Birkhäuser

Operator Theory: Advances and Applications

Volume 283

Founded in 1979 by Israel Gohberg

More information about this series at http://www.springer.com/series/4850

Manfred Möller • Vyacheslav Pivovarchik

Direct and Inverse Finite-Dimensional Spectral Problems on Graphs

Birkhäuser

Manfred Möller
School of Mathematics
University of the Witwatersrand
Johannesburg, South Africa

Vyacheslav Pivovarchik
Department of Higher Mathematics and Statistics
South-Ukrainian National Pedagogical University
Odessa, Ukraine

ISSN 0255-0156 ISSN 2296-4878 (electronic)
Operator Theory: Advances and Applications
ISBN 978-3-030-60483-7 ISBN 978-3-030-60484-4 (eBook)
https://doi.org/10.1007/978-3-030-60484-4

Mathematics Subject Classification (2010): 15A18, 15A22, 15A29, 34B45, 39A70, 47B30, 74H45, 74J25, 74K05

This book is published under the imprint Birkhäuser, www.birkhauser-science.com by the registered company Springer Nature Switzerland AG
The registered company address is: Gewerbestrasse 11, 6330 Cham, Switzerland

Contents

Preface

Vibrating strings and the sound they generate have fascinated humankind ever since homo sapiens inhabited the earth, and almost every culture has constructed musical string instruments. Hence one of the earliest subjects of Mathematical Physics was the investigation of strings, initiated in the 18th century by Euler, d'Alembert and Lagrange. Two of the earliest comprehensive accounts of the theory of sound are the monographs [69] by Helmholtz, published in 1863, and [121] by Rayleigh, published in 1877, see also [70] and [122] for later editions. In particular, Helmholtz considered a string loaded at the midpoint, and Rayleigh calculated the eigenfunctions and the eigenfrequencies as a function of the mass position along the string. Of the many results published in this field, we just mention three more recent ones. In [36] the eigenfrequencies of a loaded string with a concentrated mass were found. In [103] the authors have studied by means of numerical calculations and experimentally the eigenfunctions and the corresponding eigenfrequencies of a string-mass chain. In [41] the theoretically calculated frequencies of strings are compared with experimental data.

Mathematicians prefer to deal first with idealized simplified models, and hence M.G. Krein introduced the notion of Stieltjes string, which is a massless thread bearing point (concentrated) masses, which he called beads. A Stieltjes string is also called string-mass chain (see [103]). In his works M.G. Krein considered small transverse vibrations of Stieltjes strings admitting infinitely many beads accumulating at one of the ends of the string. However, in this monograph we restrict our attention to the case of a finite number of beads. The same model is used for vibrations of point masses connected by springs (see, e.g., [120], [62], [63], [93], [123], [37]). This model is also known in the theory of synthesis of electrical circuits (see, e.g., [66], [130], [140], [146], and in particular the Cauer method [33]). Indeed, such strings are widely used as simple models in physics (see, e.g., [84], [54], [55] and [64] where beads are connected with threads of small density). The problem of vibrations of a tree of Stieltjes strings was first considered in [60].

For the case of a finite number of beads the following spectral problems were considered in [59]: the Dirichlet-Dirichlet problem, i.e., the problem with the Dirichlet boundary conditions at both ends, and the Neumann-Dirichlet problem, i.e., the problem with the Neumann condition at the left end and the Dirichlet condition at the right end of the string. Physically, the Dirichlet condition means that the end is fixed and the Neumann condition describes a free end. It was shown that the eigenvalues of each of these problems are simple, positive and interlace. The corresponding inverse problem of recovering the masses of the beads and the lengths of the intervals into which the beads divide the string by the spectra of Dirichlet-Dirichlet and Neumann-Dirichlet problems and the total length of the string was also completely solved: it was shown that for two sequences of positive numbers to be the spectra of a Dirichlet-Dirichlet and a Neumann-Dirichlet problem, respectively, it is necessary and sufficient that the sequences interlace. The algorithm of recovering the string data consists of expanding the quotient of the characteristic polynomials of the Dirichlet-Dirichlet and the Neumann-Dirichlet problem into a continued fraction. This continued fraction algorithm originates in papers of Cauer [33] and Fry [58]. It was also shown in [59] that the solution of the inverse problem is unique.

The investigation of spectra for (over)damped dynamical systems was started in [47], represented as matrix polynomials of degree 2 in the eigenvalue parameter. This investigation was motivated by electrical circuits, but also applies to damped Stieltjes strings. The inverse problem, i.e., the problem of recovering the parameters of the string from the spectrum of its vibrations and from the total length of the string, for a Stieltjes string with a finite number of beads with point-wise (i.e., one-dimensional) damping at the right end and the left end free was solved in [4]. Also this problem can be reduced to the problem of damped oscillators considered in [133], [134]. Another approach to inverse problem for damped finite-dimensional system was developed in [85] where the given data included not only eigenvalues but also the so-called Jordan pairs. An inverse problem for a Stieltjes string with damping at the midpoint was solved in [18].

Direct and inverse spectral problems on graphs is a rapidly developing branch of analysis. Usually, the Sturm-Liouville equation, the Dirac equation or the Stieltjes string equation is considered on the edges of a graph subject to matching and boundary conditions at the vertices. This can be described as a spectral problem for an operator (see, e.g., [50], [119], [42], [117]) which is selfadjoint under certain conditions (see, e.g., [29], [77]). The justification of such models can be found in [83]. There are different approaches to inverse problems on graphs: firstly, recovering the form of a metric graph using spectral or scattering data (see, e.g., [10], [67]), and secondly, the form of the graph is known a priori and the aim is to recover the potentials of the Sturm-Liouville equation on the edges (see, e.g., [107], [108], [143]) or the mass distribution (see, e.g., [17]), using spectral or scattering data.

The tradition of using Dirichlet and Neumann spectra to solve an inverse problem on a graph came from the classical results of Borg [13], Levitan and Gasymov [89], and Marchenko [92] for inverse spectral problems on a finite interval. It is possible to consider two boundary value problems, one with Neumann condition at a certain pendant vertex, and one with Dirichlet condition at this vertex (see [27], [143]), and to use their spectra as the given data for solving the inverse problem. However, it is possible (see, e.g., [86]) to introduce analogues of Neumann and Dirichlet conditions at an interior vertex chosen as the root. Such problems were considered in [107], [108], and earlier for the star graph of two edges, i.e., for the three spectra problem, in [106], [61], [72]. Analogous problems generated by the equations of Stieltjes strings were considered in [17] and [18].

In this monograph we present a comprehensive study of direct and inverse spectral problems on Stieltjes strings and on graphs of Stieltjes strings. The boundary conditions at the pendant vertices of the graph will be Dirichlet or Neumann conditions while at interior vertices we apply continuity conditions and the condition of balance of forces (which is the Kirchhoff condition in electrical circuits synthesis terms). Sometimes we may allow mixed conditions, also called Robin conditions, at pedant vertices. The direct problem is to describe the eigenvalues, whereas the inverse problem is to determine the data of the Stieltjes string or graph of Stieltjes strings from the eigenvalues. These eigenvalue problems can be described by matrices, that is, roughly said, by a problem

$$(\lambda^2 M - i\lambda K - A)Y = 0,$$

where A, K, M are square matrices, λ is the (complex) eigenvalue parameter, and λ is an eigenvalue if the above equation has a nonzero solution Y. M will normally be a diagonal matrix with real positive entries in the diagonal, and A will be selfadjoint. For undamped problems, $K = 0$, we will replace λ^2 with z, and we will usually call the numbers z characteristic values to distinguish them from the eigenvalues λ. In the undamped case, the spectral problem is then represented by

$$(zM - A)Y = 0,$$

and the characteristic values are therefore real. We also consider some spectral problems with point-wise damping (viscous friction), in which case $K \neq 0$. The eigenvalues of such problems lie in the upper half-plane. Of course, a number λ or z is an eigenvalue or characteristic value if and only if the corresponding determinant of the matrix problem has the value zero. Hence the main tasks in the study of eigenvalues and characteristic values are to investigate when matrices are singular and to investigate the zeros of characteristic polynomials. We will also study Sturm-Liouville problems on graphs for which the eigenvalues share properties with the characteristic values of Stieltjes string problems on these graphs and where the techniques developed for graphs of Stieltjes strings apply.

Now we briefly outline the contents of this monograph.

In Chapter 1 we consider boundary value problems which describe small transverse vibrations of Stieltjes strings, the corresponding spectral problems, their characteristic values, and the inverse problems of recovering the string data from the characteristic values. In Section 1.1 we give a brief historical overview on the continued fraction approach in the synthesis of electrical circuits.

In Section 1.2 we consider the class of Stieltjes functions which occur in mechanics. We introduce the notion of a Stieltjes string with n beads and describe equations of its motion. Cauer-Fry polynomials are introduced to express the characteristic values of boundary value problems on Stieltjes strings as the zeros of Cauer-Fry polynomials. The Stieltjes functions are realized as quotients of two Cauer-Fry polynomials. At each end of the string, Dirichlet, Neumann or Robin conditions may be imposed. We present each of these problems as a spectral problem for an $n \times n$ linear matrix pencil $z \mapsto zM - A$, where M is the diagonal matrix describing the masses of the beads and A is a tridiagonal Jacobi matrix. The characteristic polynomials for each of the Dirichlet-Dirichlet, Dirichlet-Neumann, Neumann-Dirichlet, Neumann-Neumann, Robin-Dirichlet, Robin-Neumann spectral problems are obtained as particular Cauer-Fry polynomials. These are polynomials of degree n, and their zeros, which are the characteristic values of the corresponding problems, are simple and positive, except for the smallest zero of the Neumann-Neumann problem, which is 0. Also, interlacing properties of the characteristic values of these problems are shown. Furthermore, the Lagrange identity and the Sturm oscillation theorem for Stieltjes strings are proved.

In Section 1.3 we consider various inverse problems for Stieltjes strings. First we deal with the two spectra inverse problem. The direct problem has $2n+1$ degrees of freedom, namely, the masses of the n beads and the lengths of the subintervals into which the string is divided by the beads. On the other hand, the two spectra consist of $2n$ arbitrary positive numbers (degrees of freedom) if none of the boundary problems is the Neumann-Neumann problem, and of $2n-1$ arbitrary positive numbers if one of the boundary problems is the Neumann-Neumann problem. It is shown that given any two interlacing sequences of positive numbers which have in total $2n$ or $2n-1$ terms, respectively, uniquely determine the string data for the Dirichlet-Dirichlet and Dirichlet-Neumann problem, or Dirichlet-Neumann and Neumann-Neumann problems with the given sequences as positive characteristic values, respectively, if one or two of the string data is prescribed. The prescribed data can be, for example, the total length of the string, the distances of an outermost bead from the corresponding end of the string and the total mass of the beads, or the distances of the outermost beads from the corresponding end of the string. It is also shown that not necessarily unique solutions of the inverse problem for the Dirichlet-Dirichlet and Neumann-Neumann or Dirichlet-Neumann and Neumann-Dirichlet conditions, respectively, exist. In the four spectra problem four finite sequences of positive real numbers are given, and the questions is if these sequences can be identified as certain characteristic values of the Dirichlet-Dirichlet, Dirichlet-Neumann, Neumann-Dirichlet and Neumann-Neumann problems for a Stieltjes string. We prove that the four spectra problem has a unique solution if the sequences have certain interlacing properties and if an arbitrary length of the string is prescribed.

In Section 1.4 we compare the characteristic values of a spectral problem of a Stieltjes string with the characteristic values of problems on two substrings into which an interior point divides the string. We also solve the so-called three spectra problem: given characteristic values of spectral problems on the substrings and on the whole string together with the lengths of the substrings, find the data of the string. We consider a Stieltjes string analogue of the Hochstadt-Lieberman theorem. Here we prove, roughly speaking, that the knowledge of masses of the beads lying in the left part of the string and the intervals between them together with the spectrum of a problem on the whole interval uniquely determines the masses of the other beads and the intervals between them.

Chapter 2 deals with spectral problems which arise in the description of small transverse vibrations of a star graph composed by Stieltjes strings. We consider two different boundary value problems for a star graph of Stieltjes strings with continuity and Kirchhoff conditions at the interior vertex. In each of our two main results we impose conditions on two sequences of real numbers necessary and sufficient to be the spectra of the Dirichlet and the Neumann problem on a star graph of g Stieltjes strings; in the first case the root lies at the central vertex, in the second case the root is at a pendant vertex. In both cases we establish a constructive method to recover the values of the masses, including the central one, and lengths of the subintervals between them. This method uses the representation of rational functions with interlacing zeros and poles by (possibly branching) continued fractions. If the root is a pendant vertex, then the spectra of the Dirichlet and the Neumann problems together with the total length of the main edge uniquely determine the values of the masses and lengths of the subintervals between them of the main edge. The remaining inverse problem on the subgraph of $g-1$ edges may be viewed as an inverse problem with root at the central vertex to which our first result applies.

In Section 2.1 we consider the direct and the inverse spectral problem for the case when the root is the central vertex of the star graph with g edges. That is, we impose Dirichlet boundary conditions at all pendant vertices, while at the central vertex we consider Kirchhoff and continuity conditions for the Neumann problem and Dirichlet conditions for the Dirichlet problem (in which case the whole problem decouples into g separate Dirichlet-Dirichlet problems). We allow a mass M to be placed at the central vertex, so that the Dirichlet problem may be viewed as the limit $M \to \infty$ of the Neumann problem. We investigate the spectra of the corresponding Neumann and Dirichlet problems and their relation to each other, including monotonicity in terms of the central mass M. We prove that the two spectra interlace non-strictly, and if they have a characteristic value z in common, then its multiplicity $n_{D,g}(z)$ as a Dirichlet eigenvalue and its multiplicity $n_{N,g}(z)$ as a Neumann eigenvalue satisfy $n_{D,g}(z) = n_{N,g}(z) + 1$. We then show that the necessary conditions are also sufficient for the solution of the inverse problem: given two sequences satisfying these conditions and the total lengths $l_1, \ldots, l_g$ of all strings, we construct a mass distribution so that the corresponding star graph of Stieltjes strings with root at the central vertex has these two sequences as Neumann and Dirichlet characteristic values. Since we do not assume strict interlacing of the sequences, this solution need not be unique. The recovering procedure, based on the decomposition of Stieltjes functions into continued fractions, is constructive. The solution of the inverse problem provides sufficient conditions for the existence of a star-patterned matrix with two given sequences being the spectra of the matrix and its first principal submatrix.

In Section 2.2 symmetric Stieltjes strings are identified as particular star graphs of Stieltjes strings with two edges.

In Section 2.3 we consider the direct and the inverse spectral problem for the case when the root is one of the pendant vertices of the star graph with g edges. That is, we impose Dirichlet boundary conditions at all other pendant vertices, Kirchhoff and continuity conditions at the central vertex, where again a mass M may be placed, and at the pendant vertex chosen as root the Neumann condition for the Neumann problem and the Dirichlet condition for the Dirichlet problem. We prove that the two spectra interlace non-strictly and if they have a characteristic value z in common, then its multiplicity $n_{D,g}(z)$ as a Dirichlet characteristic value and its multiplicity $n_{N,g}(z)$ as a Neumann characteristic value satisfy the inequalities $n_{D,g}(z) \leq g-1$, $n_{N,g}(z) \leq g-1$, and $n_{D,g}(z)+n_{N,g}(z) \leq 2g-3$. We also establish a relation of the spectral polynomials of the above Dirichlet and Neumann problems with the boundary value problems for the star subgraph with $g-1$ edges obtained from the original graph by deleting the main edge (i.e., the edge incident with the root). We then show that the necessary conditions are also sufficient for the solution of the inverse problem: given two sequences satisfying these conditions together with the lengths of the g strings, we construct a mass distribution so that the corresponding star graph of Stieltjes strings with root at a pendant vertex has these two sequences as Neumann and Dirichlet characteristic values. Moreover, we show that the two spectra and the total length of the main edge uniquely determine the masses of the beads and the lengths of the intervals between them on this main edge; the mass distribution on the other edges cannot be uniquely determined. The recovering procedure is based on the decomposition of the quotient of the characteristic functions of the Dirichlet and the Neumann boundary value problem, which is a Stieltjes function, into branching continued fractions. In fact, the coefficients at the non-branching part of this expansion are the uniquely determined masses and the subintervals between them on the main string, while the mass distribution on the $g-1$ other edges may be recovered by our first inverse theorem and the corresponding algorithm. An example in Section 2.4 illustrates that our method, in fact, allows to construct all solutions of the inverse problem.

In Chapter 3 we consider spectral problems related to damped vibrations of Stieltjes strings and star graphs of Stieltjes strings. More precisely, we assume that the movement of at least one bead is subject to damping. In Section 3.1 we describe the spectra of small transverse vibrations of a Stieltjes string with n beads, with both ends fixed (Dirichlet-Dirichlet boundary value problem) or with the left end fixed and the right end free (Dirichlet-Neumann boundary value problem). We prove that the problem has $2n$ eigenvalues, counted with multiplicity, which lie in the closed upper half-plane and are symmetric with respect to the imaginary axis. All eigenvalues lie in the open upper half-plane if at least one of the outermost beads is subject to damping. In Section 3.2 we consider the inverse problem for the case where only the bead closest to the right end of the string is damped. Given a sequence of $2n$ complex numbers in the open upper half-plane whose terms are symmetric with respect to the imaginary axis, the total length of the string, and the distance of the rightmost bead to the right end of the string, the inverse Dirichlet-Dirichlet problem and Dirichlet-Neumann problem, respectively, have a unique solution, i.e., there is a unique Stieltjes string whose corresponding sequence of the eigenvalues is the given sequence. Relations between the characteristic functions of the Dirichlet-Dirichlet problem and the Dirichlet-Neumann problem are derived in Section 3.3.

In Sections 3.4 and 3.5 inverse problems are considered for Stieltjes strings with a damped bead at one endpoint and Robin conditions at the other endpoint.

Direct and inverse problems on Stieltjes strings with damping at an interior point are considered in Section 3.6 and, more generally, on star graphs of Stieltjes strings with damping at the central vertex are considered in Section 3.7.

In Chapter 4 we generalize the results from Chapter 2 to trees. An arbitrary vertex v_0 of the tree is chosen as the root. At interior vertices, we will impose continuity and Kirchhoff conditions, which we call generalized Neumann conditions, or we will impose generalized Dirichlet conditions. Although we are mainly interested in the case that, except for the root, the boundary conditions at pendant vertices are Dirichlet conditions and the conditions at interior vertices are generalized Neumann conditions, it is convenient for auxiliary results to allow (generalized) Dirichlet conditions or (generalized) Neumann conditions at each of the vertices of the tree. The characteristic values of the boundary value problem are the zeros of the corresponding characteristic polynomial. In Section 4.1, characteristic polynomials for boundary eigenvalue problems on trees are defined and basic properties are established. Cutting trees into subtrees is useful to study these characteristic polynomials, and in Section 4.2, relations between characteristic polynomials of trees and complementary subtrees are obtained. Letting n be the total number of beads on the tree, it is shown that the characteristic polynomial has degree n, its zeros are all real and nonnegative, and 0 is a zero if and only if Neumann conditions are imposed at all pendant vertices. In Sections 4.3 and 4.4 we consider the Dirichlet and Neumann problems which are defined by Dirichlet or (generalized) Neumann conditions at the root, respectively, Dirichlet conditions at all other vertices which are pendant, and generalized Neumann conditions at all other vertices which are interior. In Section 4.3 it is shown that the characteristic values of the Neumann problem and the Dirichlet problem interlace, but in general not strictly. Upper bounds for the multiplicities of the characteristic values are obtained in terms of the structure of the tree. In Section 4.4 the inverse problem is considered for any given rooted tree in case of strict interlacing of the characteristic values. It is shown that for any $2n$ positive numbers and any given lengths of the edges of the tree, with finitely many exceptions, there are n beads on the tree such that the interlacing characteristic values of the Dirichlet and Neumann problems on the tree are the given $2n$ positive numbers. The positions of the beads on the tree and their masses are not uniquely determined by the characteristic values.

In Chapter 5 we consider spectral problems for general graphs of Stieltjes strings. In Section 5.1 we give a linear matrix polynomial representation $z \mapsto zM - A$ of this problem with a diagonal matrix M with nonnegative entries and a symmetric matrix A with real entries. It is shown that if each edge carries at least one bead, then values z for which $zM - A$ is a singular matrix satisfy $z \geq 0$. In Section 5.2 an often more convenient matrix polynomial $z \mapsto \Phi(z)$ of smaller size, but which is not necessarily linear in z, is introduced. It is shown that the characteristic polynomial $\det \Phi(z)$ is a constant multiple of $\det(zM - A)$. In Section 5.3 this characteristic polynomial is expressed in terms of characteristic polynomials for problems on subgraphs. As particular cases, series connection of graphs is considered in Section 5.4, and parallel connection of graphs is considered in Section

5.5. In Section 5.6 (sharp) upper bounds for the multiplicities of characteristic values are obtained. These bounds depend on the shape of the graph and possibly on the numbers of (generalized) Dirichlet conditions at pendant vertices and at interior vertices. In particular, cyclically connected graphs, trees and generalized quasi-trees, and problems with only generalized Neumann conditions at interior vertices are considered. In Section 5.7, graphs of symmetric Stieltjes strings are considered. Properties of the characteristic values are obtained, and it is shown that the characteristic values uniquely determine the distribution of the beads.

In Chapter 6 we consider self-adjoint differential operators on graphs. In Section 6.1 we briefly recall the definition of selfadjoint systems of differential operators on a bounded interval. We then consider in Section 6.2 selfadjoint differential operators on metric graphs, where all edges of any such graph are supposed to have the same length. Here Neumann or Dirichlet conditions are imposed at the pendant vertices and continuity and Kirchhoff conditions at the interior vertices. In subsequent sections of Chapter 6 it will be assumed that the potential is the same on each edge and that the potential is symmetric. In Sections 6.3 and 6.4 the characteristic values of the boundary value problem on the graph are described as the zeros of entire functions of the form $S_1^s(P \circ S_2)$, where s is an integer and P is a polynomial which are determined by the structure of the graph and the boundary conditions at pendant vertices, and S_1 and S_2 are the characteristic functions of the identical Sturm-Liouville problems on the edges with normalized Dirichlet and Neumann conditions. For the tetrahedral graph, the octahedral graph, the cube graph and for star graphs, the characteristic polynomials are calculated explicitly. In Section 6.5 the particular case of perfect binary trees is considered. In this case, explicit formulas for the polynomial P mentioned earlier in this paragraph are obtained. Finally, in Section 6.6 properties of the eigenvalues are obtained for star graphs with finite edges as well as with infinite edges.

In three appendices we provide some background material to the extent it is needed in the main body of this monograph.

In Appendix A we present definitions and results with proofs on some polynomials and rational functions such as rational Nevanlinna functions, rational Stieltjes functions and Hermite-Biehler polynomials.

Appendix B contains a brief summary on results for linear and quadratic matrix pencils as well as Schur complements.

In Appendix C basic definitions and properties in graph theory are given. Then more particular advanced results which are essential in the main body of this monograph are investigated for cyclically connected graphs, quasi-trees and adjacency and incidence matrices.

Chapter 1

Stieltjes strings

1.1 Stieltjes functions in electrical circuits

An important role in the synthesis of electrical circuits is played by the 'scalar impedance function' $Z(\lambda)$ from which the impedance may be derived by writing $\lambda = i\omega$ (see [28], [58], [33]), where ω is the cyclic frequency of a sine current.

It is known, see [28, Abstract, p. 1], that $Z(\lambda)$ is a 'positive real' function, that is,

(i) $Z(\lambda)$ is a rational function which is real for real values of λ;

(ii) the real part of $Z(\lambda)$ is positive when the real part of λ is positive.

O. Brune also proved that for each positive real function $Z(\lambda)$ a circuit can be found containing resistances R_j, capacities C_j and inductivities L_j such that the given $Z(\lambda)$ is the impedance function of this circuit. Necessary and sufficient conditions for a function to be positive real are given in [28, p. 33]:

(i) no poles lie in the open right half-plane;

(ii) pure imaginary poles have positive real finite residues;

(iii) $\operatorname{Re} Z(i\omega) \geq 0$ for real ω.

It should be mentioned that a particular case corresponding to lossless circuits consisting only of C_j's and L_j's was considered earlier by R. Foster [56].

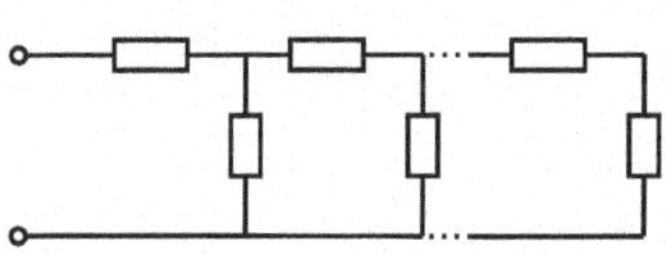

Figure 1.1

M. Möller, V. Pivovarchik, *Direct and Inverse Finite-Dimensional Spectral Problems on Graphs*, Operator Theory: Advances and Applications 283, https://doi.org/10.1007/978-3-030-60484-4_1

As far as we know, the first who used expansions of rational functions into continued fractions to describe physical phenomena was W. Cauer [33]. In [58] the point of departure was a very simple observation regarding the ladder-type network shown in Figure 1.1. The ordinary laws of circuit theory lead immediately to a continued fraction expansion for the scalar impedance function $Z(\lambda)$ (see [58, (1)]).

For example, if all the series elements in Figure 1 are resistances R_j, $j = 1, \ldots, n$, and all the shunt elements are capacities C_j, $j = 1, \ldots, n$, indexed from right to left, then (see [33, (20)] and [58, (2)])

$$Z(\lambda) = R_n + \cfrac{1}{C_n\lambda + \cfrac{1}{R_{n-1} + \cfrac{1}{C_{n-1}\lambda + \ddots + \cfrac{1}{R_1 + \cfrac{1}{C_1\lambda}}}}}. \tag{1.1.1}$$

If again all the series elements are resistances R_j, $j = 1, \ldots, n$ but all the shunt elements are inductances L_j, $j = 1, \ldots, n$, then (see [33, (26)])

$$Z(\lambda) = R_n + \cfrac{1}{M_n z + \cfrac{1}{R_{n-1} + \cfrac{1}{M_{n-1} z + \ddots + \cfrac{1}{R_1 + \cfrac{1}{M_1 z}}}}}, \tag{1.1.2}$$

where $z = \lambda^{-1}$ and $M_j = L_j^{-1}$, $j = 1, \ldots, n$. Infinite continued fractions of the form

$$\cfrac{1}{a_1 z + \cfrac{1}{a_2 + \cfrac{1}{a_3 z + \ddots + \cfrac{1}{a_{2n} + \cfrac{1}{a_{2n-1} z} + \ddots}}}} \tag{1.1.3}$$

with positive coefficients a_j, $j = 1, 2, \ldots$, were introduced by T.-J. Stieltjes in [128] and are hence called Stieltjes continued fractions.

We quote the following assessment of the culmination of classical network synthesis from [32, p. 7].

"Darlington's Theorem: Any positive real function $Z(\lambda)$ admits a realization as the input impedance of a lossless (frequency-dependent) two-port terminated in a positive (frequency-independent) resistor R.

In other words, the set of positive real rational functions $Z(\lambda)$ admits a linear fractional parametrization

$$Z(\lambda) = \frac{A(\lambda)R + B(\lambda)}{C(\lambda)R + D(\lambda)}, \quad \mathbf{T}(\lambda) = \begin{pmatrix} A(\lambda) & B(\lambda) \\ C(\lambda) & D(\lambda) \end{pmatrix}$$

where $\mathbf{T}(\lambda)$ sweeps through the set of rational chain matrices of lossless two-ports. The numerous applications of this lossless embedding of passive impedances stems from its canonical separation of thermodynamical aspects (dissipation of energy) and Hamiltonian concepts (storage of energy, dynamics, holonomic constraints or frequency dependence) within the class of linear passive systems. The theorem was apparently devised by W. Cauer [34], G. Cocci [39], and S. Darlington [45], all independently of the other. Cauer's proof seems to be not only the shortest (it requires only half a printed page, [34] (p. 232]) but also the most elegant and forward-looking. Instead of dealing with positive real impedances $Z(\lambda)$, Cauer applies a bilinear transform

$$T(\lambda) = \frac{Z\lambda) + 1}{Z(\lambda) - 1} := \frac{g(\lambda)}{h(\lambda)}$$

and discusses all rational solutions of a (spectral) factorization of the rational (and for $\lambda^2 = -\omega^2$ positive) function

$$T(\lambda)T(-\lambda) - 1 := \frac{f(\lambda)f(-\lambda)}{h(\lambda)h(-\lambda)}.$$

The polynomials f, g, h, in which $g(\lambda)$ is Hurwitz, are called Betriebskenngrößen by Cauer because they are the key quantities in the design of reactance filters whose preassigned characteristics are specified in terms of insertion loss (Betriebsdämpfung). Cauer's polynomials f, g, h still survive (in the very same notation!) in Belevitch's celebrated canonic form of the scattering matrix of a real lossless twoport [9] p. 428 (note 8)."

Generalizations of Darlington's method to the infinite-dimensional case can be found in [4].

1.2 Boundary value problems for Stieltjes strings

1.2.1 Boundary value problems

Following M.G. Krein, a Stieltjes string of length l is a massless string bearing n beads of masses $m_1, \dots, m_n$ in its interior, where $n \in \mathbb{N}$. Selecting one of the endpoints and calling it the left endpoint, the beads are indexed in such a way that the distance from the left endpoint increases with the index. Denoting the distance from the left endpoint to the first bead by l_0, the distance between the j-th bead and the $j+1$-th bead by l_j, $j = 1, \dots, n-1$, and the distance from the last bead to the right endpoint by l_n, it is clear that $\sum_{k=0}^{n} l_k = l$. In other words, the beads with masses $m_1, \dots, m_n$ divide the string into $n+1$ threads (massless strings) of length $l_0, \dots, l_n$.

When a Stieltjes string with the location of the beads on the string and the masses of the beads is given, we will briefly call this a *distribution of beads* (on the string). In particular, the positive numbers $m_1, \dots, m_n$ and $l_0, \dots, l_n$ determine a unique distribution of beads. In later chapters, we will consider (particular) graphs of Stieltjes strings, and the notion distribution of beads would apply there accordingly. Sometimes it will be convenient to allows strings without beads, i.e., $n = 0$.

The string is assumed to be stretched by a stretching force equal to 1. Taking into account that on the threads the string has zero density, the general solution of the differential equation of the string is a linear function of the position variable for each of the $n+1$ threads. Hence small transverse vibrations of such a Stieltjes string are described by the three term recurrence relations

$$\frac{v_k(t) - v_{k-1}(t)}{l_{k-1}} - \frac{v_{k+1}(t) - v_k(t)}{l_k} + m_k v_k''(t) = 0 \quad (k = 1, \dots, n), \tag{1.2.1}$$

where $v_k(t)$ is the transverse displacement of the k-th bead for $k = 1, \dots, n$ at time t and $v_0(t)$ and $v_{n+1}(t)$ are the transverse displacements of the left and right endpoints of the string, respectively. The Dirichlet-Dirichlet problem is generated by the equations (1.2.1) and the boundary conditions

$$v_0(t) = 0, \ v_{n+1}(t) = 0, \tag{1.2.2}$$

and the Neumann-Dirichlet problem is generated by the equations (1.2.1) and boundary conditions

$$v_0(t) = v_1(t), \ v_{n+1}(t) = 0. \tag{1.2.3}$$

Substituting the separation of variables ansatz $v_k(t) = u_k e^{i\lambda t}$ into (1.2.1), we obtain

$$\frac{u_k - u_{k-1}}{l_{k-1}} - \frac{u_{k+1} - u_k}{l_k} - m_k \lambda^2 u_k = 0 \quad (k = 1, \dots, n, \ \lambda \in \mathbb{C}), \tag{1.2.4}$$

for the amplitudes u_k of the vibrations of the beads and the endpoints of the string, $k = 0, \dots, n+1$, see [59, Supplement II, (13)].

Figure 1.2 illustrates the problem:

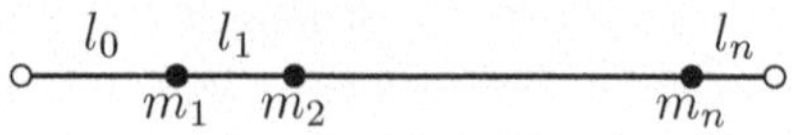

Figure 1.2

Remark 1.2.1. The aim is to find nontrivial solution $(u_0, u_1, \ldots, u_{n+1})$ of (1.2.4) which satisfy certain boundary conditions, one at the left endpoint and one at the right endpoint. The values of $\lambda \in \mathbb{C}$ for which such nontrivial solutions exist are called *eigenvalues* of the corresponding boundary value problem, and the set of all eigenvalues is called the *spectrum* of the boundary value problem. Since the eigenvalue parameter λ occurs quadratic in (1.2.4), it is convenient to replace λ^2 with z. Then we call a number $z \in \mathbb{C}$ a *characteristic value* of the corresponding boundary value problem if this problem has a nontrivial solution. Clearly, a number $\lambda \in \mathbb{C}$ is an eigenvalue if and only if $z = \lambda^2$ is a characteristic value, and in that case also $-\lambda$ is an eigenvalue. Therefore the eigenvalues are uniquely determined by the characteristic values, and in this chapter we will therefore mainly consider characteristic values. That is, when referring to the recursion relation (1.2.4) we will often replace λ^2 with z.

The boundary conditions (1.2.2) or (1.2.3) would give $u_0 = 0$ or $u_0 = u_1$, respectively, but it is convenient to admit more general boundary condition at the left endpoint of the form

$$u_0 = cu_1 \ (c \in \mathbb{R}). \tag{1.2.5}$$

Similarly, at the right endpoint we admit the boundary condition

$$u_{n+1} = du_n \ (d \in \mathbb{R}). \tag{1.2.6}$$

For $c, d \in \mathbb{R}$ consider the tridiagonal $n \times n$ matrix

$$A_c^d = \begin{pmatrix} \frac{1-c}{l_0} + \frac{1}{l_1} & -\frac{1}{l_1} & 0 & \cdots & \cdots & \cdots & 0 \\ -\frac{1}{l_1} & \frac{1}{l_1} + \frac{1}{l_2} & -\frac{1}{l_2} & 0 & \cdots & \cdots & 0 \\ 0 & -\frac{1}{l_2} & \frac{1}{l_2} + \frac{1}{l_3} & -\frac{1}{l_3} & \ddots & & \vdots \\ \vdots & & \ddots & \ddots & \ddots & \ddots & \vdots \\ \vdots & & & \ddots & \ddots & \ddots & 0 \\ \vdots & & & & -\frac{1}{l_{n-2}} & \frac{1}{l_{n-2}} + \frac{1}{l_{n-1}} & -\frac{1}{l_{n-1}} \\ 0 & \cdots & \cdots & \cdots & 0 & -\frac{1}{l_{n-1}} & \frac{1}{l_{n-1}} + \frac{1-d}{l_n} \end{pmatrix} \tag{1.2.7}$$

in case $n > 1$ and $A_c^d = \left(\frac{1-c}{l_0} + \frac{1-d}{l_1}\right)$ in case $n = 1$.

The following result is easily verified.

Proposition 1.2.2. *Let $c, d \in \mathbb{R}$, $z \in \mathbb{C}$, and $Y = (u_1, \ldots, u_n)^\mathsf{T} \in \mathbb{C}^n \setminus \{0\}$. Define u_0 and u_{n+1} by (1.2.5) and (1.2.6), respectively. Then (1.2.4) is satisfied if and only if*

$$(A_c^d - zM)Y = 0,$$

where $M = \operatorname{diag}(m_1, \ldots, m_n)$.

By Proposition 1.2.2, problem (1.2.4), (1.2.5), (1.2.6) has a nontrivial solution if and only if $\det(A_c^d - zM) = 0$. The polynomials defined by $z \mapsto \det(A_c^d - zM)$ and $\lambda \mapsto \det(A_c^d - \lambda^2 M)$ and any of their nonzero multiples will be called *characteristic polynomial* of (1.2.4), (1.2.5), (1.2.6). It will be clear from the context which of those polynomials we mean. Hence the eigenvalues and characteristic values of (1.2.4), (1.2.5), (1.2.6) are the zeros of the characteristic polynomial. In this chapter, the eigenvalues will always occur in pairs $(\lambda, -\lambda)$ and it is therefore sufficient and more convenient to consider the characteristic values $z = \lambda^2$ instead.

Proposition 1.2.3. *Let $c, d \in (-\infty, 1]$. Then the matrix A_c^d is symmetric and positive semidefinite, and all eigenvalues of the spectral problem for $A_c^d - zM$ with spectral parameter z are nonnegative and simple. The number 0 is an eigenvalue if and only if $c = 1$ and $d = 1$.*

Proof. The symmetry of A_c^d is obvious. For the vectors $Y = (u_1, \dots, u_n)^\mathsf{T} \neq 0$ and $Y^* = (\overline{u_1}, \dots, \overline{u_n})$ we calculate

$$\begin{aligned} Y^* A_c^d Y &= \sum_{j=1}^{n-1} \frac{1}{l_j} |u_j|^2 + \sum_{j=1}^{n-1} \frac{1}{l_j} |u_{j+1}|^2 + \frac{1-c}{l_0} |u_1|^2 + \frac{1-d}{l_n} |u_n|^2 \\ &\quad - \sum_{j=1}^{n-1} \frac{1}{l_j} (\overline{u_j} u_{j+1} + u_j \overline{u_{j+1}}) \\ &= \frac{1-c}{l_0} |u_1|^2 + \frac{1-d}{l_n} |u_n|^2 + \sum_{j=1}^{n-1} \frac{1}{l_j} |u_{j+1} - u_j|^2, \end{aligned}$$

which means that A_c^d is positive semidefinite and that A_c^d is positive definite when $c \neq 1$ or $d \neq 1$. Since $A_1^1 (1, \dots, 1)^\mathsf{T} = 0$, it follows that 0 is an eigenvalue of the spectral problem when $c = 1$ and $d = 1$. Furthermore, the lower left $(n-1) \times (n-1)$ submatrix of $A_c^d - zM$ is upper triangular with nonzero diagonal elements, which implies that for all $z \in \mathbb{C}$ the rank of $A_c^d - zM$ is at least $n - 1$. Therefore all of its eigenvalues are simple. An application of Proposition B.2.1 completes the proof. □

1.2.2 Cauer-Fry polynomials

For properties of characteristic values beyond those found in Proposition 1.2.3 we are going to use a different representation of the characteristic polynomials. When $u_1 = 0$, then (1.2.5) would give $u_0 = 0$, and a recursive substitution into (1.2.4) would lead to the trivial solution

$$0 = u_0 = u_1 = \cdots = u_{n+1}.$$

Therefore, when searching for nontrivial solutions we may assume $u_1 \neq 0$. Then, for each $z \in \mathbb{C}$ and each $u_1 \neq 0$, (1.2.4) together with (1.2.5) has a unique solution which can be written in the form

$$u_k = R_{2k-2}(z)u_1 \quad (k = 0, \dots, n+1). \tag{1.2.8}$$

Substituting (1.2.8) into (1.2.4), we arrive at

$$\frac{R_{2k-2}(z) - R_{2k-4}(z)}{l_{k-1}} - \frac{R_{2k}(z) - R_{2k-2}(z)}{l_k} = m_k z R_{2k-2}(z) \tag{1.2.9}$$

for $k = 1, \dots, n$. Introducing

$$R_{2k-1}(z) = \frac{R_{2k}(z) - R_{2k-2}(z)}{l_k} \quad (k = 0, \dots, n), \tag{1.2.10}$$

(1.2.9) is equivalent to

$$R_{2k-1}(z) = R_{2k-3}(z) - m_k z R_{2k-2}(z) \quad (k = 1, \dots, n). \tag{1.2.11}$$

The polynomials R_k $(k = -2, \dots, 2n)$ will be called the Cauer-Fry polynomials of (1.2.4), (1.2.5).

In order to solve the three term recurrence relations (1.2.10) and (1.2.11) we observe that (1.2.8) for $k = 1$ gives the natural initial value $R_0(z) = 1$. Then condition (1.2.5) gives $R_{-2}(z) = c$. To account for the dependence on c we will also write $R_k(\cdot, c)$ for the Cauer-Fry polynomials satisfying the initial conditions

$$R_0(z, c) = 1, \tag{1.2.12}$$
$$R_{-2}(z, c) = c. \tag{1.2.13}$$

Then the recurrence relations (1.2.10) and (1.2.11) with the initial values (1.2.12) and (1.2.13) have a unique solution $R_k(z, c)$ for $R_k(z)$, $k = -1, 1, 2, \dots, 2n$. Indeed,

$$R_{-1}(z, c) = \frac{1 - c}{l_0}, \tag{1.2.14}$$

and for $R_1(z, c), \dots, R_{2n}(z, c)$, the equation (1.2.11) is used for the odd indices whereas (1.2.10) is used for the even indices. It is easy to see by induction that $R_{2k-1}(\cdot, c)$ and $R_{2k}(\cdot, c)$ are polynomials of degree k for $k = 0, \dots, n$. Equation (1.2.11) for $z = 0$ and (1.2.14) imply that

$$R_{2n-1}(0, c) = R_{2n-3}(0, c) = \dots = R_{-1}(0, c) = \frac{1 - c}{l_0}. \tag{1.2.15}$$

Then (1.2.10) and (1.2.15) lead to

$$\begin{aligned} R_{2k}(0, c) &= \sum_{j=1}^{k} l_j R_{2j-1}(0, c) + R_0(0, c) \\ &= 1 + \frac{1 - c}{l_0} \sum_{j=1}^{k} l_j \quad (k = 0, \dots, n). \end{aligned} \tag{1.2.16}$$

In particular,

$$R_{2n}(0,c) = c + (1-c)\frac{l}{l_0}. \tag{1.2.17}$$

Lemma 1.2.4. *The recursion for the Cauer-Fry polynomials can be written as*

$$\begin{pmatrix} R_{2k-1}(z,c) \\ R_{2k}(z,c) \end{pmatrix} = \begin{pmatrix} 1 & -m_k z \\ l_k & 1 - l_k m_k z \end{pmatrix} \begin{pmatrix} R_{2k-3}(z,c) \\ R_{2k-2}(z,c) \end{pmatrix} \quad (k = 1, \dots, n). \tag{1.2.18}$$

Proof. From (1.2.10) it follows that

$$\begin{aligned} \begin{pmatrix} R_{2k-1}(z,c) \\ R_{2k}(z,c) \end{pmatrix} &= \begin{pmatrix} R_{2k-1}(z,c) \\ l_k R_{2k-1}(z,c) + R_{2k-2}(z,c) \end{pmatrix} \\ &= \begin{pmatrix} 1 & 0 \\ l_k & 1 \end{pmatrix} \begin{pmatrix} R_{2k-1}(z,c) \\ R_{2k-2}(z,c) \end{pmatrix}, \end{aligned}$$

and (1.2.11) gives

$$\begin{aligned} \begin{pmatrix} R_{2k-1}(z,c) \\ R_{2k-2}(z,c) \end{pmatrix} &= \begin{pmatrix} R_{2k-3}(z,c) - m_k z R_{2k-2}(z,c) \\ R_{2k-2}(z,c) \end{pmatrix} \\ &= \begin{pmatrix} 1 & -m_k z \\ 0 & 1 \end{pmatrix} \begin{pmatrix} R_{2k-3}(z,c) \\ R_{2k-2}(z,c) \end{pmatrix}. \end{aligned}$$

Multiplying the two coefficient matrices completes the proof. □

Lemma 1.2.4 immediately gives

$$\begin{aligned} \frac{R_{2k}(z,c)}{R_{2k-1}(z,c)} &= \frac{l_k R_{2k-3}(z,c) + (1 - l_k m_k z) R_{2k-2}(z,c)}{R_{2k-3}(z,c) - m_k z R_{2k-2}(z,c)} \\ &= l_k + \cfrac{1}{-m_k z + \cfrac{1}{\cfrac{R_{2k-2}(z,c)}{R_{2k-3}(z,c)}}} \end{aligned}$$

for $k = 2, \dots, n$. A recursive substitution and the initial conditions (1.2.12), (1.2.14) lead to the continued fraction expansion

$$\frac{R_{2n}(z,c)}{R_{2n-1}(z,c)} = l_n + \cfrac{1}{-m_n z + \cfrac{1}{l_{n-1} + \cfrac{1}{-m_{n-1} z + \ddots + \cfrac{1}{l_1 + \cfrac{1}{-m_1 z + \cfrac{1-c}{l_0}}}}}}. \tag{1.2.19}$$

Proposition 1.2.5. *Let $c \in (-\infty, 1]$. Then $R_{2n}(\cdot, c)$ and $R_{2n-1}(\cdot, c)$ have n simple zeros, which will be denoted by $(\nu_j(c))_{j=1}^n$ and $(\mu_j(c))_{j=1}^n$, respectively, where $\nu_j(c) < \nu_{j+1}(c)$ and $\mu_j(c) < \mu_{j+1}(c)$ $(j = 1, \dots, n-1)$. The zeros of $R_{2n}(\cdot, c)$ and $R_{2n-1}(\cdot, c)$ interlace as follows:*

$$0 \le \mu_1(c) < \nu_1(c) < \mu_2(c) < \cdots < \mu_n(c) < \nu_n(c), \tag{1.2.20}$$

with $\mu_1(c) = 0$ if and only if $c = 1$.

Proof. Some of the stated properties have already been shown in Proposition 1.2.3. However, since $R_{2n}(\cdot, c)$ and $R_{2n-1}(\cdot, c)$ are polynomials of degree n and since (1.2.19) and Lemma A.3.6 show that $\frac{R_{2n}(\cdot,c)}{R_{2n-1}(\cdot,c)}$ is a rational $\mathcal{S}$-function with n poles, there is no cancellation of zeros and poles. Hence the poles of $\frac{R_{2n}(\cdot,c)}{R_{2n-1}(\cdot,c)}$ are the zeros of $R_{2n-1}(\cdot, c)$, whereas its zeros are the zeros of $R_{2n}(\cdot, c)$. Therefore all statements of this proposition follows from Lemma A.3.5. □

Proposition 1.2.6. *Let $c, d \in \mathbb{R}$. Then a characteristic polynomial of* (1.2.4), (1.2.5), (1.2.6) *is*

$$(1-d)R_{2n}(\cdot, c) + dl_n R_{2n-1}(\cdot, c). \tag{1.2.21}$$

Proof. By construction of the Cauer-Fry polynomials, a nontrivial solution $(u_k)_{k=0}^{n+1}$ of (1.2.4), (1.2.5) is $(R_{2k}(z,c))_{k=-1}^n$. Then

$$\begin{aligned} u_{n+1} - du_n &= (1-d)u_{n+1} + d(u_{n+1} - u_n) \\ &= (1-d)R_{2n}(z,c)u_1 + dl_n R_{2n-1}(z,c)u_1. \end{aligned}$$

Therefore, the (simple) zeros of the polynomials $z \mapsto \det(A_c^d - zM)$ and (1.2.21) of degree n coincide (note that $R_{2n}(\cdot, c)$ and $l_n R_{2n-1}(\cdot, c)$ have the same leading coefficient by (1.2.10)). Hence these two polynomials are constant multiples of each other, which completes the proof in view of the definition of a characteristic polynomial. □

Remark 1.2.7.

1. It is clear that the zeros of $R_{2n-1}(\cdot, c)$, that is, the poles of the continued fraction expansion on the right-hand side of (1.2.19), do not depend on l_n.
2. If $c = 1$, then the zeros of $R_{2n}(\cdot, 1)$ as well as the zeros of $R_{2n-1}(\cdot, 1)$ do not depend on l_0 since the right-hand side of (1.2.19) is independent of l_0.
3. Let the string data $(m_k)_{k=1}^n$ and $(l_k)_{k=0}^n$ be given, let $\eta > 0$, and define $\widetilde{m}_k = \eta^{-1} m_k$ $(k = 1, \dots, n)$ and $\widetilde{l}_k = \eta l_k$ $(k = 0, \dots, n)$ with the corresponding Cauer-Fry polynomials $\widetilde{R}_{2n}(\cdot, c)$ and $\widetilde{R}_{2n-1}(\cdot, c)$. Then it is clear that

$$\frac{\widetilde{R}_{2n}(\cdot, c)}{\widetilde{R}_{2n-1}(\cdot, c)} = \eta \frac{R_{2n}(\cdot, c)}{R_{2n-1}(\cdot, c)}.$$

In particular, the zeros of $\widetilde{R}_{2n}(\cdot, c)$ and of $R_{2n}(\cdot, c)$ coincide, and the zeros of $\widetilde{R}_{2n-1}(\cdot, c)$ and of $R_{2n-1}(\cdot, c)$ coincide.

4. For all $c \in \mathbb{R}$, the identity $R_k(\cdot, c) = (1-c)R_k(\cdot, 0) + cR_k(\cdot, 1)$ holds when $k = -1, 2$ in view of (1.2.12) and (1.2.14), and hence the identity is also true for $k = 1, \dots, 2n$ in view of the recursions (1.2.11) and (1.2.10).

5. The coefficients of the Cauer-Fry polynomials $R_j(\cdot, c)$ depend continuously on $m_1, \dots, m_n$, $l_0, \dots, l_n$ and c, and for $j \geq 1$, their degree is positive and independent of these parameters. Due to Lemma A.6.5 it follows for $j = 1, \dots, 2n$ that the zeros of $R_j(\cdot, c)$ depend continuously on $m_1, \dots, m_n$, $l_0, \dots, l_n$ and c.

1.2.3 Dirichlet and Neumann boundary conditions

Dirichlet boundary condition at the left endpoint

Here we consider the case that the left end of the string is clamped, that is,

$$v_0(t) = 0. \tag{1.2.22}$$

This gives the Dirichlet condition

$$u_0 = 0, \tag{1.2.23}$$

which is (1.2.5) with $c = 0$.

Dirichlet-Dirichlet problem. If the right end of the string is clamped, then the Dirichlet condition

$$u_{n+1} = 0 \tag{1.2.24}$$

is obtained, which is (1.2.6) with $d = 0$.

Proposition 1.2.8. 1. *$R_{2n}(\cdot, 0)$ is the characteristic polynomial of problem* (1.2.4) *with Dirichlet-Dirichlet conditions* (1.2.23), (1.2.24).

2. *The zeros of $R_{2n}(\cdot, 0)$ are real, positive and simple. Denoting these zeros by ν_k $(k = 1, \dots, n)$ with $\nu_k < \nu_{k+1}$ for $k = 1, \dots, n-1$, it follows that*

$$R_{2n}(z, 0) = \frac{l}{l_0} \prod_{k=1}^{n} \left(1 - \frac{z}{\nu_k}\right). \tag{1.2.25}$$

Proof. Statement 1 is Proposition 1.2.6 for $c = 0$ and $d = 0$. Then statement 2 follows from Proposition 1.2.5 if we observe that the left-hand side and the right-hand side of (1.2.25) coincide for $z = 0$ in view of (1.2.17). □

Dirichlet-Neumann problem. Here the right end of the string can move freely in the direction orthogonal to the equilibrium position of the string, that is,

$$v_{n+1}(t) = v_n(t). \tag{1.2.26}$$

This gives the Neumann condition

$$u_{n+1} = u_n, \tag{1.2.27}$$

which is (1.2.6) with $d = 1$.

Proposition 1.2.9. 1. $R_{2n-1}(\cdot, 0)$ *is the characteristic polynomial of problem* (1.2.4) *with Dirichlet-Neumann conditions* (1.2.23), (1.2.27).
2. *The zeros of* $R_{2n-1}(\cdot, 0)$ *are real, positive and simple. Denoting these zeros by* μ_k $(k = 1, \dots, n)$ *with* $\mu_k < \mu_{k+1}$ *for* $k = 1, \dots, n-1$, *it follows that*

$$R_{2n-1}(z, 0) = \frac{1}{l_0} \prod_{k=1}^{n} \left(1 - \frac{z}{\mu_k}\right). \tag{1.2.28}$$

Proof. Statement 1 is Proposition 1.2.6 for $c = 0$ and $d = 1$ since we can omit the factor l_n. Then statement 2 follows from Proposition 1.2.5 if we observe that the left-hand side and the right-hand side of (1.2.28) coincide for $z = 0$ in view of (1.2.15). □

Neumann boundary condition at the left endpoint

Now assume that the left end of the string can move freely in the direction orthogonal to the equilibrium position of the string. This gives the Neumann condition

$$u_0 = u_1, \tag{1.2.29}$$

that is, (1.2.5) with $c = 1$.

Neumann-Dirichlet problem. If the right end of the string is clamped, then the Dirichlet condition (1.2.24) is obtained.

Since the Neumann-Dirichlet problem is identical to the Dirichlet-Neumann problem with reversed indexing of the beads and endpoints, Proposition 1.2.9 and (1.2.17) lead to

Proposition 1.2.10. 1. $R_{2n}(\cdot, 1)$ *is the characteristic polynomial of problem* (1.2.4) *with Neumann-Dirichlet conditions* (1.2.24), (1.2.29).

2. *The zeros of* $R_{2n}(\cdot, 1)$ *are real, positive and simple. Denoting these zeros by* ξ_k $(k = 1, \dots, n)$ *with* $\xi_k < \xi_{k+1}$ *for* $k = 1, \dots, n-1$, *it follows that*

$$R_{2n}(z, 1) = \prod_{k=1}^{n} \left(1 - \frac{z}{\xi_k}\right). \tag{1.2.30}$$

Neumann-Neumann problem. If the right end of the string can move freely in the direction orthogonal to the equilibrium position of the string, then the Neumann condition (1.2.27) is obtained.

Proposition 1.2.11. 1. *$R_{2n-1}(\cdot,1)$ is the characteristic polynomial of problem* (1.2.4) *with Neumann-Neumann conditions* (1.2.24), (1.2.29).

2. *The zeros of $R_{2n-1}(\cdot,1)$ are real, nonnegative and simple, and $R_{2n-1}(0,1) = 0$. Denoting the zeros by ζ_k, $k = 1,\dots,n$ with $\zeta_k < \zeta_{k+1}$ for $k = 1,\dots,n-1$, it follows that $\zeta_1 = 0$ and*

$$R_{2n-1}(z,1) = -\left(\sum_{k=1}^{n} m_k\right) z \prod_{k=2}^{n}\left(1-\frac{z}{\zeta_k}\right). \tag{1.2.31}$$

Proof. Statement 1 is Proposition 1.2.6 for $c = 1$ and $d = 1$ since we can omit the factor l_n. Then statement 2 follows from Proposition 1.2.5 and the fact that the derivatives on the left-hand side and the right-hand side of (1.2.31) coincide for $z = 0$. Indeed, differentiating (1.2.11) and observing (1.2.16) gives

$$R'_{2k-1}(0,1) = R'_{2k-3}(0,1) - m_k \quad (k = 1,\dots,n).$$

A recursive substitution and $R'_{-1}(0,1) = 0$, see (1.2.14), show that

$$R'_{2n-1}(0,1) = -\sum_{k=1}^{n} m_k. \qquad \square$$

Proposition 1.2.12. *Let $R_k^-(\cdot,c)$ be the Cauer-Fry polynomials for the same string, but with opposite orientation. Then*

$$l_n R^-_{2n}(\cdot,0) = l_0 R_{2n}(\cdot,0), \tag{1.2.32}$$
$$l_n R^-_{2n-1}(\cdot,0) = R_{2n}(\cdot,1), \tag{1.2.33}$$
$$R^-_{2n}(\cdot,1) = l_0 R_{2n-1}(\cdot,0), \tag{1.2.34}$$
$$R^-_{2n-1}(\cdot,1) = R_{2n-1}(\cdot,1). \tag{1.2.35}$$

Proof. The zeros of the polynomials $R_{2n}(\cdot,0)$ and $R^-_{2n}(\cdot,0)$ are the characteristic values of the Dirichlet-Dirichlet problem for the original string as well as the string with opposite orientation. Hence these two polynomials are multiples of each other. Then (1.2.32) follows from (1.2.17).

The zeros of the polynomials $R_{2n}(\cdot,1)$ and $R^-_{2n-1}(\cdot,0)$ are the characteristic values of the Neumann-Dirichlet problem for the original string and the Dirichlet-Neumann problem for the string with opposite orientation, respectively. But these two problems are the same and hence these two polynomials are multiples of each other. Then (1.2.33) follows from (1.2.15) and (1.2.17).

The identity (1.2.34) is (1.2.33) for the string with opposite orientation.

The zeros of the polynomials $R_{2n-1}(\cdot,1)$ and $R^{-}_{2n-1}(\cdot,1)$ are the characteristic values of the Neumann-Neumann problem for the original string as well as the string with opposite orientation. Hence these two polynomials are multiples of each other. Then (1.2.35) follows from (1.2.31). □

Comparison of the characteristic values

Lemma 1.2.13. *The following interlacing holds:*

$$0 < \mu_1 < \nu_1 < \cdots < \mu_n < \nu_n; \tag{1.2.36}$$

$$0 < \xi_1 < \nu_1 < \cdots < \xi_n < \nu_n; \tag{1.2.37}$$

$$0 = \zeta_1 < \mu_1 < \cdots < \zeta_n < \mu_n; \tag{1.2.38}$$

$$0 = \zeta_1 < \xi_1 < \cdots < \zeta_n < \xi_n. \tag{1.2.39}$$

Proof. The interlacings (1.2.36) and (1.2.39) are (1.2.20) for $c = 0$ and $c = 1$, respectively. Interchanging the ends of the string will interchange the rôles of μ_k and ξ_k, and hence also (1.2.37) and (1.2.38) hold. □

For two sets X and Y of real numbers, the notation $X < Y$ will mean that $x < y$ for all $x \in X$ and all $y \in Y$. It will be convenient to put $\nu_0 := 0$ and $\zeta_{n+1} := \nu_n$. Then (1.2.36)–(1.2.39) is equivalent to $\zeta_1 = 0$ and

$$\{\nu_{k-1}, \zeta_k\} < \{\mu_k, \xi_k\} < \{\nu_k, \zeta_{k+1}\} \quad (k = 1, \ldots, n). \tag{1.2.40}$$

1.2.4 Robin-Dirichlet and Robin-Neumann problems

In this subsection, a Robin boundary condition

$$\frac{v_1(t) - v_0(t)}{l_0} - \gamma v_0(t) = 0 \tag{1.2.41}$$

with some parameter $\gamma > 0$ is imposed at the left end of the string. Physically, this means that the left end of the string is prolonged by a thread that is fixed at distance γ^{-1} to the left of the string; another interpretation is that Robin conditions simulate an elastic support at the endpoint of the string (see [35, p. 15]). Substituting the separation of variables ansatz $v_k(t) = u_k e^{i\lambda t}$ (1.2.41), we obtain

$$\frac{u_1 - u_0}{l_0} - \gamma\, u_0 = 0. \tag{1.2.42}$$

The boundary condition (1.2.42) can be written in the form

$$u_0 = \frac{u_1}{1 + \gamma l_0} =: \gamma_0 u_1. \tag{1.2.43}$$

Clearly, (1.2.43) is of the form (1.2.5) with $c = \gamma_0 \in (0,1)$. In particular, (1.2.15) and (1.2.17) give

$$R_{2n-1}(0,\gamma_0) = R_{2n-3}(0,\gamma_0) = \cdots = R_{-1}(0,\gamma_0) = \frac{1-\gamma_0}{l_0} = \frac{\gamma}{1+\gamma l_0}, \tag{1.2.44}$$

$$R_{2n}(0,\gamma_0) = \gamma_0 + (1-\gamma_0)\frac{l}{l_0} = \frac{1+\gamma l}{1+\gamma l_0}. \tag{1.2.45}$$

Robin-Dirichlet problem. If the right end of the string is clamped, then the Dirichlet condition (1.2.24) is obtained.

Robin-Neumann problem. If the right end of the string can move freely in the direction orthogonal to the equilibrium position of the string, then the Neumann condition (1.2.27) is obtained.

Theorem 1.2.14. 1. $R_{2n}(\cdot,\gamma_0)$ *is the characteristic polynomial of problem* (1.2.4) *with Robin-Dirichlet conditions* (1.2.42), (1.2.24).

2. $R_{2n-1}(\cdot,\gamma_0)$ *is the characteristic polynomial of problem* (1.2.4) *with Robin-Neumann conditions* (1.2.42), (1.2.27).

3. *The zeros of* $R_{2n}(\cdot,\gamma_0)$ *as well as the zeros of* $R_{2n-1}(\cdot,\gamma_0)$ *are real and simple, and the sequence* $(\nu_k^{(\gamma)})_{k=1}^n$ *of the zeros of* $R_{2n}(\cdot,\gamma_0)$ *and the sequence* $(\mu_k^{(\gamma)})_{k=1}^n$ *of the zeros of* $R_{2n-1}(\cdot,\gamma_0)$ *can be indexed in such a way that*

$$0 < \mu_1^{(\gamma)} < \nu_1^{(\gamma)} < \cdots < \mu_n^{(\gamma)} < \nu_n^{(\gamma)}. \tag{1.2.46}$$

Proof. Statement 1 and 2 are Proposition 1.2.6 for $c = \gamma_0$ and $d = 0$ or $d = 1$, respectively. Statement 3 follows from Proposition 1.2.5. ☐

Remark 1.2.15. We have mentioned at the beginning of this subsection that the Robin condition can be considered as a Dirichlet condition for a prolonged string. Indeed, defining

$$\widehat{l}_0 = l_0 + \frac{1}{\gamma}, \quad \widehat{u}_0 = u_0 - \frac{u_1 - u_0}{l_0\gamma},$$

a straightforward calculation gives

$$\frac{u_1 - \widehat{u}_0}{\widehat{l}_0} = \frac{u_1 - u_0}{l_0}.$$

It is easy to see that $c \in (0,1)$ can be written as $c = \gamma_0$ for $\gamma = \frac{1}{l_0}(\frac{1}{c} - 1) > 0$. Together with the corresponding procedure at the right end it therefore follows that the problem (1.2.4), (1.2.5), (1.2.6) with $c, d \in (0,1)$ is equivalent to the Dirichlet-Dirichlet problem for a string which is prolonged by $l_0\frac{c}{1-c}$ on the left and by $l_n\frac{d}{1-d}$ on the right.

1.2.5 Lagrange identity and Sturm oscillation theorem

The following theorem is the discrete analogue of of the Lagrange identity in the theory of differential equations.

Theorem 1.2.16 (Lagrange identity). *For all* $c_1, c_2 \in \mathbb{R}$ *and* $k = 1, \ldots, n$,

$$R_{2k-1}(z, c_1)R_{2k}(z, c_2) - R_{2k}(z, c_1)R_{2k-1}(z, c_2) = \frac{c_2 - c_1}{l_0}. \tag{1.2.47}$$

Proof. The recurrence relations (1.2.10) and (1.2.11) give for $k = 2, \ldots, n$ that

$$\begin{aligned} &R_{2k-1}(z, c_1)R_{2k}(z, c_2) - R_{2k}(z, c_1)R_{2k-1}(z, c_2) \\ &\qquad = R_{2k-1}(z, c_1)R_{2k-2}(z, c_2) - R_{2k-2}(z, c_1)R_{2k-1}(z, c_2) \\ &\qquad = R_{2k-3}(z, c_1)R_{2k-2}(z, c_2) - R_{2k-2}(z, c_1)R_{2k-3}(z, c_2). \end{aligned}$$

Proceeding this way and using (1.2.12) and (1.2.14) we arrive at

$$\begin{aligned} &R_{2k-1}(z, c_1)R_{2k}(z, c_2) - R_{2k}(z, c_1)R_{2k-1}(z, c_2) \\ &\qquad = R_{-1}(z, c_1)R_0(z, c_2) - R_0(z, c_1)R_{-1}(z, c_2) = \frac{c_2 - c_1}{l_0}. \end{aligned}$$ □

Given three of the values on the left-hand side of the Lagrange identity allows to find the fourth one, provided the other factor in the corresponding product is not zero. The following two examples show that one can find string problems where both factors have the value 0 at the right endpoint for some positive z.

Example 1.2.17. Let $n \geq 3$. Then there is a string with n beads such that

$$R_{2n}(2, 0) = 0, \quad R_{2n-1}(2, 0) = 1, \quad R_{2n}(2, 1) = 1, \quad R_{2n-1}(2, 1) = 0. \tag{1.2.48}$$

One can choose the string with $l_k = 1$ $(k = 0, \ldots, n)$ and $m_k = 1$ $(k = 1, \ldots, n)$ except for the following string data:

(i) if $n = 4\widehat{n} - 1$ $(\widehat{n} \in \mathbb{N})$, then $m_{2\widehat{n}} = 2$;

(ii) if $n = 4\widehat{n}$ $(\widehat{n} \in \mathbb{N})$, then $l_{2\widehat{n}} = l_{2\widehat{n}+2} = \frac{1}{2}$;

(iii) if $n = 4\widehat{n} + 1$ $(\widehat{n} \in \mathbb{N})$, then $m_{2\widehat{n}-1} = m_{2\widehat{n}} = m_{2\widehat{n}+2} = m_{2\widehat{n}+3} = \frac{1}{2}$;

(iv) if $n = 4\widehat{n} + 2$ $(\widehat{n} \in \mathbb{N})$, then $l_{2\widehat{n}-1} = l_{2\widehat{n}} = l_{2\widehat{n}+2} = l_{2\widehat{n}+3} = \frac{3}{2}$.

Proof. Defining

$$\widehat{T}(a, b) := \begin{pmatrix} 1 & -2b \\ a & 1 - 2ab \end{pmatrix}, \quad T_n := \widehat{T}(l_n, m_n) \cdots \widehat{T}(l_1, m_1),$$

it follows from Lemma 1.2.4 that

$$\begin{pmatrix} R_{2n-1}(2, c) \\ R_{2n}(2, c) \end{pmatrix} = T_n \begin{pmatrix} R_{-1}(2, c) \\ R_0(2, c) \end{pmatrix} = T_n \begin{pmatrix} 1 - c \\ 1 \end{pmatrix},$$

where we have used that $l_0 = 1$. Then (1.2.48) will hold if and only if

$$T_n = \begin{pmatrix} 1 & 0 \\ -1 & 1 \end{pmatrix}. \tag{1.2.49}$$

Observing

$$\widehat{T}(1,1)^2 = \begin{pmatrix} 1 & -2 \\ 1 & -1 \end{pmatrix}\begin{pmatrix} 1 & -2 \\ 1 & -1 \end{pmatrix} = \begin{pmatrix} -1 & 0 \\ 0 & -1 \end{pmatrix}$$

and writing $n = 4(\widehat{n}-1) + n_0$ with $n_0 \in \{3,4,5,6\}$ it follows from $l_k = 1$ and $m_k = 1$ for $k = 1,\dots,2\widehat{n}-2$ and $k = 2\widehat{n}+n_0-1,\dots,n$ that

$$T_n = (\widehat{T}(1,1))^{2(\widehat{n}-1)}\widetilde{T}_{n_0}(\widehat{T}(1,1))^{2(\widehat{n}-1)} = (-I)^{\widehat{n}-1}\widetilde{T}_{n_0}(-I)^{\widehat{n}-1} = \widetilde{T}_{n_0},$$

where

$$\widetilde{T}_{n_0} = \widehat{T}(l_{2\widehat{n}+n_0-2}, m_{2\widehat{n}+n_0-2})\dots\widehat{T}(l_{2\widehat{n}-1}, m_{2\widehat{n}-1}).$$

Hence (1.2.49) is equivalent to

$$\widetilde{T}_{n_0} = \begin{pmatrix} 1 & 0 \\ -1 & 1 \end{pmatrix} \quad (n_0 = 3,4,5,6). \tag{1.2.50}$$

A straightforward calculation shows that (1.2.50) is true for the data as given by (i)–(iv) and the remaining parameters being 1. □

Example 1.2.18. Let $n \in \mathbb{N} \setminus \{2\}$. Then there is a string with n beads such that

$$R_{2n}(2,0) = 1, \quad R_{2n-1}(2,0) = 0, \quad R_{2n}(2,1) = 0, \quad R_{2n-1}(2,1) = -1. \tag{1.2.51}$$

One can choose the string with $l_k = 1$ $(k = 0,\dots,n)$ and $m_k = 1$ $(k = 1,\dots,n)$ except for the following string data:

(i) if $n = 4\widehat{n}-1$ $(\widehat{n} \in \mathbb{N})$, then $l_{2\widehat{n}+1} = 3$, $m_{2\widehat{n}-1} = \frac{3}{2}$ and $m_{2\widehat{n}} = 2$;

(ii) if $n = 4\widehat{n}$ $(\widehat{n} \in \mathbb{N})$, then $l_{2\widehat{n}+2} = 2$ and $m_{2\widehat{n}+1} = \frac{3}{2}$;

(iii) if $n = 4\widehat{n}+1$ $(\widehat{n} \in \mathbb{N} \cup \{0\})$, then $m_{2\widehat{n}+1} = \frac{1}{2}$;

(iv) if $n = 4\widehat{n}+2$ $(\widehat{n} \in \mathbb{N})$, then $m_{2\widehat{n}-1} = m_{2\widehat{n}} = m_{2\widehat{n}+3} = m_{2\widehat{n}+4} = \frac{1}{2}$.

Proof. With the notation and reasoning from Example 1.2.17 it it clear that (1.2.51) holds if and only if

$$T_n = \begin{pmatrix} 1 & -1 \\ 1 & 0 \end{pmatrix}, \tag{1.2.52}$$

and that this holds if and only if

$$\widetilde{T}_{n_0} = \begin{pmatrix} 1 & -1 \\ 1 & 0 \end{pmatrix} \quad (n_0 = 1,3,4,6). \tag{1.2.53}$$

A straightforward calculation shows that (1.2.53) is true for the data as given by (i)–(iv) and the remaining parameters being 1. □

Define a partition $X = \{0 = x_0 < x_1 < \cdots < x_n < x_{n+1} = l\}$ of $[0, l]$ by

$$x_k = x_{k-1} + l_{k-1} \quad (k = 1, \ldots, n). \tag{1.2.54}$$

Then

$$x_n + l_n = \sum_{j=0}^{n-1} l_j + l_n = l = x_{n+1},$$

and hence (1.2.54) also holds for $k = n + 1$. Therefore

$$l_k = x_{k+1} - x_k \quad (k = 0, \ldots, n),$$

and (1.2.4) can be written in the form

$$\frac{u_k - u_{k-1}}{x_k - x_{k-1}} - \frac{u_{k+1} - u_k}{x_{k+1} - x_k} - m_k z u_k = 0 \quad (k = 1, \ldots, n). \tag{1.2.55}$$

Let $z \in \mathbb{C}$ and let $u = (u_0, \ldots, u_{n+1})$ be a displacement vector given by a solution of the recurrence relation (1.2.55). It is easy to see that any two consecutive entries u_k and u_{k+1} $(k = 0, \ldots, n)$ uniquely determine u. In particular, if $u \neq 0$, then $u_k \neq 0$ or $u_{k+1} \neq 0$ for all $k = 0, \ldots, n$. Clearly, the set of solutions u is a two-dimensional vector space. The corresponding movement of the Stieltjes string under the separation ansatz is described by the real or imaginary part of $\phi(x)e^{i\lambda t}$, when λ is a real parameter with $z = \lambda^2$ and x is the distance of a point on the string in equilibrium position from the left endpoint. The amplitude function ϕ is piecewise linear and continuous, and its graph is given by the line segments connecting (x_k, u_k) and (x_{k+1}, u_{k+1}) for $k = 0, \ldots, n$. Observe that the function ϕ is uniquely determined by the partition X and the displacement vector u, and we therefore also denote ϕ by $\phi_{X,u}$.

The following figures show two examples of the amplitude function with $n = 6$, $l_0 = 13$, $l_1 = 18$, $l_2 = 11$, $l_3 = 20$, $l_4 = 22$, $l_5 = 15$, $l_6 = 19$, $m_1 = 1$, $m_2 = 2$, $m_3 = 1$, $m_4 = 2$, $m_5 = 1$, $m_6 = 2$, $\lambda = 0.25$, where the dotted line is the equilibrium position of the string:

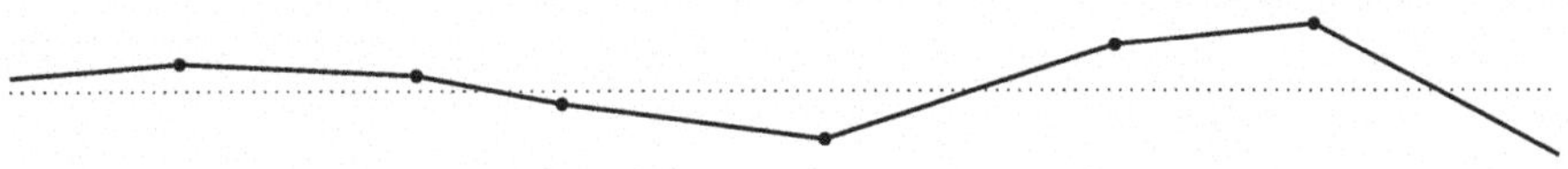

Figure 1.3: $u_0 = 1$ and $u_1 = 2$

Figure 1.4: $u_0 = 0$ and $u_1 = 1$

A point $x \in [0,l]$ is called a *node of u* if $\phi(x) = 0$. A node $x \in (0,l)$ is called an *internal node*. For example, in each of the above two examples there are three internal nodes.

It is convenient to consider nodes as partition points of the Stieltjes string. To this end we generalize Stieltjes strings to strings which may have massless partition points. Hence let $m \geq n$ and let $\widetilde{X} = \{0 = \widetilde{x}_0 < \widetilde{x}_1 < \cdots < \widetilde{x}_m < \widetilde{x}_{m+1}\}$ be a partition of $[0,l]$ with $X \subset \widetilde{X}$. This means $\widetilde{x}_0 = x_0$, $\widetilde{x}_{m+1} = x_{n+1}$, and that there is a strictly increasing sequence $(\tau_k)_{k=0}^{n+1}$ such that such that $x_k = \widetilde{x}_{\tau_k}$ for $k = 0,\dots,n+1$. Then we define $\widetilde{m}_{\tau_k} = m_k$ for $k = 1,\dots,n$ and $\widetilde{m}_k = 0$ if $j \in \{1,\dots,m\}$ but $j \neq \tau_k$ for all $k = 1,\dots,n$, and consider the recurrence relation

$$\frac{\widetilde{u}_k - \widetilde{u}_{k-1}}{\widetilde{x}_k - \widetilde{x}_{k-1}} - \frac{\widetilde{u}_{k+1} - \widetilde{u}_k}{\widetilde{x}_{k+1} - \widetilde{x}_k} - \widetilde{m}_k z \widetilde{u}_k = 0 \quad (k = 1,\dots,m). \tag{1.2.56}$$

The recurrence relations (1.2.55) and (1.2.56) have unique solutions $u = (u_0,\dots,u_{n+1})$ and $\widetilde{u} = (\widetilde{u}_0,\dots,\widetilde{u}_{n+1})$ subject to initial conditions for u_0, u_1 and $\widetilde{u}_0$, $\widetilde{u}_1$, respectively. Recall that we denote the corresponding piecewise linear continuous functions on $[0,l]$ by $\phi_{X,u}$, and $\phi_{\widetilde{X},\widetilde{u}}$, respectively.

Lemma 1.2.19. *Let X, u and $\widetilde{X}$ be as above and let $\widetilde{u} = (\widetilde{u}_0,\dots,\widetilde{u}_{m+1})$ be the solution of* (1.2.56) *with the initial values $\widetilde{u}_0 = u_0$ and $\widetilde{u}_1 = \widetilde{u}_0 + \frac{\widetilde{x}_1}{x_1}(u_1 - u_0)$. Then $\phi_{\widetilde{X},\widetilde{u}} = \phi_{X,u}$.*

Proof. By induction on $k = 1,\dots,m+1$ we will show that the two functions $\phi_{\widetilde{X},\widetilde{u}}$ and $\phi_{X,u}$ coincide on the interval $[0,x_k]$ for $k = 1,\dots,n$. Since

$$\frac{\widetilde{u}_1 - \widetilde{u}_0}{\widetilde{x}_1} = \frac{u_1 - u_0}{x_1},$$

both $\phi_{\widetilde{X},\widetilde{u}}$ and $\phi_{X,u}$ are linear on $[0,\widetilde{x}_1]$ with the same initial point $\phi_{\widetilde{X},\widetilde{u}}(0) = u_0 = \phi_{X,u}(0)$ and the same slope. This proves the induction base if $\widetilde{x}_1 = x_1$. Otherwise, $\tau_1 > 1$, and (1.2.56) for $j = 1,\dots,\tau_1 - 1$ gives

$$\frac{\widetilde{u}_{j+1} - \widetilde{u}_j}{\widetilde{x}_{j+1} - \widetilde{x}_j} = \frac{\widetilde{u}_j - \widetilde{u}_{j-1}}{\widetilde{x}_j - \widetilde{x}_{j-1}},$$

that is, all line segments of the graph of $\phi_{\widetilde{X},\widetilde{u}}$ between 0 and x_1 have the same slope, and we already know that this slope equals the slope of $\phi_{X,u}$ on $[0,x_1]$. This shows that $\phi_{\widetilde{X},\widetilde{u}}$ coincides with $\phi_{X,u}$ on $[0,x_1]$ also when $\widetilde{x}_1 \neq x_1$.

Now let $k \in \{1,\dots,n\}$ and assume that $\phi_{\widetilde{X},\widetilde{u}}$ coincides with $\phi_{X,u}$ on $[0,x_k]$. Then

$$\begin{aligned}
\frac{\widetilde{u}_{\tau_k+1} - \widetilde{u}_{\tau_k}}{\widetilde{x}_{\tau_k+1} - \widetilde{x}_{\tau_k}} &= \frac{\widetilde{u}_{\tau_k} - \widetilde{u}_{\tau_k-1}}{\widetilde{x}_{\tau_k} - \widetilde{x}_{\tau_k-1}} - \widetilde{m}_{\tau_k} z \widetilde{u}_{\tau_k} \\
&= \frac{u_k - u_{k-1}}{x_k - x_{k-1}} - m_k z u_k \\
&= \frac{u_{k+1} - u_k}{x_{k+1} - x_k},
\end{aligned}$$

where we have used that by induction hypothesis the slopes of $\phi_{\widetilde{X},\widetilde{u}}$ and $\phi_{X,u}$ coincide on $[x_{k-1}, x_k]$ and thus on $[\widetilde{x}_{\tau_k-1}, x_k]$. Hence $\phi_{\widetilde{X},\widetilde{u}}$ and $\phi_{X,u}$ have the same slope on $[x_k, \widetilde{x}_{\tau_k+1}]$ and the same initial value $u_k = \widetilde{u}_{\tau_k}$ at $x_k = \widetilde{x}_{\tau_k}$. Proceeding as in the first part of the proof it follows that $\phi_{\widetilde{X},\widetilde{u}}$ and $\phi_{X,u}$ coincide on $[x_k, x_{k+1}]$. □

Theorem 1.2.20. *Let z_1 and z_2 be nonnegative real numbers satisfying $z_2 \geq z_1$ and let $u = (u_0, \ldots, u_{n+1})$ and $w = (w_0, \ldots, w_{n+1})$ be solutions of the recurrence relation* (1.2.4) *with $\lambda^2 = z_1$ and $\lambda^2 = z_2$, respectively. If $z_1 = z_2$ let these solutions be linearly independent. Then between any two nodes of u there exists at least one node of w.*

Proof. In view of Lemma 1.2.19 we may assume without loss of generality that the nodes of both u and w are included as partition points, that is, we allow $m_k = 0$ for some k. By (1.2.4) we have

$$\frac{(u_k - u_{k-1})w_k}{l_{k-1}} - \frac{(u_{k+1} - u_k)w_k}{l_k} - m_k z_1 u_k w_k = 0$$

and

$$\frac{u_k(w_k - w_{k-1})}{l_{k-1}} - \frac{u_k(w_{k+1} - w_k)}{l_k} - m_k z_2 u_k w_k = 0$$

for $k = 1, \ldots, n$, and therefore

$$\begin{aligned}\frac{(u_k - u_{k-1})w_k}{l_{k-1}} - \frac{(u_{k+1} - u_k)w_k}{l_k} - \frac{u_k(w_k - w_{k-1})}{l_{k-1}} + \frac{u_k(w_{k+1} - w_k)}{l_k}&\\ - m_k(z_1 - z_2)u_k w_k = 0,&\end{aligned}$$

which simplifies to

$$-\frac{u_{k-1}w_k}{l_{k-1}} - \frac{u_{k+1}w_k}{l_k} + \frac{u_k w_{k-1}}{l_{k-1}} + \frac{u_k w_{k+1}}{l_k} - m_k(z_1 - z_2)u_k w_k = 0. \tag{1.2.57}$$

Now assume that x_{k_1} and x_{k_2}, $0 \leq k_1 < k_2 \leq n+1$, are adjacent nodes of u. Since $u \neq 0$ we must have $k_2 \neq k_1 + 1$, so that $k_1 + 1 \leq k_2 - 1$. Again by Lemma 1.2.19 and its proof we may remove all massless partition points between x_{k_1} and x_{k_2}, and the same argument shows that there is an integer k_0 with $k_1 < k_0 < k_2$ and $m_{k_0} > 0$. Adding equations (1.2.57) with $k = k_1 + 1, \ldots, k_2 - 1$ we obtain

$$-\frac{u_{k_2}w_{k_2-1}}{l_{k_2-1}} + \frac{u_{k_2-1}w_{k_2}}{l_{k_2-1}} + \frac{u_{k_1+1}w_{k_1}}{l_{k_1}} - \frac{u_{k_1}w_{k_1+1}}{l_{k_1}} = (z_1 - z_2)\sum_{k=k_1+1}^{k_2-1} m_k u_k w_k.$$

Assume that w has no node between x_{k_1} and x_{k_2}. Without loss of generality we may assume that $u_k > 0$ and $w_k > 0$ for $k = k_1 + 1, \ldots k_2 - 1$, replacing u and w with their negatives, if necessary. Then $u_{k_1} = 0 = u_{k_2}$ leads to

$$\frac{u_{k_2-1}w_{k_2}}{l_{k_2-1}} + \frac{u_{k_1+1}w_{k_1}}{l_{k_1}} = (z_1 - z_2)\sum_{k=k_1+1}^{k_2-1} m_k u_k w_k \leq (z_1 - z_2)m_{k_0}u_{k_0}w_{k_0}. \tag{1.2.58}$$

By construction and assumption, the left-hand side of (1.2.58) is nonnegative. But the right-hand side is negative if $z_1 < z_2$, so that we have arrived at a contradiction in this case. Finally, let $z_1 = z_2$. Assume that $w_{k_1} = 0$. Since u and w are uniquely determined by u_{k_1}, u_{k_1+1} and w_{k_1}, w_{k_1+1} via the linear recurrence relation (1.2.4), it would follow that $w = w_{k_1+1}u_{k_1+1}^{-1}u$, which would contradict the linear independence of u and w. Hence $w_{k_1} > 0$, which shows that the left-hand side of (1.2.58) is positive, whereas the right-hand side is zero. Therefore also in this case we have a contradiction, and the proof of the theorem is complete. □

Corollary 1.2.21. *Let $u = (u_0, \dots, u_{n+1})$ and $w = (w_0, \dots, w_{n+1})$ be linearly independent solutions of the recurrence relation* (1.2.4). *Then between any two adjacent nodes of u there exists exactly one node of w.*

Proof. By Theorem 1.2.20 we know that between two adjacent nodes of u, say at y_1 and y_2, there is at least one node of w. If there were at least two such nodes, then interchanging the rôles of u and w would give a node y_3 of u between these two nodes of w, and y_3 would lie between y_1 and y_2, contradicting the fact that y_1 and y_2 are adjacent nodes of u. □

Theorem 1.2.22. *Let $n \in \mathbb{N}$ and $c, d \in [0, 1]$. Then an eigenvector of the k-th eigenvalue ($k = 1, \dots, n$) of the boundary value problem* (1.2.4), (1.2.5), (1.2.6) *possesses $k - 1$ internal nodes.*

Proof. Consider a nontrivial solution vector u with amplitude function ϕ of the recurrence relation (1.2.4). Recall that the graph of ϕ consists of $n+1$ line segments, none of which is in equilibrium position. Hence each line segment can contribute at most one node, which means that the number of nodes of u is at most $n + 1$.

Next consider the initial conditions $u_0 = 0$, $u_1 = 1$ and for $z \in [0, \infty)$ denote the amplitude function of the corresponding solution vector $u = u(z)$ by $\phi_D(\cdot, z)$. Let $\kappa_D(z)$ be the number of internal nodes of $u(z)$. We already know that the total number of nodes is at most $n + 1$, and since 0 is a node in this case, it follows that $\kappa_D(z) \le n$ for all $z \in [0, \infty)$. Recall from Proposition 1.2.8 that the characteristic values of the Dirichlet problem are denoted by ν_k ($k = 1, \dots, n$) and satisfy $0 < \nu_k < \nu_{k+1}$ ($k = 1, \dots, n - 1$). Since 0 and l are nodes of $u(\nu_n)$, it follows that $\kappa_D(\nu_n) \le n - 1$.

Now let $0 \le z < y$. Since 0 is a node of $u(z)$, $u(z)$ has at least $\kappa_D(z) + 1$ nodes and therefore $u(y)$ has at least $\kappa_D(z)$ internal nodes by Theorem 1.2.20. This shows that κ_D is a nondecreasing function on $[0, \infty)$. Furthermore, for each $k = 1, \dots, n$, both endpoints are nodes of $u(\nu_k)$, and therefore $u(\nu_k)$ has $\kappa_D(\nu_k)+2$ nodes, and another application of Theorem 1.2.20 gives that $\kappa_D(\nu_k) + 1 \le \kappa_D(z)$ for all $z > \nu_k$. In particular

$$0 \le \kappa_D(\nu_1) < \kappa_D(\nu_2) < \cdots < \kappa_D(\nu_n) \le n - 1.$$

But this immediately gives $\kappa_D(\nu_k) = k - 1$ for $k = 1, \dots, n$, that is, the statement of the Corollary for the Dirichlet-Dirichlet problem has been proved.

Next we observe that the above reasoning completely determines the function κ_D on $[0, \infty)$, namely, $\kappa_D(z) = 0$ for $z \in [0, \nu_1]$, $\kappa_D(z) = k$ for $z \in (\nu_k, \nu_{k+1}]$ $(k = 1, \dots, n)$, and $\kappa_D(z) = n$ for $z \in (\nu_n, \infty)$. Then (1.2.36) proves the statement of this corollary for the Dirichlet-Neumann eigenvalues μ_k $(k = 1, \dots, n)$ and by symmetry also for the Neumann-Dirichlet eigenvalues ξ_k $(k = 1, \dots, n)$.

For the Neumann-Neumann eigenvalues ζ_k $(k = 1, \dots, n)$ we first observe that $\zeta_1 = 0$, and in that case the amplitude function is constant and has therefore no zeros, i.e., $\kappa_N(\zeta_1) = 0$, where $\kappa_N(z)$ $(z \in [0, \infty))$ is the number of internal nodes of the solution $u = u(z)$ of (1.2.4) with $u_0 = u_1 = 1$. Next we observe that $u_{n+1} = R_{2n}(\cdot, 1)$ has simple zeros by Proposition 1.2.10, so that u_{n+1} changes sign at the Neumann-Dirichlet eigenvalues ξ_k $(k = 1, \dots, n)$. Since u is constant at $z = 0$, we have that $R_{2n}(0, 1) = 1 > 0$. Hence $R_{2n}(z, 1) > 0$ for $z < \xi_1$ and $(-1)^{k-1} R_{2n}(z, 1) > 0$ for $\xi_{k-1} < z < \xi_k$ $(k = 2, \dots, n)$. Using the notation with possibly massless internal nodes, let $\widetilde{u}_k$ be such a node. Then (1.2.56) gives

$$\frac{\widetilde{u}_{k-1}}{\widetilde{x}_k - \widetilde{x}_{k-1}} + \frac{\widetilde{u}_{k+1}}{\widetilde{x}_{k+1} - \widetilde{x}_k} = 0,$$

which shows that $\widetilde{u}_{k-1}$ and $\widetilde{u}_{k+1}$ have opposite signs. In other words, u changes sign at each internal node. Therefore the parity of the number of internal nodes equals the sign of u_{n+1}. The above reasoning and (1.2.39) lead to

$$(-1)^{\kappa_N(\zeta_k)} = (-1)^{k-1} \quad (k = 2, \dots, n). \tag{1.2.59}$$

On the other hand, Corollary 1.2.21 shows that

$$\kappa_D(z) \le \kappa_N(z) \le \kappa_D(z) + 1$$

for all $z \in [0, \infty)$. Recalling that κ_D is nondecreasing, it follows in view of (1.2.38) that

$$\kappa_D(\mu_{k-1}) \le \kappa_D(\zeta_k) \le \kappa_N(\zeta_k) \le \kappa_D(\zeta_k) + 1 \le \kappa_D(\mu_k) + 1.$$

Inserting $\kappa_D(\mu_j) = j - 1$ leads to

$$k - 2 \le \kappa_N(\zeta_k) \le k.$$

Together with (1.2.59) the required identity $\kappa_N(\zeta_k) = k - 1$ follows.

So far we have proved the theorem for $c, d \in \{0, 1\}$. If one or both of c, d belong to $(0, 1)$, say $d \in (0, 1)$ and $c \in \{0, 1\}$, then according to Remark 1.2.15 we prolong the string with partition $X = (x_0, \dots, x_n, x_{n+1})$ to a string with partition $\widehat{X} = (x_0, \dots, x_n, \widehat{x}_{n+1})$ with $\widehat{x}_{n+1} > x_{n+1}$ so that the boundary condition (1.2.6) becomes the Dirichlet condition at the right end of the prolonged string. We have already shown that the amplitude function $\phi_{\widehat{X}, \widehat{u}}$ for an eigenvector $\widehat{u}$ corresponding to the k-th eigenvalue has $k-1$ internal nodes. Then $\phi_{\widehat{X}, \widehat{u}}|_{[x_0, x_{n+1}]}$ is an eigenvector $\phi_{X,u}$ of the boundary value problem on the original string corresponding to the k-th eigenvalue. Since $\phi_{\widehat{X}, \widehat{u}}$ is linear on the interval $[x_n, \widehat{x}_{n+1}]$ with $\phi_{\widehat{X}, \widehat{u}}(\widehat{x}_{n+1}) = 0$, the function $\phi_{\widehat{X}, \widehat{u}}$ has no internal nodes in $[x_n, \widehat{x}_{n+1})$. Therefore, $\phi_{X,u}$ has the same number of internal nodes as $\phi_{\widehat{X}, \widehat{u}}$, that is, $k - 1$ nodes. The same reasoning can be applied to the remaining cases. □

1.3 Inverse problems for Stieltjes strings

1.3.1 Two-spectra inverse problems

Theorem 1.3.1. *Let $n \in \mathbb{N}$ and $l > 0$ be given together with two sequences of real numbers $(\nu_k)_{k=1}^n$ and $(\mu_k)_{k=1}^n$ which have the interlacing property (1.2.36).*

Then there exists a unique pair of sequences of positive numbers $(m_k)_{k=1}^n$ and $(l_k)_{k=0}^n$ with $\sum\limits_{k=0}^n l_k = l$ such that the characteristic values of the Dirichlet-Dirichlet problem (1.2.4), (1.2.23), (1.2.24) are ν_k $(k = 1, \dots, n)$ and such that the characteristic values of the Dirichlet-Neumann problem (1.2.4), (1.2.23), (1.2.27) are μ_k $(k = 1, \dots, n)$.

Proof. The function f defined by

$$f(z) = l\frac{\prod\limits_{k=1}^n \left(1 - \dfrac{z}{\nu_k}\right)}{\prod\limits_{k=1}^n \left(1 - \dfrac{z}{\mu_k}\right)} \tag{1.3.1}$$

is a rational $\mathcal{S}$-function by Corollary A.3.3. In view of Lemma A.3.5 there are positive numbers $l_0, \dots, l_n$ and positive numbers $m_1, \dots, m_n$ such that

$$f(z) = l_n + \cfrac{1}{-m_n z + \cfrac{1}{l_{n-1} + \cfrac{1}{-m_{n-1} z + \ddots + \cfrac{1}{l_1 + \cfrac{1}{-m_1 z + \cfrac{1}{l_0}}}}}}. \tag{1.3.2}$$

Clearly,

$$l = f(0) = \sum_{k=0}^n l_k .$$

Defining the functions $R_k(\cdot, 0)$ by (1.2.10), (1.2.11), (1.2.12), (1.2.13) with $c = 0$ and with the above values of l_k and m_k, it follows from (1.2.19) and (1.3.2) that

$$f(z) = \frac{R_{2n}(z, 0)}{R_{2n-1}(z, 0)}. \tag{1.3.3}$$

But the n zeros of $R_{2n}(\cdot, 0)$ are the characteristic values of the problem (1.2.4), (1.2.23), (1.2.24) by Proposition 1.2.8, and the n zeros of $R_{2n-1}(\cdot, 0)$ are the characteristic values of the problem (1.2.4), (1.2.23), (1.2.27) by Proposition 1.2.9. This shows that the two constructed problems have the required characteristic values.

Since the rational $\mathcal{S}$-function f is uniquely determined by its zeros, its poles and $f(0)$, which are given data, the uniqueness part of Lemma A.3.5 and the fact that $\frac{R_{2n}(\cdot,0)}{R_{2n-1}(\cdot,0)}$ has the representation (1.2.19) show that the sequences $(m_k)_{k=1}^n$ and $(l_k)_{k=0}^n$ with $\sum\limits_{k=0}^{n} l_k = l$ are uniquely determined by the given data. □

Theorem 1.3.2. *Let $n \in \mathbb{N}$ and $l > 0$ be given together with two sequences of real numbers $(\xi_k)_{k=1}^n$ and $(\nu_k)_{k=1}^n$ which have the interlacing property* (1.2.37).

Then there exists a unique pair of sequences of positive numbers $(m_k)_{k=1}^n$ and $(l_k)_{k=0}^n$ with $\sum\limits_{k=0}^{n} l_k = l$ such that the characteristic values of the Dirichlet-Dirichlet problem (1.2.4), (1.2.23), (1.2.24) *are ν_k $(k = 1, \ldots, n)$ and such that the characteristic values of the Neumann-Dirichlet problem* (1.2.4), (1.2.29), (1.2.24) *are ξ_k $(k = 1, \ldots, n)$.*

Proof. By Theorem 1.3.1 there is a unique pair of sequences of positive numbers $(\widehat{m}_k)_{k=1}^n$ and $(\widehat{l}_k)_{k=0}^n$ with $\sum\limits_{k=0}^{n} \widehat{l}_k = l$ such that the problem

$$\frac{\widehat{u}_k - \widehat{u}_{k-1}}{\widehat{l}_{k-1}} - \frac{\widehat{u}_{k+1} - \widehat{u}_k}{\widehat{l}_k} - \widehat{m}_k z \widehat{u}_k = 0 \quad (k = 1, \ldots, n), \tag{1.3.4}$$

$\widehat{u}_0 = 0$, $\widehat{u}_{n+1} = 0$ has the characteristic values ν_k $(k = 1, \ldots, n)$ and such that the problem (1.3.4), $\widehat{u}_0 = 0$, $\widehat{u}_{n+1} = \widehat{u}_n$ has the characteristic values ξ_k $(k = 1, \ldots, n)$. Putting $m_k = \widehat{m}_{n+1-k}$ $(k = 1, \ldots, n)$, $l_k = \widehat{l}_{n-k}$ $(k = 0, \ldots, n)$, $u_k = \widehat{u}_{n+1-k}$ $(k = 0, \ldots, n+1)$, (1.3.4) becomes (1.2.4), and therefore the numbers ν_k $(k = 1, \ldots, n)$ are the characteristic values of (1.2.4), (1.2.23), (1.2.24), whereas the numbers ξ_k $(k = 1, \ldots, n)$ are the characteristic values of (1.2.4), (1.2.29), (1.2.24).

The uniqueness is clear from the uniqueness in Theorem 1.3.1 and the fact that the relationship between $(m_k)_{k=1}^n$, $(l_k)_{k=0}^n$, $(u_k)_{k=0}^{n+1}$ and $(\widehat{m}_k)_{k=1}^n$, $(\widehat{l}_k)_{k=0}^n$, $(\widehat{u}_k)_{k=0}^{n+1}$ is bijective. □

Theorem 1.3.3. *Let $n \in \mathbb{N}$, $m > 0$ and $l_0 > 0$ be given together with two sequences of real numbers $(\xi_k)_{k=1}^n$ and $(\zeta_k)_{k=1}^n$ which have the interlacing property* (1.2.39).

Then there exists a unique pair of sequences of positive numbers $(m_k)_{k=1}^n$ and $(l_k)_{k=1}^n$ with $\sum\limits_{k=1}^{n} m_k = m$ such that the characteristic values of the Neumann-Dirichlet problem (1.2.4), (1.2.29), (1.2.24) *are ξ_k $(k = 1, \ldots, n)$ and such that the characteristic values of the Neumann-Neumann problem* (1.2.4), (1.2.29), (1.2.27) *are ζ_k $(k = 1, \ldots, n)$.*

Proof. The function f defined by

$$f(z) = -\frac{1}{m} \frac{\prod_{k=1}^{n}\left(1 - \frac{z}{\xi_k}\right)}{z \prod_{k=2}^{n}\left(1 - \frac{z}{\zeta_k}\right)} \tag{1.3.5}$$

is a rational $\mathcal{S}$-function by Corollary A.3.3. In view of Lemma A.3.5 there are positive numbers $l_1, \dots, l_n$ and positive numbers $m_1, \dots, m_n$ such that

$$f(z) = l_n + \cfrac{1}{-m_n z + \cfrac{1}{l_{n-1} + \cfrac{1}{-m_{n-1}z + \ddots + \cfrac{1}{l_1 + \cfrac{1}{-m_1 z}}}}}. \tag{1.3.6}$$

Defining the functions $R_k(\cdot, 1)$ by (1.2.10), (1.2.11), (1.2.12), (1.2.13) with $c = 1$ and with the above values of l_k and m_k, it follows from (1.2.19) and (1.3.6) that

$$f(z) = \frac{R_{2n}(z, 1)}{R_{2n-1}(z, 1)}. \tag{1.3.7}$$

Propositions 1.2.10 and 1.2.11 show that $\sum_{k=1}^{n} m_k = m$. But the n zeros of $R_{2n}(\cdot, 1)$ are the characteristic values of problem (1.2.4), (1.2.29), (1.2.24) by Proposition (1.2.8), whereas the zeros of $R_{2n-1}(\cdot, 1)$ are the characteristic values of problem (1.2.4), (1.2.29), (1.2.27) by Proposition (1.2.10). This shows that the two constructed problems have the required spectrum.

Since the rational $\mathcal{S}$-function f is uniquely determined by its zeros, its poles and $\lim_{z\to 0} z f(z)$, which are given data, the uniqueness part of Lemma A.3.5 and the fact that $\frac{R_{2n}(\cdot,1)}{R_{2n-1}(\cdot,1)}$ is an $\mathcal{S}$-function show that the sequences $(m_k)_{k=1}^{n}$ and $(l_k)_{k=1}^{n}$ with $\sum_{k=1}^{n} m_k = m$ are uniquely determined by the given data. □

The following theorem is a variant of Theorem 1.3.3 with symmetric assumptions with respect to the prescribed lengths of the intervals, whereas the total mass of the beads is then determined by the given data.

Theorem 1.3.4. *Let $n \in \mathbb{N}$, $l_0 > 0$ and $l_n > 0$ be given together with two sequences of real numbers $(\xi_k)_{k=1}^{n}$ and $(\zeta_k)_{k=1}^{n}$ which have the interlacing property* (1.2.39).

Then there exists a unique pair of sequences of positive numbers $(m_k)_{k=1}^{n}$ and $(l_k)_{k=1}^{n-1}$ such that the characteristic values of the Neumann-Dirichlet problem (1.2.4), (1.2.29), (1.2.24) *are ξ_k $(k = 1, \dots, n)$ and such that the characteristic values of the Neumann-Neumann problem* (1.2.4), (1.2.29), (1.2.27) *are ζ_k $(k = 1, \dots, n)$.*

Proof. The function f defined by

$$f(z) = l_n \frac{\prod_{k=1}^{n} (z - \xi_k)}{z \prod_{k=2}^{n} (z - \zeta_k)} \tag{1.3.8}$$

is a rational $\mathcal{S}$-function by Corollary A.3.3 and satisfies $\lim_{z\to-\infty} f(z) = l_n$. In view of Lemma A.3.5 there are positive numbers $l_1, \dots, l_{n-1}$ and positive numbers $m_1, \dots, m_n$ such that

$$f(z) = l_n + \cfrac{1}{-m_n z + \cfrac{1}{l_{n-1} + \cfrac{1}{-m_{n-1} z + \ddots + \cfrac{1}{l_1 + \cfrac{1}{-m_1 z}}}}}. \tag{1.3.9}$$

We can now continue as in the proof of Theorem 1.3.3 to complete the proof of this theorem. □

Theorem 1.3.5. *Let $n \in \mathbb{N}$, $m > 0$ and $l_n > 0$ be given together with two sequences of real numbers $(\mu_k)_{k=1}^n$ and $(\zeta_k)_{k=1}^n$ which have the interlacing property* (1.2.38).

Then there exists a unique pair of sequences of positive numbers $(m_k)_{k=1}^n$ and $(l_k)_{k=1}^n$ with $\sum_{k=1}^{n} m_k = m$ such that the characteristic values of the Dirichlet-Neumann problem (1.2.4), (1.2.23), (1.2.27) *are μ_k $(k = 1, \dots, n)$ and the characteristic values of the Neumann-Neumann problem* (1.2.4), (1.2.29), (1.2.27) *are ζ_k $(k = 1, \dots, n)$.*

Proof. By Theorem 1.3.3 there is a unique pair of sequences of positive numbers $(\widehat{m}_k)_{k=1}^n$ and $(\widehat{l}_k)_{k=1}^n$ with $\sum_{k=1}^{n} \widehat{m}_k = m$ such that problem

$$\frac{\widehat{u}_k - \widehat{u}_{k-1}}{\widehat{l}_{k-1}} - \frac{\widehat{u}_{k+1} - \widehat{u}_k}{\widehat{l}_k} - \widehat{m}_k z \widehat{u}_k = 0 \quad (k = 1, \dots, n), \tag{1.3.10}$$

$\hat{u}_0 = \widehat{u}_1$, $\widehat{u}_{n+1} = 0$ has the characteristic values μ_k $(k = 1, \dots, n)$ and such that problem (1.3.10), $\widehat{u}_0 = \widehat{u}_1$, $\widehat{u}_{n+1} = \widehat{u}_n$ has the characteristic values ζ_k $(k = 1, \dots, n)$. Note that a value for $\widehat{l}_0$ is not needed here since $\widehat{u}_0 = \widehat{u}_1$. Putting $m_k = \widehat{m}_{n+1-k}$ $(k = 1, \dots, n)$, $l_k = \widehat{l}_{n-k}$ $(k = 0, \dots, n-1)$, $u_k = \widehat{u}_{n+1-k}$ $(k = 0, \dots, n+1)$, (1.3.4) becomes (1.2.4), and therefore the numbers μ_k $(k = 1, \dots, n)$ are the characteristic values of (1.2.4), (1.2.23), (1.2.27), whereas the numbers ζ_k $(k = 1, \dots, n)$ are the characteristic values of (1.2.4), (1.2.29), (1.2.27).

The uniqueness is clear from the uniqueness in Theorem 1.3.3 and the fact that the relationship between $(m_k)_{k=1}^n$, $(l_k)_{k=0}^{n-1}$, $(u_k)_{k=0}^{n+1}$ and $(\widehat{m}_k)_{k=1}^n$, $(\widehat{l}_k)_{k=1}^n$, $(\widehat{u}_k)_{k=0}^{n+1}$ is bijective. □

Applying Theorem 1.3.4 while interchanging the values of l_0 and l_n gives

Theorem 1.3.6. *Let $n \in \mathbb{N}$, $l_0 > 0$ and $l_n > 0$ be given together with two sequences of real numbers $(\mu_k)_{k=1}^n$ and $(\zeta_k)_{k=1}^n$ which interlace as in (1.2.38).*

Then there exists a unique pair of sequences of positive numbers $(m_k)_{k=1}^n$ and $(l_k)_{k=1}^{n-1}$ such that the characteristic values of the Dirichlet-Neumann problem (1.2.4), (1.2.23), (1.2.27) are μ_k $(k = 1, \dots, n)$ and such that the characteristic values of the Neumann-Neumann problem (1.2.4), (1.2.29), (1.2.27) are ζ_k $(k = 1, \dots, n)$.

Theorem 1.3.7. *Let $n \in \mathbb{N}$. For two nondecreasing sequences $(\nu_k)_{k=1}^n$ and $(\zeta_k)_{k=1}^n$ of real numbers to be sequences of the characteristic values of the Dirichlet-Dirichlet and the Neumann-Neumann problems generated by the same sequences $(m_k)_{k=1}^n$ and $(l_k)_{k=0}^n$ it is necessary and sufficient that the following conditions are satisfied:*

1. $\zeta_1 = 0$, $\nu_{k-1} < \nu_k$, $\zeta_{k-1} < \zeta_k$ $(k = 2, \dots, n)$.
2. *The sequences are interlaced as follows:*
   - i) $\zeta_k < \nu_k$ $(k = 1, \dots, n)$,
   - ii) $\nu_k < \zeta_{k+2}$ $(k = 1, \dots, n-2)$.

Proof. The necessity of these conditions follows from (1.2.37) and (1.2.39). To prove the sufficiency of these conditions we define the polynomials S_1 and S_2 by

$$S_1(z) = \prod_{k=1}^{n} \left(1 - \frac{z}{\nu_k}\right), \tag{1.3.11}$$

$$S_2(z) = -z \prod_{k=2}^{n} \left(1 - \frac{z}{\zeta_k}\right). \tag{1.3.12}$$

Let $a_1 = \zeta_1 = 0$, $a_k = \max\{\nu_{k-1}, \zeta_k\}$ $(k = 2, \dots, n)$, $b_k = \min\{\nu_k, \zeta_{k+1}\}$ $(k = 1, \dots, n-1)$, $b_n = \nu_n$. The conditions 1 and 2 show that

$$a_1 < b_1 \le a_2 < b_2 \le \cdots \le a_n < b_n,$$

which gives mutually disjoint intervals (a_k, b_k) $(k = 1, \dots, n)$. The function S_1 is positive on $(-\infty, \nu_1)$ and changes sign at each ν_k $(k = 1, \dots, n)$, and the function S_2 is positive on $(-\infty, \zeta_1)$ and changes sign at each ζ_k $(k = 1, \dots, n)$. Therefore, observing in particular that

$$(a_k, b_k) = (\nu_{k-1}, \nu_k) \cap (\zeta_k, \zeta_{k+1}) \quad \text{for } k = 2, \dots, n-1,$$

it follows that $S_1 S_2$ is negative on each of the intervals (a_k, b_k) $(k = 1, \dots, n)$. For each $k \in \{1, \dots, n\}$ let $c_k \in (a_k, b_k)$ (e.g., the midpoint of the interval) and choose a positive number C such that $C < -S_1(c_k)S_2(c_k)$ for $k = 1, \ldots, n$. We put

$$T(z) = C + S_1(z)S_2(z). \tag{1.3.13}$$

Then $T(a_k) = C = T(b_k)$ and $T(c_k) < 0$. Therefore T has at least two zeros in each of the n intervals (a_k, b_k) $(k = 1, \ldots, n)$. On the other hand, T is a polynomial of degree $2n$, so that the number of zeros in each of the intervals (a_k, b_k) $(k = 1, \ldots, n)$ is exactly two. For $k = 1, \ldots, n$ let μ_k be the zero of T in (a_k, c_k) and let ξ_k be the zero of T in (c_k, b_k).

Therefore the sequences $(\nu_k)_{k=1}^n$ and $(\mu_k)_{k=1}^n$ satisfy the assumptions of Theorem 1.3.1. Choosing now $l > 0$ arbitrary, it follows from Theorem 1.3.1 that there are sequences of positive numbers $(m_k)_{k=1}^n$ and $(l_k)_{k=0}^n$ with $\sum\limits_{k=0}^{n} l_k = l$ such that the characteristic values of problem (1.2.4), (1.2.23), (1.2.24) are ν_k $(k = 1, \ldots, n)$ and such that the characteristic values of problem (1.2.4), (1.2.23), (1.2.27) are μ_k $(k = 1, \ldots, n)$.

Since S_1 and $R_{2n}(\cdot, 0)$ have the same zeros by Proposition 1.2.8 and since $S_1(0) = 1 = \frac{l_0}{l} R_{2n}(0, 0)$ by (1.3.11) and (1.2.25), it follows that $S_1 = \frac{l_0}{l} R_{2n}(\cdot, 0)$. In view of Proposition 1.2.11 it remains to show that the zeros of S_2 and $R_{2n-1}(\cdot, 1)$ coincide. We recall the Lagrange identity, Theorem 1.2.16,

$$R_{2n-1}(z, 0)R_{2n}(z, 1) - R_{2n}(z, 0)R_{2n-1}(z, 1) = \frac{1}{l_0}. \tag{1.3.14}$$

Since T and $R_{2n-1}(\cdot, 0)$ have zeros at μ_k $(k = 1, \ldots, n)$, it follows from (1.3.13) and (1.3.14) that

$$C + \frac{l_0}{l} R_{2n}(\mu_k, 0)S_2(\mu_k) = 0 \quad \text{and} \quad \frac{1}{l_0} + R_{2n}(\mu_k, 0)R_{2n-1}(\mu_k, 1) = 0$$

for $k = 1, \ldots n$. This leads to $S_2(\mu_k) = lCR_{2n-1}(\mu_k, 1)$ $(k = 1, \ldots, n)$. Furthermore, $S_2(0) = 0$ and $R_{2n-1}(0, 1) = 0$, and hence the polynomial $S_2 - lCR_{2n}(\cdot, 1)$ has at least $n+1$ zeros. Therefore it is the zero polynomial, that is, $S_2 = lCR_{2n}(\cdot, 1)$ and it follows that problem (1.2.4), (1.2.29), (1.2.27) has the characteristic values ζ_k $(k = 1 \ldots n)$. □

Theorem 1.3.8. *Let $n \in \mathbb{N}$. For two nondecreasing sequences $(\mu_k)_{k=1}^n$ and $(\xi_k)_{k=1}^n$ of positive real numbers to be sequences of the characteristic values of the Dirichlet-Neumann and the Neumann-Dirichlet problems generated by the same sequences $(m_k)_{k=1}^n$ and $(l_k)_{k=0}^n$ it is necessary and sufficient that the following conditions are satisfied:*

1. $\mu_{k-1} < \mu_k$, $\xi_{k-1} < \xi_k$ $(k = 2, \ldots, n)$.
2. *The sequences are interlaced as follows:*
   $\mu_{k-1} < \xi_k$, $\xi_{k-1} < \mu_k$ $(k = 2, \ldots, n)$.
3. *All zeros of the polynomial S are real, where*

$$S(z) = \prod_{k=1}^{n} \left(1 - \frac{z}{\mu_k}\right) \prod_{k=1}^{n} \left(1 - \frac{z}{\xi_k}\right) - 1.$$

Proof. To prove the necessity of these conditions we observe that properties 1 and 2 follow from (1.2.36) and (1.2.37). By (1.2.28), (1.2.30) and Theorem 1.2.16,

$$S(z) = l_0 R_{2n-1}(z,0)R_{2n}(z,1) - 1 = l_0 R_{2n}(z,0)R_{2n-1}(z,1).$$

But the zeros of $R_{2n}(\cdot,0)$ and $R_{2n-1}(\cdot,1)$ are real by Propositions 1.2.8 and 1.2.11. Therefore also property 3 holds.

To prove the sufficiency of these conditions, let $a_k = \min\{\mu_k, \xi_k\}$ and $b_k = \max\{\mu_k, \xi_k\}$ $(k = 1,\dots,n)$. Then $a_k \le b_k$ $(k = 1,\dots,n)$, and, in view of properties 1 and 2, $b_{k-1} < a_k$ $(k = 2,\dots,n)$. Clearly,

$$T(z) := S(z) + 1 = \prod_{k=1}^{n}\left(1 - \frac{z}{\mu_k}\right)\prod_{k=1}^{n}\left(1 - \frac{z}{\xi_k}\right)$$

has the following properties: $T(z) > 0$ when $z < a_1$, $b_{k-1} < z < a_k$ $(k = 2,\dots,n)$, $b_n < z$, and $T(z) < 0$ when $a_k < z < b_k$ $(k = 1,\dots,n)$. Hence there are numbers d_k $(k = 1,\dots,n)$ with $a_k \le d_k \le b_k$ such that $T'(d_k) = 0$ and numbers $c_k \in (b_{k-1}, a_k)$ for $k = 2,\dots,n$ such that $T'(c_k) = 0$. Since T' is a polynomial of degree $2n-1$, T' has no other zeros. In particular, T and therefore also S are strictly increasing on (b_{k-1}, c_k) and strictly decreasing on (c_k, a_k) for $k = 2,\dots,n$, strictly decreasing on $(-\infty, a_1)$ and strictly increasing on (b_n,∞). Observing that $S(a_k) = S(b_k) = -1$ $(k = 1,\dots,n)$, it follows that $\zeta_1 = 0$ is the only zero of S on $(-\infty, a_1)$, that S has at most two zeros, counted with multiplicity, on each interval (b_{k-1}, a_k) $(k = 2,\dots,n)$, that S has exactly one simple zero ν_n on (b_n,∞) since $\lim\limits_{z\to\infty} S(z) = \infty$, and that $S(z) \le -1$ for $a_k \le z \le b_k$ $(k = 1,\dots,n)$. By property 3, S has exactly $2n$ real zeros, and therefore S has exactly two zeros, which we will denote by ν_{k-1} and ζ_k, in each of the intervals (b_{k-1}, a_k) $(k = 2,\dots,n)$. For definiteness, we choose $\nu_{k-1} \le \zeta_k$.

It follows that $\mu_k \le b_k < \nu_k$ $(k = 1,\dots,n)$ and $\nu_{k-1} < a_k \le \mu_k$ $(k = 2,\dots,n)$. Therefore the sequences $(\nu_k)_{k=1}^n$ and $(\mu_k)_{k=1}^n$ satisfy the assumptions of Theorem 1.3.1. Choosing now $l > 0$, it follows from Theorem 1.3.1 that there are sequences of positive numbers $(m_k)_{k=1}^n$ and $(l_k)_{k=0}^n$ with $\sum\limits_{k=0}^{n} l_k = l$ such that the characteristic values of problem (1.2.4), (1.2.23), (1.2.24) are ν_k $(k = 1,\dots,n)$ and such that the characteristic values of problem (1.2.4), (1.2.23), (1.2.27) are μ_k $(k = 1,\dots,n)$.

Defining

$$S_3(z) = \sum_{k=1}^{n}\left(1 - \frac{z}{\xi_k}\right)$$

it remains to show that the zeros of S_3 and $R_{2n}(\cdot,1)$ coincide. For $k = 1,\dots,n$ we have in view of (1.2.28) that

$$R_{2n-1}(\nu_k,0)S_3(\nu_k) = \frac{1}{l_0}T(\nu_k) = \frac{1}{l_0},$$

and by Theorem 1.2.16,

$$R_{2n-1}(\nu_k,0)R_{2n}(\nu_k,1)=\frac{1}{l_0}\,.$$

This shows that $S_3(\nu_k)=R_{2n}(\nu_k,1)$ $(k=1,\dots,n)$. Finally, $S_3(0)=1=R_{2n}(0,1)$ by (1.2.30). Therefore $S_3=R_{2n}(\cdot,1)$ since both polynomials have degree n and since they coincide at $n+1$ points. Hence problem (1.2.4), (1.2.29), (1.2.24) has the characteristic values ξ_k $(k=1,\dots,n)$. □

Theorem 1.3.9. *Let $n\in\mathbb{N}$ and let two sequences of real numbers $(\widehat{\nu}_k)_{k=1}^n$ and $(\widehat{\mu}_k)_{k=1}^n$ be given which have the interlacing property (1.2.46). Then there is a number $\eta>0$ such that the following statement is true:*
For $l>0$ and $\gamma>0$ there exists a pair of sequences of positive numbers $(m_k)_{k=1}^n$ and $(l_k)_{k=0}^n$ with $\sum\limits_{k=0}^{n}l_k=l$ such that the characteristic values of the Robin-Dirichlet problem (1.2.4), (1.2.42), (1.2.24) are $\widehat{\nu}_k$ $(k=1,\dots,n)$ and such that the characteristic values of Robin-Neumann problem (1.2.4), (1.2.42), (1.2.27) are $\widehat{\mu}_k$ $(k=1,\dots,n)$ if and only if $l\gamma>\eta$. In case such a pair of sequences exists, this pair is unique.

Proof. The function f defined by

$$f(z)=\frac{\prod\limits_{k=1}^{n}\left(1-\dfrac{z}{\widehat{\nu}_k}\right)}{\prod\limits_{k=1}^{n}\left(1-\dfrac{z}{\widehat{\mu}_k}\right)}\tag{1.3.15}$$

is a rational $\mathcal{S}_0$-function by Corollary A.3.3. In view of Lemma A.3.5 there are positive numbers $a_0,\dots,a_n$ and positive numbers $b_1,\dots,b_n$ such that

$$f(z)=a_n+\cfrac{1}{-b_nz+\cfrac{1}{a_{n-1}+\cfrac{1}{-b_{n-1}z+\ddots+\cfrac{1}{a_1+\cfrac{1}{-b_1z+\cfrac{1}{a_0}}}}}}\,.\tag{1.3.16}$$

Clearly,

$$1=f(0)=\sum_{k=0}^{n}a_k\,,$$

and in particular $a_k<1$ for $k=0,\dots,n$. In view of (1.2.19), (1.2.44) and (1.2.45) the Cauer-Fry polynomials for a solution would have to satisfy

$$\frac{R_{2n}(\cdot,\gamma_0)}{R_{2n-1}(\cdot,\gamma_0)}=\left(l+\frac{1}{\gamma}\right)f.$$

Since the continued fraction representations are unique, it follows from Remark 1.2.7, part 3, that

$$l_k = \left(l + \frac{1}{\gamma}\right) a_k \ (k = 1, \dots, n), \quad \frac{l_0}{1 - \gamma_0} = \left(l + \frac{1}{\gamma}\right) a_0,$$
$$m_k = \left(l + \frac{1}{\gamma}\right)^{-1} b_k \quad (k = 1, \dots, n).$$

Hence there is a solution, which would then be unique, if and only if all l_k and all m_k satisfying the above identities were positive, i.e., if and only if $l_0 > 0$. But since

$$\frac{l_0}{1 - \gamma_0} = \frac{l_0}{1 - \dfrac{1}{1 + \gamma l_0}} = \frac{1 + \gamma l_0}{\gamma} = \frac{1}{\gamma} + l_0,$$

we arrive at

$$l_0 = \left(l + \frac{1}{\gamma}\right) a_0 - \frac{1}{\gamma}.$$

Hence $l_0 > 0$ if and only if $\gamma l + 1 > \frac{1}{a_0}$, and the proof is complete by setting

$$\eta = \frac{1}{a_0} - 1.$$

We finally observe that $a_0 < 1$ implies that $\eta > 0$. □

1.3.2 Four-spectra inverse problem

In this subsection we use parts of the four sequences of characteristic values $(\nu_k)_{k=1}^n$, $(\mu_k)_{k=1}^n$, $(\xi_k)_{k=1}^n$ and $(\zeta_k)_{k=1}^n$ together with some other parameters to recover the sequences $(m_k)_{k=1}^n$ and $(l_k)_{k=0}^n$.

The main theorem of this subsection is

Theorem 1.3.10. *Let $n \geq 2$ be a natural number and let four increasing sequences of natural numbers $(k_j)_{j=1}^{n_1}$, $(p_j)_{j=1}^{n_2}$, $(r_j)_{j=1}^{n_3}$, $(s_j)_{j=1}^{n_4}$ be given such that $k_{n_1} = n$,*

$$n_1 + n_2 = n, \quad \{k_j : j = 1, \dots, n_1\} \cup \{p_j - 1 : j = 1, \dots, n_2\} = \{1, \dots, n\},$$
$$n_3 + n_4 = n, \quad \{r_j : j = 1, \dots, n_3\} \cup \{s_j : j = 1, \dots, n_4\} = \{1, \dots, n\}.$$

Let a constant $l > 0$ be given together with four sequences of positive real numbers $(\nu_{k_j})_{j=1}^{n_1}$, $(\zeta_{p_j})_{j=1}^{n_2}$, $(\mu_{r_j})_{j=1}^{n_3}$, $(\xi_{s_j})_{j=1}^{n_4}$ such that the sequences $(\eta_k)_{k=1}^n$ and $(\gamma_k)_{k=1}^n$ defined by

$$(\eta_{k_j})_{j=1}^{n_1} = (\nu_{k_j})_{j=1}^{n_1}, \quad (\eta_{p_j-1})_{j=1}^{n_2} = (\zeta_{p_j})_{j=1}^{n_2}, \tag{1.3.17}$$

and

$$(\gamma_{r_j})_{j=1}^{n_3} = (\mu_{r_j})_{j=1}^{n_3}, \quad (\gamma_{s_j})_{j=1}^{n_4} = (\xi_{s_j})_{j=1}^{n_4}, \tag{1.3.18}$$

interlace:

$$\gamma_1 < \eta_1 < \gamma_2 < \eta_2 < \cdots < \gamma_n < \eta_n. \tag{1.3.19}$$

Then there exist two unique sequences of positive numbers $(m_k)_{k=1}^n$, $(l_k)_{k=0}^n$ *with* $\sum_{k=0}^n l_k = l$ *such that*

1. ν_{k_j} $(j = 1, \ldots, n_1)$ *are characteristic values of the Dirichlet-Dirichlet problem* (1.2.4), (1.2.23), (1.2.24),
2. μ_{r_j} $(j = 1, \ldots, n_3)$ *are characteristic values of the Dirichlet-Neumann problem* (1.2.4), (1.2.23), (1.2.27),
3. ξ_{s_j} $(j = 1, \ldots, n_4)$ *are characteristic values of the Neumann-Dirichlet problem* (1.2.4), (1.2.29), (1.2.24),
4. ζ_{p_j} $(j = 1, \ldots, n_2)$ *are characteristic values of the Neumann-Neumann problem* (1.2.4), (1.2.29), (1.2.27).
5. *In each case, if the characteristic values are indexed in increasing order, then the indices of the given sequences are the corresponding indices in the full sequence of characteristic values; e.g.,* ν_{k_j} *is the* k_j*-th characteristic value of the Dirichlet-Dirichlet problem.*

Proof. Due to (1.3.19) and Theorem 1.3.1 we conclude that there exist unique sequences of positive numbers $(\widetilde{m}_k)_{k=1}^n$ and $(\widetilde{l}_k)_{k=0}^n$ which together with the given l generate problem (1.2.4), (1.2.23), (1.2.24) with the characteristic values $(\eta_k)_{k=1}^n$ and problem (1.2.4), (1.2.23), (1.2.27) with the characteristic values $(\gamma_k)_{k=1}^n$. Let $(\widetilde{\xi}_k)_{k=1}^n$ be the increasing sequence of the characteristic values of the Neumann-Dirichlet problem (1.2.4), (1.2.29), (1.2.24) generated by $(\widetilde{m}_k)_{k=1}^n$, $(\widetilde{l}_k)_{k=0}^n$ and let $(\widetilde{\zeta}_k)_{k=1}^n$ be the increasing sequence of the characteristic values of Neumann-Neumann problem (1.2.4), (1.2.29), (1.2.27) generated by $(\widetilde{m}_k)_{k=1}^n$, $(\widetilde{l}_k)_{k=0}^n$. From (1.2.40) it follows that

$$\{\eta_{k-1}, \widetilde{\zeta}_k\} < \{\gamma_k, \widetilde{\xi}_k\} < \{\eta_k, \widetilde{\zeta}_{k+1}\} \qquad (k = 1, \ldots, n), \tag{1.3.20}$$

where $\eta_0 = 0$ and $\widetilde{\zeta}_{n+1} = \eta_n$. Augment the given sequence $(\nu_{k_j})_{j=1}^{n_1}$ to the sequence $(\nu_k)_{k=1}^n$ by setting

$$(\nu_{p_j-1})_{j=1}^{n_2} = (\widetilde{\zeta}_{p_j})_{j=1}^{n_2} \tag{1.3.21}$$

and augment the given sequence $(\mu_{r_j})_{j=1}^{n_3}$ to the sequence $(\mu_k)_{k=1}^n$ by setting

$$(\mu_{s_j})_{j=1}^{n_4} = (\widetilde{\xi}_{s_j})_{j=1}^{n_4}. \tag{1.3.22}$$

Since $\mu_k = \gamma_k$ or $\mu_k = \widetilde{\xi}_k$ $(k = 1, \ldots, n)$ by (1.3.18) and (1.3.22), we have

$$\min\{\gamma_k, \widetilde{\xi}_k\} \leq \mu_k \leq \max\{\gamma_k, \widetilde{\xi}_k\} \quad (k = 1, \ldots, n). \tag{1.3.23}$$

From $\nu_k = \eta_k$ or $\nu_k = \widetilde{\zeta}_{k+1}$ ($k = 1, \dots, n$) by (1.3.17) and (1.3.21), it follows that

$$\min\{\eta_k, \widetilde{\zeta}_{k+1}\} \le \nu_k \le \max\{\eta_k, \widetilde{\zeta}_{k+1}\} \quad (k = 1, \dots, n). \tag{1.3.24}$$

Hence (1.2.40), (1.3.23) and (1.3.24) show that $\mu_k < \nu_k$ ($k = 1, \dots, n$) and $\nu_k < \mu_{k+1}$ ($k = 1, \dots, n-1$). Therefore the sequences $(\nu_k)_{k=1}^n$ and $(\mu_k)_{k=1}^n$ satisfy (1.2.36). By Theorem 1.3.1 there exist two sequences $(m_k)_{k=1}^n$ and $(l_k)_{k=0}^n$ of positive numbers which satisfy $\sum_{k=0}^n l_k = l$ and generate problem (1.2.4), (1.2.23), (1.2.24) with the sequence of characteristic values $(\nu_k)_{k=1}^n$ and problem (1.2.4), (1.2.23), (1.2.27) with the sequence of characteristic values $(\mu_k)_{k=1}^n$.

Statements 1 and 2 are clear from the construction of $(\nu_k)_{k=1}^n$ and $(\mu_k)_{k=1}^n$.

To prove statements 3 and 4 consider the Lagrange identity (1.2.47) for the problem generated by $(\widetilde{m}_k)_{k=1}^n$, $(\widetilde{l}_k)_{k=0}^n$ as well as $(m_k)_{k=1}^n$, $(l_k)_{k=0}^n$:

$$\begin{aligned}\widetilde{l}_0 \widetilde{R}_{2n-1}(z,0)\widetilde{R}_{2n}(z,1) - \widetilde{l}_0 \widetilde{R}_{2n}(z,0)\widetilde{R}_{2n-1}(z,1) = 1,\\ l_0 R_{2n-1}(z,0)R_{2n}(z,1) - l_0 R_{2n}(z,0)R_{2n-1}(z,1) = 1,\end{aligned}$$

which leads to

$$\begin{aligned}P(z) &:= \widetilde{l}_0 \widetilde{R}_{2n-1}(z,0)\widetilde{R}_{2n}(z,1) - l_0 R_{2n-1}(z,0)R_{2n}(z,1)\\ &= \widetilde{l}_0 \widetilde{R}_{2n}(z,0)\widetilde{R}_{2n-1}(z,1) - l_0 R_{2n}(z,0)R_{2n-1}(z,1).\end{aligned} \tag{1.3.25}$$

For $j = 1, \dots, n_3$ we have $\mu_{r_j} = \gamma_{r_j}$, $R_{2n-1}(\mu_{r_j}, 0) = 0$ and $\widetilde{R}_{2n-1}(\gamma_{r_j}, 0) = 0$, and for $j = 1, \dots, n_4$ we have $\mu_{s_j} = \widetilde{\xi}_{s_j}$, $R_{2n-1}(\mu_{s_j}, 0) = 0$ and $\widetilde{R}_{2n}(\widetilde{\xi}_{s_j}, 1) = 0$. This shows that $P(\mu_k) = 0$ for $k = 1, \dots, n$. For $j = 1, \dots, n_1$ we have $\nu_{k_j} = \eta_{k_j}$, $R_{2n}(\nu_{k_j}, 0) = 0$ and $\widetilde{R}_{2n}(\eta_{k_j}, 0) = 0$, and for $j = 1, \dots, n_2$ we have $\nu_{p_j-1} = \widetilde{\zeta}_{p_j}$, $R_{2n}(\nu_{p_j-1}, 0) = 0$ and $\widetilde{R}_{2n-1}(\widetilde{\zeta}_{p_j}, 1) = 0$. This shows that $P(\nu_k) = 0$ for $k = 1, \dots, n$. Furthermore, $R_{2n-1}(0,1) = 0$ and $\widetilde{R}_{2n-1}(0,1) = 0$ gives $P(0) = 0$. Thus P is a polynomial of degree at most $2n$ which has $2n+1$ distinct zeros ν_k ($k = 1, \dots, n$), μ_k ($k = 1, \dots, n$) and 0. Hence $P = 0$.

Therefore the zeros of $\widetilde{R}_{2n-1}(\cdot, 0)\widetilde{R}_{2n}(\cdot, 1)$ and $R_{2n-1}(\cdot, 0)R_{2n}(\cdot, 1)$ coincide. Above we have already identified n of the common zeros, which are the zeros μ_k ($k = 1, \dots, n$) of $R_{2n-1}(\cdot, 0)$. The remaining zeros can be identified as the increasing sequence $(\widehat{\xi}_k)_{k=1}^n$ of the zeros of $R_{2n}(\cdot, 1)$ on the one hand or the numbers $\widetilde{\xi}_{r_j}$ ($j = 1, \dots, n_3$) and $\xi_{s_j} = \gamma_{s_j}$ ($j = 1, \dots, n_4$). Hence the numbers ξ_{s_j} ($j = 1, \dots, n_4$) are characteristic values of the Neumann-Dirichlet problem for the sequences $(m_k)_{k=1}^n$ and $(l_k)_{k=0}^n$. This completes the proof of statement 3.

For the proof of statement 4 we use that the zeros of $\widetilde{R}_{2n}(\cdot, 0)\widetilde{R}_{2n-1}(\cdot, 1)$ and $R_{2n}(\cdot, 0)R_{2n-1}(\cdot, 1)$ coincide because $P = 0$ in (1.3.25). We have already identified n of the common zeros, which are the zeros ν_k ($k = 1, \dots, n$) of $R_{2n}(\cdot, 0)$. The remaining zeros can be identified as the increasing sequence $(\widehat{\zeta}_k)_{k=1}^n$ of the zeros of $R_{2n-1}(\cdot, 1)$ on the one hand or the numbers $\zeta_{p_j} = \eta_{p_j-1}$ ($j = 1, \dots, n_2$), 0, and

$\widetilde{\zeta}_{k_j+1}$ $(j = 1, \ldots, n_1 - 1)$. Hence the numbers ζ_{p_j} $(j = 1, \ldots, n_2)$ are characteristic numbers of the Neumann-Neumann problem for the sequences $(m_k)_{k=1}^n$ and $(l_k)_{k=0}^n$. This completes the proof of statement 4.

To prove uniqueness, let two pairs of sequences $(m_k^{(q)})_{k=1}^n$ and $(l_k^{(q)})_{k=0}^n$ $(q = 1, 2)$ be given for which the corresponding characteristic polynomials and characteristic values are also denoted by superscripts (1) and (2) and where $2n$ of the characteristic values are given according to (1.3.17)–(1.3.19). Below we will also refer to these characteristic values as characteristic values of problems (1) and (2), respectively. Let the sequences $(\widetilde{m}_k)_{k=1}^n$ and $(\widetilde{l}_k)_{k=0}^n$ be defined as at the beginning of this proof. As in (1.3.25), the Lagrange identity leads to

$$\begin{aligned} P^{(q)}(z) &:= \widetilde{l}_0 \widetilde{R}_{2n-1}(z,0)\widetilde{R}_{2n}(z,1) - l_0^{(q)} R_{2n-1}^{(q)}(z,0) R_{2n}^{(q)}(z,1) \\ &= \widetilde{l}_0 \widetilde{R}_{2n}(z,0)\widetilde{R}_{2n-1}(z,1) - l_0^{(q)} R_{2n}^{(q)}(z,0) R_{2n-1}^{(q)}(z,1) \end{aligned} \tag{1.3.26}$$

for $q = 1, 2$.

For $j = 1, \ldots, n_3$, the numbers $\gamma_{r_j} = \mu_{r_j}$ are characteristic values of the Dirichlet-Neumann problem for (1) and (2), and therefore $R_{2n-1}^{(q)}(\mu_{r_j}, 0) = 0$ and $\widetilde{R}_{2n-1}(\gamma_{r_j}, 0) = 0$. For $j = 1, \ldots, n_4$, the numbers $\gamma_{s_j} = \xi_{s_j}$ are characteristic values of the Neumann–Dirichlet problem for (1) and (2), and therefore $R_{2n}^{(q)}(\xi_{s_j}, 1) = 0$ and $\widetilde{R}_{2n-1}(\gamma_{r_j}, 0) = 0$. This shows that $P^{(q)}(\gamma_k) = 0$ for $k = 1, \ldots, n$. For $j = 1, \ldots, n_1$, the numbers $\eta_{k_j} = \nu_{k_j}$ are characteristic values of the Dirichlet-Dirichlet problem for (1) and (2), and therefore $R_{2n}^{(q)}(\nu_{k_j}, 0) = 0$ and $\widetilde{R}_{2n}(\eta_{k_j}, 0) = 0$. For $j = 1, \ldots, n_2$, the numbers $\eta_{p_j-1} = \zeta_{p_j}$ are characteristic values of the Neumann-Neumann problem for (1) and (2), and therefore $R_{2n-1}^{(q)}(\zeta_{p_j}, 1) = 0$ and $\widetilde{R}_{2n}(\eta_{p_j-1}, 0) = 0$. This shows that $P^{(q)}(\eta_k) = 0$ for $k = 1, \ldots, n$. Furthermore, $R_{2n-1}^{(q)}(0, 1) = 0$ and $\widetilde{R}_{2n-1}(0, 1) = 0$, so that $P^{(q)}(0) = 0$. Thus $P^{(q)}$ is a polynomial of degree at most $2n$ which has $2n + 1$ zeros γ_k $(k = 1, \ldots, n)$, η_k $(k = 1, \ldots, n)$, and 0. Hence $P^{(q)} = 0$ for $q = 1, 2$. We obtain that $l_0^{(1)} R_{2n-1}^{(1)}(\cdot, 0) R_{2n}^{(1)}(\cdot, 1) = l_0^{(2)} R_{2n-1}^{(2)}(\cdot, 0) R_{2n}^{(2)}(\cdot, 1)$ and $l_0^{(1)} R_{2n}^{(1)}(\cdot, 0) R_{2n-1}^{(1)}(\cdot, 1) = l_0^{(2)} R_{2n}^{(2)}(\cdot, 0) R_{2n-1}^{(2)}(\cdot, 1)$. But this implies that

$$\begin{aligned} (\mu_k^{(1)})_{k=1}^n \cup (\xi_k^{(1)})_{k=1}^n &= (\mu_k^{(2)})_{k=1}^n \cup (\xi_k^{(2)})_{k=1}^n, \\ (\nu_k^{(1)})_{k=1}^n \cup (\zeta_k^{(1)})_{k=1}^n &= (\nu_k^{(2)})_{k=1}^n \cup (\zeta_k^{(2)})_{k=1}^n, \end{aligned}$$

with, say, nondecreasing order in the concatenations (unions) of the sequences. From (1.2.40) we know that

$$\begin{aligned} \{0\} = \{\zeta_1^{(q)}\} < \{\mu_1^{(q)}, \xi_1^{(q)}\} < \{\nu_1^{(q)}, \zeta_2^{(q)}\} < \{\mu_2^{(q)}, \xi_2^{(q)}\} \\ < \cdots < \{\nu_{n-1}^{(q)}, \zeta_n^{(q)}\} < \{\mu_n^{(q)}, \xi_n^{(q)}\} < \{\nu_n^{(q)}\} \end{aligned}$$

for $q = 1, 2$. This implies that

$$\{\mu_k^{(1)}, \xi_k^{(1)}\} = \{\mu_k^{(2)}, \xi_k^{(2)}\} \quad (k = 1, \dots, n),$$
$$\{\nu_k^{(1)}, \zeta_{k+1}^{(1)}\} = \{\nu_k^{(2)}, \zeta_{k+1}^{(2)}\} \quad (k = 1, \dots, n-1), \quad \nu_n^{(1)} = \nu_n^{(2)}.$$

For $k = 1, \dots, n$, either $k = r_j$ or $k = s_j$ for some index j, and it follows that $\mu_k^{(1)} = \mu_{r_j} = \mu_k^{(2)}$ or $\xi_k^{(1)} = \xi_{s_j} = \xi_k^{(2)}$ is given, and therefore both identities hold for $k = 1, \dots, n$. This shows that the Dirichlet-Neumann as well as the Neumann-Dirichlet characteristic values coincide for both sets of string data. Similarly, for $k = 1, \dots, n-1$, either $k = k_j$ or $k = p_j - 1$ for some index j, and it follows that $\nu_k^{(1)} = \nu_{k_j} = \nu_k^{(2)}$ or $\zeta_{k+1}^{(1)} = \zeta_{p_j} = \zeta_{k+1}^{(2)}$ is given, and therefore both identities hold for $k = 1, \dots, n-1$. This shows that the Dirichlet-Dirichlet as well as the Neumann-Neumann characteristic values coincide for both sets of string data. The uniqueness part of Theorem 1.3.1 then shows that $m_k^{(1)} = m_k^{(2)}$ $(k = 1, \dots, n)$ and $l_k^{(1)} = l_k^{(2)}$ $(k = 0, \dots, n)$.

Next we are going to prove statement 5 for the Dirichlet-Dirichlet and the Dirichlet-Neumann problems. We already know that the sequences $(\nu_k)_{k=1}^n$ and $(\mu_k)_{k=1}$ are increasing sequences of the characteristic values of the Dirichlet-Dirichlet and the Dirichlet-Neumann problem, respectively. By definition of these two sequences it follows that ν_{k_j} $(j = 1, \dots, n_1)$ is the k_j-th eigenvalue of the Dirichlet-Dirichlet problem and that μ_{r_j} $(j = 1, \dots, n_3)$ is the r_j-th eigenvalue of the Dirichlet-Neumann problem.

To prove statement 5 for the Neumann-Dirichlet problem and the Neumann-Neumann problem, we are going to use the notation from the existence part of this proof. First observe that the inequality (1.2.40) for the string data $(l_k)_{k=0}^n$ and $(m_k)_{k=1}^n$ is

$$\{\nu_{k-1}, \widehat{\zeta}_k\} < \{\mu_k, \widehat{\xi}_k\} < \{\nu_k, \widehat{\zeta}_{k+1}\} \quad (k = 1, \dots, n). \tag{1.3.27}$$

From (1.3.20), (1.3.24) and (1.3.27) we conclude that

$$\widetilde{\xi}_k < \min\{\eta_k, \widetilde{\zeta}_{k+1}\} \le \nu_k < \widehat{\xi}_{k+1} \quad (k = 1, \dots, n-1),$$
$$\widehat{\xi}_{k-1} < \nu_{k-1} \le \max\{\eta_{k-1}, \widetilde{\zeta}_k\} < \min\{\gamma_k, \widetilde{\xi}_k\} \le \widetilde{\xi}_k \quad (k = 2, \dots, n).$$

Hence

$$\widetilde{\xi}_1 < \widehat{\xi}_2, \quad \widehat{\xi}_{n-1} < \widetilde{\xi}_n, \quad \widehat{\xi}_{k-1} < \widetilde{\xi}_k < \widehat{\xi}_{k+1} \quad (k = 2, \dots, n-1).$$

Since $\widetilde{\xi}_{r_j} \in \{\widehat{\xi}_1, \dots, \widehat{\xi}_n\}$ $(j = 1, \dots, n_3)$, it follows from the above inequalities that $\widetilde{\xi}_{r_j} = \widehat{\xi}_{r_j}$ $(j = 1, \dots, n_3)$. Therefore $\{\xi_{s_j} : j = 1, \dots, n_4\} = \{\widehat{\xi}_{s_j} : j = 1, \dots, n_4\}$, and since both $(\xi_{s_j})_{j=1}^4$ and $(\widehat{\xi}_{s_j})_{j=1}^4$ are increasing sequences, we obtain that $\xi_{s_j} = \widehat{\xi}_{s_j}$ $(j = 1, \dots, n_4)$. Hence ξ_{s_j} $(j = 1, \dots, n_4)$ is the s_j-th characteristic value of the Neumann-Dirichlet problem.

Observing that $\mu_{r_j} = \gamma_{r_j}$ $(j = 1, \ldots, n_3)$ by (1.3.18) and that $\mu_{s_j} = \widetilde{\xi}_{s_j}$ $(j = 1, \ldots, n_4)$ by (1.3.22), the inequalities (1.3.20) show that

$$\mu_k \le \max\{\gamma_k, \widetilde{\xi}_k\} < \eta_k < \min\{\gamma_{k+1}, \widetilde{\xi}_{k+1}\} \le \mu_{k+1} \quad (k = 1, \ldots, n-1).$$

Together with (1.3.27) and (1.3.17) it follows for $j = 1, \ldots . n_2$ that

$$\widehat{\zeta}_{p_j-1} < \mu_{p_j-1} < \eta_{p_j-1} = \zeta_{p_j} < \mu_{p_j} < \widehat{\zeta}_{p_j+1},$$

which gives

$$\widehat{\zeta}_{p_j-1} < \zeta_{p_j} < \widehat{\zeta}_{p_j+1} \quad (j = 1, \ldots, n_2).$$

Hence ζ_{p_j} $(j = 1, \ldots, n_2)$ is the p_j-th characteristic value of the Neumann-Neumann problem. □

Remark 1.3.11. 1. In the proof of Theorem 1.3.10 we have applied Theorem 1.3.1 twice to first find the data of an auxiliary string and then to find the data of the string which solves the four spectra problem. The first application can be replaced by using the Lagrange identity and Lagrange interpolation to find the characteristic polynomials $\widetilde{R}_{2n}(\cdot, 1)$ and $\widetilde{R}_{2n-1}(\cdot, 1)$. This approach may be easier for explicit calculations, see Example 1.3.12 below.

2. We have already mentioned above that we have used Theorem 1.3.1 to prove Theorem 1.3.10. However, Theorem 1.3.10 can be seen as a generalization of Theorems 1.3.1 and 1.3.2. To this end we observe that in the formulation and the proof of Theorem 1.3.10 we only require explicitly $k_{n_1} = n$, so that $n_1 > 0$. In order to have really parts of four spectra occurring, $n \ge 2$ is needed. However, the statements and proofs still hold if one only assumes that $n \ge 1$ and allows n_2, n_3 or n_4 to have the value 0. In particular, Theorem 1.3.1 is the particular case $n_2 = n_4 = 0$, and Theorem 1.3.2 is the particular case $n_2 = n_3 = 0$.

3. For Theorems 1.3.1, 1.3.2, 1.3.3, 1.3.4, 1.3.5, 1.3.6, 1.3.9, 1.3.10 it easily follows from Lemma A.6.5 and Corollary A.6.7 that the distribution of the beads depends continuously on the given sequences and the other given data, such as l, m, l_0 as they occur in the particular theorem. Furthermore, the number η constructed in the proof of Theorem 1.3.9 depends continuously on the given data of that theorem.

Example 1.3.12. Given $n = 3$, $\xi_1 = 1$, $\zeta_2 = 2$, $\mu_2 = 3$, $\nu_2 = 4$, $\mu_3 = 5$, $\nu_3 = 6$ and $l = 1$, find the distribution of the beads.

According to Theorem 1.3.10 we set $\gamma_1 = \xi_1 = 1$, $\eta_1 = \zeta_2 = 2$, $\gamma_2 = \mu_2 = 3$, $\eta_2 = \nu_2 = 4$, $\gamma_3 = \mu_3 = 5$, $\eta_3 = \nu_3 = 6$ and observe that (1.3.19) is satisfied. Then (1.2.25) and (1.2.28) give

$$\widetilde{R}_6(z, 0) = \frac{1}{\widetilde{l}_0}\left(1 - \frac{z}{2}\right)\left(1 - \frac{z}{4}\right)\left(1 - \frac{z}{6}\right),$$

$$\widetilde{R}_5(z, 0) = \frac{1}{\widetilde{l}_0}(1 - z)\left(1 - \frac{z}{3}\right)\left(1 - \frac{z}{5}\right).$$

Now using the Lagrange identity and Lagrange interpolation we arrive at

$$\widetilde{R}_6(z,1) = -\frac{1}{3}(2z^3 - 18z^2 + 37z - 3),$$
$$\widetilde{R}_5(z,1) = -\frac{16}{15}z(2z^2 - 12z + 13).$$

In the notation of the proof of Theorem 1.3.10, the zeros of $\widetilde{R}_6(\cdot,1)$ are $\widetilde{\xi}_1 = 3 - \sqrt{8.5}$, $\widetilde{\xi}_2 = 3$ and $\widetilde{\xi}_3 = 3 + \sqrt{8.5}$, while the zeros of $\widetilde{R}_5(\cdot,1)$ are $\widetilde{\zeta}_1 = 0$, $\widetilde{\zeta}_2 = 3 - \sqrt{2.5}$ and $\widetilde{\zeta}_3 = 3 + \sqrt{2.5}$. Then it follows that $\nu_1 = \widetilde{\zeta}_2 = 3 - \sqrt{2.5}$ and $\mu_1 = \widetilde{\xi}_1 = 3 - \sqrt{8.5}$. This gives

$$R_6(z,0) = \frac{1}{l_0}\left(1 - \frac{z}{3-\sqrt{2.5}}\right)\left(1 - \frac{z}{4}\right)\left(1 - \frac{z}{6}\right),$$
$$R_5(z,0) = \frac{1}{l_0}\left(1 - \frac{z}{3-\sqrt{8.5}}\right)\left(1 - \frac{z}{3}\right)\left(1 - \frac{z}{5}\right).$$

The continued fraction expansion (1.2.19) of the quotient $\frac{R_{2n}(z,0)}{R_{2n-1}(z,0)}$ gives, up to 5 decimal places $l_0 \approx 0.52115$, $l_1 \approx 0.34848$, $l_2 \approx 0.09314$, $l_3 \approx 0.03723$, $m_1 \approx 1.27635$, $m_2 \approx 4.53523$, $m_3 = 8.05508$.

1.4 Stieltjes strings with one interior condition

1.4.1 The problem and its spectra

In this section we consider a Stieltjes string S with n interior beads as in Section 1.2 and with an additional partition point P between two beads. The movement of the string at P will be subject to certain restrictions. We may now consider this string

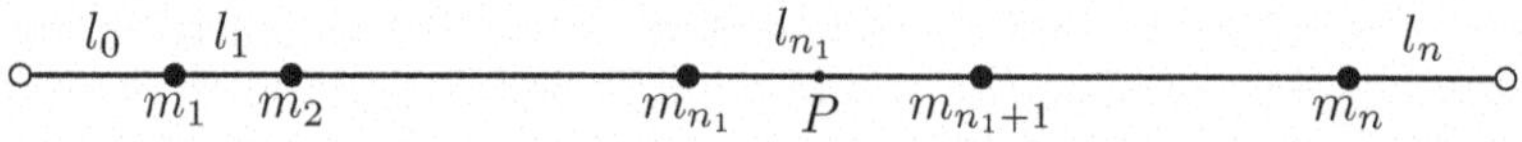

Figure 1.5

as being composed of two strings, S_1 and S_2, where S_1 starts at the left endpoint and terminates at P and where S_2 starts at the right endpoint and also terminates at P. Let n_j $(j = 1, 2)$ be the number of beads on S_j. Then $n = n_1 + n_2$. Denoting the masses, lengths between the beads and amplitudes by the superscript (1) for the substring S_1 and by the superscript (2) for the substring S_2, we arrive at the notation $l_k^{(1)} = l_k$ $(k = 0, \dots, n_1 - 1)$, $m_k^{(1)} = m_k$ $(k = 1, \dots, n_1)$, $l_k^{(2)} = l_{n-k}$ $(k = 0, \dots, n_2 - 1)$, $m_k^{(2)} = m_{n+1-k}$ $(k = 1, \dots, n_2)$. The interval of length l_{n_1} between the beads at position n_1 and $n_1 + 1$ is divided by P into two intervals of lengths $l_{n_1}^{(1)}$ and $l_{n_2}^{(2)}$ and thus $l_{n_1}^{(1)} + l_{n_2}^{(2)} = l_{n_1}$. Figures 1.5 and 1.6 illustrate

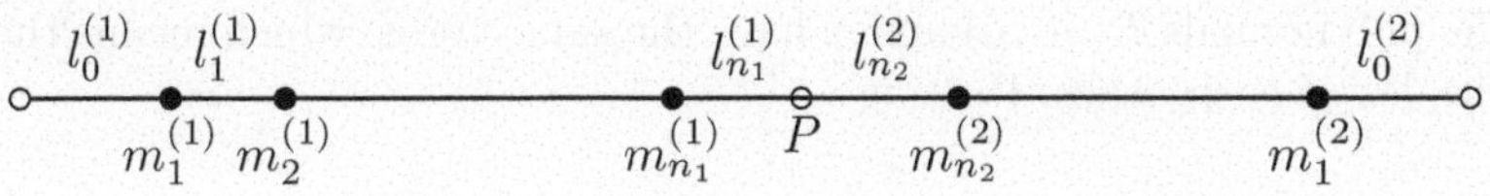

Figure 1.6

this situation. The amplitudes of the string S at the beads and at the endpoints of S are denoted by u_k ($k = 0, \ldots, n+1$), whereas the corresponding amplitudes for the substrings S_1 and S_2 are denoted by $u_k^{(1)}$ ($k = 0, \ldots, n_1+1$) and $u_k^{(2)}$ ($k = 0, \ldots, n_2+1$), respectively. The connectedness of the string at P gives the natural interface condition $u_{n_1+1}^{(1)} = u_{n_2+1}^{(2)}$ at the partition point P. By Lemma 1.2.19 the recurrence relation (1.2.4) is equivalent to

$$\frac{u_k^{(j)} - u_{k+1}^{(j)}}{l_k^{(j)}} + \frac{u_k^{(j)} - u_{k-1}^{(j)}}{l_{k-1}^{(j)}} - m_k^{(j)}\lambda^2 u_k^{(j)} = 0, \quad (k = 1, \ldots, n_j;\, j = 1, 2), \tag{1.4.1}$$

$$u_{n_1+1}^{(1)} = u_{n_2+1}^{(2)}, \tag{1.4.2}$$

$$\frac{u_{n_1+1}^{(1)} - u_{n_1}^{(1)}}{l_{n_1}^{(1)}} + \frac{u_{n_2+1}^{(2)} - u_{n_2}^{(2)}}{l_{n_2}^{(2)}} = 0. \tag{1.4.3}$$

We impose Dirichlet boundary conditions at the ends of the string:

$$u_0^{(j)} = 0 \quad (j = 1, 2). \tag{1.4.4}$$

By Proposition 1.2.8, the characteristic function of this Dirichlet-Dirichlet problem for the whole string is $R_{2n}(\cdot, 0)$. On the other hand, for the substrings with the Dirichlet conditions (1.4.4) the definition (1.2.8) gives, setting again $z = \lambda^2$ as in Section 1.2,

$$u_{n_1+1}^{(1)} = R_{2n_1}^{(1)}(z,0)u_1^{(1)}, \quad u_{n_2+1}^{(2)} = R_{2n_2}^{(2)}(z,0)u_1^{(2)}.$$

Due to (1.4.2), (1.2.10) and (1.4.3) we have

$$R_{2n_1}^{(1)}(z,0)u_1^{(1)} = R_{2n_2}^{(2)}(z,0)u_1^{(2)},$$
$$R_{2n_1-1}^{(1)}(z,0)u_1^{(1)} + R_{2n_2-1}^{(2)}(z,0)u_1^{(2)} = 0.$$

This linear system has a nontrivial solution $(u_1^{(1)}, u_1^{(2)})$ if and only if the coefficient matrix is singular. Therefore, the characteristic values of problem (1.4.1)–(1.4.4) coincide with the zeros of the polynomial

$$\phi = R_{2n_1}^{(1)}(\cdot,0)R_{2n_2-1}^{(2)}(\cdot,0) + R_{2n_1-1}^{(1)}(\cdot,0)R_{2n_2}^{(2)}(\cdot,0). \tag{1.4.5}$$

The polynomials $R_{2n}(\cdot,0)$ and ϕ have the same zeros, which means that they are multiples of each other. Putting

$$l^{(j)}=\sum_{k=0}^{n_j} l_k^{(j)} \quad (j=1,2) \tag{1.4.6}$$

it follows from (1.2.15) and (1.2.17) that

$$\phi(0)=\frac{l^{(1)}}{l_0^{(1)}}\cdot\frac{1}{l_0^{(2)}}+\frac{1}{l_0^{(1)}}\cdot\frac{l^{(2)}}{l_0^{(2)}}=\frac{l}{l_0 l_n},$$

and therefore

$$\phi=\frac{1}{l_n}R_{2n}(\cdot,0)\,. \tag{1.4.7}$$

Hence (1.4.5) can be written in the form

$$l_0^{(1)}R_{2n}(\cdot,0)=\Big[l_0^{(1)}R_{2n_1}^{(1)}(\cdot,0)\Big]\Big[l_0^{(2)}R_{2n_2-1}^{(2)}(\cdot,0)\Big]+\Big[l_0^{(1)}R_{2n_1-1}^{(1)}(\cdot,0)\Big]\Big[l_0^{(2)}R_{2n_2}^{(2)}(\cdot,0)\Big]. \tag{1.4.8}$$

Lemma 1.4.1. *The functions*

$$\frac{\phi}{R_{2n_1-1}^{(1)}(\cdot,0)R_{2n_2-1}^{(2)}(\cdot,0)} \quad \textit{and} \quad \frac{R_{2n_1}^{(1)}(\cdot,0)R_{2n_2}^{(2)}(\cdot,0)}{\phi}$$

are $\mathcal{S}_0$-functions.

Proof. In view of (1.2.19) and by Lemmas A.3.6 and A.3.5, both terms on the right-hand side of

$$\frac{\phi}{R_{2n_1-1}^{(1)}(\cdot,0)R_{2n_2-1}^{(2)}(\cdot,0)}=\frac{R_{2n_1}^{(1)}(\cdot,0)}{R_{2n_1-1}^{(1)}(\cdot,0)}+\frac{R_{2n_2}^{(2)}(\cdot,0)}{R_{2n_2-1}^{(2)}(\cdot,0)}$$

are $\mathcal{S}_0$-functions. Therefore, the left-hand side is also an $\mathcal{S}_0$-function. The same reasoning and Lemma A.2.4 show that

$$-\frac{\phi}{R_{2n_1}^{(1)}(\cdot,0)R_{2n_2}^{(2)}(\cdot,0)}=-\frac{R_{2n_2-1}^{(2)}(\cdot,0)}{R_{2n_2}^{(2)}(\cdot,0)}-\frac{R_{2n_1-1}^{(1)}(\cdot,0)}{R_{2n_1}^{(1)}(\cdot,0)}$$

is a rational Nevanlinna function. Furthermore, in view of (1.4.7) and Proposition 1.2.8, the real polynomials $R_{2n_1}^{(1)}(\cdot,0)$, $R_{2n_2}^{(2)}(\cdot,0)$ and ϕ are positive at 0 and have no negative zeros. Hence all three functions and therefore also the quotient $\frac{R_{2n_1}^{(1)}(\cdot,0)R_{2n_2}^{(2)}(\cdot,0)}{\phi}$ are positive on $(-\infty,0]$, which together with another application of Lemma A.2.4 proves that this quotient is an $\mathcal{S}_0$-function. □

Proposition 1.4.2. 1. *If $z \in \mathbb{R}$ is a zero of at least two of the three functions $R^{(1)}_{2n_1}(\cdot,0)$, $R^{(2)}_{2n_2}(\cdot,0)$, ϕ, then z is a zero of all three functions.*

2. *If $z \in \mathbb{R}$ is a zero of at least two of the three functions $R^{(1)}_{2n_1-1}(\cdot,0)$, $R^{(2)}_{2n_2-1}(\cdot,0)$, ϕ, then z is a zero of all three functions.*

Proof. 1. Assume that $R^{(1)}_{2n_1}(z,0) = 0 = R^{(2)}_{2n_2}(z,0)$. Since $\frac{R^{(1)}_{2n_1}(\cdot,0)R^{(2)}_{2n_2}(\cdot,0)}{\phi}$ is an $\mathcal{S}_0$-function by Lemma 1.4.1, it has only simple zeros, and therefore $\phi(z) = 0$ follows. Now assume that $R^{(1)}_{2n_1}(z,0) = 0 = \phi(z)$. Then $R^{(1)}_{2n_1-1}(z,0)R^{(2)}_{2n_2}(z,0) = 0$ by (1.4.5). But from either (1.2.36) or the Lagrange identity (1.2.47) we conclude from $R^{(1)}_{2n_1}(z,0) = 0$ that $R^{(1)}_{2n_1-1}(z,0) \neq 0$. Hence $R^{(2)}_{2n_2}(z,0) = 0$. The remaining case that $R^{(2)}_{2n_2}(z,0) = 0 = \phi(z)$ follows by interchanging the indices 1 and 2.

2. This is similar to part 1. We only have to observe that also the poles of $\mathcal{S}_0$-functions are simple. □

Together with the Dirichlet-Dirichlet problem (1.4.1)–(1.4.4), i.e., (1.2.4), (1.2.23), (1.2.24), for the string S, we will consider boundary value problems associated with the strings S_1 and S_2. If we clamp the string at the point P, we obtain Dirichlet-Dirichlet problems for $j = 1$ and $j = 2$, that is,

$$\frac{u^{(j)}_k - u^{(j)}_{k+1}}{l^{(j)}_k} + \frac{u^{(j)}_k - u^{(j)}_{k-1}}{l^{(j)}_{k-1}} - m^{(j)}_k \lambda^2 u^{(j)}_k = 0, \quad (k = 1,\dots,n_j;\ j = 1,2), \tag{1.4.9}$$

$$u^{(j)}_{n_j+1} = 0, \tag{1.4.10}$$

$$u^{(j)}_0 = 0. \tag{1.4.11}$$

Furthermore, we consider the Dirichlet-Neumann problems for the strings S_1 and S_2, that is with

$$u^{(j)}_{n_j+1} = u^{(j)}_{n_j} \tag{1.4.12}$$

instead of (1.4.10).

Remark 1.4.3. In view of Propositions 1.2.8 and 1.2.9, the characteristic values ν_k $(k = 1,\dots,n)$ of the problem (1.4.1)–(1.4.4), and, for $j = 1,2$, the characteristic values $\nu^{(j)}_k$ $(k = 1,\dots,n_j)$ of problem (1.4.9)–(1.4.11) as well as the characteristic values $\mu^{(j)}_k$ $(k = 1,\dots,n_j)$ of problem (1.4.9), (1.4.11), (1.4.12) are real, positive and simple. Therefore, the characteristic values will be indexed in increasing order, that is, for admissible indices $k < k'$ and $j = 1,2$ we have $\nu_k < \nu_{k'}$, $\nu^{(j)}_k < \nu^{(j)}_{k'}$ and $\mu^{(j)}_k < \mu^{(j)}_{k'}$, respectively. In view of (1.4.5) and Corollary A.6.7, these characteristic values depend continuously on the distribution of the beads.

Theorem 1.4.4. *The characteristic values defined in Remark* 1.4.3 *have the following properties:*

1. $\nu_1 < \nu^{(1)}_1$ *and* $\nu_1 < \nu^{(2)}_1$.

2. *For each* $k = 2, \dots, n$ *the following alternative holds: either the interval* (ν_1, ν_k) *contains exactly* $k-1$ *terms of the concatenated sequence* $(\nu_j^{(1)})_{j=1}^{n_1} \cup (\nu_j^{(2)})_{j=1}^{n_2}$ *and* $\nu_k \notin \{\nu_j^{(1)}\}_{j=1}^{n_1} \cup \{\nu_j^{(2)}\}_{j=1}^{n_2}$, *or the interval* (ν_1, ν_k) *contains exactly* $k-2$ *terms of the concatenated sequence* $(\nu_j^{(1)})_{j=1}^{n_1} \cup (\nu_j^{(2)})_{k=1}^{n_2}$ *and* $\nu_k \in \{\nu_j^{(1)}\}_{j=1}^{n_1} \cap \{\nu_j^{(2)}\}_{j=1}^{n_2}$.
3. *For each* $k = 1, \dots, n$ *the following alternative holds: either the interval* $(0, \nu_k)$ *contains exactly* k *terms of the concatenated sequence* $(\mu_j^{(1)})_{j=1}^{n_1} \cup (\mu_j^{(2)})_{j=1}^{n_2}$ *and* $\nu_k \notin \{\mu_j^{(1)}\}_{j=1}^{n_1} \cup \{\mu_j^{(2)}\}_{j=1}^{n_2}$, *or the interval* $(0, \nu_k)$ *contains exactly* $k-1$ *terms of the concatenated sequence* $(\mu_j^{(1)})_{j=1}^{n_1} \cup (\mu_j^{(2)})_{j=1}^{n_2}$ *and* $\nu_k \in \{\mu_j^{(1)}\}_{j=1}^{n_1} \cap \{\mu_j^{(2)}\}_{j=1}^{n_2}$.
4. *For each* $k = 1, \dots, n$ *the following alternative holds: either* $\nu_k \notin \{\nu_j^{(1)}\}_{j=1}^{n_1} \cup \{\mu_j^{(2)}\}_{j=1}^{n_2}$, *and then the interval* $(0, \nu_k)$ *contains* $k-1$ *or* k *terms of the concatenated sequence* $(\nu_j^{(1)})_{j=1}^{n_1} \cup (\mu_j^{(2)})_{j=1}^{n_2}$, *or*
$$\nu_k \in \left(\{\nu_j^{(1)}\}_{j=1}^{n_1} \cup \{\mu_j^{(2)}\}_{j=1}^{n_2}\right) \setminus \left(\nu_k \in \{\nu_j^{(1)}\}_{j=1}^{n_1} \cap \{\mu_j^{(2)}\}_{j=1}^{n_2}\right),$$
and then the interval $(0, \nu_k)$ *contains* $k-2$ *or* $k-1$ *terms of the concatenated sequence* $(\nu_j^{(1)})_{j=1}^{n_1} \cup (\mu_j^{(2)})_{j=1}^{n_2}$.

Proof. 1. Let $\psi = \frac{R_{2n_1}^{(1)}(\cdot,0) R_{2n_2}^{(2)}(\cdot,0)}{\phi}$ and recall that ϕ is a positive multiple of $R_{2n}(\cdot, 0)$, see (1.4.7). Each pole of ψ is a zero of ϕ, and each zero of ϕ is either a pole or a zero of ψ by Proposition 1.4.2, part 1. Since ψ is an $\mathcal{S}_0$-function by Lemma 1.4.1, its smallest pole must be smaller than its smallest zero. Therefore the smallest zero ν_1 of ϕ must be a pole of ψ, and all zeros of ψ must be larger than ν_1. This shows that $\nu_1 < \nu_1^{(1)}$ and $\nu_1 < \nu_1^{(2)}$.

2. By statement 1, ν_1 is a pole of ψ, and by Proposition 1.4.2, ν_k is either a pole or a zero of ψ. Let $0 \le k' \le k-2$ be the number of zero-pole cancellations in the interval (ν_1, ν_k) of the quotient representation of ψ. Hence ψ has $k-2-k'$ poles in (ν_1, ν_k). Since the poles and the zeros of ψ interlace, it follows that ψ has $k-1-k'$ zeros in (ν_1, ν_k) if ν_k is a pole of ψ and $k-2-k'$ zeros in (ν_1, ν_k) if ν_k is a zero of ψ. In view of Proposition 1.4.2, this results in the following alternative: either $R_{2n_1}^{(1)}(\nu_k, 0) \ne 0$, $R_{2n_2}^{(2)}(\nu_k, 0) \ne 0$ and $R_{2n_1}^{(1)}(\cdot, 0) R_{2n_2}^{(2)}(\cdot, 0)$ has $k-1$ zeros in (ν_1, ν_k), or $R_{2n_1}^{(1)}(\nu_k, 0) = 0 = R_{2n_2}^{(2)}(\nu_k, 0)$ and $R_{2n_1}^{(1)}(\cdot, 0) R_{2n_2}^{(2)}(\cdot, 0)$ has $k-2$ zeros in (ν_1, ν_k). In view of the notation introduced in Remark 1.4.3, this completes the proof of statement 2.

3. Let $\omega = \frac{\phi}{R_{2n_1-1}^{(1)}(\cdot,0) R_{2n_2-1}^{(2)}(\cdot,0)}$. By Remark 1.4.3, all zeros of $R_{2n}(\cdot, 0)$, $R_{2n_1-1}^{(1)}(\cdot, 0)$ and $R_{2n_2-1}^{(2)}(\cdot, 0)$ are positive real numbers. By Proposition 1.4.2, part 2, ν_k is either a pole or a zero of ω. Let $0 \le k' \le k-1$ be the number

of zero-pole cancellations within the interval $(0, \nu_k)$ in the quotient representation of ω. Hence ω has $k-1-k'$ zeros in $(0, \nu_k)$. Since the poles and the zeros of ω interlace and since the smallest pole of ω is smaller than any zeros of ω, it follows that ω has $k-k'$ poles in $(0, \nu_k)$ if ν_k is a zero of ω and $k-1-k'$ poles in $(0, \nu_k)$ if ν_k is a pole of ω. In view of Proposition 1.4.2, this results in the following alternative: either $R^{(1)}_{2n_1-1}(\nu_k, 0) \neq 0$, $R^{(2)}_{2n_2-1}(\nu_k, 0) \neq 0$ and $R^{(1)}_{2n_1-1}(\cdot, 0)R^{(2)}_{2n_2-1}(\cdot, 0)$ has k zeros in $(0, \nu_k)$, or $R^{(1)}_{2n_1-1}(\nu_k, 0) = 0 = R^{(2)}_{2n_2-1}(\nu_k, 0)$ and $R^{(1)}_{2n_1-1}(\cdot, 0)R^{(2)}_{2n_2-1}(\cdot, 0)$ has $k-1$ zeros in $(0, \nu_k)$. In view of the notation introduced in Remark 1.4.3, this completes the proof of statement 3.

4. First let us consider the case that $\nu_k \notin \{\nu_j^{(1)}\}_{j=1}^{n_1} \cup \{\mu_j^{(2)}\}_{j=1}^{n_2}$. We know from statement 3 that $\nu_k \notin \{\mu_j^{(1)}\}_{j=1}^{n_1} \cup \{\mu_j^{(2)}\}_{j=1}^{n_2}$ in this case, and therefore the interval $(0, \nu_k)$ contains k terms of $(\mu_j^{(1)})_{j=1}^{n_1} \cup (\mu_j^{(2)})_{j=1}^{n_2}$. From $\mu_j^{(1)} < \nu_j^{(1)} < \mu_{j+1}^{(1)}$ for $j = 1, \dots, n_1 - 1$, see (1.2.36), it follows that the interval $(0, \nu_k)$ contains $k-1$ or k terms of $(\nu_j^{(1)})_{j=1}^{n_1} \cup (\mu_j^{(2)})_{j=1}^{n_2}$.

If $\nu_k \in \{\nu_j^{(1)}\}_{j=1}^{n_1} \cup \{\mu_j^{(2)}\}_{j=1}^{n_2}$, then it follows from statements 2 and 3 that $\nu_k \in \{\nu_j^{(1)}\}_{j=1}^{n_1} \cap \{\nu_j^{(2)}\}_{j=1}^{n_2}$ or $\nu_k \in \{\mu_j^{(1)}\}_{j=1}^{n_1} \cap \{\mu_j^{(2)}\}_{j=1}^{n_2}$. If $\nu_k \in \{\nu_j^{(1)}\}_{j=1}^{n_1} \cap \{\nu_j^{(2)}\}_{j=1}^{n_2}$, then, in view of statements 1 and 2, the interval $(0, \nu_k)$ contains $k-2$ terms of the sequence $(\nu_j^{(1)})_{j=1}^{n_1} \cup (\nu_j^{(2)})_{j=1}^{n_2}$. Then the interlacing (1.2.36) shows that the interval $(0, \nu_k)$ contains $k-2$ or $k-1$ terms of the concatenated sequence $(\nu_j^{(1)})_{j=1}^{n_1} \cup (\mu_j^{(2)})_{j=1}^{n_2}$. If $\nu_k \in \{\mu_j^{(1)}\}_{j=1}^{n_1} \cap \{\mu_j^{(2)}\}_{j=1}^{n_2}$, then the interval $(0, \nu_k)$ contains $k-1$ terms of the sequence $(\mu_j^{(1)})_{j=1}^{n_1} \cup (\mu_j^{(2)})_{j=1}^{n_2}$ by statement 3. Again we see that the interval $(0, \nu_k)$ contains $k-2$ or $k-1$ terms of the sequence $(\nu_j^{(1)})_{j=1}^{n_1} \cup (\mu_j^{(2)})_{j=1}^{n_2}$.

Finally, if $\nu_k \in \{\nu_j^{(1)}\}_{j=1}^{n_1}$, then $\nu_k \in \{\nu_j^{(2)}\}_{j=1}^{n_2}$ by statement 2, and $\nu_k \notin \{\mu_j^{(2)}\}_{j=1}^{n_2}$ follows. Similarly, if $\nu_k \in \{\mu_j^{(2)}\}_{j=1}^{n_2}$, then $\nu_k \in \{\mu_j^{(1)}\}_{j=1}^{n_1}$ by statement 3, and $\nu_k \notin \{\nu_j^{(1)}\}_{j=1}^{n_1}$ follows. □

1.4.2 The inverse problem

First we consider the problem of finding the sequences of string data $(m_k^{(1)})_{k=1}^{n_1}$, $(m_k^{(2)})_{k=1}^{n_2}$, $(l_k^{(1)})_{k=0}^{n_1}$, and $(l_k^{(2)})_{k=0}^{n_2}$ from the sequences of the characteristic values $(\nu_k)_{k=1}^{n_1+n_2}$, $(\nu_k^{(1)})_{k=1}^{n_1}$, $(\nu_k^{(2)})_{k=1}^{n_2}$, the length of the string S and the length of the string S_1.

Theorem 1.4.5. *Let n_1 and n_2 be positive integers and let $l > 0$ and $\widehat{l} \in (0, l)$. Let three strictly increasing sequences of positive real numbers $(\nu_k)_{k=1}^{n_1+n_2}$, $(\nu_k^{(1)})_{k=1}^{n_1}$ and $(\nu_k^{(2)})_{k=1}^{n_2}$ be given. Let the concatenated sequence $(\theta_k)_{k=1}^{n_1+n_2} = (\nu_k^{(1)})_{k=1}^{n_1} \cup (\nu_k^{(2)})_{k=1}^{n_2}$ be indexed such that $\theta_k \leq \theta_{k'}$ if $k < k'$. We require that these sequences satisfy the following conditions:*

(i) *The sequences* $(\nu_k)_{k=1}^{n_1+n_2}$ *and* $(\theta_k)_{k=1}^{n_1+n_2}$ *interlace as follows:*

$$\nu_1 < \theta_1 \leq \cdots \leq \nu_{n_1+n_2} \leq \theta_{n_1+n_2}. \tag{1.4.13}$$

(ii) *If* $\theta_k = \nu_{k+1}$ *or* $\nu_{k+1} = \theta_{k+1}$ *for some* $k \in \{1, \ldots, n_1 + n_2 - 1\}$, *then* $\theta_k = \nu_{k+1} = \theta_{k+1}$.

Then there exist sequences of positive real numbers $(m_k^{(1)})_{k=1}^{n_1}$, $(m_k^{(2)})_{k=1}^{n_2}$, $(l_k^{(1)})_{k=0}^{n_1}$, *and* $(l_k^{(2)})_{k=0}^{n_2}$ *such that* $\sum\limits_{k=0}^{n_1} l_k^{(1)} = \widehat{l}$, $\sum\limits_{k=0}^{n_2} l_k^{(2)} = l - \widehat{l}$, *which generate the Dirichlet-Dirichlet problem* (1.4.1)–(1.4.4) *on the Stieltjes string* S, *and, for* $j = 1, 2$, *the Dirichlet-Dirichlet problems* (1.4.9)–(1.4.11) *on the strings* S_j, *with the sequences of characteristic values* $(\nu_k)_{k=1}^{n_1+n_2}$, $(\nu_k^{(1)})_{k=1}^{n_1}$ *and* $(\nu_k^{(2)})_{k=1}^{n_2}$, *respectively. The sequences* $(m_k^{(1)})_{k=1}^{n_1}$, $(m_k^{(2)})_{k=1}^{n_2}$, $(l_k^{(1)})_{k=0}^{n_1}$, *and* $(l_k^{(2)})_{k=0}^{n_2}$ *are uniquely determined by the given data if and only if all inequalities in* (1.4.13) *are strict.*

Proof. We introduce the polynomials

$$P_0(z) = l \prod_{k=1}^{n_1+n_2} \left(1 - \frac{z}{\nu_k}\right), \tag{1.4.14}$$

$$Q_1(z) = \widehat{l} \prod_{k=1}^{n_1} \left(1 - \frac{z}{\nu_k^{(1)}}\right), \tag{1.4.15}$$

and

$$Q_2(z) = (l - \widehat{l}) \prod_{k=1}^{n_2} \left(1 - \frac{z}{\nu_k^{(2)}}\right). \tag{1.4.16}$$

For all $k \in \{1, \ldots, n_1 + n_2 - 1\}$ with $\theta_k = \nu_{k+1} = \theta_{k+1}$ choose a number $\gamma(\nu_{k+1}) \in (0, 1)$. Define

$$\beta_k^{(1)} = \begin{cases} \dfrac{P_0(\nu_k^{(1)})}{Q_2(\nu_k^{(1)})} & \text{if } Q_2(\nu_k^{(1)}) \neq 0 \\[2ex] \gamma(\nu_k^{(1)}) \dfrac{P_0'(\nu_k^{(1)})}{Q_2'(\nu_k^{(1)})} & \text{if } Q_2(\nu_k^{(1)}) = 0 \end{cases} \qquad (k = 1, \ldots, n_1), \tag{1.4.17}$$

and let P_1 be the unique polynomial of degree at most n_1 satisfying

$$P_1(0) = 1, \quad P_1(\nu_k^{(1)}) = \beta_k^{(1)} \ (k = 1, \ldots, n_1), \tag{1.4.18}$$

namely, the Lagrange interpolation polynomial

$$P_1(z) = \sum_{k=1}^{n_1} \frac{\beta_k^{(1)} z}{\nu_k^{(1)}} \prod_{p=1,\, p \neq k}^{n_1} \frac{(z - \nu_p^{(1)})}{(\nu_k^{(1)} - \nu_p^{(1)})} + \prod_{k=1}^{n_1} \frac{\nu_k^{(1)} - z}{\nu_k^{(1)}}.$$

Let $k \in \{1, \ldots, n_1\}$. If $Q_2(\nu_k^{(1)}) \neq 0$, then $P_0(\nu_k^{(1)}) \neq 0$ by assumption (ii). There is a number $j \geq k$ such that $\nu_k^{(1)} = \theta_j$. In view of (1.4.13), P_0 has j simple zeros in the interval $(0, \theta_j)$. Therefore $(-1)^j P_0(\nu_k^{(1)}) > 0$. By definition, $\nu_m^{(2)} < \theta_j$ holds exactly for $m = 1, \ldots, j-k$. Hence $(-1)^{j-k} Q_2(\nu_k^{(1)}) > 0$. If $Q_2(\nu_k^{(1)}) = 0$, then $P_0(\nu_k^{(1)}) = 0$ by (1.4.13). There is a number $j > k$ such that $\nu_k^{(1)} = \nu_j = \nu_{j-k}^{(2)}$. Hence it follows that $(-1)^j P_0'(\nu_k^{(1)}) > 0$ and $(-1)^{j-k} Q_2'(\nu_k^{(1)}) > 0$. From the above inequalities and (1.4.17) we can now conclude that

$$P_1(0) > 0, \quad (-1)^k P_1(\nu_k^{(1)}) > 0 \quad (k = 1, \ldots, n_1).$$

Consequently, the zeros $\mu_k^{(1)}$ $(k = 1, \ldots, n_1)$ of P_1 are all positive and interlace with the zeros $\nu_k^{(1)}$ of Q_1:

$$0 < \mu_1^{(1)} < \nu_1^{(1)} < \cdots < \mu_{n_1}^{(1)} < \nu_{n_1}^{(1)}.$$

Thus, due to $\frac{Q_1(0)}{P_1(0)} = \widehat{l} > 0$ we conclude that $\frac{Q_1}{P_1}$ is an $\mathcal{S}_0$-function.

By Lemma A.3.5 the $\mathcal{S}_0$-function $\frac{Q_1}{P_1}$ can be expanded into continued fractions

$$\frac{Q_1(z)}{P_1(z)} = l_{n_1}^{(1)} + \cfrac{1}{-m_{n_1}^{(1)} z + \cfrac{1}{l_{n_1-1}^{(1)} + \cfrac{1}{-m_{n_1-1}^{(1)} z + \ddots + \cfrac{1}{l_1^{(1)} + \cfrac{1}{-m_1^{(1)} z + \cfrac{1}{l_0^{(1)}}}}}}}. \tag{1.4.19}$$

where $(l_k^{(1)})_{k=0}^{n_1}$ and $(m_k^{(1)})_{k=1}^{n_1}$ are sequences of positive numbers which we identify with the lengths of the threads and the masses of the beads we are looking for. From (1.4.15), (1.4.18) and (1.4.19) we obtain

$$\widehat{l} = \frac{Q_1(0)}{P_1(0)} = l_{n_1}^{(1)} + l_{n_1-1}^{(1)} + \cdots + l_1^{(1)} + l_0^{(1)}.$$

Comparing (1.4.19) with (1.2.19) we conclude that

$$R_{2n_1}^{(1)}(z, 0) = T_1 Q_1(z), \quad R_{2n_1-1}^{(1)}(z, 0) = T_1 P_1(z), \tag{1.4.20}$$

where $T_1 \neq 0$ is a constant and $R_{2n_1}^{(1)}(\cdot, 0)$, $R_{2n_1-1}^{(1)}(\cdot, 0)$ are the Cauer-Fry polynomials associated with these two sequences.

In the same way as above we construct the interpolation polynomial P_2 of degree at most n_2 which satisfies

$$P_2(0) = 1, \quad P_2(\nu_k^{(2)}) = \beta_k^{(2)} \quad (k = 1, \ldots, n_2), \tag{1.4.21}$$

where

$$\beta_k^{(2)} = \begin{cases} \dfrac{P_0(\nu_k^{(2)})}{Q_1(\nu_k^{(2)})} & \text{if } Q_1(\nu_k^{(2)}) \neq 0 \\ \left[1-\gamma(\nu_k^{(2)})\right] \dfrac{P_0'(\nu_k^{(2)})}{Q_1'(\nu_k^{(2)})} & \text{if } Q_1(\nu_k^{(2)}) = 0 \end{cases} \qquad (k = 1, \ldots, n_2). \qquad (1.4.22)$$

As for P_1 we conclude that

$$P_2(0) > 0, \quad (-1)^k P_2(\nu_k^{(2)}) > 0 \ (k = 1, \ldots, n_2).$$

Consequently, the zeros $\mu_k^{(2)}$ of P_2 are all positive and interlace with the zeros $\nu_k^{(2)}$ of Q_2:

$$0 < \mu_1^{(2)} < \nu_1^{(2)} < \cdots < \mu_{n_2}^{(2)} < \nu_{n_2}^{(2)}.$$

Thus, due to $\frac{Q_2(0)}{P_2(0)} = l - \widehat{l} > 0$ we conclude that $\frac{Q_2}{P_2}$ is an $\mathcal{S}_0$-function.

In the same way as above we obtain sequences of positive numbers $(l_k^{(2)})_{k=0}^{n_2}$ and $(m_k^{(2)})_{k=1}^{n_2}$ and $T_2 \neq 0$ such that

$$R_{2n_2}^{(2)}(z,0) = T_2 Q_2(z), \ R_{2n_2-1}^{(2)}(z,0) = T_2 P_2(z) \qquad (1.4.23)$$

and such that $\sum\limits_{k=0}^{n_2} l_k^{(2)} = l - \widehat{l}$. Substituting (1.4.20) and (1.4.23) into (1.4.5) we obtain

$$\phi = T_1 T_2 (Q_1 P_2 + Q_2 P_1).$$

Let us compare the polynomial P_0 with the polynomial $\widehat{P}_0 := Q_1 P_2 + Q_2 P_1$. It follows from (1.4.14)–(1.4.17) and (1.4.22) that

$$\begin{aligned} P_0(0) &= l = \widehat{P}_0(0), \\ P_0(\nu_k^{(1)}) &= Q_2(\nu_k^{(1)}) P_1(\nu_k^{(1)}) = \widehat{P}_0(\nu_k^{(1)}) \quad (k = 1, \ldots, n_1), \\ P_0(\nu_k^{(2)}) &= Q_1(\nu_k^{(2)}) P_2(\nu_k^{(2)}) = \widehat{P}_0(\nu_k^{(2)}) \quad (k = 1, \ldots, n_2). \end{aligned}$$

For $j \in \{1, \ldots, n_1\}$ and $k \in \{1, \ldots, n_2\}$ with $\nu_j^{(1)} = \nu_k^{(2)}$ we calculate

$$\begin{aligned} P_0'(\nu_k^{(2)}) &= \gamma(\nu_j^{(1)}) P_0'(\nu_j^{(1)}) + [1 - \gamma(\nu_k^{(2)})] P_0'(\nu_k^{(2)}) \\ &= \beta_j^{(1)} Q_2'(\nu_j^{(1)}) + \beta_k^{(2)} Q_1'(\nu_k^{(2)}) \\ &= P_1(\nu_j^{(1)}) Q_2'(\nu_j^{(1)}) + P_2(\nu_k^{(2)}) Q_1'(\nu_k^{(2)}) \\ &= \widehat{P}_0'(\nu_k^{(2)}), \end{aligned}$$

where we have used that $Q_2(\nu_j^{(1)}) = 0 = Q_1(\nu_k^{(2)})$. Consequently, $P_0 = \widehat{P}_0$. Hence, $\phi = T_1 T_2 P_0$, and the sequence of zeros of ϕ is exactly $(\nu_k)_{k=1}^{n_1+n_2}$.

If not all inequalities in (1.4.13) are strict, then there are $j \in \{1, \dots, n_1\}$ and $k \in \{1, \dots, n_2\}$ such that $\nu_j^{(1)} = \nu_k^{(2)}$. In that case, the polynomial P_1 depends on $\gamma(\nu_j^{(1)})$, and varying $\gamma(\nu_j^{(1)})$ in the interval $(0, 1)$ we arrive at a family of distinct polynomials P_1. But then also $\frac{Q_1}{P_1}$ has a family of distinct continued fraction expansions (1.4.19), so that we arrive at distinct pairs of sequences $(m_k^{(1)})_{k=1}^{n_1}$ and $(l_k^{(1)})_{k=1}^{n_1}$.

Now assume that all inequalities in (1.4.13) are strict. We observe that, by (1.4.7) and (1.2.25), the functions $l_0^{(1)} l_0^{(2)} \phi$, $l_0^{(1)} R_{2n_1}^{(1)}(\cdot, 0)$ and $l_0^{(2)} R_{2n_2}^{(2)}(\cdot, 0)$ are uniquely determined by $(\nu_k)_{k=1}^{n_1+n_2}$, $(\nu_k^{(1)})_{k=1}^{n_1}$, $(\nu_k^{(2)})_{k=1}^{n_2}$, l and $\widehat{l}$. From (1.4.7) and (1.4.8) it follows that the values

$$l_0^{(1)} R_{2n_1-1}^{(1)}(\nu_k^{(1)}, 0) = \frac{l_0^{(1)} l_0^{(2)} \phi(\nu_k^{(1)})}{l_0^{(2)} R_{2n_2}^{(2)}(\nu_k^{(1)}, 0)} \qquad (k = 1, \dots, n_1)$$

and $l_0^{(1)} R_{2n_1-1}^{(1)}(0, 0) = 1$ are uniquely determined by the given data, which means that $l_0^{(1)} R_{2n_1-1}^{(1)}(\cdot, 0)$ and therefore also the function $\frac{R_{2n_1}^{(1)}(\cdot,0)}{R_{2n_1-1}^{(1)}(\cdot,0)}$ are uniquely determined by the given data. Hence Lemma A.3.5 and (1.2.19) show that $(l_k^{(1)})_{k=1}^{n_1-1}$ and $(m_k^{(1)})_{k=0}^{n_1-1}$ are unique. The same reasoning also show that $(l_k^{(2)})_{k=1}^{n_2-1}$ and $(m_k^{(2)})_{k=0}^{n_2}$ are unique. □

Now we consider the problem of finding the sequences $(m_k^{(1)})_{k=1}^{n_1}$, $(m_k^{(2)})_{k=1}^{n_2}$, $(l_k^{(1)})_{k=0}^{n_1}$, and $(l_k^{(2)})_{k=0}^{n_2}$ from the sequences of the characteristic values $(\nu_k)_{k=1}^{n_1+n_2}$, $(\mu_k^{(1)})_{k=1}^{n_1}$, $(\mu_k^{(2)})_{k=1}^{n_2}$ and the length of the string S.

Theorem 1.4.6. *Let n_1 and n_2 be positive integers and let $l > 0$. Let three strictly increasing sequences of positive real numbers $(\nu_k)_{k=1}^{n_1+n_2}$, $(\mu_k^{(1)})_{k=1}^{n_1}$ and $(\mu_k^{(2)})_{k=1}^{n_2}$ be given. Let the concatenated sequence $(\tau_k)_{k=1}^{n_1+n_2} = (\mu_k^{(1)})_{k=1}^{n_1} \cup (\mu_k^{(2)})_{k=1}^{n_2}$ be indexed such that $\tau_k \le \tau_{k'}$ if $k < k'$. We require that these sequences satisfy the conditions:*

(i) *The sequences $(\nu_k)_{k=1}^{n_1+n_2}$ and $(\tau_k)_{k=1}^{n_1+n_2}$ interlace as follows:*

$$\tau_1 \le \nu_1 \le \dots \le \tau_{n_1+n_2} < \nu_{n_1+n_2}. \tag{1.4.24}$$

(ii) *If $\tau_k = \nu_k$ or $\nu_k = \tau_{k+1}$ for some $k \in \{1, \dots, n_1 + n_2 - 1\}$, then $\tau_k = \nu_k = \tau_{k+1}$.*

Then there exist sequences of positive real numbers $(m_k^{(1)})_{k=1}^{n_1}$, $(m_k^{(2)})_{k=1}^{n_2}$, $(l_k^{(1)})_{k=0}^{n_1-1}$, and $(l_k^{(2)})_{k=0}^{n_2-1}$ with

$$\sum_{k=0}^{n_1-1} l_k^{(1)} + \sum_{k=0}^{n_2-1} l_k^{(2)} < l \tag{1.4.25}$$

such that for each pair of positive numbers $l_{n_1}^{(1)}$, $l_{n_2}^{(2)}$ *with*

$$l_{n_1}^{(1)} + l_{n_2}^{(2)} = l - \sum_{k=0}^{n_1-1} l_k^{(1)} + \sum_{k=0}^{n_2-1} l_k^{(2)}, \tag{1.4.26}$$

the Dirichlet-Dirichlet problem (1.4.1)–(1.4.4) *on the Stieltjes string* S, *and, for* $j = 1, 2$, *the Dirichlet-Neumann problems* (1.4.9), (1.4.11), (1.4.12) *on the Stieltjes string* S_j, *generate the three sequences of characteristic values* $(\nu_k)_{k=1}^{n_1+n_2}$, $(\mu_k^{(1)})_{k=1}^{n_1}$, $(\mu_k^{(2)})_{k=1}^{n_2}$. *The sequences of Stieltjes string data* $(m_k^{(1)})_{k=1}^{n_1}$, $(m_k^{(2)})_{k=1}^{n_2}$, $(l_k^{(1)})_{k=0}^{n_1-1}$, *and* $(l_k^{(2)})_{k=0}^{n_2-1}$ *are uniquely determined by the given data if and only if all inequalities in* (1.4.24) *are strict.*

Proof. We introduce the polynomials

$$P_0(z) = l \prod_{k=1}^{n_1+n_2} \left(1 - \frac{z}{\nu_k}\right), \tag{1.4.27}$$

$$Q_1(z) = \prod_{k=1}^{n_1} \left(1 - \frac{z}{\mu_k^{(1)}}\right), \tag{1.4.28}$$

$$Q_2(z) = \prod_{k=1}^{n_2} \left(1 - \frac{z}{\mu_k^{(2)}}\right). \tag{1.4.29}$$

For all $k \in \{1, \ldots, n_1+n_2-1\}$ with $\tau_k = \nu_k = \tau_{k+1}$ choose a number $\gamma(\nu_k) \in (0, 1)$. Define

$$\beta_k^{(1)} = \begin{cases} \dfrac{P_0(\mu_k^{(1)})}{Q_2(\mu_k^{(1)})} & \text{if } Q_2(\mu_k^{(1)}) \neq 0 \\[2ex] \gamma(\mu_k^{(1)}) \dfrac{P_0'(\mu_k^{(1)})}{Q_2'(\mu_k^{(1)})} & \text{if } Q_2(\mu_k^{(1)}) = 0 \end{cases} \qquad (k = 1, \ldots, n_1), \tag{1.4.30}$$

let $\widetilde{l} \in (0, l)$ and let P_1 be the unique polynomial of degree at most n_1 satisfying

$$P_1(0) = \widetilde{l}, \quad P_1(\mu_k^{(1)}) = \beta_k^{(1)} \quad (k = 1, \ldots, n_1), \tag{1.4.31}$$

namely, the Lagrange interpolation polynomial

$$P_1(z) = \sum_{k=1}^{n_1} \frac{\beta_k^{(1)} z}{\mu_k^{(1)}} \prod_{p=1,\, p\neq k}^{n_1} \frac{(z - \mu_p^{(1)})}{(\mu_k^{(1)} - \mu_p^{(1)})} + \widetilde{l} \prod_{k=1}^{n_1} \frac{\mu_k^{(1)} - z}{\mu_k^{(1)}}. \tag{1.4.32}$$

Let $k \in \{1, \ldots, n_1\}$. If $Q_2(\mu_k^{(1)}) \neq 0$, then $P_0(\mu_k^{(1)}) \neq 0$ by assumption (ii). There is a number $j \geq k$ such that $\mu_k^{(1)} = \tau_j$. In view of (1.4.24), P_0 has $j-1$ simple zeros in the interval $(0, \tau_j)$. Therefore $(-1)^{j-1} P_0(\mu_k^{(1)}) > 0$. By definition, $\mu_m^{(2)} < \tau_j$ holds exactly for $m = 1, \dots, j-k$. Hence $(-1)^{j-k} Q_2(\mu_k^{(1)}) > 0$. If $Q_2(\mu_k^{(1)}) = 0$, then $P_0(\mu_k^{(1)}) = 0$ by (1.4.24). There is a number $j \geq k$ such that $\mu_k^{(1)} = \nu_j = \mu_{j+1-k}^{(2)}$. Hence it follows that $(-1)^j P_0'(\mu_k^{(1)}) > 0$ and $(-1)^{j+1-k} Q_2'(\mu_k^{(1)}) > 0$. From the above inequalities and (1.4.30) we can now conclude that

$$(-1)^{k-1} \beta_k^{(1)} > 0 \quad (k = 1, \dots, n_1), \tag{1.4.33}$$

and therefore

$$P_1(0) > 0, \quad (-1)^{k-1} P_1(\mu_k^{(1)}) > 0 \quad (k = 1, \dots, n_1). \tag{1.4.34}$$

In the same way as above we construct the interpolation polynomial P_2 of degree at most n_2 which satisfies

$$P_2(0) = l - \widetilde{l}, \quad P_2(\mu_k^{(2)}) = \beta_k^{(2)} \quad (k = 1, \dots, n_2), \tag{1.4.35}$$

where

$$\beta_k^{(2)} = \begin{cases} \dfrac{P_0(\mu_k^{(2)})}{Q_1(\mu_k^{(2)})} & \text{if } Q_1(\mu_k^{(2)}) \neq 0 \\[2ex] \left[1 - \gamma(\mu_k^{(2)})\right] \dfrac{P_0'(\mu_k^{(2)})}{Q_1'(\mu_k^{(2)})} & \text{if } Q_1(\mu_k^{(2)}) = 0 \end{cases} \quad (k = 1, \dots, n_2). \tag{1.4.36}$$

As for P_1 we conclude that

$$(-1)^{k-1} \beta_k^{(2)} > 0 \quad (k = 1, \dots, n_2), \tag{1.4.37}$$

$$P_2(0) > 0, \quad (-1)^{k-1} P_2(\mu_k^{(2)}) > 0 \quad (k = 1, \dots, n_2).$$

Let us compare the polynomial P_0 with the polynomial $\widehat{P}_0 := Q_1 P_2 + Q_2 P_1$. It follows from (1.4.27)–(1.4.30) and (1.4.35) that

$$\begin{aligned} P_0(0) &= l = \widehat{P}_0(0), \\ P_0(\mu_k^{(1)}) &= Q_2(\mu_k^{(1)}) P_1(\mu_k^{(1)}) = \widehat{P}_0(\mu_k^{(1)}) \quad (k = 1, \dots, n_1), \\ P_0(\mu_k^{(2)}) &= Q_1(\mu_k^{(2)}) P_2(\mu_k^{(2)}) = \widehat{P}_0(\mu_k^{(2)}) \quad (k = 1, \dots, n_2). \end{aligned}$$

For $j \in \{1, \dots, n_1\}$ and $k \in \{1, \dots, n_2\}$ with $\mu_j^{(1)} = \mu_k^{(2)}$ we calculate

$$\begin{aligned} P_0'(\mu_k^{(2)}) &= \gamma(\mu_j^{(1)}) P_0'(\mu_j^{(1)}) + [1 - \gamma(\mu_k^{(2)})] P_0'(\mu_k^{(2)}) \\ &= \beta_j^{(1)} Q_2'(\mu_j^{(1)}) + \beta_k^{(2)} Q_1'(\mu_k^{(2)}) \\ &= P_1(\mu_j^{(1)}) Q_2'(\mu_j^{(1)}) + P_2(\mu_k^{(2)}) Q_1'(\mu_k^{(2)}) \\ &= \widehat{P}_0'(\mu_k^{(2)}), \end{aligned}$$

where we have used that $Q_2(\mu_j^{(1)}) = 0 = Q_1(\mu_k^{(2)})$. Consequently, $P_0 = \widehat{P}_0$.

It is clear from the definition of P_0, Q_1 and Q_2 that the coefficients of $(-z)^{n_1+n_2}$ of P_0, of $(-z)^{n_1}$ of Q_1 and of $(-z)^{n_2}$ of Q_2 are positive numbers α_0, α_1 and α_2, respectively. From the representation (1.4.32) we see that the coefficient of $(-z)^{n_1}$ of P_1 is

$$-\sum_{k=1}^{n_1} \frac{\beta_k^{(1)}}{\mu_k^{(1)}} \prod_{p=1,\, p\neq k}^{n_1} \frac{1}{(\mu_p^{(1)} - \mu_k^{(1)})} + \widetilde{l} \prod_{k=1}^{n_1} \frac{1}{\mu_k^{(1)}} = \delta_3 + \alpha_3 \widetilde{l},$$

where $\alpha_3 > 0$ and $\delta_3 < 0$ in view of (1.4.33). In the same way, we obtain that the coefficient of $(-z)^{n_2}$ of P_2 is $\delta_4 + \alpha_4(l - \widetilde{l})$ with $\alpha_4 > 0$ and $\delta_4 < 0$. Comparing the coefficients of $(-z)^n$ in the identity $P_0 = \widehat{P}_0$ gives

$$\alpha_0 = \alpha_1(\delta_4 + \alpha_4(l - \widetilde{l})) + \alpha_2(\delta_3 + \alpha_3 \widetilde{l}).$$

For $f(\widetilde{l}) = \alpha_2(\delta_3 + \alpha_3 \widetilde{l})$ we have $f(0) = \alpha_2\delta_3 < 0$ and $f(l) = \alpha_0 - \alpha_1\delta_4 > \alpha_0$. Hence there is $\widetilde{l} \in (0, l)$ such that $f(\widetilde{l}) = \frac{\alpha_0}{2}$. This means that $\alpha_2(\delta_3 + \alpha_3\widetilde{l}) = \frac{\alpha_0}{2}$ and $\alpha_1(\delta_4 + \alpha_4(l - \widetilde{l})) = \frac{\alpha_0}{2}$, so that the coefficients $\delta_3 + \alpha_3\widetilde{l}$ of $(-z)^{n_1}$ of P_1 and the coefficient $\delta_4 + \alpha_4(l - \widetilde{l})$ of $(-z)^{n_2}$ of P_2 are positive. We will now continue with this particular value of $\widetilde{l}$. Then we conclude that $P_1(z) \to \infty$ and $P_2(z) \to \infty$ as $z \to -\infty$, that the degree of P_1 is n_1, and that the degree of P_2 is n_2.

We now determine the location of the zeros of P_1 and P_2. From above and (1.4.34) it follows that P_1 has an even number of zeros in the intervals $(-\infty, 0)$ and $(0, \mu_1^{(1)})$ and an odd number of zeros in the intervals $(\mu_k^{(1)}, \mu_{k+1}^{(1)})$ for $k = 1, \ldots, n_1-1$. Since the degree of P_1 is n_1, P_1 has one simple zero $\nu_k^{(1)}$ $(k = 1, \ldots, n_1)$ in the interval $(\mu_k^{(1)}, \mu_{k+1}^{(1)})$ and one simple zero $\nu_{n_1}^{(1)}$ in $(\mu_{n_1}^{(1)}, \infty)$. It follows that the zeros of P_1 and Q_1 interlace as follows:

$$0 < \mu_1^{(1)} < \nu_1^{(1)} < \cdots < \mu_{n_1}^{(1)} < \nu_{n_1}^{(1)}.$$

Thus, due to $\frac{P_1(0)}{Q_1(0)} = \widetilde{l} > 0$ we conclude that $\frac{P_1}{Q_1}$ is an $\mathcal{S}_0$-function.

By Lemma A.3.5 the $\mathcal{S}_0$-function $\frac{P_1}{Q_1}$ can be expanded into continued fraction

$$\frac{P_1(z)}{Q_1(z)} = l_{n_1}^{(1)} + \cfrac{1}{-m_{n_1}^{(1)} z + \cfrac{1}{l_{n_1-1}^{(1)} + \cfrac{1}{-m_{n_1-1}^{(1)} z + \ddots + \cfrac{1}{l_1^{(1)} + \cfrac{1}{-m_1^{(1)} z + \cfrac{1}{l_0^{(1)}}}}}}}. \tag{1.4.38}$$

where $(l_k^{1)})_{k=0}^{n_1}$ and $(m_k^{(1)})_{k=1}^{n_1}$ are sequences of positive numbers which we identify with the lengths of the threads and the masses of the beads we are looking for. From (1.4.28), (1.4.31) and (1.4.38) we obtain

$$\widetilde{l} = \frac{P_1(0)}{Q_1(0)} = l_{n_1}^{(1)} + l_{n_1-1}^{(1)} + \cdots + l_1^{(1)} + l_0^{(1)}.$$

Comparing (1.4.38) with (1.2.19) we conclude that

$$R_{2n_1}^{(1)}(z,0) = T_1 P_1(z), \quad R_{2n_1-1}^{(1)}(z,0) = T_1 Q_1(z), \tag{1.4.39}$$

where $T_1 \neq 0$ is a constant and $R_{2n_1}^{(1)}(\cdot,0)$, $R_{2n_1-1}^{(1)}(\cdot,0)$ are the Cauer-Fry polynomials associated with these two sequences.

As for P_1, the zeros $\nu_k^{(2)}$ $(k = 1, \ldots, n_2)$ of P_2 interlace with the zeros of Q_2:

$$0 < \mu_1^{(2)} < \nu_1^{(2)} < \cdots < \mu_{n_2}^{(2)} < \nu_{n_2}^{(2)}.$$

Thus, due to $\frac{P_2(0)}{Q_2(0)} = l - \widetilde{l} > 0$ we conclude that $\frac{P_2}{Q_2}$ is an $\mathcal{S}_0$-function.

In the same way as above we obtain sequences of positive numbers $(l_k^{(2)})_{k=0}^{n_2}$ and $(m_k^{(2)})_{k=1}^{n_2}$ and $T_2 \neq 0$ such that

$$R_{2n_2}^{(2)}(z,0) = T_2 P_2(z), \; R_{2n_2-1}^{(2)}(z,0) = T_2 Q_2(z) \tag{1.4.40}$$

and such that $\sum\limits_{k=0}^{n_2} l_k^{(2)} = l - \widetilde{l}$. Substituting (1.4.39) and (1.4.40) into (1.4.5) we obtain

$$\phi = T_1 T_2 (Q_1 P_2 + Q_2 P_1).$$

Hence, $\phi = T_1 T_2 P_0$, and the sequence of zeros of ϕ is exactly $(\nu_k)_{k=1}^{n_1+n_2}$.

In view of Remark 1.2.7, part 1, the characteristic values $(\mu_k^{(1)})_{k=1}^{n_1}$ and $(\mu_k^{(2)})_{k=1}^{n_2}$ are independent of $l_{n_1}^{(1)}$ and $l_{n_2}^{(2)}$. Furthermore, $l_{n_1} = l_{n_1}^{(1)} + l_{n_2}^{(2)}$, see Figures 1.5 and 1.6, is uniquely determined by l, $l_k^{(1)}$ $(k = 0, \ldots, n_1 - 1)$ and $l_k^{(2)}$ $(k = 0, \ldots, n_2 - 1)$, and therefore any pair of positive numbers $l_{n_1}^{(1)}$, $l_{n_2}^{(2)}$ satisfying (1.4.26) gives the same characteristic values. This completes the existence proof.

If not all inequalities in (1.4.24) are strict, then there are $j \in \{1, \ldots, n_1\}$ and $k \in \{1, \ldots, n_2\}$ such that $\mu_j^{(1)} = \mu_k^{(2)}$. In that case, the polynomial P_1 depends on $\gamma(\mu_j^{1)})$, and varying $\gamma(\mu_j^{1)})$ in the interval $(0,1)$ we arrive at a family of distinct polynomials P_1. But then also $\frac{Q_1}{P_1}$ has a family of distinct continued fraction expansions (1.2.19), so that we arrive at distinct pairs of sequences $(m_k^{(1)})_{k=1}^{n_1}$ and $(l_k^{(1)})_{k=1}^{n_1-1}$.

Now assume that all inequalities in (1.4.24) are strict. By (1.4.7), (1.2.25) and (1.2.28), the functions $l_0^{(1)} l_0^{(2)} \phi$, $l_0^{(1)} R_{2n_1-1}^{(1)}(\cdot,0)$ and $l_0^{(2)} R_{2n_2-1}^{(2)}(\cdot,0)$ are uniquely determined by $(\nu_k)_{k=1}^{n_1+n_2}$, $(\mu_k^{(1)})_{k=1}^{n_1}$, $(\mu_k^{(2)})_{k=1}^{n_2}$ and l. From (1.4.7) and (1.4.8) it follows that the values

$$l_0^{(1)} R_{2n_1}^{(1)}(\mu_k^{(1)}, 0) = \frac{l_0^{(1)} l_0^{(2)} \phi(\mu_k^{(1)})}{l_0^{(2)} R_{2n_2-1}^{(2)}(\mu_k^{(1)}, 0)} \quad (k = 1, \ldots, n_1)$$

are uniquely determined by the given data. Then the difference of two such functions $l_0^{(1)} R_{2n_1}^{(1)}(\cdot, 0)$ and $\widetilde{l}_0^{(1)} \widetilde{R}_{2n_1}^{(1)}(\cdot, 0)$ is zero at $\mu_k^{(1)}$ $(k = 1, \ldots, n_1)$, which means that there is $c \in \mathbb{R}$ such that

$$l_0^{(1)} R_{2n_1}^{(1)}(\cdot, 0) - \widetilde{l}_0^{(1)} \widetilde{R}_{2n_1}^{(1)}(\cdot, 0) = c l_0^{(1)} R_{2n_1-1}^{(1)}(\cdot, 0).$$

Therefore, in view of $l_0^{(1)} R_{2n_1-1}^{(1)}(\cdot, 0) = \widetilde{l}_0^{(1)} \widetilde{R}_{2n_1-1}^{(1)}(\cdot, 0)$,

$$\begin{aligned} \frac{R_{2n_1}^{(1)}(\cdot, 0)}{R_{2n_1-1}^{(1)}(\cdot, 0)} - \frac{\widetilde{R}_{2n_1}^{(1)}(\cdot, 0)}{\widetilde{R}_{2n_1-1}^{(1)}(\cdot, 0)} &= \frac{l_0^{(1)} R_{2n_1}^{(1)}(\cdot, 0)}{l_0^{(1)} R_{2n_1-1}^{(1)}(\cdot, 0)} - \frac{\widetilde{l}_0^{(1)} \widetilde{R}_{2n_1}^{(1)}(\cdot, 0)}{\widetilde{l}_0^{(1)} \widetilde{R}_{2n_1-1}^{(1)}(\cdot, 0)} \\ &= \frac{l_0^{(1)} R_{2n_1}^{(1)}(\cdot, 0) - \widetilde{l}_0^{(1)} \widetilde{R}_{2n_1}^{(1)}(\cdot, 0)}{l_0^{(1)} R_{2n_1-1}^{(1)}(\cdot, 0)} \\ &= c. \end{aligned}$$

Since the representation (1.2.19) is unique by Lemma A.3.5, it follows that the sequences $(m_k^{(1)})_{k=1}^{n_1}$ and $(l_k^{(1)})_{k=0}^{n_1-1}$ are uniquely determined by the given data. In the same way, it also follows that the sequences $(m_k^{(2)})_{k=1}^{(n_2)}$ and $(l_k^{(2)})_{k=0}^{n_2-1}$ are uniquely determined by the given data. □

Remark 1.4.7. Theorems 1.4.4 and 1.4.6 deal with the direct and inverse problems, respectively, for the Dirichlet-Dirichlet problem of a string and the Dirichlet-Neumann problem of two substrings. It is easy to see that the properties 1 and 3 in Theorem 1.4.4 are equivalent to the conditions (i) and (ii) in Theorem 1.4.6. Furthermore, with $(l_k^{(1)})_{k=0}^{n_1-1}$ and $(l_k^{(2)})_{k=0}^{n_1-1}$ also l_{n_1} is uniquely determined by the given data in Theorem 1.4.6. Only the position of the partition point P between the bead with index n_1 and the bead with index n_1+1, see Figures 1.5 and 1.6, can vary according to the range for $\widetilde{l}$ given in Theorem 1.4.6. Hence the assumptions in Theorem 1.4.6 are necessary and sufficient to solve the inverse problem.

Now we consider the problem of finding the sequences $(m_k^{(1)})_{k=1}^{n_1}$, $(m_k^{(2)})_{k=1}^{n_2}$, $(l_k^{(1)})_{k=0}^{n_1}$, and $(l_k^{(2)})_{k=0}^{n_2}$ from the sequences of the characteristic values $(\nu_k)_{k=1}^{n_1+n_2}$, $(\nu_k^{(1)})_{k=1}^{n_1}$, $(\mu_k^{(2)})_{k=1}^{n_2}$ and some other parameters.

For the remainder of this subsection we will consider some subcases with different behavior regarding uniqueness. Let n_1 and n_2 be positive integers and let $l > 0$ and $\widehat{l} \in (0, l)$ and let the sequences $(\nu_k^{(1)})_{k=1}^{n_1}$, $(\mu_k^{(2)})_{k=1}^{n_2}$ and $(\nu_k)_{k=1}^{n_1+n_2}$ be given. With these given data, we will associate the polynomials

$$P_0(z) = l \prod_{k=1}^{n_1+n_2} \left(1 - \frac{z}{\nu_k}\right), \tag{1.4.41}$$

$$P_1(z) = \widehat{l} \prod_{k=1}^{n_1} \left(1 - \frac{z}{\nu_k^{(1)}}\right), \tag{1.4.42}$$

$$Q_2(z) = \prod_{k=1}^{n_2} \left(1 - \frac{z}{\mu_k^{(2)}}\right) \tag{1.4.43}$$

and

$$\Phi = P_0 - P_1 Q_2. \tag{1.4.44}$$

Clearly,

$$\Phi(0) = l - \widehat{l} > 0. \tag{1.4.45}$$

First we consider a case where the solution of the inverse problem is unique.

Theorem 1.4.8. *Let n_1 and n_2 be positive integers and let $l > 0$ and $\widehat{l} \in (0, l)$. Let three strictly increasing sequences of positive real numbers $(\nu_k)_{k=1}^{n_1+n_2}$, $(\nu_k^{(1)})_{k=1}^{n_1}$ and $(\mu_k^{(2)})_{k=1}^{n_2}$ be given. Let the concatenated sequence $(\vartheta_k)_{k=1}^{n_1+n_2} = (\nu_k^{(1)})_{k=1}^{n_1} \cup (\mu_k^{(2)})_{k=1}^{n_2}$ be indexed such that $\vartheta_k \le \vartheta_{k'}$ if $k < k'$. We require that these sequences satisfy the following conditions:*

(i) *The sequences $(\nu_k)_{k=1}^{n_1+n_2}$ and $(\vartheta_k)_{k=1}^{n_1+n_2}$ interlace as follows:*

$$\nu_1 < \vartheta_1 < \cdots < \nu_{n_1+n_2} < \vartheta_{n_1+n_2}. \tag{1.4.46}$$

(ii) $\vartheta_{n_1+n_2} = \nu_{n_1}^{(1)}$.

Then there exist unique sequences of positive real numbers $(m_k^{(1)})_{k=1}^{n_1}$, $(m_k^{(2)})_{k=1}^{n_2}$, $(l_k^{(1)})_{k=0}^{n_1}$, and $(l_k^{(2)})_{k=0}^{n_2}$ such that $\sum\limits_{k=0}^{n_1} l_k^{(1)} = \widehat{l}$, $\sum\limits_{k=0}^{n_2} l_k^{(2)} = l - \widehat{l}$, which generate the Dirichlet-Dirichlet problem (1.4.1)–(1.4.4) for the string S, the Dirichlet-Dirichlet problem (1.4.9)–(1.4.11) for the string S_1 and the Dirichlet-Neumann problem (1.4.9), (1.4.11), (1.4.12) for the string S_2 with the sequences of characteristic values $(\nu_k)_{k=1}^{n_1+n_2}$, $(\nu_k^{(1)})_{k=1}^{n_1}$ and $(\mu_k^{(2)})_{k=1}^{n_2}$, respectively.

Proof. Since $\Phi(\nu_k) = -P_1(\nu_k)Q_2(\nu_k)$, (1.4.46) implies

$$(-1)^k \Phi(\nu_k) > 0 \quad (k = 1, \ldots, n_1 + n_2).$$

Since $\Phi(\vartheta_k) = P_0(\vartheta_k)$,

$$(-1)^k \Phi(\vartheta_k) > 0 \quad (k = 1, \ldots, n_1 + n_2).$$

Hence each of the intervals $(0,\nu_1)$, (ϑ_k,ν_{k+1}), $(k=1,\dots,n_1+n_2-1)$ contains exactly one zero of Φ, and Φ has no other zeros. We denote the zero of Φ lying in $(0,\nu_1)$ by $\mu_1^{(1)}$. For $k=1,\dots,n_1+n_2-1$ denote the zero of Φ lying in (ϑ_k,ν_{k+1}) by $\nu_r^{(2)}$ if $\vartheta_k=\mu_r^{(2)}$ and by $\mu_{r+1}^{(1)}$ if $\vartheta_k=\nu_r^{(1)}$. Therefore, for both $j=1$ and $j=2$, the sequences $(\nu_k^{(j)})_{k=1}^{n_j}$ and $(\mu_k^{(j)})_{k=1}^{n_j}$ interlace as follows:

$$0<\mu_1^{(j)}<\nu_1^{(j)}<\mu_2^{(j)}<\nu_2^{(j)}<\cdots<\mu_{n_j}^{(j)}<\nu_{n_j}^{(j)}. \tag{1.4.47}$$

Thus the sequences $(\nu_k^{(j)})_{k=1}^{n_j}$ and $(\mu_k^{(j)})_{k=1}^{n_j}$ satisfy the assumptions of Theorem 1.3.1, and hence there exists a unique tuple of sequences $((m_k^{(j)})_{k=1}^{n_j},(l_k^{(j)})_{k=0}^{n_j})$ with $\sum\limits_{k=0}^{n_1}l_k^{(1)}=\widehat{l}$ and $\sum\limits_{k=0}^{n_2}l_k^{(2)}=l-\widehat{l}$, respectively, which generates Dirichlet-Dirichlet and Dirichlet-Neumann problems with the characteristic values $(\nu_k^{(1)})_{k=1}^{n_2}$ and $(\mu_k^{(2)})_{k=1}^{n_2}$, respectively.

By the above construction, the zeros of Φ and $\left[l_0^{(1)}R_{2n_1-1}^{(1)}(\cdot,0)\right]\left[l_0^{(2)}R_{2n_2}^{(2)}(\cdot,0)\right]$ coincide. Since also $\Phi(0)=l-\widehat{l}=\left[l_0^{(1)}R_{2n_1-1}^{(1)}(0,0)\right]\left[l_0^{(2)}R_{2n_2}^{(2)}(0,0)\right]$, we conclude that $\Phi=\left[l_0^{(1)}R_{2n_1-1}^{(1)}(\cdot,0)\right]\left[l_0^{(2)}R_{2n_2}^{(2)}(\cdot,0)\right]$. By (1.2.25) and (1.2.28), the polynomials P_1 and Q_2 defined by (1.4.42) and (1.4.43) have the representations

$$P_1=l_0^{(1)}R_{2n_1}^{(1)}(\cdot,0),\quad Q_2=l_0^{(2)}R_{2n_2-1}^{(2)}(\cdot,0).$$

In view of (1.4.8) it follows that $P_0=l_0^{(1)}R_{2n}(\cdot,0)$. But this shows that the given sequence $(\nu_k)_{k=1}^{n_1+n_2}$ is the sequence of the characteristic values of the Dirichlet-Dirichlet problem (1.4.1)–(1.4.4) for this string.

To prove the uniqueness, we observe that Φ is uniquely determined by the given data and that $\Phi=\left[l_0^{(1)}R_{2n_1-1}^{(1)}(\cdot,0)\right]\left[l_0^{(2)}R_{2n_2}^{(2)}(\cdot,0)\right]$ by (1.2.25), (1.2.28) and (1.4.8). In particular, the zeros of Φ are unique. Recall that (1.4.46) is assumed and that (1.4.47) holds because of (1.2.36). We will now show that the zeros of $R_{2n_2}^{(2)}(\cdot,0)$ are uniquely determined. Indeed, let $k\in\{1,\dots,n_1+n_2-1\}$ with $R_{2n_2-1}^{(2)}(\vartheta_k,0)=0$. If also $R_{2n_2-1}^{(2)}(\vartheta_{k+1},0)=0$, then the interlacing property shows that there is $z_0\in(\vartheta_k,\vartheta_{k+1})$ such that $R_{2n_2}^{(2)}(z_0,0)=0$. Since Φ has no zeros in $[\nu_{k+1},\vartheta_{k+1})$, it follows that the zero of Φ in (ϑ_k,ν_{k+1}) is a zero of $R_{2n_2}^{(2)}(\cdot,0)=0$. If, however, $R_{2n_2-1}^{(2)}(\vartheta_{k+1},0)\neq 0$, then $R_{2n_1}^{(1)}(\vartheta_{k+1},0)=0$. Hence there is a number $r\in\{1,\dots,k\}$ such that $\vartheta_k=\mu_r^{(2)}$, and, by definition of ϑ_{k+1}, $\vartheta_{k+1}=\nu_{k-r+1}^{(1)}$ follows. Therefore $R_{2n_2}^{(2)}(\cdot,0)$ has $r-1$ zeros in $(0,\vartheta_k)$ and $R_{2n_1-1}^{(1)}(\cdot,0)$ has $k-r+1$ zeros in $(0,\vartheta_{k+1})$ by (1.4.47). Since Φ has k zeros in $(0,\vartheta_k)$, it follows that all zeros of $R_{2n_1-1}^{(1)}(\cdot,0)$ in $(0,\vartheta_{k+1})$ must lie in $(0,\vartheta_k)$. Hence the zero of Φ in (ϑ_k,ν_{k+1}) is a zero of $R_{2n_2}^{(2)}(\cdot,0)$. In this way we have uniquely identified n_2 zeros of $R_{2n_2}^{(2)}(\cdot,0)$, which are all zeros of this polynomial. The remaining zeros of Φ now determine uniquely the zeros of $R^{(1)}_{2n_1-1}(\cdot,0)$. Finally, by Theorem 1.3.1, $(m^{(1)}_k)_{k=1}^{n_1}$, $(m^{(2)}_k)_{k=1}^{n_2}$, $(l^{(1)}_k)_{k=0}^{n_1}$, and $(l^{(2)}_k)_{k=0}^{n_2}$ are uniquely determined by the given data. □

It is clear that condition (i) in Theorem 1.4.8 is equivalent to requiring for each $k = 1,\dots,n_1+n_2$ that $\nu_k \neq \vartheta_k$ and that the interval $(0,\nu_k)$ contains exactly $k-1$ terms of the sequence $(\nu^{(1)}_k)_{k=1}^{n_1} \cup (\mu^{(2)}_k)_{k=1}^{n_2}$. This is a particular case of the necessary conditions for the characteristic values derived in Theorem 1.4.4, statement 4. However, these conditions are not sufficient. A solution of the inverse problem would require that $\Phi = \left[l^{(1)}_0 R^{(1)}_{2n_1-1}(\cdot,0)\right]\left[l^{(2)}_0 R^{(2)}_{2n_2}(\cdot,0)\right]$. In particular, the coefficient of $(-z)^{n_1+n_2}$ of Φ must be positive. It is easy to see that this condition is

$$l\prod_{k=1}^{n_1+n_2}\frac{1}{\nu_k} - \hat{l}\prod_{k=1}^{n_1}\frac{1}{\nu^{(1)}_k}\prod_{k=1}^{n_2}\frac{1}{\mu^{(2)}_k} > 0. \tag{1.4.48}$$

Under this and another explicit assumption, we can now give a result for the case that each interval $(0,\nu_k)$ contains k terms of the sequence $(\nu^{(1)}_k)_{k=1}^{n_1} \cup (\mu^{(2)}_k)_{k=1}^{n_2}$.

Theorem 1.4.9. *Let n_1 and n_2 be positive integers and let $l>0$ and $\widehat{l}\in(0,l)$. Let three strictly increasing sequences of positive real numbers $(\nu_k)_{k=1}^{n_1+n_2}$, $(\nu^{(1)}_k)_{k=1}^{n_1}$ and $(\mu^{(2)}_k)_{k=1}^{n_2}$ be given. Let the concatenated sequence $(\vartheta_k)_{k=1}^{n_1+n_2} = (\nu^{(1)}_k)_{k=1}^{n_1}\cup(\mu^{(2)}_k)_{k=1}^{n_2}$ be indexed such that $\vartheta_k \le \vartheta_{k'}$ if $k<k'$. We require that these sequences satisfy the following conditions:*

(i) *The sequences $(\nu_k)_{k=1}^{n_1+n_2}$ and $(\vartheta_k)_{k=1}^{n_1+n_2}$ interlace as follows:*

$$\vartheta_1 < \nu_1 < \cdots < \vartheta_{n_1+n_2} < \nu_{n_1+n_2}. \tag{1.4.49}$$

(ii) $\vartheta_1 = \mu^{(2)}_1$.

(iii) *The inequality* (1.4.48) *is satisfied.*

Then there exist unique sequences of positive real numbers $(m^{(1)}_k)_{k=1}^{n_1}$, $(m^{(2)}_k)_{k=1}^{n_2}$, $(l^{(1)}_k)_{k=0}^{n_1}$, and $(l^{(2)}_k)_{k=0}^{n_2}$ such that $\sum\limits_{k=0}^{n_1} l^{(1)}_k = \widehat{l}$, $\sum\limits_{k=0}^{n_2} l^{(2)}_k = l-\widehat{l}$, which generate the Dirichlet Dirichlet problem (1.4.1)–(1.4.4) *for the string S, the Dirichlet-Dirichlet problem* (1.4.9)–(1.4.11) *for the string S_1 and the Dirichlet-Neumann problem* (1.4.9), (1.4.11), (1.4.12) *for the string S_2 with the sequences of characteristic values $(\nu_k)_{k=1}^{n_1+n_2}$, $(\nu^{(1)}_k)_{k=1}^{n_1}$ and $(\mu^{(2)}_k)_{k=1}^{n_2}$, respectively.*

Proof. We can prove this theorem in the way we have proved Theorem 1.4.8. However, we will transform the problem to a problem with given data as in Theorem 1.4.8 and then apply conclusions in the proof of Theorem 1.4.8. Thus define

$\widetilde{n}_1 = n_2$, $\widetilde{n}_2 = n_1$,

$$\widetilde{\nu}_k = \frac{1}{\nu_{n_1+n_2-k+1}} \quad (k = 1, \ldots, n_1 + n_2),$$
$$\widetilde{\nu}_k^{(1)} = \frac{1}{\mu_{n_2-k+1}^{(2)}} \quad (k = 1, \ldots, \widetilde{n}_1),$$
$$\widetilde{\mu}_k^{(2)} = \frac{1}{\nu_{n_1-k+1}^{(1)}} \quad (k = 1, \ldots, \widetilde{n}_2),$$
$$\widetilde{\vartheta}_k = \frac{1}{\zeta_{n_1+n_2-k+1}} \quad (k = 1, \ldots, n_1 + n_2),$$

and

$$\alpha = \prod_{k=1}^{n_2} \mu_k^{(2)}, \quad \beta = \prod_{k=1}^{n_1} \nu_k^{(1)}, \quad \gamma = \prod_{k=1}^{n_1+n_2} \nu_k, \quad \widetilde{l} = \frac{l}{\gamma}, \quad \widetilde{\widehat{l}} = \frac{\widehat{l}}{\alpha\beta}.$$

From (1.4.48) it follows that $\widetilde{l} > \widetilde{\widehat{l}}$. Then $(\widetilde{\zeta}_k)_{k=1}^{n_1+n_2} = (\widetilde{\nu}_k^{(1)})_{k=1}^{\widetilde{n}_1} \cup (\widetilde{\mu}_k^{(2)})_{k=1}^{\widetilde{n}_2}$ and the sequences $(\widetilde{\nu}_k)_{k=1}^{n_1+n_2}$ and $(\widetilde{\vartheta}_k)_{k=1}^{n_1+n_2}$ interlace as in (1.4.46). Defining $\widetilde{P}_0$, $\widetilde{P}_1$, $\widetilde{Q}_2$ and $\widetilde{\Phi}$ according to (1.4.41)–(1.4.44), the proof of Theorem 1.4.8 gives unique sequences of positive numbers $(\widetilde{\nu}_k^{(2)})_{k=1}^{\widetilde{n}_2}$ and $(\widetilde{\mu}_k^{(1)})_{k=1}^{\widetilde{n}_1}$ which satisfy the interlacing (1.4.47) for the numbers with tilde and such that $\widetilde{\Phi} = \widetilde{Q}_1\widetilde{P}_2$, where

$$\widetilde{P}_2(z) = (\widetilde{l} - \widetilde{\widehat{l}}) \prod_{k=1}^{\widetilde{n}_2} \left(1 - \frac{z}{\widetilde{\nu}_k^{(2)}}\right), \quad \widetilde{Q}_1(z) = \prod_{k=1}^{\widetilde{n}_1} \left(1 - \frac{z}{\widetilde{\mu}_k^{(1)}}\right).$$

Putting

$$\nu_k^{(2)} = \frac{1}{\widetilde{\mu}_{n_2-k+1}^{(1)}} \quad (k = 1, \ldots, n_2), \quad \mu_k^{(1)} = \frac{1}{\widetilde{\nu}_{n_1-k+1}^{(2)}} \quad (k = 1, \ldots, n_1), \tag{1.4.50}$$

and

$$\delta = \prod_{k=1}^{n_1} \mu_k^{(1)}, \quad \varepsilon = \prod_{k=1}^{n_2} \nu_k^{(2)},$$

we arrive at the polynomials

$$P_2(z) = (l - \widehat{l}) \prod_{k=1}^{n_2} \left(1 - \frac{z}{\nu_k^{(2)}}\right), \quad Q_1(z) = \prod_{k=1}^{n_1} \left(1 - \frac{z}{\mu_k^{(1)}}\right).$$

Then

$$
\begin{aligned}
(-z)^{n_1+n_2}\widetilde{P}_0\left(\frac{1}{z}\right) &= \widetilde{l}\prod_{k=1}^{n_1+n_2}(-z+\nu_k) = \widetilde{l}\gamma\prod_{k=1}^{n_1}\left(1-\frac{z}{\nu_k}\right) = \gamma\frac{\widetilde{l}}{l}P_0(z),\\
(-z)^{n_2}\widetilde{P}_1\left(\frac{1}{z}\right) &= \widetilde{\widetilde{l}}\prod_{k=1}^{n_2}(-z+\mu_k^{(2)}) = \alpha\widetilde{\widetilde{l}}Q_2(z),\\
(-z)^{n_1}\widetilde{Q}_2\left(\frac{1}{z}\right) &= \prod_{k=1}^{n_1}(-z+\nu_k^{(1)}) = \frac{\beta}{\widehat{l}}P_1(z),\\
(-z)^{n_2}\widetilde{Q}_1\left(\frac{1}{z}\right) &= \prod_{k=1}^{n_2}\left(-z+\frac{1}{\widetilde{\mu}_k^{(1)}}\right) = \frac{\varepsilon}{l-\widehat{l}}P_2(z),\\
(-z)^{n_1}\widetilde{P}_2\left(\frac{1}{z}\right) &= (\widetilde{l}-\widetilde{\widetilde{l}})\prod_{k=1}^{n_1}\left(-z+\frac{1}{\widetilde{\nu}_k^{(2)}}\right) = \delta(\widetilde{l}-\widetilde{\widetilde{l}})Q_1(z).
\end{aligned}
$$

Replacing z by $\frac{1}{z}$ in the identity $\widetilde{P}_0(z) = \widetilde{Q}_1(z)\widetilde{P}_2(z) + \widetilde{P}_1(z)\widetilde{Q}_2(z)$ and then multiplying the resulting identity with $(-z)^{n_1+n_2}$, the above representations lead to

$$\gamma\frac{\widetilde{l}}{l}P_0(z) = \varepsilon\delta\frac{\widetilde{l}-\widetilde{\widetilde{l}}}{l-\widehat{l}}Q_1(z)P_2(z) + \alpha\beta\frac{\widetilde{\widetilde{l}}}{\widehat{l}}P_1(z)Q_2(z).$$

Comparing the coefficients of $(-z)^{n_1+n_2}$ in $\widetilde{P}_0(z) = \widetilde{Q}_1(z)\widetilde{P}_2(z)+\widetilde{P}_1(z)\widetilde{Q}_2(z)$ leads to $\widetilde{l}\gamma = (\widetilde{l}-\widetilde{\widetilde{l}})\varepsilon\delta + \widetilde{\widetilde{l}}\alpha\beta$, which shows that

$$\varepsilon\delta\frac{\widetilde{l}-\widetilde{\widetilde{l}}}{l-\widehat{l}} = \frac{\widetilde{l}\gamma-\widetilde{\widetilde{l}}\alpha\beta}{l-\widehat{l}} = 1.$$

Hence we arrive at $P_0 = Q_1P_2 + P_1Q_2$. For $j = 1, 2$, the zeros $(\nu_k^{(j)})_{j=1}^{n_j}$ of Q_j and $(\mu_k^{(j)})_{j=1}^{n_j}$ of P_j satisfy the interlacing property (1.4.47). In view of Theorem 1.3.1, there are string data such that $P_j = l_0^{(j)}R_{2n_j}^{(j)}(\cdot,0)$ and $Q_j = l_0^{(j)}R_{2n_j-1}^{(j)}(\cdot,0)$, and (1.4.5) gives $P_0 = l_0^{(j)}R_{2n_1+n_2}(\cdot,0)$. Furthermore,

$$\sum_{k=1}^{n_1} l_k^{(1)} = \frac{P_1(0)}{Q_1(0)} = \widehat{l}, \quad \sum_{k=1}^{n_2} l_k^{(2)} = \frac{P_2(0)}{Q_2(0)} = l-\widehat{l}.$$

Hence we have found string data which generate the given characteristic values.

To prove uniqueness, we observe that the proof of Theorem 1.4.8 shows that the sequences $(\widetilde{\mu}_k^{(1)})_{k=1}^{n_1}$ and $(\widetilde{\nu}_k^{(2)})_{k=1}^{n_2}$ are uniquely determined by the given data. But by (1.4.50) this means that the characteristic values of the Dirichlet-Neumann problems on the strings S_1 and S_2 are uniquely determined by the given data, and therefore also the string data are unique by Theorem 1.3.1. □

Remark 1.4.10. Theorems 1.4.4 and 1.4.5 deal with the direct and inverse problems, respectively, for the Dirichlet-Dirichlet problems of a string and two substrings. It is easy to see that the properties 1 and 2 in Theorem 1.4.4 are equivalent to the conditions (i) and (ii) in Theorem 1.4.5. Therefore the assumptions in Theorem 1.4.5 are necessary and sufficient to solve the inverse problem.

Remark 1.4.11. In Theorems 1.4.8 and 1.4.9, and in the cases when the distributions of the beads are unique also in Theorems 1.4.5 and 1.4.6, the distribution of the beads depends continuously on the given data in view of Lemma A.6.5 and Corollary A.6.7.

Remark 1.4.12. In Theorem 1.4.4, statement 4, we have derived properties for the characteristic values when considering the Dirichlet-Dirichlet problems on S and S_1 and the Dirichlet-Neumann problem on S_2. Theorems 1.4.8 and 1.4.9 give sufficient conditions for the solvability of the corresponding inverse problem. We have already seen that apart from the properties stated in Theorem 1.4.4, statement 4, we also need condition (1.4.48) for the existence of a solution of the inverse problem. Furthermore, all zeros of Φ must be real and positive, but this property is not guaranteed.

Indeed, consider the particular example $n_1 = 1$, $n_2 = 1$, $\nu_1 = 2$, $\nu_2 = 4$, $\vartheta_1 = 1$, $\vartheta_2 = 8$, $\widehat{l} = 1$, $l > 1$. Then (1.4.48) is clearly satisfied, and

$$\begin{aligned}\Phi(z) &= l\left(1-\frac{z}{2}\right)\left(1-\frac{z}{4}\right) - (1-z)\left(1-\frac{z}{8}\right)\\ &= \frac{1}{8}(l-1)\left[z^2 - \left(6-\frac{3}{l-1}\right)z + 8\right].\end{aligned}$$

Then Φ has no real zeros if

$$\frac{11}{2} - 3\sqrt{2} < l < \frac{11}{2} + 3\sqrt{2},$$

two negative real zeros if

$$1 < l < \frac{11}{2} - 3\sqrt{2},$$

and two positive real zeros if

$$l > \frac{11}{2} + 3\sqrt{2}.$$

Hence to guarantee positive real zeros, l has to be sufficiently large. However, this example also shows that in order to find the range of admissible l one needs to find the zeros of Φ.

In general, if $\frac{\widehat{l}}{l}$ is sufficiently close to 0, then all zeros of Φ are real, positive and simple. Indeed, since P_1Q_2 tends to 0 uniformly on compact sets if l is fixed and $\widehat{l}$ tends to 0, it follows from Rouché's theorem that for each $\varepsilon > 0$ there is $\delta > 0$ such that $\frac{\widehat{l}}{l} \in (0, \delta)$ implies that all zeros $\hat{\nu}_k$, $(k = 1, \dots n_1 + n_2)$ of Φ are real, simple and satisfy $|\hat{\nu}_k - \nu_k| < \varepsilon$. In particular, for sufficiently small $\varepsilon > 0$, all zeros of Φ are positive. If we additionally assume that $\{\nu_k\}_{k=1}^{n_1+n_2} \cap \{\vartheta_k\}_{k=1}^{n_1+n_2} = \emptyset$, then it is clear that for each $k \in \{1, \dots, n_1 + n_2\}$ the number of terms of $(\vartheta_k)_{k=1}^{n_1+n_2}$ in the interval $(0, \nu_k)$ equals the number of terms of $(\vartheta_k)_{k=1}^{n_1+n_2}$ in the interval $(0, \hat{\nu}_k)$.

Let us finally consider an example with $n_1 \geq 2$ such that

$$\nu_1 < \nu_1^{(1)} < \nu_2^{(1)} < \nu_2 < \dots .$$

Then choosing l and $\widehat{l}$ such that $\frac{\widehat{l}}{l}$ is sufficiently small, we obtain, with the notation of the previous paragraph,

$$\widehat{\nu}_1 < \nu_1^{(1)} < \nu_2^{(1)} < \widehat{\nu}_2 < \dots .$$

To solve the inverse problem, we need for all $k = 1, \dots, n+1$ that there is $j \in \{1, \dots, n_1\}$ such that $\widehat{\nu}_k = \mu_j^{(1)}$ or that there is $j \in \{1, \dots, n_2\}$ such that $\widehat{\nu}_k = \nu_j^{(2)}$. When $k = 1$, we thus have the allocation $\widehat{\nu}_1 = \mu_1^{(1)}$ or $\widehat{\nu}_1 = \nu_1^{(2)}$. But from $\widehat{\nu}_1 < \nu_1^{(1)} < \widehat{\nu}_2 < \mu_1^{(2)} < \nu_1^{(2)}$ we see that $\widehat{\nu}_1 \neq \nu_1^{(2)}$ and therefore $\widehat{\nu}_1 = \mu_1^{(1)}$. Then $\widehat{\nu}_2 = \mu_2^{(1)}$ or $\widehat{\nu}_2 = \nu_1^{(2)}$. But $\widehat{\nu}_2 < \vartheta_3 \leq \mu_1^{(2)}$ shows that $\widehat{\nu}_2 \neq \nu_1^{(2)}$, whereas $\widehat{\nu}_2 = \mu_2^{(1)}$ would lead to the contradiction $\nu_2^{(1)} < \widehat{\nu}_2 = \mu_2^{(1)}$.

The above examples and reasonings show that conditions which are necessary and sufficient to solve the inverse problem would be quite cumbersome to derive and verify. Hence we are not going to pursue this problem any further.

1.4.3 The finite-dimensional Hochstadt-Lieberman problem

In this subsection it will be shown that the distribution of the beads on a string is uniquely determined by the length of the string, by the distribution of the beads on half of the string, and by the characteristic values of the Dirichlet-Dirichlet problem. Figure 1.7 illustrates the notation used in the following theorem.

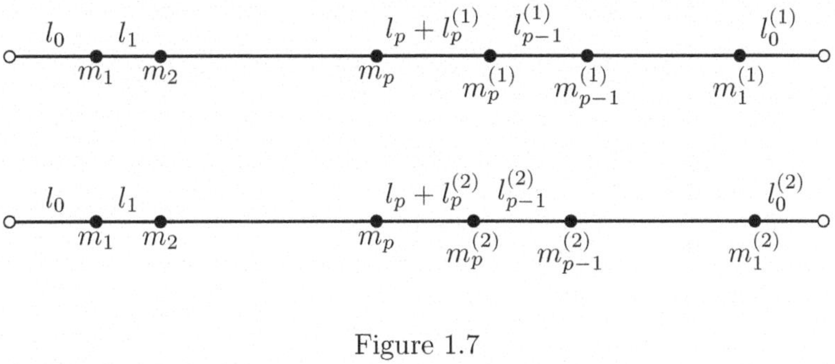

Figure 1.7

Theorem 1.4.13. *Let $p \in \mathbb{N}$ and put $n = 2p$. Let sequences of positive numbers $(m_k)_{k=1}^p$, $(l_k)_{k=0}^p$, $(m_k^{(1)})_{k=1}^p$, $(l_k^{(1)})_{k=0}^p$, $(m_k^{(2)})_{k=1}^p$, $(l_k^{(2)})_{k=0}^p$ be given such that $\sum\limits_{k=0}^p l_k^{(1)} = \sum\limits_{k=0}^p l_k^{(2)}$. For $j = 1,2$, let S_j be the string generated by $(m_k)_{k=1}^p$, $(l_k)_{k=0}^p$, $(m_k^{(j)})_{k=1}^p$, $(l_k^{(j)})_{k=0}^p$ as indicated in Figure 1.7. For $j = 1,2$, let σ_j be the sequence of the characteristic values of the Dirichlet-Dirichlet problem (1.2.4), (1.2.23), (1.2.24) for the string S_j.*

Then $\sigma_1 = \sigma_2$ implies $m_k^{(1)} = m_k^{(2)}$ for $k = 1, \dots, p$ and $l_k^{(1)} = l_k^{(2)}$ for $k = 0, \dots, p$.

Proof. The strings S_j ($j = 1,2$) are composed of two substrings $S_{j,1}$ and $S_{j,2}$, where $S_{j,1}$ is the string with the given data $(m_k)_{k=1}^p$, $(l_k)_{k=0}^p$, and $S_{j,2}$ is the string with the given data $(m_k^{(j)})_{k=1}^p$, $(l_k^{(j)})_{k=0}^p$. In particular, $S_{1,1} = S_{2,1}$. Denoting the characteristic polynomials of S_j, $S_{j,1}$ and $S_{j,2}$ by $R_k^{(j)}(\cdot,0)$, $R_k^{(j,1)}(\cdot,0)$ and $R_k^{(j,2)}(\cdot,0)$, respectively, we have from (1.4.8) that

$$R_{2n}^{(j)}(\cdot,0) = R_{2p}^{(j,1)}(\cdot,0)\left[l_0^{(j)} R_{2p-1}^{(j,2)}(\cdot,0)\right] + R_{2p-1}^{(j,1)}(\cdot,0)\left[l_0^{(j)} R_{2p}^{(j,2)}(\cdot,0)\right].$$

Since the total length of S_1 equals the total length of S_2 and since $\sigma_1 = \sigma_2$, it follows from (1.2.25) that $R_{2n}^{(1)}(\cdot,0) = R_{2n}^{(2)}(\cdot,0)$. Also, since $S_{1,1} = S_{2,1}$, it follows that $R_k^{(1,1)}(\cdot,0) = R_k^{(2,1)}(\cdot,0)$. Hence taking the differences of the above identity for $j = 1$ and $j = 2$ it follows that

$$\begin{aligned} 0 = R_{2p}^{(1,1)}(\cdot,0)&\left[l_0^{(1)} R_{2p-1}^{(1,2)}(\cdot,0) - l_0^{(2)} R_{2p-1}^{(2,2)}(\cdot,0)\right] \\ &+ R_{2p-1}^{(1,1)}(\cdot,0)\left[l_0^{(1)} R_{2p}^{(1,2)}(\cdot,0) - l_0^{(2)} R_{2p}^{(2,2)}(\cdot,0)\right]. \end{aligned}$$

Since the polynomials $R_{2p}^{(1,1)}(\cdot,0)$ and $R_{2p-1}^{(1,1)}(\cdot,0)$ do not have common zeros, see (1.2.36), it follows that $l_0^{(1)} R_{2p}^{(1,2)}(\cdot,0) - l_0^{(2)} R_{2p}^{(2,2)}(\cdot,0)$ is zero at the p zeros of $R_{2p}^{(1,1)}(\cdot,0)$. Hence there is a constant C such that

$$l_0^{(1)} R_{2p}^{(1,2)}(\cdot,0) - l_0^{(2)} R_{2p}^{(2,2)}(\cdot,0) = C R_{2p}^{(1,1)}(\cdot,0).$$

Then

$$l_0^{(1)} R_{2p-1}^{(1,2)}(\cdot,0) - l_0^{(2)} R_{2p-1}^{(2,2)}(\cdot,0) = -C R_{2p-1}^{(1,1)}(\cdot,0).$$

Evaluating the latter identity at 0 gives in view of (1.2.28) that $C = 0$. Since the lengths of S_1 and S_2 are equal and since the lengths of $S_{1,1}$ and $S_{1,2}$ are equal, also the lengths of $S_{1,2}$ and $S_{2,2}$ are equal. It follows from (1.2.25) and (1.2.28) that the characteristic values of the Dirichlet-Dirichlet and Dirichlet-Neumann problems for $S_{1,2}$ and $S_{2,2}$ coincide. An application of the uniqueness part of Theorem 1.3.1 completes the proof. □

The uniqueness result in Theorem 1.4.13 raises the following question. Let $p \in \mathbb{N}$ and put $n = 2p$. Given a sequence $(\nu_k)_{k=1}^n$ of distinct positive real numbers, two sequences of positive real numbers $(m_k)_{k=1}^p$ and $(l_k)_{k=0}^p$ and a positive real number $l > \sum\limits_{k=0}^{p} l_k$, does there exist a Stieltjes string S of total length l with n beads such that the sequence of the characteristic values of the Dirichlet-Dirichlet problem for S is $(\nu_k)_{k=1}^n$ and such that the lengths the threads and the masses of the beads of the left part of S are $(m_k)_{k=1}^p$ and $(l_k)_{k=0}^p$, respectively? The following example for the simplest case $n = 2$ shows that restrictions are needed and that these restrictions are most likely difficult to describe for $n > 2$. We will use the notation from Figure 1.7 and set $l_2 := l_0^{(1)}$ and $m_2 = m_1^{(1)}$.

Example 1.4.14. Let S be a string of length 4 with 2 beads and $l_0 = 1$, $l_1^{(1)} = 1$, $m_1 = 1$ and $0 < \nu_1 < \nu_2$. Then there are $l_2 \in (0, 2)$ and $m_2 > 0$ such that ν_1 and ν_2 are the characteristic values of the Dirichlet-Dirichlet problem on S if and only if $16 < 12(\nu_1 + \nu_2) - 9\nu_1\nu_2 < 24$.

Proof. A straightforward calculation with the given data of the string shows that

$$R_4(z, 0) = m_2(2 + l_2 - l_2^2)z^2 - (3 + 4m_2 - m_2 l_2^2)z + 4.$$

On the other hand, by (1.2.25),

$$R_4(z, 0) = 4\left(1 - \frac{z}{\nu_1}\right)\left(1 - \frac{z}{\nu_2}\right) = \frac{4}{\nu_1\nu_2}z^2 - \frac{4(\nu_1 + \nu_2)}{\nu_1\nu_2}z + 4.$$

Comparing the coefficients gives the following two equations which describe the relation between the string data and the spectral data:

$$m_2(2 + l_2 - l_2^2) = \frac{4}{\nu_1\nu_2},$$

$$3 + m_2(4 - l_2^2) = \frac{4(\nu_1 + \nu_2)}{\nu_1\nu_2}.$$

The first equation gives

$$m_2 = \frac{4}{\nu_1\nu_2(2 + l_2 - l_2^2)}, \tag{1.4.51}$$

and substituting (1.4.51) into the second equation results in

$$3 + \frac{4(2 + l_2)}{\nu_1\nu_2(1 + l_2)} = \frac{4(\nu_1 + \nu_2)}{\nu_1\nu_2}. \tag{1.4.52}$$

For any admissible values of l_2, ν_1 and ν_2, (1.4.51) gives a (unique) positive number m_2, and therefore the restrictions on ν_1 and ν_2 are completely described by (1.4.52), which can be rewritten as

$$[4(\nu_1 + \nu_2) - 3\nu_1\nu_2 - 4](1 + l_2) = 4.$$

Since $l_2 \in (0,2)$, this leads to the admissible range

$$4 < 12(\nu_1 + \nu_2) - 9\nu_1\nu_2 - 12 < 12. \qquad \square$$

1.5 Notes

M.G. Krein [78], [79] and L. de Branges [19], [20], [21], [22], [23], [24], [25], [26] considered equations occurring in the theory of small transverse vibrations of a string. They investigated a very wide class of strings (Krein strings) including those which do not possess density in its usual meaning, i.e., the mass distribution function M is not differentiable. The spectra of two boundary value problems describing small transverse vibrations of such a string with different (Neumann-Dirichlet and Dirichlet-Dirichlet) boundary conditions together with the length of the string uniquely determine the mass distribution function M of the string. This result was stated by M.G. Krein and proved by L. de Branges (see also [48], p. 252). Moreover, these authors found necessary and sufficient conditions on two sequences of real numbers to be the spectra of two boundary value problems generated by a string of this class. These conditions include strict interlacing of the two sequences (see also [75]).

However, a constructive algorithm of recovering the mass distribution function from the two spectra and the length of the string in the general case is unknown. It is known for two particular cases: firstly for the case of 'smooth' strings where the density $\rho = M'$ exists and is twice continuously differentiable and is strictly positive and secondly for Stieltjes strings.

In the first case, using the Liouville transform (see [40], p.292, [96], p.40) one can reduce the string equation to the Sturm-Liouville equation and use results of B.M. Levitan and G.M. Gasymov [89] or V.A. Marchenko [92].

In the second case, the solution of two spectra problems for Stieltjes strings is read off from the coefficients of the continued fraction expansion of a rational function whose zeros and poles are the two given spectra. This method was used by Gantmacher and Krein (see [59, p. 287]). In particular, Theorems 1.3.1 and 1.3.3 were proved in [59].

It was stated in Section 1.1 that Cauer and Fry expressed the impedance function as a Stieltjes function. For this reason the numerator and denominator of such Stieltjes functions are called Cauer-Fry polynomials. The Stieltjes string equations derived in Section 1.2 also occur in the theory of longitudinal vibrations of point masses joined by springs, see [93] and [62].

The Sturm oscillation theorem for Stieltjes strings can be found in [131].

The problem considered in Subsection 1.4.2 (based on [16]) is a finite-dimensional analogue of the three spectra problem for Sturm-Liouville equation solved in [105], [67], see also [72] and for Jacobi matrices in [95].

The result in Subsection 1.4.3 is a finite-dimensional analogue of Hochstadt-Lieberman theorem (see [71] for the original infinite-dimensional version).

Chapter 2

Vibrations of star graphs

2.1 Star graphs with root at the centre

In this section, we consider a plane star graph of g (≥ 2) Stieltjes strings, that is, g Stieltjes strings which are joined at a common vertex, called the central vertex of the star graph. In this section, the central vertex will be called the root of the star graph. At the root, a bead of mass $M > 0$ may be placed. Since the cases with and without a bead at the central vertex can often be considered simultaneously, we put $M = 0$ when there is no bead at the central vertex. We assume that this graph is stretched and study its small transverse vibrations in two different cases:

(D1) the central vertex is fixed (Dirichlet problem),

(N1) the central vertex is free to move in the direction orthogonal to the equilibrium position of the strings (Neumann problem).

We investigate the relation of the characteristic values of the Neumann problem (N1) to those of the problem (D1) which decouples completely into q Dirichlet problems on the edges of the star graph.

In the sequel, we label the edges of the star graph by $j = 1, \ldots, g$ ($g \geq 2$) and we assume that each edge is a Stieltjes string. We suppose that the j-th edge consists of $n_j + 1$ ($n_j \geq 0$) threads of length $l_k^{(j)}$ ($k = 0, \ldots, n_j$) with beads of mass $m_k^{(j)}$ ($k = 1, \ldots, n_j$) separating them, both counted from the exterior towards the centre. The length of the j-th edge is denoted by $l^{(j)} := \sum_{k=0}^{n_j} l_k^{(j)}$.

By $v_k^{(j)}(t)$ ($k = 1, \ldots, n_j$, $j = 1, \ldots, g$) we denote the transverse displacement of the k-th bead on the j-th string at time t, and by $v_0^{(j)}(t)$, $v_{n_j+1}^{(j)}(t)$ those of the ends of the j-th string. If we assume the threads to be stretched by forces each equal to 1, the Lagrange equations for the small transverse vibrations of the graph

M. Möller, V. Pivovarchik, *Direct and Inverse Finite-Dimensional Spectral Problems on Graphs*, Operator Theory: Advances and Applications 283, https://doi.org/10.1007/978-3-030-60484-4_2

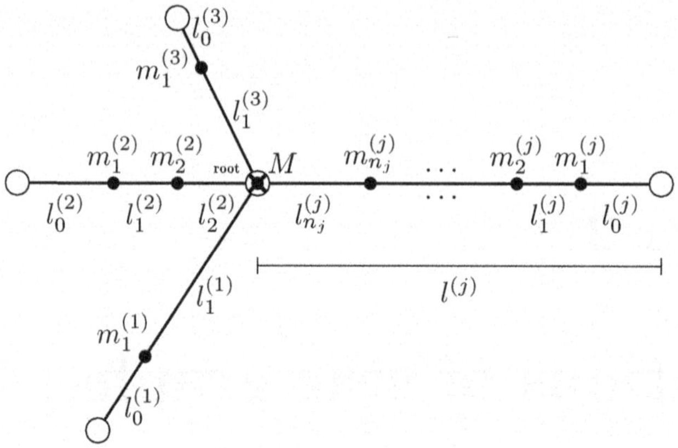

Figure 2.1: Star graph with root at the central vertex

are given by (see (1.2.1))

$$\frac{v_k^{(j)}(t) - v_{k-1}^{(j)}(t)}{l_{k-1}^{(j)}} - \frac{v_{k+1}^{(j)}(t) - v_k^{(j)}(t)}{l_k^{(j)}} + m_k^{(j)} v_k^{(j)\prime\prime}(t) = 0$$
$$(k = 1, \ldots, n_j,\ j = 1, \ldots, g).$$

At the central vertex joining the edges, the continuity of the graph requires that

$$v_{n_1+1}^{(1)}(t) = v_{n_2+1}^{(2)}(t) = \cdots = v_{n_g+1}^{(g)}(t),$$

and the balance of forces implies that

$$\sum_{j=1}^{g} \frac{v_{n_j+1}^{(j)}(t) - v_{n_j}^{(j)}(t)}{l_{n_j}^{(j)}} + M v_{n_1+1}^{(1)\prime\prime}(t) = 0.$$

In case where a bead is absent at the central vertex, the balance of forces condition may be called Kirchhoff condition because it is called so in the theory of electrical circuits. For the pendant vertices we assume the general boundary conditions

$$v_0^{(j)}(t) = c_j v_1^{(j)}(t) \quad (c_j \in \mathbb{R},\ j = 1, \ldots, g).$$

Separation of variables $v_k^{(j)}(t) = u_k^{(j)} \mathrm{e}^{\mathrm{i}\lambda t}$ leads to the following difference equations for the displacement amplitudes $u_k^{(j)}$ $(k = 0, \ldots, n_j, j = 1, \ldots, g)$ for the Neumann and Dirichlet problem:

Dirichlet problem (D1). If we clamp the central vertex, then the problem decouples and consists of the g separate problems on the edges with Dirichlet boundary condition at the central vertex end,

$$\frac{u_k^{(j)} - u_{k-1}^{(j)}}{l_{k-1}^{(j)}} - \frac{u_{k+1}^{(j)} - u_k^{(j)}}{l_k^{(j)}} - m_k^{(j)}\lambda^2 u_k^{(j)} = 0 \quad (k = 1, \ldots, n_j), \tag{2.1.1}$$

$$u_{n_j+1}^{(j)} = 0, \tag{2.1.2}$$

$$u_0^{(j)} = c_j u_1^{(j)} \tag{2.1.3}$$

for $j = 1, \ldots, g$.

Neumann problem (N1). If the central vertex is allowed to move freely in the direction orthogonal to the equilibrium position of the strings, we obtain

$$\frac{u_k^{(j)} - u_{k-1}^{(j)}}{l_{k-1}^{(j)}} - \frac{u_{k+1}^{(j)} - u_k^{(j)}}{l_k^{(j)}} - m_k^{(j)}\lambda^2 u_k^{(j)} = 0 \quad (k = 1, \ldots, n_j,\ j = 1, \ldots, g), \tag{2.1.4}$$

$$u_{n_1+1}^{(1)} = u_{n_2+1}^{(2)} = \ldots = u_{n_g+1}^{(g)}, \tag{2.1.5}$$

$$\sum_{j=1}^{g} \frac{u_{n_j+1}^{(j)} - u_{n_j}^{(j)}}{l_{n_j}^{(j)}} - M\lambda^2 u_{n_1+1}^{(1)} = 0, \tag{2.1.6}$$

$$u_0^{(j)} = c_j u_1^{(j)} \quad (j = 1, \ldots, g). \tag{2.1.7}$$

Note that the Dirichlet problem (D1) and the Neumann problem (N1) share the equations (2.1.1) and (2.1.3). If M tends to ∞, then the Neumann problem (N1) becomes the Dirichlet problem (D1). Indeed, in this case the condition (2.1.6) becomes $u_{n_1+1}^{(1)} = 0$ and, together with (2.1.5), it follows that (2.1.2) holds for $j = 1, \ldots, g$. We will denote by

$$n := \sum_{j=1}^{g} n_j$$

the number of beads on the star graph without the possible bead at the centre.

As in Chapter 1 we will replace λ^2 with z and we will call a complex number z a characteristic value of (D1) or (N1) when the corresponding problem has a nontrivial solution for that value of z. The number of linearly independent solutions, that is, the dimension of the vector space of solutions, will be called the geometric multiplicity of the characteristic value. The dimension of the corresponding solution space will be denoted by $n_{D,g}(z)$ and $n_{N,g}(z)$, respectively.

2.1.1 The direct spectral problem

In this subsection we investigate the interlacing properties and multiplicities of the characteristic values of the Dirichlet problem (D1) and the Neumann problem (N1).

For simplicity we will assume for the remainder of this section that each pendant vertex obeys either the Dirichlet condition or the Neumann condition. This assumption is in most of what follows in this section not necessary, $c_j \in [0,1]$ may suffice.

In other parts of this monograph we will make use of notation and results which include the case $g = 1$. Therefore, in this subsection, we will consider the problems (D1) and (N1) for $g \geq 1$. Problem (D1) in case $g = 1$ has been considered in Sections 1.2 and 1.3. For (N1) in case $g = 1$, the condition (2.1.5) is void, and for $M > 0$, the boundary condition (2.1.6) has not been considered in Chapter 1.

According to Subsection 1.2.1, for each $j = 1, \dots, g$, one may obtain the solutions $u_k^{(j)}$ ($k = 1, \dots, n_j + 1$) of (2.1.1) with initial condition $u_0^{(j)} = c_j u_1^{(j)}$ as in (2.1.3) successively in the form

$$u_k^{(j)} = R_{2k-2}^{(j)}(z, c_j) u_1^{(j)} \quad (k = 1, \dots, n_j + 1). \tag{2.1.8}$$

From (1.2.10) we recall the notation

$$R_{2k-1}^{(j)}(\cdot, c_j) = \frac{R_{2k}^{(j)}(\cdot, c_j) - R_{2k-2}^{(j)}(\cdot, c_j)}{l_k^{(j)}} \quad (k = 1, \dots, n_j).$$

We may therefore use all the properties of the Cauer-Fry polynomials R_k which were derived in Chapter 1. Some of the expressions involving these polynomials are quite lengthy, and since the c_j will be fixed, we will often omit these parameters in the presentations of the Cauer-Fry polynomials and write $R_k^{(j)}$ instead of $R_k^{(j)}(\cdot, c_j)$.

The discussion of the Stieltjes string problems on page 10 and Propositions 1.2.8–1.2.11 give the following

Remark 2.1.1. For $j = 1, \dots, g$, the characteristic values of problem (2.1.1)–(2.1.3) on the j-th edge are given by the zeros of the polynomial

1. $R_{2n_j}^{(j)}$ for the Dirichlet condition $u_{n_j+1}^{(j)} = 0$,
2. $R_{2n_j-1}^{(j)}$ for the Neumann condition $u_{n_j+1}^{(j)} = u_{n_j}^{(j)}$,

and all characteristic values are real, positive (except for the Neumann-Neumann problem, where the smallest characteristic value is 0) and simple and interlace according to (1.2.36)–(1.2.39).

We introduce the polynomials $\phi_{D,g}$ and $\phi_{N,g}$ defined by

$$\phi_{D,g}(z) = \prod_{j=1}^{g} R_{2n_j}^{(j)}(z), \tag{2.1.9}$$

$$\phi_{N,g}(z) = \sum_{j=1}^{g} \left[\left(R_{2n_j-1}^{(j)}(z) - \frac{M}{g} z R_{2n_j}^{(j)}(z) \right) \prod_{k=1,\, k\neq j}^{g} R_{2n_k}^{(k)}(z) \right], \tag{2.1.10}$$

and we introduce the rational functions

$$\phi_g = \frac{\phi_{D,g}}{\phi_{N,g}} \tag{2.1.11}$$

and

$$\phi^{(j)} = \frac{R^{(j)}_{2n_j}}{R^{(j)}_{2n_j-1}} \quad (j = 1, \dots, g). \tag{2.1.12}$$

Note that the polynomial $\phi_{N,g}$ may also be written as

$$\phi_{N,g}(z) = \left(\sum_{j=1}^{g} \frac{R^{(j)}_{2n_j-1}(z)}{R^{(j)}_{2n_j}(z)} - Mz \right) \phi_{D,g}(z). \tag{2.1.13}$$

Theorem 2.1.2. *The function ϕ_g has the representation*

$$\phi_g(z) = \frac{1}{\sum\limits_{j=1}^{g} \frac{1}{\phi^{(j)}(z)} - Mz} \tag{2.1.14}$$

and is an $\mathcal{S}$-function, and an $\mathcal{S}_0$-function if at least one pendant vertex is clamped.

Proof. The representation (2.1.14) of ϕ_g follows from (2.1.11), (2.1.13), (2.1.12). The representation (1.2.19) and Lemmas A.3.6 and A.3.5 show that $\phi^{(j)}$ is a rational $\mathcal{S}$-function for all $j = 1, \dots, g$ and therefore a rational Nevanlinna function. Furthermore, $\phi^{(j)}$ is a rational $\mathcal{S}_0$-function for those j for which the j-th edge is clamped at the pendant vertex. In particular, $\phi^{(j)}(z) > 0$ for $z \in (-\infty, 0)$ and $\phi^{(j)}(0) > 0$ for those j for which the j-th edge is clamped at the pendant vertex. Using Lemma A.2.4 and the fact that a finite sum of rational Nevanlinna functions is a rational Nevanlinna function, it follows that

$$z \mapsto -\sum_{j=1}^{g} \frac{1}{\phi^{(j)}(z)} + Mz$$

is a rational Nevanlinna function. Another application of Lemma A.2.4 shows that ϕ_g is a Nevanlinna function. Finally, $\phi_g(z) > 0$ for $z \in (-\infty, 0)$ and $\phi_g(0) > 0$ if at least one pendant vertex is clamped, which shows that ϕ_g is a rational $\mathcal{S}$-function or $\mathcal{S}_0$-function. □

Theorem 2.1.3. *The characteristic values of the Dirichlet problem* (D1)*, counted with multiplicity, are the zeros of $\phi_{D,g}$.*

Proof. The conditions (2.1.2) imply that $\left((u_k^{(j)})_{k=1}^{n_j} \right)_{j=1}^{g}$ is a solution of (D1) if and only if for each $j = 1, \dots, g$, $(u_k^{(j)})_{k=1}^{n_j}$ is a solution of the Dirichlet-Dirichlet or Neumann-Dirichlet problem on the j-th edge. In particular, the dimension of the solution space of problem (D1) equals the sum of the dimensions of the solution spaces on the edges, if we observe that any of these dimensions is 0 if the corresponding problem has only the trivial solution, for which all displacements are 0. Taking Remark 2.1.1 into account completes the proof. □

Theorem 2.1.4. *The characteristic values of the Neumann problem* (N1), *counted with multiplicity, are the zeros of the polynomial* $\phi_{N,g}$. *Let* $z \in \mathbb{C}$. *If* $n_{D,g}(z) = 0$, *then* $n_{N,g}(z) = 0$ *or* $n_{N,g}(z) = 1$. *If* $n_{D,g}(z) > 0$, *then* $n_{N,g}(z) = n_{D,g}(z) - 1$.

Proof. First let $g = 1$ and $z \in \mathbb{C}$. Then (N1) has a nontrivial solution if and only if

$$\phi_{N,1}(z)u_1^{(1)} = R_{2n_1-1}^{(1)}(z)u_1^{(1)} - MzR_{2n_1}^{(1)}(z)u_1^{(1)} = 0$$

for $u_1^{(1)} \neq 0$, which means that the characteristic values of (N1) are the zeros of $\phi_{N,1}$. Because $(u_0^{(1)}, u_1^{(1)}, \ldots, u_{n+1}^{(1)})$ satisfying (N1) is uniquely determined by $u_1^{(1)}$, the multiplicity of each characteristic value is 1. By (1.2.36) and (1.2.39), $R_{2n_1-1}^{(1)}$ has only simple zeros which are distinct from the zeros of $\phi_{D,1} = R_{2n_1}^{(1)}$, which are also simple. In particular, the zeros of $\phi_{N,1}$ are distinct from the zeros of $\phi_{D,1}$. Since ϕ_1 is a Nevanlinna function by Theorem 2.1.2, its poles are simple, which implies that the zeros of $\phi_{N,1}$ are simple. This completes the proof of Theorem 2.1.4 for $g = 1$.

Now let $g \geq 2$. Then the conditions (2.1.5) and (2.1.6) at the central vertex yield the following system of linear equations for $u_1^{(j)}$ $(j = 1, \ldots, g)$:

$$R_{2n_1}^{(1)}(z)u_1^{(1)} = R_{2n_2}^{(2)}(z)u_1^{(2)} = \cdots = R_{2n_g}^{(g)}(z)u_1^{(g)}, \tag{2.1.15}$$

$$\sum_{j=1}^{g} R_{2n_j-1}^{(j)}(z)u_1^{(j)} - MzR_{2n_1}^{(1)}(z)u_1^{(1)} = 0. \tag{2.1.16}$$

In view of (2.1.15), the equation (2.1.16) can be written in the form

$$\sum_{j=1}^{g} \left(R_{2n_j-1}^{(j)}(z) - \frac{M}{g} z R_{2n_j}^{(j)}(z) \right) u_j^{(1)} = 0.$$

Then the system of equations has a nontrivial solution if and only if its coefficient matrix

$$A(z) = \begin{pmatrix} R_{2n_1}^{(1)}(z) & -R_{2n_2}^{(2)}(z) & 0 & \cdots & 0 & 0 \\ 0 & R_{2n_2}^{(2)}(z) & -R_{2n_3}^{(3)}(z) & \cdots & 0 & 0 \\ \vdots & & \ddots & \ddots & \vdots & \vdots \\ 0 & 0 & 0 & \cdots & R_{2n_{g-1}}^{(g-1)}(z) & -R_{2n_g}^{(g)}(z) \\ a_{g,1}(z) & a_{g,2}(z) & a_{g,3}(z) & \cdots & a_{g,g-1}(z) & a_{g,g}(z) \end{pmatrix}$$

is singular, where

$$a_{g,j}(z) = R^{(j)}_{2n_j-1}(z) - \frac{M}{g} z R^{(j)}_{2n_j}(z) \quad (j = 1, \ldots, g).$$

Expanding the determinant of $A(z)$ with respect to the last row, it follows that $\det A(z) = \phi_{N,g}(z)$.

We still have to prove the statements about the multiplicities. Let $z \in \mathbb{C}$. Clearly, $n_{N,g}(z)$ equals the defect of the matrix $A(z)$. From Theorem 2.1.3 it follows that the number of indices $j \in \{1, \ldots, g\}$ for which $R^{(j)}_{2n_j}(z) = 0$ is $n_{D,g}(z)$. The upper right $(g-1) \times (g-1)$ submatrix $\widehat{A}(z)$ of $A(z)$ is a lower triangular matrix with diagonal entries $-R^{(2)}_{2n_2}(z), \ldots, -R^{(g)}_{2n_g}(z)$. If $n_{D,g}(z) = 0$, then it follows from Theorem 2.1.3 that these diagonal entries are nonzero. Therefore $\widehat{A}(z)$ has full rank $g - 1$, which implies $n_{N,g}(z) = 0$ or $n_{N,g}(z) = 1$. If $n_{D,g}(z) > 0$, then we will assume without loss of generality that $R^{(1)}_{2n_1}(z) = 0$, see Theorem 2.1.3. Then the first $g - 1$ entries of the first column of $\hat{A}(z)$ are 0, whereas the last entry is $a_{g,1}(z) = R_{2n_1-1}(z) \neq 0$ because the zeros of $R^{(1)}_{2n_1-1}$ and $R^{(1)}_{2n_1}$ interlace (see (1.2.36) and (1.2.39)). Hence the last row of $A(z)$ is linearly independent of the first $g-1$ rows of $A(z)$. Since $n_{D,g}(z) - 1$ of the values of $R^{(2)}_{2n_2}(z), \ldots, R^{(g)}_{2n_g}(z)$ are 0 (see Theorem 2.1.3), the rank of $\hat{A}(z)$ is $(g - 1) - (n_{D,g}(z) - 1)$. It follows that $A(z)$ has rank $g - n_{D,g}(z) + 1$, which means that $n_{N,g}(z) = n_{D,g}(z) - 1$.

Finally, it remains to verify the multiplicities of $\phi_{N,g}$. Therefore, let $z \in \mathbb{C}$ such that $n_{N,g}(z) > 0$. By Theorem 2.1.2 we know that ϕ_g is a rational $\mathcal{S}$-function. In particular, its zeros and poles are simple. If $n_{D,g}(z) = 0$, it follows that z is a simple zero of $\phi_{N,g}$, and the multiplicity of the zero z of $\phi_{N,g}$ therefore equals the multiplicity $n_{N,g}(z)$ of the characteristic value z of (N1). Now let $p = n_{D,g}(z) > 0$. Without loss of generality we may assume that $R^{(j)}_{n_j}(z) = 0$ for $j = 1, \ldots, p$ and $R^{(j)}_{n_j}(z) \neq 0$ for $j = p + 1, \ldots, g$. Since each product in the definition (2.1.10) of $\phi_{N,g}$ contains at least $p - 1$ factors $R^{(k)}_{2n_k}$ which are zero at z, it is clear that $\phi_{N,g}$ has a zero of multiplicity at least $p - 1$ at z. To prove that this order is exactly $p - 1$, we are going to show that $(\phi_{N,g})^{(p-1)}(z) \neq 0$. Firstly it is clear that

$$(\phi_{N,g})^{(p-1)}(z) = (p-1)! \sum_{j=1}^{p} \left[R^{(j)}_{2n_j-1}(z) \prod_{k=1,\, k \neq j}^{p} R^{(k)\prime}_{2n_k}(z) \right] \prod_{k=p+1}^{g} R^{(k)}_{2n_k}(z),$$

since all other summand have at least one factor $R^{(k)}_{2n_k}(z)$ with $k \leq p$. We rewrite this as

$$(\phi_{N,g})^{(p-1)}(z) = (p-1)! \left[\sum_{j=1}^{p} \frac{R^{(j)}_{2n_j-1}(z)}{R^{(j)\prime}_{2n_j}(z)} \right] \prod_{k=1}^{p} R^{(k)\prime}_{2n_k}(z) \prod_{k=p+1}^{g} R^{(k)}_{2n_k}(z). \quad (2.1.17)$$

For $j = 1, \dots, p$ we find that

$$\frac{R_{2n_j}^{(j)\prime}(z)}{R_{2n_j-1}^{(j)}(z)} = \frac{R_{2n_j}^{(j)\prime}(z)R_{2n_j-1}^{(j)}(z) - R_{2n_j}^{(j)}(z)R_{2n_j-1}^{(j)\prime}(z)}{(R_{2n_j-1}^{(j)}(z))^2} = \left(\frac{R_{2n_j}^{(j)}(z)}{R_{2n_j-1}^{(j)}(z)}\right)' > 0$$

by Lemma A.2.5. Hence the sum in (2.1.17) is positive, whereas all other factors are nonzero. Therefore the multiplicity of the zero z of $\phi_{N,g}$ equals $n_{N,g}(z)$. □

We can now prove the following relations between the characteristic values of the Dirichlet problem (D1) and the characteristic values of the Neumann problem (N1).

Theorem 2.1.5. *Let $n' = n + 1$ if $M > 0$ and $n' = n$ if $M = 0$. Then the sequence of characteristic values $(\mu_k)_{k=1}^{n'}$ of the Neumann problem* (N1) *and the sequence of the characteristic values $(\nu_k)_{k=1}^{n}$ of the Dirichlet problem* (D1) *are real and have the following properties, when indexed in nondecreasing order:*

1. $0 \le \mu_1 < \nu_1 \le \dots \le \mu_n \le \nu_n$ *and* $\nu_n < \mu_{n+1}$ *if* $M > 0$, *where* $\mu_1 = 0$ *if and only if all pendant vertices are freely moving;*
2. $\nu_{k-1} = \mu_k$ *if and only if* $\mu_k = \nu_k$ $(k = 2, \dots, n)$;
3. *the multiplicity of* ν_k *does not exceed* g $(k = 1, \dots, n)$.

Proof. From its definition it is clear that $\phi_{D,g}$ is a polynomial of degree n, and all zeros of $\phi_{D,g}$ are real and positive by Propositions 1.2.8 and 1.2.10. Hence the characteristic values of the Dirichlet problem (D1) can be indexed as indicated. From the definition of $\phi_{N,g}$ in (2.1.10) it is clear that $\phi_{N,g}$ is a polynomial of degree at most n if $M = 0$ and a polynomial of degree exactly $n + 1$ if $M > 0$. But since ϕ_g is an $\mathcal{S}$-function by Theorem 2.1.2, the degree of $\phi_{N,g}$ is at least as large as the degree of $\phi_{D,g}$, which is n. Hence all characteristic values of the Neumann problem (N1) are real and nonnegative and can be indexed as indicated, with $\mu_1 = 0$ as stated in part 1. By Lemmas A.3.5 and A.4.4, the interlacing given in statement 1 follows, however only in the weaker form $\mu_1 \le \nu_1$ and $\nu_n \le \mu_{n+1}$ in case $M > 0$. Here we note that Lemma A.4.4 becomes applicable in the case $M > 0$ if we introduce an auxiliary number $\nu_{n+1} > \mu_{n+1}$.

Since the zeros of $R_{2n_k}^{(k)}$ $(k = 1, \dots, g)$ are simple, it follows that $n_{D,g}(z) \le g$ for all $z \in \mathbb{C}$. This proves statement 3.

If $\nu_{k-1} = \mu_k$ or if $\mu_k = \nu_k$ for indices k for which these characteristic values are defined, then μ_k is a common zero of $\phi_{N,g}$ and $\phi_{D,g}$. But then $n_{D,g}(\mu_k) = n_{N,g}(\mu_k) + 1$ by Theorem 2.1.4, so that there are indices j_1 and j_2 with $1 \le j_1 < j_2 \le n$ such that

$$\mu_{j_1} < \nu_{j_1} = \mu_{j_1+1} = \dots = \mu_{j_2} = \nu_{j_2} \tag{2.1.18}$$

with $\nu_{j_2} < \mu_{j_2+1}$ in case $M > 0$ or $j_2 < n$. This proves statement 2 and the strict inequalities in statement 1. □

Corollary 2.1.6. *At least one of two neighbouring characteristic values $\mu_k < \mu_{k+1}$ $(k = 2, \ldots, n-1)$ is simple.*

Proof. If both μ_k and μ_{k+1} were not simple, it would follow from (2.1.18) that $\mu_k = \nu_k$ and $\nu_k = \mu_{k+1}$, which would give the contradiction $\mu_k = \mu_{k+1}$. □

Proposition 2.1.7. *The characteristic values $(\mu_k)_{k=1}^{n'}$ of the Neumann problem* (N1) *have the following monotonicity properties:*

1. *μ_k is a nonincreasing function of $M \in [0, \infty)$ for $k = 1, \ldots, n+1$, where we have set $\mu_{n+1} := \infty$ if $M = 0$;*
2. *$\mu_k \to \nu_{k-1}$ $(k = 2, \ldots, n+1)$ and $\mu_1 \to 0$ as $M \to \infty$.*

Proof. For the purpose of this proof, we write $\mu_k(M)$ and $\phi_{N,g}(\cdot, M)$ to indicate the dependence on $M \in [0, \infty)$. When $\nu_{k-1} = \nu_k$ for some $k \in \{2, \ldots, n\}$, then $\nu_{k-1} = \mu_k(M) = \nu_k$ for all $M \in [0, \infty)$, and the statements of this proposition are trivial.

Now let $k \in \{1, \ldots, n+1\}$ with $\nu_{k-1} \neq \nu_k$, where $\nu_0 := 0$ and $\nu_{n+1} := \infty$. Then Theorem 2.1.5, statement 2, gives that $\nu_{k-1} < \mu_k(M) < \nu_k$ for all $M \in [0, \infty)$, except for the case when $M = 0$ and $k = n+1$ where $\mu_{n+1}(0) = \nu_{n+1} = \infty$. From (2.1.13) we conclude for $M, M' \in [0, \infty)$, $M' < M$, that

$$-\frac{\phi_{N,g}(z, M)}{\phi_{D,g}(z)} = -\frac{\phi_{N,g}(z, M')}{\phi_{D,g}(z)} + (M - M')z. \tag{2.1.19}$$

Then

$$-\frac{\phi_{N,g}(\mu_k(M'), M)}{\phi_{D,g}(\mu_k(M'))} = (M - M')\mu_k(M') > 0$$

and the fact that the rational function $-\frac{\phi_{N,g}(\cdot,M)}{\phi_{D,g}}$ is a Nevanlinna function (see Theorem 2.1.2 and Lemma A.2.4) and hence increasing between its poles (see Lemma A.2.5) shows that $\nu_{k-1} < \mu_k(M) < \mu_k(M')$. This completes the proof of part 1.

Finally, to prove part 2 in case $\nu_{k-1} \neq \nu_k$, let $w \in (\nu_{k-1}, \nu_k)$. Then (2.1.19) gives

$$-\frac{\phi_{N,g}(w, M)}{\phi_{D,g}(w)} = -\frac{\phi_{N,g}(w, 0)}{\phi_{D,g}(w)} + Mw,$$

which is positive for sufficiently large M, say $M \geq M_0$. We already know that $-\frac{\phi_{N,g}(\cdot,M)}{\phi_{D,g}}$ is increasing on (ν_{k-1}, ν_k), and therefore

$$-\frac{\phi_{N,g}(z, M)}{\phi_{D,g}(z)} > 0$$

when $z \in [w, \nu_k)$ and $M \geq M_0$. Since $\phi_{N,g}(\mu_k(M), M) = 0$ and $\mu_k(M) \in (\nu_{k-1}, \nu_k)$ for $M > 0$, it follows that $\mu_k(M) \in (\nu_{k-1}, w)$ for $M \geq M_0$. This completes the proof of part 2. □

2.1.2 The inverse spectral problem

In this subsection we investigate the inverse problem, that is, the question if two sequences of nonnegative real numbers satisfying properties 1–3 in Theorem 2.1.5 can be realized as sequences of the characteristic values of the Neumann problem (N1) and the Dirichlet problem (D1) on a star graph when all pendant vertices carry the Dirichlet condition.

More precisely, suppose that $g \in \mathbb{N}$, and that a sequence of lengths $l^{(j)} > 0$ ($j = 1, \ldots, g$) as well as sequences $(\mu_k)_{k=1}^{n'}$, $(\nu_k)_{k=1}^{n}$ with $n' = n$ or $n' = n+1$ having the properties 1–3 in Theorem 2.1.5 are given. Can we determine nonnegative integers n_j, sequences of masses $(m_k^{(j)})_{k=1}^{n_j}$, a mass M and sequences of lengths $(l_k^{(j)})_{k=0}^{n_j}$ for $j = 1, \ldots, g$ so that the corresponding star graph has the sequences $(\mu_k)_{k=1}^{n'}$, $(\nu_k)_{k=1}^{n}$ as Neumann and Dirichlet characteristic values, respectively?

Theorem 2.1.8. *Let $n, g \in \mathbb{N}$ and let $n' = n$ or $n' = n + 1$. Let $(l^{(j)})_{j=1}^{g}$ be a sequence of positive real numbers and let sequences of real numbers $(\mu_k)_{k=1}^{n'}$, $(\nu_k)_{k=1}^{n}$ be given such that*

(i) $0 < \mu_1 < \nu_1 \le \cdots \le \mu_n \le \nu_n$, *and additionally* $\nu_n < \mu_{n+1}$ *if* $n' = n + 1$;

(ii) $\nu_{k-1} = \mu_k$ *if and only if* $\mu_k = \nu_k$ $(k = 2, \ldots, n)$;

(iii) *the multiplicity of each* ν_k *in* $(\nu_k)_{k=1}^{n}$ *does not exceed* g.

Then there exists a star graph of g Stieltjes strings, i.e., a sequence $(n_j)_{j=1}^{g}$ of nonnegative integers, $M \ge 0$ with $M = 0$ if and only if $n' = n$, and sequences $(m_k^{(j)})_{k=1}^{n_j}$, $(l_k^{(j)})_{k=0}^{n_j}$ $(j = 1, \ldots, g)$ of positive numbers with $n = \sum_{j=1}^{g} n_j$ and $\sum_{k=0}^{n_j} l_k^{(j)} = l^{(j)}$ such that the Neumann problem (N1), (2.1.4)–(2.1.7), *has the characteristic values $(\mu_k)_{k=1}^{n'}$ and the Dirichlet problem* (D1), (2.1.1)–(2.1.3), *has the characteristic values $(\nu_k)_{k=1}^{n}$, with $c_j = 0$ in* (2.1.3) *and* (2.1.7) *for all j. If $n \ge g$, all n_j can be chosen to be positive. If $n < g$, n of the numbers n_j can be chosen to be* 1, *and $g - n$ of the n_j must be* 0.

Proof. Let $g' = \min\{n, g\}$. Trivially, the multiplicity of ν_k in $(\nu_k)_{k=1}^{n}$ does not exceed n, so that assumption (iii) holds with g replaced by g', and we can choose a sequence of positive integers $(n_j)_{j=1}^{g'}$ such that $n = \sum_{j=1}^{g'} n_j$ and such that the sequence $(\nu_k)_{k=1}^{n}$ can be written as the concatenation

$$(\nu_k)_{k=1}^{n} = \bigcup_{j=1}^{g'} (\nu_\kappa^{(j)})_{\kappa=1}^{n_j} \tag{2.1.20}$$

with sequences $(\nu_\kappa^{(j)})_{\kappa=1}^{n_j}$ so that $\nu_\kappa^{(j)} > \nu_{\kappa'}^{(j)}$ for $\kappa > \kappa'$ and $j = 1, \ldots, g'$. For example, we may define the sequences $(\nu_\kappa^{(j)})_{\kappa=1}^{n_j}$ as follows. Writing n in the form $n = \widetilde{n} g' + \widehat{n}$ with integers $\widetilde{n} \ge 1$ and $0 \le \widehat{n} \le g' - 1$, we put $n_j = \widetilde{n} + 1$ for

$j = 1, \dots, \widehat{n}$, $n_j = \widetilde{n}$ for $j = \widehat{n}+1, \dots, g'$, and $\nu_\kappa^{(j)} = \nu_{(\kappa-1)g'+j}$ for $j = 1, \dots, g'$ and $\kappa = 1, \dots, n_j$.

We consider the rational function ψ_g defined by

$$\psi_g(z) = \left(\sum_{j=1}^{g} \frac{1}{l^{(j)}}\right) \frac{\prod_{k=1}^{n'} \left(1 - \frac{z}{\mu_k}\right)}{\prod_{k=1}^{n} \left(1 - \frac{z}{\nu_k}\right)}. \tag{2.1.21}$$

The function $\frac{1}{\psi_g}$ is a rational $\mathcal{S}_0$-function by Lemma A.3.5 since $\psi_g(0) > 0$ and since its poles and zeros, after cancellation of common factors, are all simple in view of condition (ii), interlace strictly and are ordered as in (i). In particular, each ν_k $(k = 1, \dots, n)$ is a simple pole of ψ_g, although any pole of ψ_g may be represented by ν_k with up to g different indices k.

Since $\frac{1}{\psi_g}$ is a Nevanlinna function, so is $-\psi_g$ (see Lemma A.2.4). In view of Lemma A.2.7 there are constants $A_0 \geq 0$, $A_1, \dots, A_n > 0$, and $B \in \mathbb{R}$ so that ψ_g has the partial fractions expansion

$$\psi_g(z) = -A_0 z + \sum_{k=1}^{n} \frac{A_k}{z - \nu_k} + B. \tag{2.1.22}$$

We note that the coefficients are not unique when ν_i has the same value for more than one index. But we can still choose them to be positive, say by requiring $A_j = A_k$ when $\nu_j = \nu_k$. Putting $z = 0$ in (2.1.22) we find

$$B = \sum_{j=1}^{g} \frac{1}{l^{(j)}} + \sum_{k=1}^{n} \frac{A_k}{\nu_k}.$$

We now define

$$A_\kappa^{(j)} := A_k \ \text{ if } \nu_\kappa^{(j)} = \nu_k \quad (j = 1, \dots g',\ \kappa = 1, \dots n_j,\ k = 1, \dots, n),$$

$$B_j := \frac{1}{l^{(j)}} + \sum_{\kappa=1}^{n_j} \frac{A_\kappa^{(j)}}{\nu_\kappa^{(j)}} \quad (j = 1, \dots g), \tag{2.1.23}$$

$$\psi^{(j)}(z) := \sum_{\kappa=1}^{n_j} \frac{A_\kappa^{(j)}}{z - \nu_\kappa^{(j)}} + B_j \quad (j = 1, \dots g). \tag{2.1.24}$$

Therefore we have

$$\psi_g(z) = -A_0 z + \sum_{j=1}^{g} \psi^{(j)}(z). \tag{2.1.25}$$

Now consider $j \in \{1, \dots, g'\}$. Since $A_\kappa^{(j)} = A_k > 0$, the derivative of the rational function $\psi^{(j)}$ is negative on the real axis. Observing further that $\psi^{(j)}(0) = \frac{1}{l^{(j)}} > 0$

it follows that $\psi^{(j)}$ has one simple zero $\mu_1^{(j)} > 0$ in $(0, \nu_1^{(j)})$, one simple zero $\mu_\kappa^{(j)}$ in each of the intervals $(\nu_{\kappa-1}^{(j)}, \nu_\kappa^{(j)})$ for $\kappa = 2, \dots, n_j$ and therefore no other zeros. Hence the zeros and poles of $\psi^{(j)}$ interlace as follows:

$$0 < \mu_1^{(j)} < \nu_1^{(j)} < \mu_2^{(j)} < \nu_2^{(j)} < \cdots < \mu_{n_j}^{(j)} < \nu_{n_j}^{(j)}.$$

Therefore, $\frac{1}{\psi^{(j)}}$ is an $\mathcal{S}_0$-function and by Lemma A.3.5 there exist unique sequences of positive numbers $(m_\kappa^{(j)})_{\kappa=1}^{n_j}$ and $(l_\kappa^{(j)})_{\kappa=0}^{n_j}$ such that

$$\frac{1}{\psi^{(j)}(z)} = l_{n_j}^{(j)} + \cfrac{1}{-m_{n_j}^{(j)} z + \cfrac{1}{l_{n_j-1}^{(j)} + \cfrac{1}{-m_{n_j-1}^{(j)} z + \cdots + \cfrac{1}{l_1^{(j)} + \cfrac{1}{-m_1^{(j)} z + \cfrac{1}{l_0^{(j)}}}}}}}. \tag{2.1.26}$$

From (2.1.24) and (2.1.26) with $z = 0$ we see that

$$l^{(j)} = \frac{1}{\psi^{(j)}(0)} = l_{n_j}^{(j)} + l_{n_j-1}^{(j)} + \ldots + l_1^{(j)} + l_0^{(j)},$$

and this is clearly also true when $n < g$ and $j \in \{g'+1, \dots, g\}$ if we put $l_0^{(j)} := l^{(j)}$. Hence the above sequences of masses and lengths yield a star graph of Stieltjes strings with g edges of length $l^{(j)}$ $(j = 1, \dots, g)$. For this star graph we find in view of (2.1.26) and (1.2.19) and with the notation (2.1.12) that

$$\frac{1}{\psi^{(j)}(z)} = \phi^{(j)}(z) \quad (j = 1, \dots, g).$$

We set $M = A_0$ and observing (2.1.25) and (2.1.14) we arrive at

$$\psi_g(z) = -Mz + \sum_{j=1}^{g} \frac{1}{\phi^{(j)}(z)} = \frac{1}{\phi_g(z)}. \tag{2.1.27}$$

From (2.1.9) and (1.2.25) we find

$$\begin{aligned} \phi_{D,g}(z) &= \prod_{j=1}^{g} R_{2n_j}^{(j)}(z) = \prod_{j=1}^{g} \frac{l^{(j)}}{l_0^{(j)}} \prod_{k=1}^{n_j} \left(1 - \frac{z}{\nu_k^{(j)}}\right) \\ &= \prod_{j=1}^{g} \frac{l^{(j)}}{l_0^{(j)}} \prod_{k=1}^{n} \left(1 - \frac{z}{\nu_k}\right). \end{aligned} \tag{2.1.28}$$

Hence Theorem 2.1.3 shows that the sequence of the characteristic values of the Dirichlet problem (D1) is $(\nu_k)_{k=1}^n$. Finally, in view of (2.1.11), (2.1.27) and (2.1.21),

$$\phi_{N,g}(z) = \frac{\phi_{D,g}(z)}{\phi_g(z)} = \phi_{D,g}(z)\psi_g(z)$$
$$= \left(\sum_{j=1}^g \frac{1}{l^{(j)}}\right) \prod_{j=1}^g \frac{l^{(j)}}{l_0^{(j)}} \prod_{k=1}^{n'} \left(1 - \frac{z}{\mu_k}\right), \tag{2.1.29}$$

which shows in view of Theorem 2.1.4 that the sequence of the characteristic values of the Neumann problem (N1) is $(\mu_k)_{k=1}^{n'}$, where $n' = n$ if and only if $M = 0$. □

Theorem 2.1.9. *Let the assumptions and notation be as in Theorem 2.1.8.*

1. *The number M in Theorem 2.1.8 is uniquely determined by the given data.*
2. *In addition we assume that the sequences $(\mu_k)_{k=1}^{n'}$ and $(\nu_k)_{k=1}^n$ interlace strictly and that the distribution (2.1.20) of the latter characteristic values onto the g strings is prescribed. Then the sequences $(m_k^{(j)})_{k=1}^{n_j}$ and $(l_k^{(j)})_{k=0}^{n_j}$ $(j = 1, \dots, g)$ are unique.*

Proof. Consider a solution of the inverse problem and the function ϕ_g defined in (2.1.11) associated with this solution. Then the zeros and poles of $\frac{1}{\phi_g}$ coincide with the zeros and poles of ψ_g defined in (2.1.21). Furthermore, in view of (2.1.13), (1.2.25) and (1.2.28),

$$\frac{1}{\phi_g(0)} = \frac{\phi_{N,g}(0)}{\phi_{D,g}(0)} = \sum_{j=1}^g \frac{1}{l^{(j)}} = \psi_g(0).$$

Hence ϕ_g is uniquely determined by the given data. But (2.1.14) gives

$$\psi_g(z) = \frac{1}{\phi_g(z)} = \sum_{j=1}^g \frac{1}{\phi^{(j)}(z)} - Mz, \tag{2.1.30}$$

and therefore

$$M = -\lim_{z\to\infty} \frac{\psi(z)}{z}$$

is uniquely determined by the given data. This completes the proof of statement 1.

For the proof of statement 2 we observe that the functions $\phi^{(j)}$ $(j = 1, \dots, g)$ defined in (2.1.12) are Nevanlinna functions, and hence also the functions $-\frac{1}{\phi^{(j)}}$ are Nevanlinna functions. Therefore Lemma A.2.7 gives the partial fractions representation

$$\frac{1}{\phi^{(j)}(z)} = \sum_{\kappa=1}^{n_j} \frac{a_\kappa^{(j)}}{z - \nu_\kappa^{(j)}} + b_j,$$

where $a_\kappa^{(j)} > 0$ ($j = 1, \dots, q$, $\kappa = 1, \dots, n_j$) and $b_j \in \mathbb{R}$ ($j = 1, \dots, q$). By assumption, the numbers $\nu_\kappa^{(j)}$ are uniquely allocated as terms of the sequence $(\nu_k)_{k=1}^n$, and since the ν_k are pairwise distinct, the coefficients in the partial fractions expansion (2.1.22) are uniquely determined by ψ_g. Comparing the partial fractions expansions on the left and right hand sides of (2.1.30) shows that $a_\kappa^{(j)} = A_k$ when $\nu_\kappa^{(j)} = \nu_k$. Furthermore, $\frac{1}{\psi^{(j)}(0)} = \frac{1}{l^{(j)}}$ in view of (2.1.12), (1.2.24) and (1.2.26). Therefore $\phi^{(j)}$ ($j = 1, \dots, g$) is uniquely determined by the given data, and an application of Lemma A.3.5 completes the proof. □

2.1.3 Restrictions on the multiplicities

In this subsection, let $g \geq 2$. It will also be assumed without loss of generality that the edges are labelled in such a way that $n_1 \geq n_2 \geq \cdots \geq n_g$.

We augment the notation which was used earlier in this section.

Notation 2.1.10. We denote by

1. $\Lambda_N = (\mu_k)_{k=1}^{n'}$, $\mu_k \geq \mu_{k'} > 0$ for $k > k'$, the nondecreasing sequence of the characteristic values of the Neumann problem (N1), (2.1.4)–(2.1.7), on the star graph,
2. $\Lambda_D = (\nu_k)_{k=1}^n = \bigcup\limits_{j=1}^{g} (\nu_\kappa^{(j)})_{\kappa=1}^{n_j}$, $\nu_k \geq \nu_{k'} > 0$ for $k > k'$, the nondecreasing sequence of the characteristic values of the Dirichlet problem (D1) on the star graph, where
3. $(\nu_\kappa^{(j)})_{\kappa=1}^{n_j}$, $\nu_\kappa^{(j)} > \nu_{\kappa'}^{(j)}$ for $\kappa > \kappa'$, are the characteristic values of the Dirichlet-Dirichlet or Neumann-Dirichlet problem (2.1.1)–(2.1.3) on the j-th edge for $j = 1, \dots, g$,
4. $\widetilde{\Lambda}_N = (\widetilde{\mu}_k)_{k=1}^{r_N}$, $\widetilde{\mu}_k > \widetilde{\mu}_{k'}$ for $k > k'$ the sequence of distinct characteristic values of the Neumann problem (N1),
5. $\widetilde{\Lambda}_D = (\widetilde{\nu}_k)_{k=1}^{r_D}$, $\widetilde{\nu}_k > \widetilde{\nu}_{k'}$ for $k > k'$ the set of distinct characteristic values of the Dirichlet problem (D1),
6. $(p_k(N))_{k=1}^{r_N}$ and $(p_k(D))_{k=1}^{r_D}$ the vectors of multiplicities of the characteristic values $\widetilde{\mu}_k$ ($k = 1, \dots r_N$) of problem (N1) and $\widetilde{\nu}_k$ ($k = 1, \dots, r_N$) of problem (D1).

Remark 2.1.11. 1. By definition of $p_k(N)$ and $p_k(D)$ as multiplicities, it is clear that

$$\sum_{k=1}^{r_N} p_k(N) = n', \quad \sum_{k=1}^{r_D} p_k(D) = n. \tag{2.1.31}$$

2. The number r_D of different positive Dirichlet eigenvalues satisfies $r_D \geq n_1$ since $\nu_1^{(1)} < \nu_2^{(1)} < \cdots < \nu_{n_1}^{(1)}$.

To derive necessary conditions on the multiplicities of the characteristic values of the Neumann and Dirichlet problems, we need the notion of majorization, which goes back to Muirhead [99] for the case of integers and was generalized to nonnegative numbers by Hardy, Littlewood, and Polyá (see [68, 2.18]).

Definition 2.1.12. Let $x = (x_i)_{i=1}^s$ and $y = (y_i)_{i=1}^t$ be two vectors with nonnegative entries ordered nonincreasingly, that is, $x_1 \geq x_2 \geq \cdots \geq x_s \geq 0$ and $y_1 \geq y_2 \geq \cdots \geq y_t \geq 0$. Then x is said to *majorize* y,

$$x \succ y \quad :\Longleftrightarrow \quad \sum_{i=1}^{s} x_i = \sum_{i=1}^{t} y_i, \quad \sum_{i=1}^{\tau} x_i \geq \sum_{i=1}^{\tau} y_i \quad (\tau = 1, \dots, \min\{s-1, t\}). \tag{2.1.32}$$

Remark 2.1.13. If a vector $x = (x_i)_{i=1}^s$ majorizes a vector $(y_i)_{i=1}^t$, then the number of nonzero entries of x is less or equal to the number of nonzero entries of y, i.e.,

$$x \succ y \quad \Longrightarrow \quad \#\{i \in \{1, \dots, s\} : x_i > 0\} \leq \#\{i \in \{1, \dots, t\} : y_i > 0\}.$$

In fact, denote the numbers of nonzero entries of x and y by s_0 and t_0, respectively, and assume $s_0 > t_0$. By (2.1.32) this implies

$$\sum_{i=1}^{t_0} x_i \leq \sum_{i=1}^{s} x_i = \sum_{i=1}^{t} y_i = \sum_{i=1}^{t_0} y_i \leq \sum_{i=1}^{t_0} x_i,$$

where the last inequality is true in view of $t_0 \leq \min\{s-1, t\}$. Hence we have equality everywhere. Since all x_i are nonnegative, this shows that $x_i = 0$ for $i = t_0 + 1, \dots, s$, a contradiction to the assumption.

The following elementary lemma on the inversion of the nonincreasing function $\{1, \dots, g\} \to \mathbb{N}$, $j \mapsto n_j$, will be useful.

Lemma 2.1.14. *Let $g \in \mathbb{N}$, $g \geq 2$, $(n_j)_{j=1}^g \in \mathbb{N}^g$ with $n_1 \geq n_2 \geq \cdots \geq n_g$, and set $n = \sum_{j=1}^g n_j$, $n_{g+1} := 0$. For $i = 1, \dots, n_1$ define*

$$N_i := \#\{j \in \{1, \dots, g\} : n_j \geq i\} = \max\{j \in \{1, \dots, g\} : n_j \geq i\}.$$

Then $N_1 \geq N_2 \geq \cdots \geq N_{n_1}$ and

1. $\sum_{i=1}^{n_1} N_i = n$;
2. $N_i > 1 \quad (i = 1, \dots, n_2)$, $N_i = 1 \quad (i = n_2 + 1, \dots, n_1)$;
3. $\#\{i \in \{1, \dots, n_1\} : N_i = j\} = n_j - n_{j+1} \quad (j = 1, \dots, g)$;
4. $n_j = \#\{i \in \{1, \dots, n_1\} : N_i \geq j\} = \max\{i \in \{1, \dots, n_1\} : N_i \geq j\}$ $(j = 1, \dots, g)$.

Proof. Statements 1, 2 and 3 follow from the definition of N_i, and statement 4 follows from statement 3. □

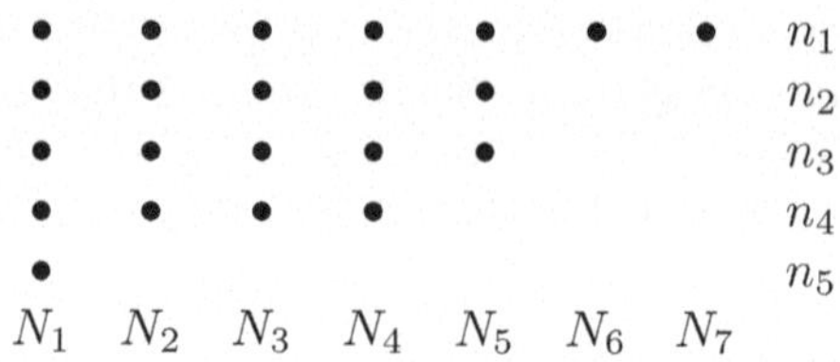

Figure 2.2

Figure 2.2 illustrates Lemma 2.1.14, where the j-th row shows the beads of the j-th string, and where n_j and N_k is the number of beads in the j-th row and k-th column, respectively.

Lemma 2.1.15. *Let $x = (x_i)_{i=1}^s$ and $y = (y_i)_{i=1}^t$ be two vectors whose entries are nonnegative integers ordered nonincreasingly, $x_1 \geq x_2 \geq \cdots \geq x_s \geq 0$ and $y_1 \geq y_2 \geq \cdots \geq y_t \geq 0$. Assume that $x_s > 0$, that $x_1 > 1$ and that $x \succ y$. Let $\widehat{s} = \max\{i \in \{1, \dots, s\} : x_i > 1\}$ and define $\widehat{x} = (\widehat{x}_i)_{i=1}^{\widehat{s}}$ by $\widehat{x}_i = x_i - 1$ $(i = 1, \dots, \widehat{s})$. Let*

$$t_1 = \min\{i \in \{1, \dots, t\} : y_i = y_s\}, \quad t_2 = \max\{i \in \{1, \dots, t\} : y_i = y_s\},$$

and define $\widehat{y} = (\widehat{y}_i)_{i=1}^t$ by

$$\begin{aligned}
&\widehat{y}_i = y_i - 1 \text{ for } i = 1, \dots, t_1 - 1,\\
&\widehat{y}_i = y_i \text{ for } i = t_1, \dots, t_1 + t_2 - s - 1,\\
&\widehat{y}_i = y_i - 1 \text{ for } i = t_1 + t_2 - s, \dots, t_2,\\
&\widehat{y}_i = y_i \text{ for } i = t_2 + 1, \dots, t.
\end{aligned}$$

Then $\widehat{x}$ and $\widehat{y}$ are nonincreasingly ordered vectors of nonnegative integers, $\widehat{x}_{\widehat{s}} > 0$, and $\widehat{x} \succ \widehat{y}$.

Proof. Since $x_1 > 1$, $\widehat{s}$ is well-defined, and $\widehat{x}_{\widehat{s}} = x_{\widehat{s}} - 1 > 0$. Clearly $\widehat{x}_i = x_i - 1 \geq x_{i+1} - 1 = \widehat{x}_{i+1}$ for $i = 1, \dots, \widehat{s} - 1$. Hence $\widehat{x}$ is nonincreasingly ordered.

By Remark 2.1.13 there is $t_0 \geq s$ such that $y_{t_0} > 0$. Then $y_s \geq y_{t_0} > 0$. Therefore, for $i = 1, \dots, t_2$, $y_i \geq y_{t_2} = y_s > 0$. This shows that $\widehat{y}_i \geq 0$ for $i = 1, \dots, t$. Next we find for $i = 1, \dots, t_1 - 2$ and $i = t_1 + t_2 - s, \dots, t_2 - 1$ that $\widehat{y}_i = y_i - 1 \geq y_{i+1} - 1 = \widehat{y}_{i+1}$. For $i = t_1, \dots, t_1 + t_s - s - 2$ and $i = t_2 + 1, \dots, t - 1$ we have $\widehat{y}_i = y_i \geq y_{i+1} = \widehat{y}_{i+1}$. In case $t_1 > 1$ and $t_2 > s$ we have $y_{t_1 - 1} \neq y_s = y_{t_1}$ and $y_{t_1-1} \geq y_{t_1}$, so that $y_{t_1-1} > y_{t_1}$. It follows that $\widehat{y}_{t_1-1} = y_{t_1-1} - 1 \geq y_{t_1} = \widehat{y}_{t_1}$. In case $t_1 > 1$ and $t_2 = s$ we have $\widehat{y}_{t_1-1} = y_{t_1-1} - 1 \geq y_{t_1} - 1 = \widehat{y}_{t_1}$. If $t_2 \neq t$, then $y_{t_2+1} \neq y_s = y_{t_2}$, and therefore $y_{t_2+1} \leq y_{t_2}$ gives $y_{t_2+1} < y_{t_2}$. Hence $\widehat{y}_{t_2+1} = y_{t_2+1} \leq y_{t_2} - 1 = \widehat{y}_{t_2}$. Altogether we have shown that $\widehat{y}$ is a nonincreasingly ordered sequence of nonnegative integers.

We have $\widehat{y}_i = y_i - 1$ for $(t_1 - 1) + (t_2 - (t_1 + t_2 - s - 1)) = s$ terms and $\widehat{y}_i = y_i$ for the remaining terms. Therefore, observing that $x_i = 1$ for $i = \widehat{s} + 1, \dots, s$, we get

$$\sum_{i=1}^{\widehat{s}} \widehat{x}_i = \sum_{i=1}^{s} x_i - s = \sum_{i=1}^{t} y_i - s = \sum_{i=1}^{t} \widehat{y}_i,$$

and, similarly for $\tau = 1, \dots, \min\{\widehat{s} - 1, t_1 - 1\}$,

$$\sum_{i=1}^{\tau} \widehat{x}_i = \sum_{i=1}^{\tau} x_i - \tau \geq \sum_{i=1}^{\tau} y_i - \tau = \sum_{i=1}^{\tau} \widehat{y}_i.$$

This completes the proof if $\widehat{s} \leq t_1$.

If $\widehat{s} > t_1$, assume that there is $\tau \in \{t_1, \dots, \widehat{s} - 1\}$ such that

$$\sum_{i=1}^{\tau} \widehat{x}_i < \sum_{i=1}^{\tau} \widehat{y}_i.$$

Choosing the smallest such τ we have

$$\sum_{i=1}^{\tau-1} \widehat{x}_i \geq \sum_{i=1}^{\tau-1} \widehat{y}_i,$$

where in case $\tau = t_1 > 1$ this holds by what we have already shown and where the case $\tau = 1$ is trivial. Taking the difference of these two inequalities shows that $\widehat{x}_\tau < \widehat{y}_\tau$, and therefore, in view of $t_1 \leq \tau \leq \widehat{s} - 1 < s \leq t_2$,

$$x_\tau \leq y_\tau = y_s.$$

For $i = \tau + 1, \dots, \widehat{s}$ it follows that $x_i \leq x_\tau \leq y_s = y_i$ and therefore $\widehat{x}_i \leq \widehat{y}_i$. Hence

$$\sum_{i=1}^{\widehat{s}} \widehat{x}_i = \sum_{i=1}^{\tau-1} \widehat{x}_i + \sum_{i=\tau}^{\widehat{s}} \widehat{x}_i < \sum_{i=1}^{\tau-1} \widehat{y}_i + \sum_{i=\tau}^{\widehat{s}} \widehat{y}_i \leq \sum_{i=1}^{t} \widehat{y}_i = \sum_{i=1}^{\widehat{s}} \widehat{x}_i,$$

which is impossible. □

Remark 2.1.16. For a vector $x = (x_i)_{i=1}^s \in \mathbb{R}^s$ we denote by $x^\downarrow = (x_i^\downarrow)_{i=1}^s \in \mathbb{R}^s$ the vector with the same entries but ordered nonincreasingly. This vector $x^\downarrow$ is uniquely determined by x and can be formally written via a permutation π of $\{1, \dots, s\}$ such that

$$x_i^\downarrow = x_{\pi(i)}, \ (i = 1, \dots, s), \quad x_1^\downarrow \geq x_2^\downarrow \geq \cdots \geq x_s^\downarrow.$$

Theorem 2.1.17. *Recalling from Notation* 2.1.10 *that r_D is the number of distinct characteristic values of the Dirichlet problem* (D1), *denote by $p^\downarrow(D) = (p_i^\downarrow(D))_{i=1}^{r_D}$ the vector of the multiplicities of the distinct characteristic values of* (D1) *in nonincreasing order, and by N_i the number of edges on which the number of beads is at least i $(i = 1, \dots, n_1)$, i. e., $N_i := \#\{j \in \{1, \dots, g\} : n_j \geq i\}$. Then*

$$(N_1, N_2, \dots, N_{n_1}) \succ (p_1^\downarrow(D), p_2^\downarrow(D), \dots, p_{r_D}^\downarrow(D)).$$

Proof. By Theorem 2.1.3 and Proposition 1.2.5 the total number of the characteristic values of (D1), counted with multiplicities, is n, which means that the sum of the multiplicities of distinct characteristic values is n. Together with Lemma 2.1.14, part 1, this implies

$$\sum_{i=1}^{r_D} p_i^{\downarrow}(D) = n = \sum_{i=1}^{n_1} N_i. \tag{2.1.33}$$

This proves the identity in (2.1.32).

On each edge of the star graph the characteristic values are simple by Proposition 1.2.5. Observing Notation 2.1.10, this implies that $r_D \geq n_1$. Hence we are going to show the inequalities in (2.1.32) for $n_1 > 1$ and $\tau = 1, \dots n_1 - 1$. For the purpose of this proof we may consider g as a parameter, and we will prove the inequalities in (2.1.32) by induction on g, where $g = 1$ is admitted. For convenience, we will write $N_1(g)$ and $p_i^{\downarrow}(D, g)$ to indicate the dependence on g. For $g = 1$ we have

$$N_i(1) = 1 = p_i^{\downarrow}(D, 1), \quad (i = 1, \dots, n_1),$$

and the inequalities in (2.1.32) are indeed equalities for all $\tau = 1, \dots, n_1 - 1$. Now let $g \geq 2$ and assume that the inequalities in (2.1.32) hold for $g - 1$. Clearly,

$$N_i(g) = \begin{cases} N_i(g-1) + 1 & \text{for } i = 1, \dots, n_g, \\ N_i(g-1) & \text{for } i = n_g + 1, \dots, n_1, \end{cases}$$

whereas for $i = 1, \dots, n_1 - 1$ the number $p_i^{\downarrow}(D, g)$ is $p_i^{\downarrow}(D, g-1)$ or $p_i^{\downarrow}(D, g-1) + 1$, where the latter case occurs for exactly n_g of these indices since the numbers $\nu_\kappa^{(g)}$ ($\kappa = 1, \dots, n_g$) are distinct. Therefore

$$\sum_{i=1}^{\tau} N_i(g) = \sum_{i=1}^{\tau} N_i(g-1) + \min\{\tau, n_g\}$$

and

$$\sum_{i=1}^{\tau} p_i^{\downarrow}(D, g) \leq \sum_{i=1}^{\tau} p_i^{\downarrow}(D, g-1) + \min\{\tau, n_g\},$$

and the inequalities in (2.1.32) for g follow from the induction hypothesis. □

Proposition 2.1.18. 1. *Let*

$$\begin{aligned} \kappa_N(j) &:= \#\{k \in \{1, \dots, r_N\} : p_k(N) = j\} \quad (j = 1, \dots, g-1), \\ \kappa_D(j) &:= \#\{k \in \{1, \dots, r_D\} : p_k(D) = j\} \quad (j = 1, \dots, g). \end{aligned} \tag{2.1.34}$$

Then

$$\kappa_N(1) = \begin{cases} \kappa_D(2) + r_D + 1 & \text{if } M > 0. \\ \kappa_D(2) + r_D & \text{if } M = 0, \end{cases}$$

$$\kappa_N(j) = \kappa_D(j+1) \quad (j = 2, \dots, g-1).$$

2. *Let $\tilde{r}_D$ be the number of distinct multiple characteristic values of the Dirichlet problem* (D1). *Then*

$$r_N = \begin{cases} r_D + \tilde{r}_D + 1 & \text{if } M > 0, \\ r_D + \tilde{r}_D & \text{if } M = 0, \end{cases} \tag{2.1.35}$$

and

$$\begin{pmatrix} p_1^{\downarrow}(D) \\ \vdots \\ \vdots \\ \vdots \\ p_{r_D}^{\downarrow}(D) \end{pmatrix} = \begin{pmatrix} p_1^{\downarrow}(N)+1 \\ \vdots \\ p_{\tilde{r}_D}^{\downarrow}(N)+1 \\ p_{\tilde{r}_D+1}^{\downarrow}(N) \\ \vdots \\ p_{r_D}^{\downarrow}(N) \end{pmatrix} \succ \begin{pmatrix} p_1^{\downarrow}(N) \\ \vdots \\ p_{\tilde{r}_D}^{\downarrow}(N) \\ p_{\tilde{r}_D+1}^{\downarrow}(N) \\ \vdots \\ p_{r_D}^{\downarrow}(N) \\ \left.\begin{matrix} 1 \\ \vdots \\ 1 \end{matrix}\right\} \tilde{r}_D \end{pmatrix} = \begin{pmatrix} p_1^{\downarrow}(N) \\ \vdots \\ \vdots \\ \vdots \\ \vdots \\ p_{r_D+\tilde{r}_D}^{\downarrow}(N) \end{pmatrix}. \tag{2.1.36}$$

3. $(N_1, N_2, \dots, N_{n_1}) \succ (p_1^{\downarrow}(N), p_2^{\downarrow}(N), \dots, p_{r_D+\tilde{r}_D}^{\downarrow}(N))$.

4. $p_j(D) \le g \ (j = 1, \dots, r_D), \quad p_j(N) \le g - 1 \ (j = 1, \dots, r_N)$.

Proof. 1. By Theorem 2.1.4 we know for each $z \in \mathbb{C}$ and $j \ge 2$ that $n_{N,g}(z) = j$ if and only if $n_{D,g}(z) = j + 1$, which proves $\kappa_N(j) = \kappa_D(j + 1)$ for $j = 2, \dots, g - 1$. Since (N1) has n' characteristic values and (D1) has n characteristic values, it follows that

$$\sum_{j=1}^{g-1} j\kappa_N(j) = n' \quad \text{and} \quad \sum_{j=1}^{g} j\kappa_D(j) = n. \tag{2.1.37}$$

Therefore

$$\begin{aligned} \kappa_N(1) &= n' - \sum_{j=2}^{g-1} j\kappa_N(j) = n' - \sum_{j=2}^{g-1} j\kappa_D(j+1) = n' - \sum_{j=3}^{g} (j-1)\kappa_D(j) \\ &= n' - n + \kappa_D(1) + 2\kappa_D(2) + \sum_{j=3}^{g} \kappa_D(j) \\ &= n' - n + \kappa_D(2) + r_D. \end{aligned}$$

2. Using the above relations for $\kappa_N(j)$ we find that

$$\begin{aligned} r_N &= \sum_{j=1}^{g-1} \kappa_N(j) = \kappa_D(2) + r_D + n' - n + \sum_{j=2}^{g-1} \kappa_D(j+1) \\ &= r_D + n' - n + \sum_{j=2}^{g} \kappa_D(j) = r_D + \widetilde{r}_D + n' - n. \end{aligned}$$

To prove (2.1.36), we first conclude from the definition of $\widetilde{r}_D$ that $p_j^\downarrow(D) > 1$ if and only if $j \le \widetilde{r}_D$. Furthermore, by Theorem 2.1.4, $p_j^\downarrow(N) = 1$ for $j = \widetilde{r}_D + 1, \dots, r_N$. But then it is clear from Theorem 2.1.4 that

$$p_j^\downarrow(D) > 1, \quad p_j^\downarrow(N) = p_j^\downarrow(D) - 1 \quad (j = 1, \dots, \widetilde{r}_D), \tag{2.1.38}$$

$$p_j^\downarrow(D) = 1, \quad p_j^\downarrow(N) = p_j^\downarrow(D) \quad (j = \widetilde{r}_D + 1, \dots, r_D), \tag{2.1.39}$$

$$p_j^\downarrow(N) = 1 \quad (j = r_D + 1, \dots, r_N). \tag{2.1.40}$$

This implies the first and the last equality in (2.1.36) if we note (2.1.35). The majorization property in (2.1.36) is obvious since the number of components where 1 is added on the left hand side is equal to $\widetilde{r}_D$ and coincides with the number of new components 1 added on the right hand side.

Statement 3 immediately follows from Theorem 2.1.17 and from (2.1.36), whereas statement 4 follows from Theorem 2.1.5. □

Theorem 2.1.19. *Let r_N be the number of distinct characteristic values of the Neumann problem* (N1)*, let $(p_i(N))_{i=1}^{r_N}$ the vector of their multiplicities, and let $p^\downarrow(N) = (p_i^\downarrow(N))_{i=1}^{r_N}$ be the corresponding vector of multiplicities in nonincreasing order, see Notation* 2.1.10*. Then:*

1. *$p_1(N) = 1$; if $p_i(N) > 1$, then $p_{i-1}(N) = p_{i+1}(N) = 1$ $(i = 2, \dots, r_N - 1)$; if $M > 0$, then $p_{r_N}(N) = 1$; if $M = 0$ and $p_{r_N}(N) > 1$, then $p_{r_N - 1}(N) = 1$;*
2. *$r_N \ge n_1 + n_2 + 1$ if $M > 0$ and $r_N \ge n_1 + n_2$ if $M = 0$;*
3. *$(N_1 - 1, N_2 - 1, \dots, N_{n_1} - 1) \succ (p_1^\downarrow(N), p_2^\downarrow(N), \dots, p^\downarrow_{r'_N - n_1}(N))$ where $r'_N = r_N - 1$ if $M > 0$ and $r'_N = r_N$ if $M = 0$.*

Proof. The statements in part 1 follow from Theorem 2.1.5, part 1, and from Corollary 2.1.6.

For statement 2 we observe that, starting counting at the first edge, each characteristic value of the Dirichlet problem on the first edge increases r_D by 1, whereas each characteristic value of the Dirichlet problem on the second edge then increases either r_D or $\widetilde{r}_D$ by 1. This shows that $r_D + \widetilde{r}_D = n_1 + n_2$ if $g = 2$, and therefore $r_D + \widetilde{r}_D \ge n_1 + n_2$ for any $g \ge 2$. Taking (2.1.35) into account completes the proof of statement 2.

To prove statement 3 we first recall that $r_D \geq n_1$ since the characteristic values for (D1) are distinct on each edge. We further note that $r'_N = r_D + \tilde{r}_D$ by (2.1.35). Therefore, in view of Proposition 2.1.18, part 3, (2.1.39) and (2.1.40),

$$\sum_{i=1}^{n_1}(N_i - 1) = \sum_{i=1}^{n_1} N_i - n_1 = \sum_{i=1}^{r_D+\tilde{r}_D} p_i^{\downarrow}(N) - n_1 = \sum_{i=1}^{r_D+\tilde{r}_D-n_1} p_i^{\downarrow}(N), \tag{2.1.41}$$

and, in view of Theorem 2.1.17 and (2.1.38), for $\tau = 1, \dots, \min\{\tilde{r}_D, n_1 - 1\}$,

$$\sum_{i=1}^{\tau}(N_i - 1) = \sum_{i=1}^{\tau} N_i - \tau \geq \sum_{i=1}^{\tau} p_i^{\downarrow}(D) - \tau = \sum_{i=1}^{\tau} p_i^{\downarrow}(N). \tag{2.1.42}$$

This completes the proof in case $\tilde{r}_D \geq n_1 - 1$. If $\tilde{r}_D < n_1 - 1$, assume there is $\tau \in \{\tilde{r}_D + 1, \dots, \min\{n_1 - 1, r_D + \tilde{r}_D - n_1\}\}$ such that

$$\sum_{i=1}^{\tau}(N_i - 1) < \sum_{i=1}^{\tau} p_i^{\downarrow}(N).$$

Letting τ be the smallest number with this property, we have

$$\sum_{i=1}^{\tau-1}(N_i - 1) \geq \sum_{i=1}^{\tau-1} p_i^{\downarrow}(N),$$

which is trivially true if $\tau = 1$. Taking the difference gives

$$N_\tau - 1 < p_\tau^{\downarrow}(N) = 1$$

in view of (2.1.39) and (2.1.40). That is, $N_\tau < 2$ and therefore $N_\tau = 1$. It follows that $N_i = 1$ for $i = \tau, \dots, n_1$. But this would lead to

$$\sum_{i=1}^{n_1}(N_i - 1) = \sum_{i=1}^{\tau}(N_i - 1) < \sum_{i=1}^{\tau} p_i^{\downarrow}(N) \leq \sum_{i=1}^{r_D+\tilde{r}_D-n_1} p_i^{\downarrow}(N),$$

which contradicts (2.1.41). □

Remark 2.1.20.

1. If $M > 0$, then the number of simple characteristic values of the Neumann problem (N1) is at least $\max\{n_1, [\frac{r_N}{2}]\} + 1$.
2. If $M = 0$, then the number of simple characteristic values of the Neumann problem (N1) is at least $\max\{n_1, [\frac{r_N+1}{2}]\}$.

Proof. Recall from Proposition 2.1.18 that the number of simple characteristic values of (N1) is denoted by $\kappa_N(1)$. Since $r_D \geq n_1$, Proposition 2.1.18 gives $\kappa_N(1) \geq n_1 + 1$ if $M > 0$ and $\kappa_N(1) \geq n_1$ if $M = 0$.

Recall from Theorem 2.1.5 that μ_1 and, in case $M > 0$, μ_{n+1} are simple. Hence it follows that $\#\{\mu_k : k = 2, \dots, n\} = r'_N - 1$. By Corollary 2.1.6, multiple characteristic values μ_k ($k = 2, \dots, n$) are separated by simple characteristic values. Therefore at least $[\frac{r'_N-1}{2}]$ of these characteristic values are simple, which gives $\kappa_N(1) \geq \frac{r_N}{2} + 1$ if $M > 0$ and $\kappa_N(1) \geq \frac{r_N+1}{2}$ if $M > 0$. □

2.1.4 The inverse problem with given numbers of beads on the edges

In this subsection we consider the inverse problem under the additional assumption that the number of beads on each edge is prescribed. Furthermore, the problems (N1) and (D1) will always to be considered with $c_j = 0$ for all j, i. e., all pendant vertices are assumed to be clamped.

Theorem 2.1.21. *Let $g \in \mathbb{N}$, $g \geq 2$, $(l^{(j)})_{j=1}^g \in (0,\infty)^g$, and $n \in \mathbb{N}$. Let $n' = n+1$ or $n' = n$. Suppose that sequences $\Lambda_N = (\mu_k)_{k=1}^{n'}$, $\Lambda_D = (\nu_k)_{k=1}^n$ of positive real numbers are given and let $(n_j)_{j=1}^g \in \mathbb{N}^g$, $n_1 \geq n_2 \geq \cdots \geq n_g$, be given with $\sum_{j=1}^g n_j = n$. Define $N_i := \#\{j \in \{1,\ldots,g\} : n_j \geq i\}$ for $i = 1,\ldots,n_1$, let r_D be the number of distinct entries in Λ_D and let $(p_k^{\downarrow}(D))_{k=1}^{r_D}$ be the vector of nonincreasingly ordered multiplicities of the distinct entries in Λ_D. Then the conditions*

(i) $0 < \mu_1 < \nu_1 \leq \cdots \leq \mu_n \leq \nu_n$ *and* $\nu_n < \mu_{n+1}$ *if* $n' = n+1$,

(ii) $\nu_{k-1} = \mu_k$ *if and only if* $\mu_k = \nu_k$ $(k = 2,\ldots,n)$,

(iii) $(N_1, N_2, \ldots, N_{n_1}) \succ (p_1^{\downarrow}(D), p_2^{\downarrow}(D), \ldots, p_{r_D}^{\downarrow}(D))$,

are necessary and sufficient for the existence of a sequence of (positive) masses $(m_k^{(j)})_{k=1}^{n_j}$, a mass $M \geq 0$, and sequences of lengths $(l_k^{(j)})_{k=0}^{n_j}$ $(j = 1,\ldots,q)$ with $\sum_{k=0}^{n_j} l_k^{(j)} = l^{(j)}$ such that the sequences of characteristic values of the spectral problems (N1) *and* (D1) *on the corresponding star graph are Λ_N and Λ_D. Then M is uniquely determined by the given data, and $M > 0$ if and only if $n' = n+1$.*

Proof. The necessity of the conditions (i) and (ii) follows from Theorem 2.1.5, and the necessity of condition (iii) follows from Theorem 2.1.17. For the sufficiency we observe that (iii) implies $g = N_1 \geq p_1^{\downarrow}(D)$. Therefore condition (iii) of Theorem 2.1.8 is satisfied. The sufficiency of the conditions (i)–(iii) then follows from Theorem 2.1.8 and its proof if we show that (2.1.20) can be satisfied, i. e., that the sequence $(\nu_k)_{k=1}^n$ can be partitioned into g subsequences $(\nu_k^{(j)})_{k=1}^{n_j}$ $(j = 1,\ldots,g)$ such that

$$\Lambda_D = (\nu_k)_{k=1}^n = \bigcup_{j=1}^g (\nu_k^{(j)})_{k=1}^{n_j} \tag{2.1.43}$$

and $\nu_k^{(j)} \neq \nu_{k'}^{(j)}$ for $k \neq k'$.

The vectors $x = (N_1, N_2, \ldots, N_{n_1})$ and $y = (p_1^{\downarrow}(D), p_2^{\downarrow}(D), \ldots, p_{r_D}^{\downarrow}(D))$ satisfy the assumption of Lemma 2.1.15 in view of condition (iii). Then it follows from Lemma 2.1.15 that numbers $\widehat{p}_j^{\downarrow}(D)$ $(j = 1,\ldots,r(D))$ can be chosen such that $\widehat{p}_j^{\downarrow}(D) = p_j^{\downarrow}(D) - 1 \geq 0$ for exactly n_1 indices j and that $\widehat{p}_j^{\downarrow}(D) = p_j^{\downarrow}(D)$ for the remaining indices and such that

$$(N_1 - 1, N_2 - 1, \ldots, N_{n_2} - 1) \succ (\widehat{p}_1^{\downarrow}(D), \widehat{p}_2^{\downarrow}(D), \ldots, \widehat{p}_{r_D}^{\downarrow}(D)).$$

Choosing the n_1 numbers $\widetilde{\nu}_j \in \widetilde{\Lambda}_D$ for which $\widehat{p}_j^{\downarrow}(D) = p_j^{\downarrow}(D) - 1$ and arranging them in increasing order, we have constructed the sequence $(\nu_k^{(1)})_{k=1}^{n_1}$.

The remaining numbers $\nu_k \in \Lambda_D$ with $\nu_k = \widetilde{\nu}_j$ have the multiplicities $\widehat{p}_j^{\downarrow}(D)$, where $\widehat{p}_j^{\downarrow}(D) = 0$ if a number with multiplicity 1 has been taken. The numbers $N_j - 1$ $(j = 1, \dots, n_2)$ correspond to the sequence $(n_j)_{j=2}^{g}$, and successive application of the above algorithm gives the sequences $(\nu_k^{(j)})_{k=1}^{n_j}$ for $j = 2, \dots, g-1$ of elements from Λ_D. In the last step we have $\widetilde{N}_i = 1$ for $i = 1, \dots, n_g$ and a vector $(\widetilde{p}_1^{\downarrow}(D), \widetilde{p}_2^{\downarrow}(D), \dots, \widetilde{p}_{r_D}^{\downarrow}(D))$ of the multiplicities of the remaining elements from Λ_D. Because the recursion preserves majorization, we have

$$(1, \dots, 1) \succ (\widetilde{p}_1^{\downarrow}(D), \widetilde{p}_2^{\downarrow}(D), \dots, \widetilde{p}_{r_D}^{\downarrow}(D)),$$

where the vector on the left has n_g components. Then $\widetilde{p}_{n_g}^{\downarrow}(D) > 0$ in view of Remark 2.1.13. Since the entries of the vector on the right-hand side are sorted in nonincreasing order and sum up to n_g, it follows that $\widetilde{p}_k^{\downarrow}(D) = 1$ for $k = 1, \dots, n_g$ and $\widetilde{p}_k^{\downarrow}(D) = 0$ for $k = n_g+1, \dots, r_D$. That means, exactly the numbers $\widetilde{\nu}_1, \dots, \widetilde{\nu}_{n_g}$ are left from Λ_D, and sorting them into an increasing sequence $(\nu_k^g)_{k=1}^{n_g}$ gives (2.1.43) with $\nu_k^{(j)} \neq \nu_{k'}^{(j)}$ for $k \neq k'$ and $j = 1, \dots, g$. □

In the next theorem we derive sufficient conditions for a sequence of numbers to be the characteristic values of the Neumann problem on a star graph.

Theorem 2.1.22. *Let $g \in \mathbb{N}$, $g \geq 2$, $(l^{(j)})_{j=1}^{g} \in (0, \infty)^g$, $n \in \mathbb{N}$, and $n' = n + 1$ or $n' = n$. Suppose that a sequence $\Lambda_N := (\mu_k)_{k=1}^{n'}$ of positive real numbers is given, and let $(n_j)_{j=1}^{g} \subset \mathbb{N}$, $n_1 \geq n_2 \geq \cdots \geq n_g$, be given with $\sum_{j=1}^{g} n_j = n$. Define $N_i := \#\{j \in \{1, \dots, g\} : n_j \geq i\}$ for $i = 1, \dots, n_1$. Denote by r_N the number of different elements $\widetilde{\mu}_k$ in the sequence $(\mu_k)_{k=1}^{n'}$, by $(p_i(N))_{i=1}^{r_N}$ the vector of their multiplicities, and by $p^{\downarrow}(N) = (p_i^{\downarrow}(N))_{i=1}^{r_N}$ the corresponding vector of multiplicities in nonincreasing order. Then the conditions*

(i) $0 < \mu_1 < \mu_2 \leq \mu_3 \leq \ldots \leq \mu_{n-2} \leq \mu_{n-1} \leq \mu_n$; $\mu_n < \mu_{n+1}$ *if* $n' = n + 1$;

(ii) *if* $p_i(N) > 1$ *then* $p_{i-1}(N) = p_{i+1}(N) = 1$ *for* $i = 2, \dots, r_N - 1$, *and if* $n' = n$ *and* $p_{r_N}(N) > 1$, *then* $p_{r_N - 1}(N) = 1$;

(iii) $(N_1 - 1, N_2 - 1, \dots, N_{n_1} - 1) \succ (p_1^{\downarrow}(N), p_2^{\downarrow}(N), \dots, p_{r'_N - n_1}^{\downarrow}(N))$ *where* $r'_N = r_N - 1$ *if* $n' = n + 1$ *and* $r'_N = r_N$ *if* $n' = n$;

are necessary and sufficient such that there exist sequences of (positive) masses $(m_k^{(j)})_{k=1}^{n_j}$ and lengths $(l_k^{(j)})_{k=0}^{n_j}$ $(j = 1, \dots, q)$ with $\sum_{k=0}^{n_j} l_k^{(j)} = l^{(j)}$ such that the spectral problem (N1) *on the corresponding star graph with central mass $M \geq 0$ has the sequence Λ_N as characteristic values. Then $M = 0$ if and only if $n' = n$.*

Proof. The necessity of the conditions has been shown in Theorems 2.1.5 and 2.1.19.

To prove sufficiency of the conditions (i)–(iii) we will first show that the conditions (i)–(iii) ensure the existence of a sequence $\Lambda_D := (\nu_k)_{k=1}^n$ such that the two sequences Λ_N and Λ_D satisfy the assumptions of Theorem 2.1.21. To this end we set $\mu_{n+1} := \infty$ in the case $n' = n$ for convenience. We will call a term of a sequence a simple value of that sequence if it occurs exactly once in that sequence and we will call it a multiple value of that sequence if it occurs more than once in that sequence. Let $k \in \{1, \dots, n\}$ and consider μ_k and μ_{k+1}. If $\mu_k < \mu_{k+1}$ and both μ_k and μ_{k+1} are simple values of Λ_N, then we choose $\nu_k \in (\mu_k, \mu_{k+1})$. If $\mu_k < \mu_{k+1}$ and at least one of μ_k and μ_{k+1} is not simple, then exactly one is a multiple value of Λ_N in view of condition (ii). In this case, if μ_k is a multiple value, we set $\nu_k = \mu_k$, and if μ_{k+1} is a multiple value, we set $\nu_k = \mu_{k+1}$. Finally, if $\mu_k = \mu_{k+1}$, we set $\nu_k = \mu_k$.

Now we are going to verify that the sequences Λ_D and Λ_N satisfy the assumptions of Theorem 2.1.21. Assumption (i) holds by construction and since μ_1 and μ_{n+1} are simple values of Λ_N.

To verify condition (ii) in Theorem 2.1.21 let $k \in \{2, \dots, n\}$. If $\nu_{k-1} = \mu_k$, then μ_k is a multiple value of Λ_N, and therefore $\nu_k = \mu_k$ by definition of ν_k. And if $\mu_k = \nu_k$, then μ_k is a multiple value of Λ_N, and therefore $\nu_{k-1} = \mu_{(k-1)+1} = \mu_k$ by definition of ν_{k-1}.

To verify condition (iii) in Theorem 2.1.21 we first find the vector $(p_k^\downarrow(D))_{k=1}^{r_D}$. We note that by construction of ν_k $(k = 1, \dots, n)$ we have $\nu_k \in \Lambda_N$ if and only if ν_k is a multiple value of Λ_N. In this case there are unique numbers $j_1, j_2 \in \{2, \dots, n\}$ such that $j_1 \neq j_2$ and $\mu_{j_1-1} < \mu_{j_1} = \nu_k = \mu_{j_2} < \mu_{j_2+1}$. By construction of ν_j it follows that $\nu_j = \nu_k = \mu_{j_1}$ if and only if $j = j_1 - 1, \dots, j_2$. Thus we have shown that the number $\widetilde{r}_D$ of multiple values of Λ_D is greater or equal than the number of multiple values of Λ_N and that $(p_1^\downarrow(N) + 1, \dots, p_{\widetilde{r}_D}^\downarrow(N) + 1)$ is a vector of the multiplicities of the distinct multiple values of Λ_D. Since this vector is in nonincreasing order, it follows that

$$(p_1^\downarrow(D), \dots, p_{\widetilde{r}_D}^\downarrow(D)) = (p_1^\downarrow(N) + 1, \dots, p_{\widetilde{r}_D}^\downarrow(N) + 1).$$

Note that r'_N denotes the number of distinct terms in the sequence $(\mu_k)_{k=1}^n$. Hence

$$r'_N = n - \sum_{i=1}^{\widetilde{r}_D} \left(p_i^\downarrow(N) - 1 \right).$$

Similarly, the number of distinct terms in Λ_D is

$$r_D = n - \sum_{i=1}^{\widetilde{r}_D} \left(p_i^\downarrow(D) - 1 \right).$$

It therefore follows that

$$r'_N = r_D + \sum_{i=1}^{\widetilde{r}_D} \left(p_i^\downarrow(D) - p_i^\downarrow(N) \right) = r_D + \widetilde{r}_D. \tag{2.1.44}$$

We have that the number of terms in the sequence $(\mu_k)_{k=1}^n$ is

$$\sum_{i=1}^{r'_N} p_i^{\downarrow}(N) = n,$$

and condition (iii) and Lemma 2.1.14 give

$$\sum_{i=1}^{r'_N - n_1} p_i^{\downarrow}(N) = \sum_{i=1}^{n_1} (N_i - 1) = \sum_{i=1}^{n_1} N_i - n_1 = n - n_1.$$

Taking the difference shows that

$$\sum_{i=r'_N - n_1 + 1}^{r'_N} p_i^{\downarrow}(N) = n_1.$$

But since $p_i^{\downarrow}(N) \geq 1$ for $i = 1, \ldots, r'_N$, it follows for $i = r'_N - n_1 + 1, \ldots, r'_N$ that $p_i^{\downarrow}(N) = 1$, which gives $\widetilde{r}_D \leq r'_N - n_1$. With the aid of (2.1.44) we conclude that

$$r_D = r'_N - \widetilde{r}_D \geq n_1.$$

Now we are able to prove condition (iii) in Theorem 2.1.21. The equality in the majorization condition holds since

$$\sum_{i=1}^{n_1} N_i = n = \sum_{i=1}^{r_D} p_i^{\downarrow}(D). \tag{2.1.45}$$

For $\tau = 1, \ldots, n_2$ we calculate

$$\sum_{i=1}^{\tau} N_i = \sum_{i=1}^{\tau} (N_i - 1) + \tau \geq \sum_{i=1}^{\tau} p_i^{\downarrow}(N) + \tau = \sum_{i=1}^{\tau} \left(p_i^{\downarrow}(N) + 1 \right) \geq \sum_{i=1}^{\tau} p_i^{\downarrow}(D),$$

and for $\tau = n_2 + 1, \ldots, n_1 - 1$ it is clear in view of (2.1.45), $r_D \geq n_1$ and $N_i = 1$ for $i = n_2 + 1, \ldots, n_1$ that

$$\begin{aligned}
\sum_{i=1}^{\tau} N_i &= \sum_{i=1}^{r_D} p_i^{\downarrow}(D) - \sum_{i=\tau+1}^{n_1} N_i \geq \sum_{i=1}^{n_1} p_i^{\downarrow}(D) - \sum_{i=\tau+1}^{n_1} N_i \\
&= \sum_{i=1}^{\tau} p_i^{\downarrow}(D) + \sum_{i=\tau+1}^{n_1} (p_i^{\downarrow}(D) - 1) \geq \sum_{i=1}^{\tau} p_i^{\downarrow}(D).
\end{aligned}$$

The statement of this theorem follows now from Theorem 2.1.21. □

2.1.5 Comparison with results for tree-patterned matrices

Interlacing conditions of finite sequences of real numbers also play a rôle in the theory of symmetric matrices. In Section C.6, interlacing results of eigenvalues of tree-patterned matrices have been proved. In this subsection we show how the above results on star graphs of Stieltjes strings can be used to prove the existence of a star-patterned symmetric matrix and submatrix with prescribed interlacing spectra, thus giving an alternative proof of one direction of Theorem C.6.2 for star-patterned matrices.

To relate the results about characteristic values of star graphs of Stieltjes strings to star-patterned matrices, we reformulate the direct Neumann and Dirichlet problems with $c_j = 0$ for all j considered in Section 2.1.1 in the case $M > 0$ as matrix eigenvalue problems.

To this end, let $\mathcal{L}$ be the $(n+1)\times(n+1)$ matrix defined by

$$\mathcal{L} := \left(\begin{array}{c|c|c|c|c}
\sum\limits_{j=1}^{g} \frac{1}{l^{(j)}_{n_j}} & 0\cdots 0\ -\frac{1}{l^{(1)}_{n_1}} & 0\cdots 0\ -\frac{1}{l^{(2)}_{n_2}} & \cdots\cdots\cdots & 0\cdots 0\ -\frac{1}{l^{(g)}_{n_g}} \\
\hline
\begin{matrix} 0 \\ \vdots \\ 0 \\ -\frac{1}{l^{(1)}_{n_1}} \end{matrix} & \mathcal{L}_1 & 0 & 0 & 0 \\
\hline
\begin{matrix} 0 \\ \vdots \\ 0 \\ -\frac{1}{l^{(2)}_{n_2}} \end{matrix} & 0 & \mathcal{L}_2 & 0 & 0 \\
\hline
\vdots & 0 & 0 & \ddots & 0 \\
\hline
\begin{matrix} 0 \\ \vdots \\ 0 \\ -\frac{1}{l^{(g)}_{n_g}} \end{matrix} & 0 & 0 & 0 & \mathcal{L}_g
\end{array}\right),$$

where the matrices $\mathcal{L}_j$ occurring in the block diagonal are $n_j \times n_j$ matrices defined by

$$\begin{pmatrix} \frac{1}{l_0^{(j)}}+\frac{1}{l_1^{(j)}} & -\frac{1}{l_1^{(j)}} & 0 & \cdots & \cdots & \cdots & 0 \\ -\frac{1}{l_1^{(j)}} & \frac{1}{l_1^{(j)}}+\frac{1}{l_2^{(j)}} & -\frac{1}{l_2^{(j)}} & 0 & \cdots & \cdots & 0 \\ 0 & -\frac{1}{l_2^{(j)}} & \frac{1}{l_2^{(j)}}+\frac{1}{l_3^{(j)}} & -\frac{1}{l_3^{(j)}} & \ddots & & \vdots \\ \vdots & & \ddots & \ddots & \ddots & \ddots & 0 \\ \vdots & & & -\frac{1}{l_{n_j-2}^{(j)}} & \frac{1}{l_{n_j-2}^{(j)}}+\frac{1}{l_{n_j-1}^{(j)}} & & -\frac{1}{l_{n_j-1}^{(j)}} \\ 0 & \cdots & \cdots & 0 & & -\frac{1}{l_{n_j-1}^{(j)}} & \frac{1}{l_{n_j-1}^{(j)}}+\frac{1}{l_{n_j}^{(j)}} \end{pmatrix}$$

for $j = 1, \dots g$. Furthermore, let $\mathcal{M}$ be the $(n+1)\times(n+1)$ diagonal matrix with diagonal elements

$$M, m_1^{(1)}, \dots, m_{n_1}^{(1)}, m_1^{(2)}, \dots, m_{n_2}^{(2)}, \dots, m_1^{(g)}, \dots, m_{n_g}^{(g)}.$$

Then the Neumann problem (N1) in (2.1.4)–(2.1.7) with $z = \lambda^2$ is the eigenvalue problem for the matrix

$$\widetilde{\mathcal{L}} := \mathcal{M}^{-\frac{1}{2}}\mathcal{L}\mathcal{M}^{-\frac{1}{2}},$$

while the Dirichlet problem (D1) given by (2.1.1)–(2.1.3) is the eigenvalue problem for the principal submatrix $\widetilde{\mathcal{L}}_{1,1}$.

The matrix $\widetilde{\mathcal{L}}$ is T-acyclic, where the tree T is derived from the star graph under consideration with the beads and vertices of the star graph being identified as the vertices of T and with the threads of the star graph being identified as the edges of T.

Theorem 2.1.8 means that, under the assumptions therein with $M > 0$, there exists a real Hermitian star-patterned $(n+1)\times(n+1)$ matrix such that its sequence of eigenvalues coincides with the sequence $(\mu_k)_{k=1}^{n+1}$ and the sequence of the eigenvalues of the submatrix obtained by deleting the first row and the first column coincides with the sequence $(\nu_k)_{k=1}^{n}$.

Thus Theorem 2.1.8 with $M > 0$ provides sufficient conditions for two sequences $(\mu_k)_{k=1}^{n+1}$ and $(\nu_k)_{k=1}^{n}$ to be the eigenvalues of a real Hermitian star-patterned matrix and its first principal submatrix, respectively. Observe that stronger assumptions are made for star-shaped string problems, where the total lengths of the strings are given, which has no meaning in graph theory since no length is associated to the edges. Furthermore, the eigenvalues of the Dirichlet problems on the strings are simple, whereas in the notation of Theorem C.6.2, the eigenvalues of the matrices $A_{\Gamma_k(j)}$ do not need to be simple.

2.2 Symmetric Stieltjes strings

In this section we consider Stieltjes strings with clamped ends and with at least one bead which are symmetric with respect to their midpoints. With the notation from Section 1.2 this means that

$$m_k = m_{n-k+1} \quad (k = 1, \dots, n_0),$$
$$l_k = l_{n-k} \quad (k = 0, \dots, n_0),$$

where $n_0 = \left\lfloor \frac{n}{2} \right\rfloor$ is the integer part of $\frac{n}{2}$.

For symmetric Stieltjes strings, the Dirichlet-Neumann problem is identical to the Neumann-Dirichlet problem, and therefore the corresponding characteristic functions are multiples of each other. More precisely, (1.2.34) gives

Lemma 2.2.1. *If the Stieltjes string is symmetric, then*

$$R_{2n-1}(\cdot, 0) = \frac{1}{l_0} R_{2n}(\cdot, 1). \tag{2.2.1}$$

Using (2.2.1) we see that the Lagrange identity (1.2.47) for a symmetric Stieltjes string attains the form

$$\frac{1}{l_0} R_{2n}^2(z, 1) - \frac{1}{l_0} = R_{2n}(z, 0) R_{2n-1}(z, 1). \tag{2.2.2}$$

Splitting the symmetric Stieltjes string S at its midpoint P into two substrings S_1 and S_2, we arrive at a special case of a star graph with two edges and root at the central vertex. Each of the two strings has n_0 beads in its interior, and if n is odd, there is a bead of mass $M = m_{n_0+1}$ at the midpoint P. We put $M = 0$ when n is even. The Dirichlet-Dirichlet problem for the string S can therefore be considered as a special case of the Neumann problem (N1) considered in Section 2.1. Since the strings S_1 and S_2 are identical, they have the same Cauer-Fry polynomials, which will be denoted by $R_k^{(1)}(\cdot, 0)$ $(k = 0, \dots, 2n_0)$. We note that $R_k^{(1)}(\cdot, 0) = R_k(\cdot, 0)$ for $k = 0, \dots, 2n_0 - 2$, but that they may differ for $k = 2n_0 - 1$ and $k = 2n_0$. However, it is important to note that the lengths of the first subinterval is l_0 for S, S_1 and S_2.

Lemma 2.2.2. *For a symmetric string,*

$$\frac{1}{l_0} R_{2n}(z, 0) = [2R_{2n_0-1}^{(1)}(z, 0) - MzR_{2n_0}^{(1)}(z, 0)] R_{2n_0}^{(1)}(z, 0). \tag{2.2.3}$$

Proof. By Theorem 2.1.4, the function on the right hand side of (2.2.3) is a characteristic function for the Dirichlet-Dirichlet problem on S, and therefore the functions on the left hand side and on the right hand side have the same (simple) zeros. Comparing the coefficients at $z = 0$ completes the proof. □

Theorem 2.2.3. *Let $n \in \mathbb{N}$. Then each increasing sequence of positive numbers $(\nu_k)_{k=1}^n$ is the sequence of the characteristic values of the Dirichlet-Dirichlet problem* (1.2.4), (1.2.23), (1.2.24) *for a symmetric Stieltjes string with n beads and of a prescribed length $l > 0$, and the sequences of the masses $(m_k)_{k=1}^n$ and the lengths $(l_k)_{k=0}^n$ are uniquely determined by $(\nu_k)_{k=1}^n$ and l.*

Proof. In the case of even n, Theorem 1.3.1 shows that there is a unique string S_1 of length $\frac{l}{2}$ with sequences of masses $(m_k^{(1)})_{k=0}^{n_0}$ and lengths $(l_k^{(1)})_{k=0}^{n_0}$ such that the sequences of the zeros of the corresponding Cauer-Fry polynomial $R_{2n_0-1}^{(1)}(\cdot,0)$ and $R_{2n_0}^{(1)}(\cdot,0)$ are $(\nu_{2k-1})_{k=1}^{n_0}$ and $(\nu_{2k})_{k=1}^{n_0}$, respectively. Putting $m_k = m_{n+1-k} = m_k^{(1)}$ for $k = 1,\dots,n_0$, $l_k = l_{n-k} = l_k^{(1)}$ for $k = 0,\dots,n_0-1$ and $l_{n_0} = 2l_{n_0}^{(1)}$ we arrive at a symmetric string S, with S_1 being one of the two identical copies obtained by splitting S at the midpoint. Since

$$\frac{1}{l_0} R_{2n}(\cdot,0) = 2R_{2n_0-1}^{(1)}(\cdot,0)R_{2n_0}^{(1)}(\cdot,0) \tag{2.2.4}$$

by Lemma 2.2.2, the sequence $(\nu_k)_{k=1}^n$ is the sequence of characteristic values of the string S. Due to (2.2.4) and the interlacing property (1.2.36), the zeros of $R_{2n_0-1}^{(1)}(\cdot,0)$ and $R_{2n_0}^{(1)}(\cdot,0)$ are uniquely determined by the given sequence and since l is given, also the length $\frac{l}{2}$ of S_1 is given. Hence the uniqueness statement in Theorem 1.3.1 shows that S_1, and therefore also the symmetric string S, is uniquely determined by the given data.

In the case of odd n, we define

$$f_1(z) := \prod_{k=1}^{n_0}(\nu_{2k} - z), \quad f_2(z) := \prod_{k=1}^{n_0+1}(\nu_{2k-1} - z), \quad h(z) := f_2(z) + zf_1(z).$$

Since the leading term of f_1 is $(-z)^{n_0}$ and the leading term of f_2 is $(-z)^{n_0+1}$, it follows that the degree of the polynomial h is at most n_0. Furthermore, the fact that $(\nu_k)_{k=1}^n$ is an increasing sequence of positive numbers shows for $k = 1,\dots,n_0$ that $(-1)^k h(\nu_{2k-1}) = (-1)^k \nu_{2k-1} f_1(\nu_{2k-1}) < 0$ and $(-1)^k h(\nu_{2k}) = (-1)^k f_2(\nu_{2k}) > 0$, the function h has a zero $\tau_k \in (\nu_{2k-1},\nu_{2k})$ $(k = 1,\dots,n_0)$. Hence h has degree n_0, and its simple zeros τ_k $(k = 1,\dots,n_0)$ satisfy

$$0 < \tau_1 < \nu_2 < \tau_2 < \nu_4 < \dots < \tau_{n_0} < \nu_{2n_0}.$$

By Theorem 1.3.1 there is a unique string S_1 of length $\frac{l}{2}$ with sequences of masses $(m_k^{(1)})_{k=0}^{n_0}$ and lengths $(l_k^{(1)})_{k=0}^{n_0}$ such that the sequences of the zeros of the corresponding Cauer-Fry polynomials $R_{2n_0-1}^{(1)}(\cdot,0)$ and $R_{2n_0}^{(1)}(\cdot,0)$ are $(\tau_k)_{k=1}^{n_0}$ and $(\nu_{2k})_{k=1}^{n_0}$, respectively. Putting $m_k = m_{n+1-k} = m_k^{(1)}$ for $k = 1,\dots,n_0$, $l_k = l_{n-k} = l_k^{(1)}$ for $k = 0,\dots,n_0$ and

$$m_{n_0+1} = \frac{2}{l}\prod_{k=1}^{n_0}\nu_{2k}\prod_{k=1}^{n_0+1}\frac{1}{\nu_{2k-1}}, \tag{2.2.5}$$

we arrive at a symmetric string S, with S_1 being one of the two identical copies obtained by splitting S at the midpoint. It is clear that

$$R^{(1)}_{2n_0}(z,0) = \frac{l}{2l_0}\frac{f_1(z)}{f_1(0)}, \quad R^{(1)}_{2n_0-1}(z,0) = \frac{1}{l_0}\frac{h(z)}{h(0)}.$$

It follows that

$$\begin{aligned} 2R^{(1)}_{2n_0-1}(z,0) - m_{n_0+1}zR^{(1)}_{2n_0}(z,0) &= \frac{1}{l_0}\frac{h(z)}{h(0)} - m_{n_0+1}z\frac{l}{2l_0}\frac{f_1(z)}{f_1(0)} \\ &= \frac{1}{l_0 h(0)}\left(h(z) - m_{n_0+1}l\frac{h(0)}{2f_1(0)}zf_1(z)\right) \\ &= \frac{f_2(z)}{l_0 h(0)}. \end{aligned}$$

In view of Lemma 2.2.2 it follows that the generated string problem has the given sequence $(\nu_k)_{k=1}^n$ as the sequence of its characteristic values, and the string has length l.

For the uniqueness, we observe that the zeros of $R^{(1)}_{2n_0}(\cdot,0)$ are necessarily the numbers ν_{2k} $(k=1,\dots,n)$ since the function $\frac{(R^{(1)}_{2n_0})^2(\cdot,0)}{R_{2n}(\cdot,0)}$ is an $\mathcal{S}_0$-function by Lemma 2.2.2 and by Theorem 2.1.2. Furthermore, $R_{2n}(0,0) = \frac{l}{l_0}$ and $R^{(1)}_{2n_0}(0,0) = \frac{l}{2l_0}$ show that

$$\frac{R_{2n}(z,0)}{l_0(R^{(1)}_{2n_0})^2(z,0)} = 2\frac{R^{(1)}_{2n_0-1}(z,0)}{R^{(1)}_{2n_0}(z,0)} - m_{n_0+1}z$$

is uniquely determined by the given data. Thus m_{n_0+1} and the zeros of $R^{(1)}_{2n_0-1}(\cdot,0)$ are unique. Together with the uniqueness of the length of S_1 another application of Theorem 1.3.1 completes the proof. □

Remark 2.2.4. In view of (2.2.5), the mass of the central bead for odd n is explicitly given by the characteristic values and the length of the string. Conversely, given the length of the string and the mass of the central bead, (2.2.5) gives an upper bound for ν_1 and a lower bound for ν_n: $\nu_1 \le \frac{2}{lm_{n_0+1}} \le \nu_n$, where equality occurs if and only if $n=1$.

2.3 Star graphs with root at a pendant vertex

In this section we consider a plane star graph of g (≥ 2) Stieltjes strings joined at the central vertex where a bead of mass $M > 0$ may be placed. One pendant vertex is chosen to be the root of the star graph. If there is no bead at the central vertex, then we put $M = 0$. We suppose that the star graph is stretched and study its small transverse vibrations with continuity conditions and balance of forces at the central vertex in two different cases:

(D2) the root is fixed (Dirichlet problem),

(N2) the root is free to move in the direction orthogonal to the equilibrium position of the strings (Neumann problem).

We investigate the characteristic values of both problems and their relations to each other in order to be able to establish necessary and sufficient conditions for the solution of the corresponding inverse problem. Observe that both (D2) and (N2) are special cases of the Neumann problem (N1). However, to compare the characteristic values of (D2) and (N2), the approach differs from that taken in Section 2.1.

We are going to use the notation for the star graph which was introduced in Section 2.1. In the sequel, the string incident to the root is called the main edge of the string. The main edge will be labelled with $j = g$, and therefore the other edges of the star graph will be labelled with $j = 1, \ldots, g-1$.

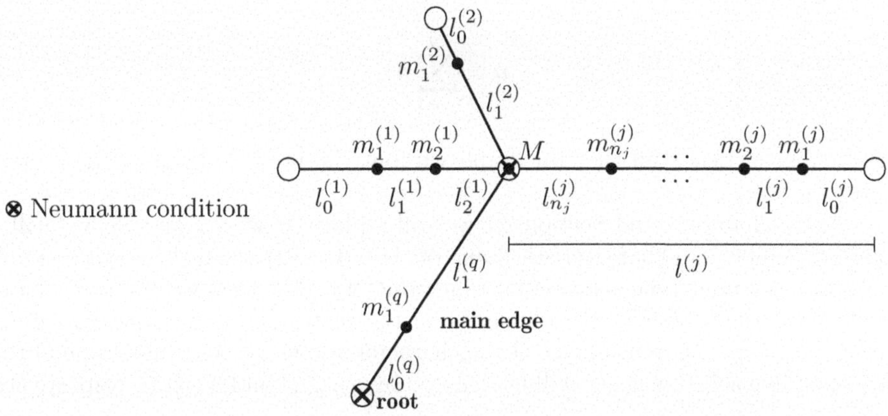

Figure 2.3: Star graph with root at a pendant vertex

Under the assumption that the threads are stretched by forces each equal to 1, we have derived in Section 2.1 the following difference equations for the amplitudes $u_k^{(j)}$:

$$\frac{u_k^{(j)} - u_{k-1}^{(j)}}{l_{k-1}^{(j)}} - \frac{u_{k+1}^{(j)} - u_k^{(j)}}{l_k^{(j)}} - m_k^{(j)}\lambda^2 u_k^{(j)} = 0 \quad (k = 1, \ldots, n_j,\ j = 1, \ldots, g), \tag{2.3.1}$$

and the following conditions at the vertices which are not the root:

$$u_{n_1+1}^{(1)} = u_{n_2+1}^{(2)} = \ldots = u_{n_g+1}^{(g)}, \tag{2.3.2}$$

$$\sum_{j=1}^{g} \frac{u_{n_j+1}^{(j)} - u_{n_j}^{(j)}}{l_{n_j}^{(j)}} - M\lambda^2 u_{n_1+1}^{(1)} = 0, \tag{2.3.3}$$

$$u_0^{(j)} = c_j u_1^{(j)} \quad (j = 1, \ldots, g-1). \tag{2.3.4}$$

For the main edge, the boundary condition at the root is:

Dirichlet problem (D2). If we clamp the root, we have to consider (2.3.1)–(2.3.4) together with

$$u_0^{(g)} = 0. \tag{2.3.5}$$

Neumann problem (N2). If the root is allowed to move freely, we have to consider (2.3.1)–(2.3.4) together with

$$u_0^{(g)} = u_1^{(g)}. \tag{2.3.6}$$

As in Section 2.1 we will denote by

$$n := \sum_{j=1}^{g} n_j$$

the number of beads on the star graph without the possible bead at the central vertex.

As in Chapter 1 and Section 2.1 we will replace λ^2 with z and we will call a complex number z a characteristic value of (D2) or (N2) when the corresponding problem has a nontrivial solution for that value of z. The number of linearly independent solutions, that is, the dimension of the vector space of solutions, will be called the geometric multiplicity of the characteristic value. The dimension of the corresponding solution space will be denoted by $\tilde{n}_{D,g}(z)$ and $\tilde{n}_{N,g}(z)$, respectively.

2.3.1 The direct spectral problem

In this subsection we investigate the relations of the characteristic values of the Dirichlet problem (D2) with those of the Neumann problem (N2). Defining

$$\begin{aligned}\phi_g(z,c) := R_{2n_g}^{(g)}(z,c) \sum_{j=1}^{g-1} \left[R_{2n_j-1}^{(j)}(z) \prod_{k=1,\ k\neq j}^{g-1} R_{2n_k}^{(k)}(z) \right] \\ + R_{2n_g-1}^{(g)}(z,c) \prod_{k=1}^{g-1} R_{2n_k}^{(k)}(z) - MzR_{2n_g}^{(g)}(z,c) \prod_{k=1}^{g-1} R_{2n_k}^{(k)}(z)\end{aligned} \tag{2.3.7}$$

it is clear that Theorem 2.1.4 and (2.1.10) with $R_k^{(g)}(z)$ replaced by $R_k^{(g)}(z,c)$ for $k = 2n_g$, $k = 2n_g - 1$, $c = 0$, $c = 1$ lead to the following result.

Theorem 2.3.1. 1. *The characteristic values of the Dirichlet problem* (D2), *counted with multiplicity, are the zeros of the polynomial* $\phi_g(\cdot,0)$.

2. *The characteristic values of the Neumann problem* (N2), *counted with multiplicity, are the zeros of the polynomial* $\phi_g(\cdot,1)$.

Proposition 2.3.2. *Let* $\phi_{D,g-1}$, $\phi_{N,g-1}$ *be defined as in* (2.1.9), (2.1.10) *for the subgraph of the* $g-1$ *edges that are not the main edge. Then*

$$\phi_g(\cdot,c) = R^{(g)}_{2n_g}(\cdot,c)\phi_{N,g-1} + R^{(g)}_{2n_g-1}(\cdot,c)\phi_{D,g-1} \quad (c=0,1), \tag{2.3.8}$$

$$\phi_{D,g-1} = l_0^{(g)}\left(R^{(g)}_{2n_g}(\cdot,1)\phi_g(\cdot,0) - R^{(g)}_{2n_g}(\cdot,0)\phi_g(\cdot,1)\right), \tag{2.3.9}$$

$$\phi_{N,g-1} = l_0^{(g)}\left(-R^{(g)}_{2n_g-1}(\cdot,1)\phi_g(\cdot,0) + R^{(g)}_{2n_g-1}(\cdot,0)\phi_g(\cdot,1)\right). \tag{2.3.10}$$

Proof. The identities (2.3.8) for $c=0$ and $c=1$ follow immediately from the definitions of $\phi_g(\cdot,c)$, $\phi_{D,g-1}$ and $\phi_{N,g-1}$. Writing (2.3.8) in matrix form, we have

$$\begin{pmatrix}\phi_g(\cdot,0)\\ \phi_g(\cdot,1)\end{pmatrix} = \begin{pmatrix}R^{(g)}_{2n_g-1}(\cdot,0) & R^{(g)}_{2n_g}(\cdot,0)\\ R^{(g)}_{2n_g-1}(\cdot,1) & R^{(g)}_{2n_g}(\cdot,1)\end{pmatrix}\begin{pmatrix}\phi_{D,g-1}\\ \phi_{N,g-1}\end{pmatrix}. \tag{2.3.11}$$

Then

$$\begin{aligned}\begin{pmatrix}\phi_{D,g-1}\\ \phi_{N,g-1}\end{pmatrix} &= \begin{pmatrix}R^{(g)}_{2n_g-1}(\cdot,0) & R^{(g)}_{2n_g}(\cdot,0)\\ R^{(g)}_{2n_g-1}(\cdot,1) & R^{(g)}_{2n_g}(\cdot,1)\end{pmatrix}^{-1}\begin{pmatrix}\phi_g(\cdot,0)\\ \phi_g(\cdot,1)\end{pmatrix}\\ &= l_0^{(g)}\begin{pmatrix}R^{(g)}_{2n_g}(\cdot,1) & -R^{(g)}_{2n_g}(\cdot,0)\\ -R^{(g)}_{2n_g-1}(\cdot,1) & R^{(g)}_{2n_g-1}(\cdot,0)\end{pmatrix}\begin{pmatrix}\phi_g(\cdot,0)\\ \phi_g(\cdot,1)\end{pmatrix}\end{aligned}$$

due to the Lagrange identity (1.2.47). This proves (2.3.9) and (2.3.10). □

Theorem 2.3.3. *The function*

$$\frac{\phi_g(\cdot,0)}{\phi_g(\cdot,1)}$$

is an $\mathcal{S}$*-function and an* $\mathcal{S}_0$*-function if at least one of the pendant vertices which is distinct from the root is clamped, and can be represented as a continued fraction*

$$l_0^{(g)}\frac{\phi_g(z,0)}{\phi_g(z,1)} = l_0^{(g)} + \cfrac{1}{-m_1^{(g)}z + \cfrac{1}{l_1^{(g)} + \cfrac{1}{-m_2^{(g)}z + \cdots + \cfrac{1}{-m_{n_g}^{(g)}z + \cfrac{1}{l_{n_g}^{(g)} + \phi_{g-1}(z)}}}}}, \tag{2.3.12}$$

where ϕ_{g-1} *is defined as in* (2.1.11) *for the subgraph of the* $g-1$ *edges that are not the main edge.*

Proof. Let

$$\Psi_k = \begin{pmatrix} R_{2k-1}^{(g)}(\cdot,0) & R_{2k}^{(g)}(\cdot,0) \\ R_{2k-1}^{(g)}(\cdot,1) & R_{2k}^{(g)}(\cdot,1) \end{pmatrix} \quad (k = 0,\ldots,n_g).$$

In view of Lemma 1.2.4 we have

$$\Psi_k = \Psi_{k-1}\Theta_k \quad (k = 1,\ldots,n_g),$$

where

$$\Theta_k(z) = \begin{pmatrix} 1 & l_k^{(g)} \\ -m_k^{(g)}z & 1 - l_k^{(g)}m_k^{(g)}z \end{pmatrix} \quad (k = 1,\ldots,n_g,\ z \in \mathbb{C}).$$

Hence it follows in view of (2.3.11) that

$$\begin{pmatrix} \phi_g(\cdot,0) \\ \phi_g(\cdot,1) \end{pmatrix} = \Psi_{n_g}\begin{pmatrix} \phi_{D,g-1} \\ \phi_{N,g-1} \end{pmatrix} = \Psi_0\Theta_1\cdots\Theta_{n_g}\begin{pmatrix} \phi_{D,g-1} \\ \phi_{N,g-1} \end{pmatrix}.$$

Let

$$\begin{pmatrix} f_1 \\ f_2 \end{pmatrix} := \Theta_1\cdots\Theta_{n_g}\begin{pmatrix} \phi_{D,g-1} \\ \phi_{N,g-1} \end{pmatrix}.$$

From (2.1.13) and Theorem 2.1.2 and recalling Definition A.4.1 it follows that $(\phi_{D,g-1},\phi_{N,g-1}) \in \mathcal{S}^0$ as well as $\frac{\phi_{D,g-1}}{\phi_{N,g-1}} \in \mathcal{S}$. A recursive application of Lemma A.4.6 shows that $(f_1,f_2) \in \mathcal{S}^0$ with $\frac{f_1}{f_2} \in \mathcal{S}$. Finally,

$$\begin{pmatrix} \phi_g(\cdot,0) \\ \phi_g(\cdot,1) \end{pmatrix} = \Psi_0\begin{pmatrix} f_1 \\ f_2 \end{pmatrix} = \begin{pmatrix} \frac{1}{l_0^{(g)}} & 1 \\ 0 & 1 \end{pmatrix}\begin{pmatrix} f_1 \\ f_2 \end{pmatrix} = \begin{pmatrix} \frac{1}{l_0^{(g)}}f_1 + f_2 \\ f_2 \end{pmatrix}$$

shows that

$$\frac{\phi_g(\cdot,0)}{\phi_g(\cdot,1)} = \frac{1}{l_0^{(g)}}\frac{f_1}{f_2} + 1$$

belongs to the class $\mathcal{S}$. The representation (2.3.12) now follows by a recursive application of (A.4.1) and from (2.1.11). In the above reasoning we may replace $\mathcal{S}$ with $\mathcal{S}_0$ if at least one of the pendant vertices which is distinct from the root is clamped. □

Proposition 2.3.4. *For all $z \in \mathbb{R}$,*

$$\min\{\widetilde{n}_{D,g}(z),\widetilde{n}_{N,g}(z)\} = \min\{n_{D,g-1}(z),n_{N,g-1}(z)\} \tag{2.3.13}$$

and $\widetilde{n}_{D,g}(z)+\widetilde{n}_{N,g}(z) \le 2g-3$. Furthermore, if $\widetilde{n}_{D,g}(z) \ge 1$ and $\widetilde{n}_{N,g}(z)\} \ge 1$, then $n_{N,g-1}(z) = \min\{\widetilde{n}_{D,g}(z),\widetilde{n}_{N,g}(z)\}$ and $n_{D,g-1}(z) = \min\{\widetilde{n}_{D,g}(z),\widetilde{n}_{N,g}(z)\} + 1$.

Proof. By Theorem 2.3.1, (2.3.9) and (2.3.10), it follows that

$$\min\{\widetilde{n}_{D,g}(z), \widetilde{n}_{N,g}(z)\} \leq \min\{n_{D,g-1}(z), n_{N,g-1}(z)\},$$

while (2.3.8) yields

$$\min\{\widetilde{n}_{D,g}(z), \widetilde{n}_{N,g}(z)\} \geq \min\{n_{D,g-1}(z), n_{N,g-1}(z)\}.$$

This proves (2.3.13). If $\widetilde{n}_{D,g}(z) \geq 1$ and $\widetilde{n}_{N,g}(z)\} \geq 1$, then $n_{D,g-1}(z) \geq 1$, and Theorem 2.1.4 gives $n_{D,g-1}(z) = n_{N,g-1}(z) + 1$. Hence (2.3.13) implies that $n_{N,g-1}(z) = \min\{\widetilde{n}_{D,g}(z), \widetilde{n}_{N,g}(z)\}$ and $n_{D,g-1}(z) = \min\{\widetilde{n}_{D,g}(z), \widetilde{n}_{N,g}(z)\} + 1$.

Since $\frac{\phi_g(\cdot,0)}{\phi_g(\cdot,1)}$ is a Nevanlinna function by Theorem 2.3.3, its zeros and poles are simple and interlace. Therefore $|\widetilde{n}_{D,g}(z) - \widetilde{n}_{N,g}(z)| \leq 1$, which gives that

$$\widetilde{n}_{D,g}(z)+\widetilde{n}_{N,g}(z) \leq 2\min\{\widetilde{n}_{D,g}(z), \widetilde{n}_{N,g}(z)\}+1 = 2\min\{n_{D,g-1}(z), n_{N,g-1}(z)\}+1.$$

If $n_{D,g-1}(z) = 0$, then it follows that

$$\widetilde{n}_{D,g}(z) + \widetilde{n}_{N,g}(z) \leq 1 \leq 2g-3,$$

whereas if $n_{D,g-1}(z) > 0$, then

$$\widetilde{n}_{D,g}(z) + \widetilde{n}_{N,g}(z) \leq 2n_{D,g-1}(z) - 1 \leq 2(g-1) - 1 = 2g-3$$

in view of Theorem 2.1.4. □

Theorem 2.3.5. *Let $n' = n$ if $M = 0$ and let $n' = n+1$ if $M > 0$. Then both the Neumann problem* (N2) *and the Dirichlet problem* (D2) *have n' characteristic values, counted with multiplicity. The nondecreasing sequence of the characteristic values $(\lambda_k)_{k=1}^n$ of the Neumann problem* (N2) *and the nondecreasing sequence of the characteristic values $(\mu_k)_{k=1}^n$ of the Dirichlet problem* (D2) *have the following properties:*

1. $0 < \lambda_1 < \mu_1 \leq \lambda_2 \leq \cdots \leq \lambda_{n'} \leq \mu_{n'}$, *where $\lambda_1 = 0$ if and only if all pendant vertices distinct from the root are freely moving;*
2. *the multiplicities of λ_k and μ_k do not exceed $g-1$;*
3. *if $\mu_k = \lambda_k$ or $\mu_k = \lambda_{k+1}$, then the sum of multiplicities of μ_k and λ_k or μ_k and λ_{k+1} does not exceed $2g-3$;*
4. *if $\mu_k = \lambda_k$ or $\mu_k = \lambda_{k+1}$, then μ_k is a zero of $\phi_{g-1} = \frac{\phi_{D,g-1}}{\phi_{N,g-1}}$.*

Proof. From Theorem 2.1.5 we know that (N1) has n' characteristic values, and therefore also (D2) and (N2) have n' characteristic values, counted with multiplicity. Since $\frac{\phi_g(\cdot,0)}{\phi_g(\cdot,1)}$ belongs to the class $\mathcal{S}_0$ by Theorem 2.3.3, Lemmas A.3.5 and A.4.4 prove statement 1, except for the strict inequality $\lambda_1 < \mu_1$ therein.
2. If the multiplicity of a characteristic value λ_k or μ_k is 1, then it does not exceed

$g-1$ since $g \geq 2$. Now assume that $z = \mu_k = \mu_{k+1}$ or $z = \lambda_k = \lambda_{k+1}$. Then $z = \lambda_{k+1}$ or $z = \mu_k$, which implies that $\widetilde{n}_{D,g}(z) \geq 1$ and $\widetilde{n}_{N,g}(z) \geq 1$. We have already noted in the proof of Proposition 2.3.4 that $|\widetilde{n}_{D,g}(z) - \widetilde{n}_{N,g}(z)| \leq 1$, which gives

$$\max\{\tilde{n}_{D,g}(z), \tilde{n}_{N,g}(z)\} \leq \min\{\tilde{n}_{D,g}(z), \tilde{n}_{N,g}(z)\} + 1 = n_{D,g-1}(z) \leq g-1$$

in view of Proposition 2.3.4 and Theorem 2.1.5, part3.

Statement 3 is part of Proposition 2.3.4, and statement 4 is immediately clear since $n_{D,g-1}(\lambda_k) > n_{N,g-1}(\lambda_k)$ by Proposition 2.3.4.

To complete the proof of statement 1, it remains to be shown that $\lambda_1 < \mu_1$. By proof of contradiction, assume that $\lambda_1 = \mu_1$. In view of (2.3.12), the n_g-th tail of $l_0^{(g)} \frac{\phi(\cdot,0)}{\phi(\cdot,1)}$ is $f_{n'-n_g} := l_{n_g}^{(g)} + \phi_{g-1}$. By statement 4 we have $\phi_{g-1}(\mu_1) = 0$, and we can conclude that $f_{n'-n_g}(\mu_1) = l_{n_g}^{(g)} > 0$. Since $\phi_{g-1} \in \mathcal{S}$ by Theorem 2.1.2, we know by Proposition A.3.2 that ϕ_{g-1} has at least one pole which is smaller than μ_1. Let z_0 be the largest pole of ϕ_{g-1} which is smaller than μ_1. Clearly, z_0 is also a pole of $f_{n'-n_g}$. By Lemma A.2.5, the function ϕ_{g-1} is increasing on $(z_0, \mu_1]$, and therefore also $f_{n'-n_g}$ is increasing on $(z_0, \mu_1]$. Hence there is $z_1 \in (z_0, \mu_1)$ such that $f_{n'-n_g}(z_1) = 0$. But this contradicts Lemma A.3.5, part 5, which states that any zero of the $n' - n_g$-th tail of $l_0^{(g)} \frac{\phi(\cdot,0)}{\phi(\cdot,1)}$ must be larger than the smallest zero of $l_0^{(g)} \frac{\phi(\cdot,0)}{\phi(\cdot,1)}$. This contradiction proves that $\lambda_1 < \mu_1$. □

One more property of the direct problem is needed for the inverse problem.

Proposition 2.3.6. *If the number of poles of* $\frac{\phi_g(\cdot,0)}{\phi_g(\cdot,1)}$ *is* n_g, *then* $M = 0$ *and* $n_g = n$.

Proof. The assumption, the uniqueness of the representation (A.3.8) of $\mathcal{S}$-functions and (2.3.12) show that ϕ_{g-1} is constant. The representation (2.1.14) of ϕ_{g-1} implies that $M = 0$ and that $\sum_{j=1}^{g-1} \frac{1}{\phi^{(j)}}$ is constant. But the functions $\phi^{(j)}$ belong to $\mathcal{S}_0$, and it follows from Lemma A.2.5 that each function $\frac{1}{\phi^{(j)}}$ is nonincreasing on $(-\infty, 0)$. Hence each function $\phi^{(j)}$ must be constant, which means $n_j = 0$ $(j = 1, \ldots, g-1)$. This shows that $n = n_g$. □

2.3.2 The inverse spectral problem

In this subsection we investigate the inverse problem, that is, the question if two sequences of positive real numbers satisfying properties 1–4 in Theorem 2.3.5 can be realized as sequences of the characteristic values of the Neumann problem (N2) and the Dirichlet problem (D2) on a star graph in the case that all pendant vertices distinct from the root are clamped.

More precisely, suppose that $g \in \mathbb{N}$ $(g \geq 2)$ is fixed and a set of lengths $l^{(j)} > 0$ $(j = 1, \ldots, g)$ as well as sequences $(\mu_k)_{k=1}^n$, $(\lambda_k)_{k=1}^n$ of positive real numbers are given. Under which conditions can we determine numbers $n_j \in \mathbb{N}_0$ $(j = 1, \ldots, g)$, the mass M of a possible bead at the central vertex, sequences of masses $(m_k^{(j)})_{k=1}^{n_j}$, and sequences of lengths $(l_k^{(j)})_{k=0}^{n_j}$ so that the corresponding star graph of Stieltjes strings has the sequences $(\mu_k)_{k=1}^n$ and $(\lambda_k)_{k=1}^n$ as characteristic values of the Dirichlet problem (D2) and the Neumann problem (N2), respectively?

Lemma 2.3.7. *Let $g \in \mathbb{N}$ $(g \geq 2)$, $l^{(j)} > 0$, $(j = 1, \ldots, g)$, $n \in \mathbb{N}$. Suppose that $(\lambda_k)_{k=1}^n$, $(\mu_k)_{k=1}^n$ are sequences of real numbers such that $(\lambda_k)_{k=1}^n \neq (\mu_k)_{k=1}^n$ and*

$$0 < \lambda_1 \leq \mu_1 \leq \lambda_2 \leq \cdots \leq \lambda_n \leq \mu_n, \tag{2.3.14}$$

and let

$$\Phi(z) := \gamma \frac{\prod\limits_{k=1}^{n}\left(1 - \frac{z}{\mu_k}\right)}{\prod\limits_{k=1}^{n}\left(1 - \frac{z}{\lambda_k}\right)}, \quad \gamma := l^{(g)} + \left(\sum_{k=1}^{g-1} \frac{1}{l^{(k)}}\right)^{-1}. \tag{2.3.15}$$

Then there exist unique $p \in \mathbb{N}$, $n_g \in \mathbb{N}_0$ with $n_g \leq p$ and $a_k > 0$ $(k = 0, \ldots, p)$, $b_k > 0$ $(k = 1, \ldots, p)$, such that

$$\sum_{k=0}^{n_g-1} a_k < l^{(g)}, \quad \sum_{k=0}^{n_g} a_k \geq l^{(g)} \tag{2.3.16}$$

with

$$a_{n_g}^1 := \sum_{k=0}^{n_g} a_k - l^{(g)} \geq 0,$$

and Φ has the representation

$$\Phi(z) = a_0 + \cfrac{1}{-b_1 z + \cfrac{1}{a_1 + \cfrac{1}{-b_2 z + \cdots + \cfrac{1}{a_{n_g-1} + \cfrac{1}{-b_{n_g} z + \cfrac{1}{a_{n_g} - a_{n_g}^1 + \widehat{f}_{n_g}(z)}}}}}} \tag{2.3.17}$$

with

$$\widehat{f}_{n_g}(z) := a_{n_g}^1 + \cfrac{1}{-b_{n_g+1} z + \cfrac{1}{a_{n_g+1} + \cfrac{1}{-b_{n_g+2} z + \cdots + \cfrac{1}{a_{p-1} + \cfrac{1}{-b_p z + \cfrac{1}{a_p}}}}}}. \tag{2.3.18}$$

Proof. In view of $(\lambda_k)_{k=1}^n \neq (\mu_k)_{k=1}^n$ and (2.3.14), Φ has at least one pole, its smallest pole is positive and less than its smallest zero, the poles and zeros of Φ are simple and interlace, and the number p of poles of Φ equals the number of zeros of Φ. Hence Φ is an $\mathcal{S}_0$-function with $\lim_{z\to-\infty}\Phi(z) > 0$. By Lemma A.3.5 there are unique $a_k > 0$ $(k = 0, \dots, p)$, $b_k > 0$ $(k = 1, \dots, p)$ such that

$$\Phi(z) = a_0 + \cfrac{1}{-b_1 z + \cfrac{1}{a_1 + \cfrac{1}{-b_2 z + \dots + \cfrac{1}{a_{p-1} + \cfrac{1}{-b_p z + \cfrac{1}{a_p}}}}}}.$$

Since we have $\sum\limits_{k=0}^{p} a_k = \Phi(0) = \gamma > l^{(g)}$, there exists a unique $n_g \in \mathbb{N}_0$ such that (2.3.16) holds and all claims follow. □

The next theorem shows that the properties obtained in Subsection 2.3.1 are also sufficient to find a star graph of Stieltjes strings which have these properties.

Theorem 2.3.8. *Let $g \in \mathbb{N}$, $g \geq 2$, $l^{(j)} > 0$ $(j = 1, \dots, g)$, $n \in \mathbb{N}$, and let λ_k $(k = 1, \dots, n)$, μ_k $(k = 1, \dots, n)$ be real numbers such that*

(i) $0 < \lambda_1 < \mu_1 \leq \lambda_2 \leq \dots \leq \lambda_n \leq \mu_n$;

(ii) *the multiplicities of λ_k in the sequence $(\lambda_k)_{k=1}^n$ and of μ_k in the sequence $(\mu_k)_{k=1}^n$ do not exceed $g - 1$;*

(iii) *if $\mu_k = \lambda_k$ or $\mu_k = \lambda_{k+1}$, then the sum of multiplicities of μ_k and λ_k or μ_k and λ_{k+1} does not exceed $2g - 3$.*

Let n_g and $\widehat{f}_{n_g}$ be defined as in Lemma 2.3.7. Further assume:

(iv) *if $\mu_k = \lambda_k$ (or $\mu_k = \lambda_{k+1}$), then $\widehat{f}_{n_g}(\mu_k) = 0$;*

(v) *if $\widehat{f}_{n_g}$ is constant, then $n = n_g$.*

Then there exists a star graph of g Stieltjes strings, i. e., numbers $n_j \in \mathbb{N}_0$ $(j = 1, \dots, g-1)$, masses $m_k^{(j)} > 0$ $(j = 1, \dots, g,\ k = 1, \dots, n_j)$, $M \in [0, \infty)$, and lengths $l_k^{(j)} > 0$ $(j = 1, \dots, g,\ k = 0, \dots, n_j)$ with $\sum_{k=0}^{n_j} l_k^{(j)} = l^{(j)}$ $(j = 1, \dots, g)$, and

$$\begin{cases} M = 0, \quad n = \sum\limits_{j=1}^{g} n_j, & \text{if } a_{n_g}^1 > 0,\\ M > 0, \quad n = 1 + \sum\limits_{j=1}^{g} n_j & \text{if } a_{n_g}^1 = 0,\end{cases}$$

with $a^1_{n_g}$ *as defined in Lemma* 2.3.7, *so that Dirichlet problem* (D2) *has the characteristic values* μ_k $(k = 1, \ldots, n)$ *and the Neumann problem* (N2) *has the characteristic values* λ_k $(k = 1, \ldots, n)$.

Proof. Due to assumptions 1 and 2, the given data yield integers $p \in \mathbb{N}$, $n_g \in \mathbb{N}_0$ and the functions Φ and $\widehat{f}_{n_g}$ as in Lemma 2.3.7. The star graph we search for will be constructed as follows. For the main edge we have already defined the number n_g of beads on this edge and with the coefficients from the continued fraction expansion of Φ given by (2.3.17) we let $m_k^{(g)} := b_k$ $(k = 1, \ldots, n_g)$ be their masses, $l_k^{(g)} := a_k$ $(k = 0, \ldots, n_g - 1)$ and

$$l_{n_g}^{(g)} := a_{n_g} - a^1_{n_g} = a_{n_g} - \sum_{k=0}^{n_g} a_k + l^{(g)} = l^{(g)} - \sum_{k=0}^{n_g-1} a_k > 0 \tag{2.3.19}$$

be the lengths of the threads between them, while the function $\widehat{f}_{n_g}$ from (2.3.18) will be used to construct the subgraph of the other $g-1$ edges using the inverse Theorem 2.1.8. Observe that the definition of the $l_k^{(g)}$ and the identity in (2.3.19) shows that $\sum_{k=0}^{n_g} l_k^{(g)} = l^{(g)}$.

For any such subgraph with total lengths $l^{(j)}$ $(j = 1, \ldots, g-1)$ of its $g-1$ strings we know from (1.2.25) and (1.2.28) that

$$\phi^{(j)}(0) = \frac{R^{(j)}_{2n_j}(0)}{R^{(j)}_{2n_j-1}(0)} = l^{(j)},$$

and then (2.1.14) and (2.3.15) show that

$$\begin{aligned}\phi_{g-1}(0) &= \left(\sum_{j=1}^{g-1} \frac{1}{\phi^{(j)}(0)}\right)^{-1} = \left(\sum_{j=1}^{g-1} \frac{1}{l_j}\right)^{-1} \\ &= \gamma - l^{(g)}.\end{aligned}$$

On the other hand, (2.3.18) and (2.3.19) give

$$\begin{aligned}\widehat{f}_{n_g}(0) &= a^1_{n_g} + \sum_{k=n_g+1}^{p} a_k = \sum_{k=0}^{p} a_k - l^{(g)} = \Phi(0) - l^{(g)} \\ &= \gamma - l^{(g)}.\end{aligned}$$

Altogether, we obtain

$$\phi_{g-1}(0) = \hat{f}_{n_g}(0) > 0. \tag{2.3.20}$$

First consider the case $n_g = p$, which is the case that $\widehat{f}_{n_g}$ is constant. But then $n = n_g$ by assumption (v). We put $M = 0$ and $n_j = 0$ $(j = 1, \ldots, g-1)$.

Then ϕ_{g-1} is constant, and (2.3.20) gives $\phi_{g-1} = \widehat{f}_{n_g}$. Together with (2.3.19) it follows that (2.3.17) is identical to (2.3.12). Since Φ has $n_g = n$ poles and zeros, it therefore follows that the characteristic values of (D2) and (N2) are given by the sequences $(\mu_k)_{k-1}^n$ and $(\lambda_k)_{k=1}^n$.

Now consider the case that $n_g < p$. Then $\widehat{f}_{n_g}$ is not constant, and from the continued fraction expansion (2.3.18) of $\widehat{f}_{n_g}$ it follows that $\widehat{f}_{n_g}$ is an S_0-function; moreover, in view of Lemmas A.3.6 and A.3.5, $\widehat{f}_{n_g}$ is the quotient of two polynomials h_{1,n_g} and h_{2,n_g},

$$\widehat{f}_{n_g} = \frac{h_{1,n_g}}{h_{2,n_g}}, \quad \deg h_{1,n_g} = \begin{cases} p - n_g & \text{if } a_{n_g}^1 > 0, \\ p - n_g - 1 & \text{if } a_{n_g}^1 = 0, \end{cases} \quad \deg h_{2,n_g} = p - n_g.$$

By Lemma A.3.5, the zeros and poles of $\widehat{f}_{n_g}$, i.e., the zeros of h_{1,n_g} and h_{2,n_g}, interlace strictly.

Since Φ has p poles and zeros, there are sequences $(\widehat{\lambda}_k)_{k=1}^p$, $(\widehat{\mu}_k)_{k=1}^p$ and $(\gamma_k)_{k=1}^{n-p}$ such that $(\lambda_k)_{k=1}^n = (\widehat{\lambda}_k)_{k=1}^p \cup (\gamma_k)_{k=1}^{n-p}$, $(\mu_k)_{k=1}^n = (\widehat{\mu}_k)_{k=1}^p \cup (\gamma_k)_{k=1}^{n-p}$, $(\widehat{\lambda}_k)_{k=1}^p$ is the sequence of the poles of Φ and $(\widehat{\mu}_k)_{k=1}^p$ is the sequence of the zeros of Φ. Now set

$$\widetilde{h}_{1,n_g}(z) := h_{1,n_g}(z) \prod_{k=1}^{n-p} (z - \gamma_k), \quad \widetilde{h}_{2,n_g}(z) := h_{2,n_g}(z) \prod_{k=1}^{n-p} (z - \gamma_k).$$

It follows that

$$\deg \widetilde{h}_{1,n_g} = \begin{cases} n - n_g & \text{if } a_{n_g}^1 > 0, \\ n - n_g - 1 & \text{if } a_{n_g}^1 = 0, \end{cases} \quad \deg \widetilde{h}_{2,n_g} = n - n_g.$$

For convenience we introduce the notation $\widetilde{n}_{h_1} = \deg \widetilde{h}_{1,n_g}$ and $\widetilde{n}_{h_2} = \deg \widetilde{h}_{2,n_g}$. Denote the nondecreasing sequence of the zeros of $\widetilde{h}_{1,n_g}$ by $(\tau_k)_{k=1}^{\widetilde{n}_{h_1}}$, and denote the nondecreasing sequence of the zeros of $\widetilde{h}_{2,n_g}$ by $(\theta_k)_{k=1}^{\widetilde{n}_{h_2}}$. Then these two sequences satisfy the interlacing property

$$0 < \theta_1 \le \tau_1 \le \theta_2 \le \cdots \le \theta_{\widetilde{n}_{h_1}} \le \tau_{\widetilde{n}_{h_1}} \text{ and } \tau_{\widetilde{n}_{h_1}} \le \theta_{\widetilde{n}_{h_2}} \text{ if } a_{n_g}^1 = 0,$$

see Lemma A.4.4. We are going to show that these sequences satisfy the assumptions of Theorem 2.1.8 with $g-1$ instead of g; more precisely, θ_k will take the role of μ_k in Theorem 2.1.8 and τ_k the role of ν_k in Theorem 2.1.8, and we will have $M = 0$ if $a_{n_g}^1 > 0$ and $M > 0$ if $a_{n_g}^1 = 0$.

Statement A. If $\tau_k = \theta_k$ or $\tau_k = \theta_{k+1}$, then the maximal interlacing sequence of those zeros of $\widetilde{h}_{1,n_g}$ and $\widetilde{h}_{2,n_g}$ which equal τ_k is of the form $\tau_r = \theta_{r+1} = \cdots = \theta_s = \tau_s$ for some $r < s$.

Indeed, from the assumption it follows that $\tau_k = \gamma_j$ for some $j \in \{1, \dots, n' - p\}$ since h_{1,n_g} and h_{2,n_g} do not have common zeros. But $\gamma_j = \lambda_t$ with $\lambda_t = \mu_t$ or

$\lambda_t = \mu_{t+1}$ for some index t, and it follows from assumption (iv) that $\widehat{f}_{n_g}(\tau_k) = 0$. Hence the maximal interlacing sequence of those zeros of $\widetilde{h}_{1,n_g}$ and $\widetilde{h}_{2,n_g}$ which equal τ_k must start and end with zeros τ_r and τ_s of $\widetilde{h}_{1,n_g}$. This completes the verification of statement A.

Observing the above interlacing properties and that statement A implies $\theta_1 < \tau_1$ and $\tau_{\widetilde{n}_{h_1}} < \theta_{\widetilde{n}_{h_2}}$ if $a^1_{n_g} > 0$, it follows that the sequences satisfy the interlacing conditions in Theorem 2.1.8 (i).

Condition (ii) in Theorem 2.1.8 holds because of statement A.

To verify condition (iii) of Theorem 2.1.8 consider some τ_k and let κ be its multiplicity. If $\kappa = 1$, then $\kappa \le g-1$ since $g \ge 2$. Now let $\kappa > 1$. By statement A, the multiplicity of θ_s in the sequence $(\theta_k)_{k=1}^{\widetilde{n}_h}$ is $\kappa - 1$. With $\tau_k = \gamma_j$ it therefore follows that the multiplicity of γ_j in the sequence $(\gamma_k)_{k=1}^{n-p}$ is $\kappa - 1$. This implies that the sum of the multiplicity of γ_j in the sequence $(\lambda_k)_{k=1}^{n}$ and of the multiplicity of γ_j in the sequence $(\mu_k)_{k=1}^{n}$ is $2\kappa - 2$ or $2\kappa - 1$. By assumption (iii) it follows that $2\kappa - 2 \le 2g - 3$, which proves $\kappa \le g - 1$.

By Theorem 2.1.8 there is a star graph of $g-1$ Stieltjes strings, i.e., a sequence $(n_j)_{j=1}^{g-1}$ of positive integers, $M \ge 0$ with $M = 0$ if and only if $a^1_{n_g} > 0$, and sequences $(m_k^{(j)})_{k=1}^{n_j}$, $(l_k^{(j)})_{k=0}^{n_j}$ $(j = 1, \dots, g-1)$ of positive numbers with $\widetilde{n}_{h_1} = \sum_{j=1}^{g-1} n_j$ and $\sum_{k=0}^{n_j} l_k^{(j)} = l^{(j)}$ such that the Neumann problem (N1) for this star graph has the characteristic values $(\theta_k)_{k=1}^{\widetilde{n}_{h_1}}$ and the Dirichlet problem (D1) for this star graph has the characteristic values $(\tau_k)_{k=1}^{\widetilde{n}_{h_2}}$. Hence $\widehat{f}_{n_g}$ and ϕ_{g-1} for this star graph have the same poles and zeros, and therefore $\widehat{f}_{n_g} = \phi_{g-1}$ in view of (2.3.20).

With the notation of this proof it follows with (2.1.28) and (2.1.29) that the identities

$$\phi_{D,g-1}(z) = \prod_{j=1}^{g-1} \frac{l^{(j)}}{l_0^{(j)}} \prod_{k=1}^{\widetilde{n}_{h_1}} \left(1 - \frac{z}{\tau_k}\right),$$

$$\phi_{N,g-1}(z) = \left(\sum_{j=1}^{g} \frac{1}{l^{(j)}}\right) \prod_{j=1}^{g-1} \frac{l_j}{l_0^{(j)}} \prod_{k=1}^{\widetilde{n}_{h_2}} \left(1 - \frac{z}{\theta_k}\right).$$

hold. The functions h_{1,n_g} and h_{2,n_g} are only unique up to a constant factor. But the polynomials $\phi_{D,g-1}$ and $\widetilde{h}_{1,n_g}$ have the same zeros, counted with multiplicity, and the polynomials $\phi_{N,g-1}$ and $\widetilde{h}_{2,n_g}$ have the same zeros, counted with multiplicity. We may therefore assume that $\phi_{D,g-1} = \widetilde{h}_{1,n_g}$, which implies in view of (2.1.11), (2.3.20) and the definitions of h_{1,n_g}, h_{2,n_g}, $\widetilde{h}_{1,n_g}$ and $\widetilde{h}_{2,n_g}$ that $\phi_{N,g-1} = \widetilde{h}_{2,n_g}$.

Let $\phi_g(\cdot, c)$ $(c = 0, 1)$ be the functions defined in (2.3.7) with respect to these string data. Then (2.3.8) gives for $c = 0, 1$ that

$$\begin{aligned}\phi_g(z, c) &= R^{(g)}_{2n_g}(\cdot, c)\widetilde{h}_{2,n_g} + R^{(g)}_{2n_g-1}(\cdot, c)\widetilde{h}_{1,n_g}\\ &= \left(R^{(g)}_{2n_g}(\cdot, c)h_{2,n_g} + R^{(g)}_{2n_g-1}(\cdot, c)h_{1,n_g}\right)\prod_{k=1}^{n-p}(z - \gamma_k).\end{aligned}$$

From Theorem 2.3.5 we know that both $\phi_g(\cdot, 0)$ and $\phi_g(\cdot, 1)$ are polynomials of degree n. Therefore, for $c = 0, 1$, $R^{(g)}_{2n_g}(\cdot, c)h_{2,n_g} + R^{(g)}_{2n_g-1}(\cdot, c)h_{1,n_g}$ is a polynomial of degree p. But from (2.3.12) and (2.3.17) we know that

$$l_0^{(g)}\frac{\phi_g(\cdot, 0)}{\phi_g(\cdot, 1)} = \Phi \tag{2.3.21}$$

has p (simple) zeros, which we denoted by $\widehat{\mu}_k$ $(k = 1, \ldots, p)$ and p (simple) poles, which we denoted by $\widehat{\lambda}_k$ $(k = 1, \ldots, p)$. Therefore the sequence $(\widehat{\mu}_k)_{k=1}^p$ is the sequence of the zeros of $R^{(g)}_{2n_g}(\cdot, 0)h_{2,n_g} + R^{(g)}_{2n_g-1}(\cdot, 0)h_{1,n_g}$ and the sequence $(\widehat{\lambda}_k)_{k=1}^p$ is the sequence of the zeros of $R^{(g)}_{2n_g}(\cdot, 1)h_{2,n_g} + R^{(g)}_{2n_g-1}(\cdot, 1)h_{1,n_g}$. Thus $(\mu_k)_{k=1}^n = (\widehat{\mu}_k)_{k=1}^p \cup (\gamma_k)_{k=1}^{n-p}$ is the sequence of the zeros of $\phi_g(\cdot, 0)$ and $(\lambda_k)_{k=1}^n = (\widehat{\lambda}_k)_{k=1}^p \cup (\gamma_k)_{k=1}^{n-p}$ is the sequence of the zeros of $\phi_g(\cdot, 1)$, that is, for the constructed data of the star graph, $(\mu_k)_{k=1}^n$ is the sequence of the characteristic values of the Dirichlet problem (D2), whereas $(\lambda_k)_{k=1}^n$ is the sequence of the characteristic values of the Neumann problem (N2). □

Corollary 2.3.9. *The characteristic values of the Dirichlet problem* (D2) *and of the Neumann problem* (N2) *together with the total length l_g and the length $l_0^{(g)}$ of the main edge uniquely determine the Stieltjes string data on the main edge, i. e., the number n_g of beads, their masses $m_k^{(g)}$ and the lengths $l_k^{(g)}$ $(k = 1, \ldots, n_g)$.*

Proof. The poles and zeros of $\frac{\phi_g(\cdot,0)}{\phi_g(\cdot,1)}$ are uniquely determined by the given characteristic values, and hence this function is uniquely determined up to a constant multiple. Prescribing $l_0^{(g)}$ therefore shows that the continued fraction expansion (2.3.12) is uniquely determined. Then n_g given by (2.3.16) is unique, and all the other data of the main string are now uniquely given by the continued fraction expansion (2.3.12). □

Remark 2.3.10.

1. If all inequalities in assumption (i) in Theorem 2.3.8 are strict, then assumption (ii) is satisfied, and the assumptions (iii) and (iv) are void. Therefore the assumptions (ii)–(iv) can be omitted in this case.
2. In the notation of Lemma 2.3.7, let $\widehat{a}_k = \gamma^{-1}a_k$ $(k = 0, \ldots, p)$. That is, the numbers $\widehat{a}_k$ are the numbers a_k in case $\gamma = 1$ and are therefore uniquely

determined by $(\mu_k)_{k=1}^n$ and $(\lambda_k)_{k=1}^n$. For convenience we put

$$\widehat{l} = \left(\sum_{j=1}^{g-1} l^{(j)}\right)^{-1}, \quad \widetilde{a}_r = \sum_{k=0}^{r} \widehat{a}_k \quad (r = 0, \ldots, p).$$

Then $0 < \widetilde{a}_0 < \cdots < \widetilde{a}_p = 1$,

$$l^{(g)} - \sum_{k=0}^{r} a_k = l^{(g)} - (l^{(g)} + \widehat{l})\widetilde{a}_r = l^{(g)}(1 - \widetilde{a}_r) - \widehat{l}\,\widetilde{a}_r \quad (r = 0, \ldots, p),$$

and therefore, for $r = 0, \ldots, p-1$,

$$\sum_{k=0}^{r} a_k < l^{(g)} \Leftrightarrow l^{(g)}(1 - \widetilde{a}_r) > \widehat{l}\widetilde{a}_r$$
$$\Leftrightarrow l^{(g)} > \widehat{l}\frac{\widetilde{a}_r}{1 - \widetilde{a}_r} =: c_r$$

as well as

$$\sum_{k=0}^{r} a_k \geq l^{(g)} \Leftrightarrow l^{(g)} \leq c_r.$$

We introduce the intervals $I_0 = (0, c_0]$, $I_r = (c_{r-1}, c_r]$ $(r = 1, \ldots, p-1)$, and $I_p = (c_{p-1}, \infty)$. Then $n_g = r$ if and only if $l^{(g)} \in I_r$. Hence, for any two sequences $(\mu_k)_{k=1}^n$ and $(\lambda_k)_{k=1}^n$ satisfying the assumptions of Lemma 2.3.7 and any numbers $r \in \{0, \ldots, p\}$ and $l^{(j)} > 0$ $(j = 1, \ldots, g-1)$ one can always find a number $l^{(g)} > 0$ such that $n_g = r$.

3. In order to find a star graph realization with at least one bead on each of its g strings it is necessary and sufficient that $n_g > 0$ and, in view of Theorem 2.1.8 and the proof of Theorem 2.3.8, that $\widetilde{n}_{h_1} \geq g - 1$. By part 2 above the largest possible value of $\widetilde{n}_{h_1}$ with $n_{h_1} > 0$ is $n - 1$, which is obtained if (and only if) l_g is in the interior of I_1. Therefore $n \geq g$ is necessary and sufficient to find a realization with at least one bead on each of the g edges of the star graph.

2.4 Examples

In this section we illustrate the inverse Theorems 2.1.8 and 2.3.8 and their constructive proofs by means of a simple example.

Example 2.4.1. Construct all star graphs with root at the central vertex with 2 edges and edge lengths $l^{(1)} = 2$, $l^{(2)} = 1$ such that the characteristic values $(\mu_k)_{k=1}^2$ of the Neumann problem (N1) and the characteristic values $(\nu_k)_{k=1}^2$ of the Dirichlet problem (D1) are given by

$$\mu_1 = 1, \quad \nu_1 = \mu_2 = \nu_2 = 2. \tag{2.4.1}$$

Solution. In the notation of Theorem 2.1.8 we have $n = g = 2$. It is easy to see that the numbers μ_1, μ_2, ν_1, ν_2 satisfy the assumptions of Theorem 2.1.8, and hence there exist such star graphs.

In order to find all of them we first observe that $M = 0$ by Theorem 2.1.5. From Theorem 2.1.3 we know that the characteristic values of (D1) are the zeros of $\phi_{D,2}$, and (2.1.9) and (1.2.25) therefore shows $n_1 = n_2 = 1$ and

$$R_2^{(j)}(z) = \frac{l^{(j)}}{l_0^{(j)}}\left(1 - \frac{z}{2}\right) \quad (j = 1, 2)$$

for any solution of the problem. In view of (1.2.28) and Lemma 1.2.13,

$$R_1^{(j)}(z) = \frac{1}{l_0^{(j)}}\left(1 - \frac{z}{\mu_1^{(j)}}\right) \quad (j = 1, 2)$$

with $0 < \mu_1^{(j)} < 2$ is necessary. Then

$$\phi^{(j)}(z) = \frac{R_2^{(j)}(z)}{R_1^{(j)}(z)} = (3 - j)\frac{1 - \frac{z}{2}}{1 - \frac{z}{\mu_1^{(j)}}} \quad (j = 1, 2)$$

due to $l^{(1)} = 2$ and $l^{(2)} = 1$. In view of (2.1.11) and Theorems 2.1.2, 2.1.3 and 2.1.4, the function

$$\frac{1}{\phi^{(1)}} + \frac{1}{\phi^{(2)}} = \frac{1}{\phi_2}$$

has a simple zero at 1 and a simple pole at 2. But

$$\begin{aligned}\frac{1}{\phi^{(1)}(z)} + \frac{1}{\phi^{(2)}(z)} &= \frac{1}{1 - \frac{z}{2}}\left[\frac{1}{2}\left(1 - \frac{z}{\mu_1^{(1)}}\right) + \left(1 - \frac{z}{\mu_1^{(2)}}\right)\right] \\ &= \frac{1}{1 - \frac{z}{2}}\frac{1}{2\mu_1^{(1)}\mu_1^{(2)}}\left[3\mu_1^{(1)}\mu_1^{(2)} - \left(2\mu_1^{(1)} + \mu_1^{(2)}\right)z\right], \end{aligned} \tag{2.4.2}$$

which shows that

$$3\mu_1^{(1)}\mu_1^{(2)} = 2\mu_1^{(1)} + \mu_1^{(2)} \tag{2.4.3}$$

is necessary. Putting $a = \frac{2}{\mu_1^{(1)}} - 1$, the necessary conditions can be written as

$$\mu_1^{(1)} = \frac{2}{a + 1}, \quad \mu_1^{(2)} = \frac{4}{5 - a}, \quad a \in (0, 3).$$

Taking Theorem 1.3.1 into account, it follows that under these conditions there is a unique pair of strings for each $a \in (0, 3)$. The string data are found from the continued fraction expansions (1.3.2). For $j = 1, 2$ we have

$$\phi^{(j)}(z) = \frac{(3-j)\mu_1^{(j)}}{2} \frac{z-2}{z-\mu_1^{(j)}}$$
$$= \frac{(3-j)\mu_1^{(j)}}{2} + \frac{1}{-\frac{2z}{(3-j)\mu_1^{(j)}\left(2-\mu_1^{(j)}\right)} + \frac{2}{(3-j)\left(2-\mu_1^{(j)}\right)}},$$

which gives

$$l_0^{(j)} = \frac{(3-j)\left(2-\mu_1^{(j)}\right)}{2}, \quad l_1^{(j)} = \frac{(3-j)\mu_1^{(j)}}{2}, \quad m_1^{(j)} = \frac{2}{(3-j)\mu_1^{(j)}\left(2-\mu_1^{(j)}\right)}$$

for $j = 1, 2$. Hence the locations and masses of the beads for all star graphs with root at the central vertex and 2 edges having the Neumann and Dirichlet characteristic values (2.4.1) are given by $M = 0$ and

$$\begin{aligned} l_0^{(1)} &= \frac{2a}{a+1}, \quad l_1^{(1)} = \frac{2}{a+1}, \quad m_1^{(1)} = \frac{(a+1)^2}{4a}, \\ l_0^{(2)} &= \frac{3-a}{5-a}, \quad l_1^{(2)} = \frac{2}{5-a}, \quad m_1^{(2)} = \frac{(5-a)^2}{4(3-a)}, \end{aligned} \tag{2.4.4}$$

where $a \in (0, 3)$ is a free parameter.

Example 2.4.2. Find all star graphs with root at a pendant vertex with 3 edges and edge lengths $l^{(1)} = 2$, $l^{(2)} = 1$, $l^{(3)} = 2$, $l_0^{(3)} = 1$ so that the characteristic values $(\lambda_k)_{k=1}^3$ of the Neumann problem (N2) and the characteristic values $(\mu_k)_{k=1}^3$ of the Dirichlet problem are given by

$$\lambda_1 = 0.5, \quad \lambda_2 = 1.5, \quad \lambda_3 = 2, \qquad \mu_1 = 1, \quad \mu_2 = \mu_3 = 2? \tag{2.4.5}$$

Solution. First we check if the assumptions of Theorem 2.3.8 are satisfied. To this end we note that $n = g = 3$. Since

$$0 < \lambda_1 < \mu_1 < \lambda_2 < \mu_2 = \lambda_3 = \mu_3,$$

(i) is satisfied. The multiplicity of all λ_k is 1, whereas the multiplicity of the μ_k is 1 or 2. Therefore (ii) is satisfied. Also (iii) holds since the maximal such multiplicity is $3 = 2g - 3$. In order to verify assumptions (iv) and (v) we first have to find n_3 and $\widehat{f}_{n_3}$. Note that γ and Φ defined by (2.3.15) with the representation (2.3.17) are

$$\gamma = l_3 + \left(\frac{1}{l_1} + \frac{1}{l_2}\right)^{-1} = \frac{8}{3},$$

$$\Phi(z) = \frac{8}{3}\frac{(1-z)\left(1-\frac{z}{2}\right)}{(1-2z)\left(1-\frac{2z}{3}\right)} = \frac{z^2-3z+2}{z^2-2z+\frac{3}{4}} = 1 + \cfrac{1}{-z+\cfrac{1}{\frac{4}{3}+\cfrac{1}{-3z+\cfrac{1}{\frac{1}{3}}}}}.$$

Hence $a_0 = 1$, $a_1 = \frac{4}{3}$ and so $a_0 = 1 < l^{(3)} = 2 < 1 + \frac{4}{3} = a_0 + a_1$. Thus, by (2.3.16), it follows that $n_3 = 1$. Moreover, we have $a_{n_3}^1 = a_0 + a_1 - l^{(3)} = \frac{1}{3}$, $a_{n_3} - a_{n_3}^1 = \frac{4}{3} - \frac{1}{3} = 1$ and hence, by (2.3.17), (2.3.18),

$$\Phi(z) = 1 + \cfrac{1}{-z+\cfrac{1}{1+\widehat{f}_1(z)}}, \quad \widehat{f}_1(z) = \frac{1}{3} + \cfrac{1}{-3z+\frac{1}{\frac{1}{3}}} = \frac{1}{3}\frac{(2-z)}{(1-z)}. \tag{2.4.6}$$

Since $\widehat{f}_1(\mu_3) = \widehat{f}_1(2) = 0$, assumption (iv) holds, and (v) is trivially true since $\widehat{f}_1$ is not constant. Therefore a graph as required *does* exist by Theorem 2.3.8.

In order to construct such graphs, we observe that (2.3.21), (2.3.12) and (2.4.6) give that $l_0^{(3)} = l_1^{(3)} = 1$, $m_1^{(3)} = 1$ and $\phi_2 = \widehat{f}_1$. Observing that $a_1^1 \neq 0$, it also follows that $M = 0$. In view of (2.1.14) we thus have

$$\frac{1}{\phi^{(1)}(z)} + \frac{1}{\phi^{(2)}(z)} = \frac{1}{\widehat{f}_1(z)} = 3\,\frac{z-1}{z-2}.$$

But this is exactly (2.4.2) satisfying (2.4.3), and hence the star graphs with the data

$$M = 0, \quad l_0^{(3)} = l_1^{(3)} = 1, \quad m_1^{(3)} = 1, \quad (2.4.4) \tag{2.4.7}$$

are exactly those graphs which have the desired characteristic values.

2.5 Notes

In this chapter we have dealt with star graphs of Stieltjes strings, which are special cases of trees and graphs of Stieltjes strings, which will be considered in Chapters 4 and 5, respectively. However, we have decided to first devote a whole chapter to this special type of trees for two reasons. Firstly, no graph theory is needed for the results of Chapter 2. Secondly, most of the results on inverse problems in this chapter are not special cases of results in Chapters 4 or 5.

Inverse problem on star graphs were first considered for the Sturm-Liouville equations in [107] for $g = 3$ and in [108] for arbitrary $g \geq 3$. The inverse spectral problem for Stieltjes string equations on a star graph without a bead at the central vertex and with strict interlacing of the given spectra was solved in [18]. In all these papers the central vertex was considered as the root, i. e., the spectra of boundary value problems with generalized Dirichlet and Neumann conditions at the central vertex were used as the given data to solve the inverse problem of reconstructing the Stieltjes string data on the edges. In this monograph we have extended those results by allowing pendant vertices to be clamped or freely moving.

The full solution of the inverse spectral problem generated by the Stieltjes string equations on a star graph rooted at the central vertex was given in [114], where also an inverse spectral problem generated by the Stieltjes string equations on a star graph rooted at a pendant vertex was considered. In these papers the numbers of beads on the edges were not prescribed in the given data. The inverse spectral problem generated by the Stieltjes string equations on a star graph rooted at the central vertex with prescribed numbers of beads on the edges was solved in [116].

The inverse problem where not just the sequence of all Neumann eigenvalues, but the sequences of the Dirichlet eigenvalues for all edges are given was considered in [87] for trees with $M = 0$ at the root. In the case when the tree is a star graph, Theorem 2.1.8 is an immediate consequence of [87, Theorem 3.3].

In [46] a three spectra inverse problem was considered for the following three spectra: the first spectrum is the spectrum of a problem on a star graph of three edges with the Dirichlet boundary conditions at the pendant vertices and the generalized Neumann conditions at the centre; the second is the spectrum of the Dirichlet-Dirichlet problem on the union of two of the graphs edges; the third is the spectrum of the Dirichlet-Neumann problem on the third edge.

The presentation in Subsection 2.1.5 generalizes results for star-patterned matrices, in particular [73, Theorem 9] (see also [47], [101]).

Many properties of the characteristic functions and characteristic values of string problems on star graphs are determined by the star graph itself and shared by various string problems, with finite or infinite spectrum, see [117], [49].

Chapter 3

Damped vibrations of Stieltjes strings and star graphs

3.1 The direct spectral problem

As in Subsection 1.2.1 we consider a Stieltjes string bearing n beads in its interior. But in this section these beads and the ends of the string will be subject to possible viscous friction.

From Subsection 1.2.1 we recall the notation $m_k > 0$ $(k = 1, \dots, n)$ for the masses of the beads, $l_k > 0$ $(k = 0, \dots, n)$ for the lengths of the threads and $l = \sum_{k=0}^{n} l_k$ for the total length of the string (see Figure 1.2). We denote by $\alpha_k \geq 0$ the coefficient of damping (viscous friction) of the k-th bead $(k = 1, \dots, n)$.

As in Chapter 1, the string is assumed to be stretched by a stretching force equal to 1. Then small transverse vibrations of such a Stieltjes string are described by the three term recurrence relations

$$\frac{v_k(t) - v_{k+1}(t)}{l_k} + \frac{v_k(t) - v_{k-1}(t)}{l_{k-1}} + m_k v_k''(t) + \alpha_k v_k'(t) = 0 \quad (k = 1, \dots, n), \tag{3.1.1}$$

where $v_k(t)$ is the transverse displacement of the k-th bead for $k = 1, \dots, n$ at time t and $v_0(t)$ and $v_{n+1}(t)$ are the transverse displacements of the left and right endpoints of the string, respectively.

The Dirichlet-Dirichlet problem is generated by equations (3.1.1) and boundary conditions

$$v_0(t) = 0, \ v_{n+1}(t) = 0, \tag{3.1.2}$$

and the Dirichlet-Neumann problem is generated by equations (3.1.1) and boundary conditions

$$v_0(t) = 0, \ v_{n+1}(t) = v_n(t). \tag{3.1.3}$$

M. Möller, V. Pivovarchik, *Direct and Inverse Finite-Dimensional Spectral Problems on Graphs*, Operator Theory: Advances and Applications 283, https://doi.org/10.1007/978-3-030-60484-4_3

Substituting $v_k(t) = u_k e^{i\lambda t}$ we obtain

$$\frac{u_k - u_{k+1}}{l_k} + \frac{u_k - u_{k-1}}{l_{k-1}} - \lambda^2 m_k u_k + i\lambda\alpha_k u_k = 0 \quad (k = 1, \dots, n) \tag{3.1.4}$$

with Dirichlet-Dirichlet boundary conditions

$$u_0 = 0, \ u_{n+1} = 0 \tag{3.1.5}$$

or Dirichlet-Neumann boundary conditions

$$u_0 = 0, \ u_{n+1} = u_n. \tag{3.1.6}$$

Similar to (1.2.5) we consider the initial condition

$$u_0 = cu_1 \quad (c \in \mathbb{R}) \tag{3.1.7}$$

and write

$$u_k = S_{2k-2}(\lambda, c)u_1 \quad (k = 0, \dots, n+1). \tag{3.1.8}$$

Defining $S_{2k-1}(\lambda, c)$ implicitly by

$$S_{2k}(\lambda, c) = l_k S_{2k-1}(\lambda, c) + S_{2k-2}(\lambda, c) \quad (k = 0, \dots, n), \tag{3.1.9}$$

and observing that then (3.1.4) can be written as

$$S_{2k-1}(\lambda, c) = S_{2k-3}(\lambda, c) + (-m_k\lambda^2 + i\alpha_k\lambda)S_{2k-2}(\lambda, c) \quad (k = 1, \dots, n), \tag{3.1.10}$$

we see that (3.1.10), (3.1.9) is a three term recurrence relation with the natural initial condition $S_0(\lambda, c) = 1$ and the initial condition (3.1.7), which gives $S_{-2}(\lambda, c) = c$. Because S_{-1} can be found from S_0 and S_{-2}, we have the three-term recurrence relation (3.1.10), (3.1.9) for $S_j(\lambda, c)$ $(j = 1, \dots, 2n)$ with the initial terms

$$S_{-1}(\lambda, c) = \frac{1-c}{l_0}, \quad S_0(\lambda, c) = 1. \tag{3.1.11}$$

It is immediately clear from the recurrence relation and the initial conditions that $S_{2k}(\cdot, c)$ and $S_{2k-1}(\cdot, c)$ are polynomials of degree $2k$ for $k = 0, \dots, n$, with the exception of $S_{-1}(\cdot, 1) = 0$.

We complement the condition (3.1.7) for $c \geq 0$ with

$$u_{n+1} = du_n \tag{3.1.12}$$

for $d \geq 0$, we recall the matrix $A_{c,d}$ from (1.2.7) and we put

$$M = \operatorname{diag}(m_1, \dots, m_n), \quad K = \operatorname{diag}(\alpha_1, \dots, \alpha_n). \tag{3.1.13}$$

The quadratic matrix pencil T_c^d is defined by

$$T_c^d(\lambda) = \lambda^2 M - i\lambda K - A_c^d. \tag{3.1.14}$$

The following result is easily verified.

Proposition 3.1.1. *Let $c, d \geq 0$, $\lambda \in \mathbb{C}$, and $Y = (u_1, \dots, u_n)^\mathsf{T} \in \mathbb{C}^n \setminus \{0\}$. Define u_0 and u_{n+1} by* (3.1.7) *and* (3.1.12)*, respectively. Then* (3.1.4) *is satisfied if and only if $T_c^d(\lambda)Y = 0$.*

A complex number λ is called an eigenvalue of the problem (3.1.4), (3.1.7), (3.1.12), if this system of equations has a nontrivial solution, which happens if and only if there is a solution with $u_1 \neq 0$. Recalling the definitions from Section B.1, this means that the eigenvalues of the problem (3.1.4), (3.1.7), (3.1.12) are the eigenvalues of the matrix pencil T_c^d. Hence we make the following

Definition 3.1.2. The zeros of the function $\det T_c^d$, counted with multiplicity, are called the eigenvalues of the problem (3.1.4), (3.1.7), (3.1.12).

It is evident from (3.1.14) that the problem (3.1.4), (3.1.7), (3.1.12) has $2n$ eigenvalues, counted with multiplicity.

Proposition 3.1.3. *Let $c, d \geq 0$. Then*

$$\det T_c^d(\lambda) = (-1)^n \left(\prod_{j=1}^{n} \frac{1}{l_j} \right) [S_{2n}(\lambda, c) - dS_{2n-2}(\lambda, c)] \tag{3.1.15}$$

$$= (-1)^n \left(\prod_{j=1}^{n} \frac{1}{l_j} \right) [(1-d)S_{2n}(\lambda, c) + dl_k S_{2n-1}(\lambda, c)] \quad (\lambda \in \mathbb{C}).$$

Proof. We will prove the first identity by induction. The second identity is then a direct consequence of (3.1.9).

For integers $1 \leq k \leq n$ let ${}^kT_c^d(\lambda)$ be the upper left $k \times k$ submatrix of $T_c^d(\lambda)$. Also let $d' = d$ if $k = n$ and $d' = 0$ if $k < n$. We are going to prove by induction on k that

$$\det {}^kT_c^d(\lambda) = (-1)^k \left(\prod_{j=1}^{k} \frac{1}{l_j} \right) [S_{2k}(\lambda, c) - d' S_{2k-2}(\lambda, c)] \quad (\lambda \in \mathbb{C}). \tag{3.1.16}$$

For $k = 1$ we calculate

$$\begin{aligned} \det {}^1T_c^d(\lambda) &= m_1\lambda^2 - i\alpha_1\lambda - \frac{1-c}{l_0} - \frac{1-d'}{l_1} \\ &= \left(m_1\lambda^2 - i\alpha_1\lambda - \frac{1-d'}{l_1} \right) S_0(\lambda, c) - S_{-1}(\lambda, c) \\ &= -S_1(\lambda, c) - \frac{1}{l_1} S_0(\lambda, c) + \frac{d'}{l_1} S_0(\lambda, c) \\ &= -\frac{1}{l_1} [S_2(\lambda, c) - d' S_0(\lambda, c)], \end{aligned}$$

which proves (3.1.16) for $k = 1$.

Now let $k \geq 2$ and assume (3.1.16) holds for all smaller indices. When $k \geq 3$, then Proposition B.3.7 and the induction hypothesis give

$$\begin{aligned}
\det{}^k T_c^d(\lambda) &= \left(m_k\lambda^2 - i\alpha_k\lambda - \frac{1}{l_{k-1}} - \frac{1-d'}{l_k}\right)\det{}^{k-1}T_c^d(\lambda) - \frac{1}{l_{k-1}^2}\det{}^{k-2}T_c^d(\lambda)\\
&= \left(m_k\lambda^2 - i\alpha_k\lambda - \frac{1}{l_{k-1}} - \frac{1-d'}{l_k}\right)(-1)^{k-1}\left(\prod_{j=1}^{k-1}\frac{1}{l_j}\right)S_{2k-2}(\lambda,c)\\
&\quad - \frac{1}{l_{k-1}^2}(-1)^k\left(\prod_{j=1}^{k-2}\frac{1}{l_j}\right)S_{2k-4}(\lambda,c).
\end{aligned}$$

In view of (3.1.16) for $k = 1$ and $S_0(\lambda, c) = 1$, this is also true when $k = 2$. Using (3.1.9) and (3.1.10), we find

$$\begin{aligned}
(-1)^k\left(\prod_{j=1}^{k} l_j\right)\det{}^k T_c^d &= l_k(-m_k\lambda^2 + i\alpha_k\lambda)S_{2k-2}(\lambda,c) + \frac{l_k}{l_{k-1}}S_{2k-2}(\lambda,c)\\
&\quad + (1-d')S_{2k-2}(\lambda,c) - \frac{l_k}{l_{k-1}}S_{2k-4}(\lambda,c)\\
&= l_k\left(S_{2k-1}(\lambda,c) - S_{2k-3}(\lambda,c)\right) + l_k S_{2k-3}(\lambda,c)\\
&\quad + S_{2k-2}(\lambda,c) - d'S_{2k-2}(\lambda,c)\\
&= S_{2k}(\lambda,c) - d'S_{2k-2}(\lambda,c).
\end{aligned}$$

Hence (3.1.16) has been shown, and (3.1.15) is the case $k = n$. □

The eigenvalues of the Dirichlet-Dirichlet problem (3.1.4), (3.1.5), i.e., the eigenvalues of the matrix pencil T_0^0, counted with algebraic multiplicity, will be denoted by $(\nu_k)_{k=-n,k\neq0}^n$. The eigenvalues of the Dirichlet-Neumann problem (3.1.4), (3.1.6), i.e., the eigenvalues of the matrix pencil T_0^1, counted with algebraic multiplicity, will be denoted by $(\mu_k)_{k=-n,k\neq0}^n$.

Remark 3.1.4. Since the coefficients of $S_j(\cdot, c)$ $(j = 1, \ldots, 2n)$ depend continuously on the distribution of the beads and on c, Corollary A.6.7 shows that the zeros of $\det T_c^d$ depend continuously on the distribution of the beads and on c and d.

Theorem 3.1.5. 1. *The sequences* $(\nu_k)_{k=-n,k\neq0}^n$ *and* $(\mu_k)_{k=-n,k\neq0}^n$ *are the zeros of* $S_{2n}(\cdot, 0)$ *and* $S_{2n-1}(\cdot, 0)$*, counted with multiplicity, respectively, and*

$$\{\nu_k\}_{k=-n,k\neq0}^n \cap \{\mu_k\}_{k=-n,k\neq0}^n = \emptyset. \tag{3.1.17}$$

2. *The sequences* $(\nu_k)_{k=-n,k\neq0}^n$ *and* $(\mu_k)_{k=-n,k\neq0}^n$ *can be indexed in such a way that for all* $k \in \{\pm1, \ldots, \pm n\}$, $\nu_{-k} = -\overline{\nu_k}$, $\mu_{-k} = -\overline{\mu_k}$ *if* μ_k *or* ν_k *is not pure imaginary, respectively.*

3. *For all* $k \in \{\pm 1, \dots, \pm n\}$, $\nu_k \neq 0$, $\mu_k \neq 0$, $\operatorname{Im} \mu_k \geq 0$ *and* $\operatorname{Im} \nu_k \geq 0$. *If* $\alpha_1 > 0$, *or* $\alpha_n > 0$, *or* $\alpha_j > 0$ *and* $\alpha_{j+1} > 0$ *for some* $j \in \{2, \dots, n-2\}$, *then* $\operatorname{Im} \mu_k > 0$ *and* $\operatorname{Im} \nu_k > 0$ *for all* $k \in \{\pm 1, \dots, \pm n\}$.

Proof. 1. The first statement is clear from Proposition 3.1.3. Suppose that $S_{2n}(\cdot, 0)$ and $S_{2n-1}(\cdot, 0)$ have a common zero, say λ_0. Substitution into (3.1.9) with $k = n$ gives $S_{2n-2}(\lambda_0, 0) = 0$. Then using (3.1.10) we arrive at $S_{2n-3}(\lambda_0, 0) = 0$ and so on. Finally, we obtain $S_0(\lambda_0, 0) = 0$, a contradiction.

Statement 2 follows from Lemma B.3.2.

3. By Proposition 1.2.3, the matrices A_0^0 and A_0^1 are positive definite. Hence all eigenvalues of T_0^0 and T_0^1 lie in the closed upper half-plane by Lemma B.3.5, part 1. They are also nonzero since A_0^0 and A_0^1 are invertible.

Now assume additionally that $\alpha_1 > 0$, or $\alpha_n > 0$, or $\alpha_j > 0$ and $\alpha_{j+1} > 0$ for some $j \in \{2, \dots, n-2\}$. Let $Y = (u_1, \dots, u_n) \in \mathbb{C}^n$ with $KY = 0$. Then $u_1 = 0$ or $u_n = 0$ or $u_k = 0$ and $u_{k+1} = 0$. For each (real) λ and $d = 0$ or $d = 1$, the matrix $\lambda^2 M - A_0^d$ is a tridiagonal matrix whose elements in the diagonals below and above the main diagonal are nonzero. Then it is easy to see that $(\lambda^2 M - A_0^d)Y = 0$ would lead to $Y = 0$. Therefore $Y \neq 0$ and $KY = 0$ give $(\lambda^2 M - A_0^d)Y \neq 0$, and an application of Lemma B.3.5, part 3, completes the proof. □

Corollary 3.1.6. *If* $\alpha_1 > 0$, *or* $\alpha_n > 0$, *or* $\alpha_j > 0$ *and* $\alpha_{j+1} > 0$ *for some* $j \in \{2, \dots, n-2\}$, *then the polynomials* $S_{2n}(\cdot, 0)$ *and* $S_{2n-1}(\cdot, 0)$ *are symmetric Hermite-Biehler polynomials.*

Proof. Since T_0^0 and T_0^1 do not have real eigenvalues by Theorem 3.1.5, part 3, it follows from Proposition B.3.6 that $\det T_0^0$ and $\det T_0^1$ belong to the SHB class. An application of Proposition 3.1.3 completes the proof. □

Proposition 3.1.7. *The eigenvalues of the Dirichlet-Dirichlet and Dirichlet-Neumann problems satisfy the following identities:*

$$l \prod_{k=-n,\, k\neq 0}^{n} \mu_k = l_n \prod_{k=-n,\, k\neq 0}^{n} \nu_k, \tag{3.1.18}$$

$$\sum_{k=-n,\, k\neq 0}^{n} \mu_k = \sum_{k=-n,\, k\neq 0}^{n} \nu_k, \tag{3.1.19}$$

and both sides of (3.1.19) *are pure imaginary.*

Proof. It is evident that, with the notation from Subsection 1.2.1, $S_j(0,0) = R_j(0,0)$, so that $S_{2n}(0,0) = \frac{l}{l_0}$ in view of (1.2.17) and $S_{2n-1}(0,0) = \frac{1}{l_0}$ in view of (1.2.15). Hence

$$S_{2n}(\lambda, 0) = \frac{l}{l_0} \prod_{k=-n,\, k\neq 0}^{n} \left(1 - \frac{\lambda}{\nu_k}\right), \quad S_{2n-1}(\lambda, 0) = \frac{1}{l_0} \prod_{k=-n,\, k\neq 0}^{n} \left(1 - \frac{\lambda}{\mu_k}\right). \tag{3.1.20}$$

Observing that $S_{2n}(\cdot,0) - l_n S_{2n-1}(\cdot,0) = S_{2n-2}(\cdot,0)$ is a polynomial of degree $2n-2$, a comparison of the coefficients for λ^{2n} and λ^{2n-1} in the representations of $S_{2n}(\cdot,0)$ and $S_{2n-1}(\cdot,0)$ in (3.1.20) leads to

$$\frac{l}{l_0}\prod_{k=-n,\,k\neq 0}^{n}\frac{1}{\nu_k} = \frac{l_n}{l_0}\prod_{k=-n,\,k\neq 0}^{n}\frac{1}{\mu_k}$$

and

$$\frac{l}{l_0}\sum_{j=-n,\,j\neq 0}^{n}\prod_{k=-n,\,k\neq 0,j}^{n}\frac{1}{\nu_k} = \frac{l_n}{l_0}\sum_{j=-n,\,j\neq 0}^{n}\prod_{k=-n,\,k\neq 0,j}^{n}\frac{1}{\mu_k}.$$

Rearranging the first equation gives (3.1.18). Substituting the first equation into the second equation we arrive at (3.1.19). It follows from Theorem 3.1.5, part 1, that $\nu_{-k}+\nu_k$ and $\mu_{-k}+\mu_k$ are pure imaginary for $k=1,\dots,n$, and hence both sides of (3.1.19) are pure imaginary. □

Theorem 3.1.8. *Let $(\nu_k)_{k=-n,\,k\neq 0}^{n}$ and $(\mu_k)_{k=-n,\,k\neq 0}^{n}$ be two sequences of complex numbers and let l be a positive number. For $(\nu_k)_{k=-n,\,k\neq 0}^{n}$ and $(\mu_k)_{k=-n,\,k\neq 0}^{n}$ to be the eigenvalues of problems (3.1.4), (3.1.5) and (3.1.4), (3.1.6), respectively, with $m_k>0$, $\alpha_k\geq 0$ $(k=1,\dots,n)$, $l_k>0$ $(k=0,\dots,n)$ and $\sum_{k=0}^{n} l_k = l$ it is necessary and sufficient that the product $l\prod\limits_{k=-n,k\neq 0}^{n}\left(1-\frac{\lambda}{\nu_k}\right)\left(1-\frac{\lambda}{\mu_k}\right)^{-1}$ can be represented as a continued fraction of the form:*

$$\begin{aligned} & l\prod_{k=-n,k\neq 0}^{n}\left(1-\frac{\lambda}{\nu_k}\right)\left(1-\frac{\lambda}{\mu_k}\right)^{-1} \\ &= a_n + \cfrac{1}{b_n\lambda^2 + ic_n\lambda + \cfrac{1}{a_{n-1} + \cfrac{1}{b_{n-1}\lambda^2 + ic_{n-1}\lambda + \cdots + \cfrac{1}{a_1 + \cfrac{1}{b_1\lambda^2 + ic_1\lambda + \cfrac{1}{a_0}}}}}} \end{aligned} \tag{3.1.21}$$

with $a_k>0$ $(k=0,\dots,n)$, $b_k<0$, $c_k\geq 0$ $(k=1,\dots,n)$.

Proof. Let $(\nu_k)_{k=-n,\,k\neq 0}^{n}$ and $(\mu_k)_{k=-n,\,k\neq 0}^{n}$ be the spectra of problems (3.1.4), (3.1.5) and (3.1.4), (3.1.6), respectively. Then

$$l\prod_{k=-n,k\neq 0}^{n}\left(1-\frac{\lambda}{\nu_k}\right)\left(1-\frac{\lambda}{\mu_k}\right)^{-1} = \frac{S_{2n}(\lambda,0)}{S_{2n-1}(\lambda,0)},$$

see (3.1.20). Using (3.1.10) and (3.1.9) we obtain (3.1.21), if we formally replace $m_k z$ in (1.2.19) with $m_k\lambda^2 - i\alpha_k\lambda$ $(k = 1, \dots, n)$. In this representation, $a_k = l_k$ $(k = 0, \dots, n)$, $b_k = -m_k$, $c_k = \alpha_k$ $(k = 1, \dots, n)$.

Conversely, if (3.1.21) holds, then the two spectral problems with $l_k = a_k$ $(k = 0, \dots, n)$, $m_k = -b_k$, $\alpha_k = c_k$ $(k = 1, \dots, n)$ satisfy $\sum_{k=0}^{n} l_k = l$, which is (3.1.21) at $\lambda = 0$, and their spectra are the zeros and poles of the left-hand side of (3.1.21). □

Corollary 3.1.9. *The eigenvalues μ_k $(k = \pm 1, \dots, \pm n)$ and ν_k $(k = \pm 1, \dots, \pm n)$ together with the given total length l uniquely determine m_k, α_k $(k = 1, \dots, n)$ and l_k $(k = 0, \dots, n)$.*

Proof. The left-hand side of (3.1.21) is uniquely determined by the given data. It is easy to see that the right-hand-side of (3.1.21) is of the form $a_n + r_n(\lambda)$, where $r_n(\lambda) \to 0$ as $n \to \infty$. Hence a_n is uniquely determined. Subtracting a_n from the right-hand side of (3.1.21) and then taking the inverse, one arrives at an expression of the form $b_n\lambda^2 + ic_n\lambda + \widetilde{r}_n(\lambda)$, where $\widetilde{r}_n(\lambda) \to \frac{1}{a_{n-1}}$ as $\lambda \to \infty$. This shows that b_n and c_n are uniquely determined. Continuing in this way, one sees that all coefficients in the continued fractions expansion are uniquely determined. But these are the numbers m_k, α_k and l_k, which proves the stated uniqueness. □

3.2 Dirichlet-Dirichlet and Dirichlet-Neumann inverse problems for a string with point-wise damping

In this section we will find sufficient conditions for a sequence of $2n$ complex numbers to be the sequence of eigenvalues of a string with n beads where exactly the n-th bead is subject to viscous friction.

Theorem 3.2.1. *Let $n \in \mathbb{N}$, $l > 0$ and $l_n \in (0, l)$ be given together with the sequence $(\nu_k)_{k=-n,\, k\neq 0}^{n}$ of complex numbers which satisfy the conditions:*

(i) *$\operatorname{Im} \nu_k > 0$ for $k = \pm 1, \dots, \pm n$;*

(ii) *$\nu_{-k} = -\overline{\nu_k}$ for not pure imaginary ν_{-k}.*

Let DP be either the Dirichlet-Dirichlet problem (3.1.4), (3.1.5) or the Dirichlet-Neumann problem (3.1.4), (3.1.6).

Then there exists a unique Stieltjes string, i.e., a unique sequence of positive numbers $(l_k)_{k=0}^{n-1}$ with $\sum\limits_{k=0}^{n-1} l_k = l - l_n$, a unique sequence of positive numbers $(m_k)_{k=1}^{n}$ and a unique positive number α which generate DP with $\alpha_k = 0$ for $k \leq n-1$ and $\alpha_n = \alpha$ such that its eigenvalues coincide with the sequence $(\nu_k)_{k=-n,\, k\neq 0}^{n}$.

It is convenient to have the initial condition at the left endpoint generalized to $u_0 = cu_1$ with $c \in [0,1)$. Therefore, consider

$$u_0 = cu_1, \; u_{n+1} = 0 \quad (c \in [0,1)) \tag{3.2.1}$$

and

$$u_0 = cu_1, \; u_{n+1} = u_n \quad (c \in [0,1)). \tag{3.2.2}$$

Then Theorem 3.2.1 is the special case $c = 0$ of the following theorem.

Theorem 3.2.2. *Let $n \in \mathbb{N}$, $\widetilde{l} > 0$, $l_n \in (0, \widetilde{l})$ and $c \in [0,1)$ be given together with the sequence $(\nu_k)_{k=-n,\,k\neq 0}^n$ of complex numbers which satisfy the conditions:*

(i) $\operatorname{Im} \nu_k > 0$ *for* $k = \pm 1, \ldots, \pm n$;

(ii) $\nu_{-k} = -\overline{\nu_k}$ *for not pure imaginary* ν_{-k}.

Let BP *be either the problem (3.1.4), (3.2.1) or the problem (3.1.4), (3.2.2).*

Then there exists a unique Stieltjes string, i.e., a unique sequence of positive numbers $(l_k)_{k=0}^{n-1}$ with $\dfrac{l_0}{1-c} + \sum\limits_{k=1}^{n} l_k = \widetilde{l}$, a unique sequence of positive numbers $(m_k)_{k=1}^n$ and a unique positive number α which generate BP *with $\alpha_k = 0$ for $k \leq n-1$ and $\alpha_n = \alpha$ such that its eigenvalues coincide with the sequence $(\nu_k)_{k=-n,\,k\neq 0}^n$. Amongst these numbers, only l_0 depends on c, and $\frac{l_0}{1-c}$ is independent of c.*

Proof. In view of assumptions (i) and (ii), Proposition A.5.4 shows that the polynomial Φ defined by

$$\Phi(\lambda) = \prod_{k=-n,\; k\neq 0}^{n} \left(1 - \frac{\lambda}{\nu_k}\right) \tag{3.2.3}$$

is of SHB class. Hence there are real polynomials $\widetilde{P}$ and $\widetilde{Q}$ such that

$$\Phi(\lambda) = \widetilde{P}(\lambda^2) + i\lambda\widetilde{Q}(\lambda^2), \tag{3.2.4}$$

see (A.5.1). Proposition A.5.4 gives

$$\widetilde{P}(0) = \Phi(0) = 1 \quad \text{and} \quad \widetilde{Q}(0) = \sum_{k=-n,k\neq 0}^{n} \frac{i}{\nu_k} > 0. \tag{3.2.5}$$

For convenience we introduce a parameter τ with $\tau = l_n^{-1}$ for the Dirichlet condition at the right endpoint and $\tau = 0$ for the Neumann condition at the right endpoint. Setting

$$\alpha = \widetilde{Q}(0)\left(\frac{1}{\widetilde{l} - l_n} + \tau\right) \tag{3.2.6}$$

it follows from (3.2.5) that

$$\alpha = \left(\frac{1}{\widetilde{l} - l_n} + \tau\right) \sum_{k=-n,k\neq 0}^{n} \frac{i}{\nu_k} > 0. \tag{3.2.7}$$

From

$$\Phi(\lambda) \underset{|\lambda|\to\infty}{=} \lambda^{2n} \prod_{j=-n,j\neq 0}^{n} \frac{1}{\nu_j} - \lambda^{2n-1} \sum_{k=-n,k\neq 0}^{n} \prod_{j=-n,j\neq 0,k}^{n} \frac{1}{\nu_j} + O(\lambda^{2n-2})$$

it follows that

$$\widetilde{P}(z) \underset{|z|\to\infty}{=} z^n \prod_{j=-n,j\neq 0}^{n} \frac{1}{\nu_j} + O(z^{n-1}),$$

$$\widetilde{Q}(z) \underset{|z|\to\infty}{=} iz^{n-1} \sum_{k=-n,k\neq 0}^{n} \prod_{j=-n,j\neq 0,k}^{n} \frac{1}{\nu_j} + O(z^{n-2}).$$

Therefore

$$\frac{\widetilde{P}(z)}{z\widetilde{Q}(z)} \underset{|z|\to\infty}{=} \left(\sum_{k=-n,k\neq 0}^{n} i\nu_k \right)^{-1} + O(z^{-1}). \tag{3.2.8}$$

Then

$$m_n := -\alpha \lim_{|z|\to\infty} \frac{\widetilde{P}(z)}{z\widetilde{Q}(z)} \tag{3.2.9}$$

is real and positive since $\operatorname{Re}(i\nu_k) = -\operatorname{Im}\nu_k < 0$ for all $k = \pm 1, \dots, \pm n$.

For real numbers η, ζ we define the polynomial $\phi(\cdot, \eta, \zeta)$ by

$$\phi(\lambda, \eta, \zeta) = \widetilde{P}(\lambda^2) + \alpha^{-1}(\eta\lambda^2 - \zeta)\widetilde{Q}(\lambda^2) + i\lambda\alpha^{-1}\widetilde{Q}(\lambda^2).$$

Since Φ is of SHB class, it follows from Theorem A.5.2 that the polynomial $\phi(\cdot, 0, 0)$ is also of SHB class. Since ϕ is a polynomial in all variables, its zeros in the λ-plane are piecewise analytic and continuous functions of η and of ζ, respectively, see [96, Theorem 9.1.1]. The zeros do not cross the real axis when η moves from 0 to m_n and ζ moves from 0 to τ if $\tau = l_n^{-1}$ and when η moves from 0 to m_n and $\zeta = 0$ is fixed if $\tau = 0$. Otherwise, there would be $\lambda \in \mathbb{R}$, $\eta \in [0, m_n]$ and $\zeta \in [0, \tau]$ such that

$$\widetilde{P}(\lambda^2) + \alpha^{-1}(\eta\lambda^2 - \zeta)\widetilde{Q}(\lambda^2) = 0 \quad \text{and} \quad \lambda\alpha^{-1}\widetilde{Q}(\lambda^2) = 0.$$

If $\lambda \neq 0$, then $\widetilde{Q}(\lambda^2) = 0$ and, consequently, $\widetilde{P}(\lambda^2) = 0$ and $\Phi(\lambda) = 0$, which contradicts condition (i). If $\lambda = 0$, then $0 = \widetilde{P}(0) - \alpha^{-1}\zeta\widetilde{Q}(0) = 1 - \alpha^{-1}\zeta\widetilde{Q}(0)$, which is impossible for $\zeta \in [0, \tau]$ due to (3.2.6).

In view of the representation

$$\phi(\lambda, \eta, \zeta) \underset{|\lambda|\to\infty}{=} \widetilde{P}(\lambda^2)\left[1 + \alpha^{-1}\eta\frac{\lambda^2\widetilde{Q}(\lambda^2)}{\widetilde{P}(\lambda^2)}\right] + O(\lambda^{2n-1})$$

it follows from (3.2.8) and (3.2.9) that the coefficient of λ^{2n} of $\phi(\cdot, \eta, \zeta)$ is a nonzero multiple of $1 - \alpha^{-1}\eta\left(\alpha m_n^{-1}\right) = 1 - \eta m_n^{-1}$. Therefore the degree of the polynomial

$\phi(\cdot,\eta,\zeta)$ is $2n$ for each $\eta \in [0, m_n)$ and each $\zeta \in [0,\tau]$. Since the polynomial $\widetilde{Q}$ has degree $n-1$, it is clear that the degree of $\phi(\cdot, m_n, \zeta)$ is $2n-1$ for each $\zeta \in [0,\tau]$. This means that the zeros of $\phi(\cdot,\eta,\zeta)$ do not come from infinity when η moves from 0 to m_n and ζ moves from 0 to τ. Therefore, $\phi(\cdot,\eta,\zeta)$ is of SHB class for each $\zeta \in [0,\tau]$ and each $\eta \in [0, m_n]$. In view of Theorem A.5.5 this implies that the function $\widetilde{R}(\cdot,\tau)$ defined by

$$\widetilde{R}(z,\tau) = \frac{\alpha^{-1}\widetilde{Q}(z)}{\widetilde{P}(z) + (m_n z - \tau)\alpha^{-1}\widetilde{Q}(z)} \tag{3.2.10}$$

is a rational $\mathcal{S}_0$-function, where the polynomial in the numerator has degree $n-1$. Since the degree of the polynomial in the denominator is at most $n-1$ and since a rational $\mathcal{S}$-function has at least as may poles as zeros, see Corollary A.3.3, also the polynomial in the denominator has degree $n-1$. Then, according to Lemma A.3.5, $\widetilde{R}(\cdot,\tau)$ has the representation

$$\widetilde{R}(z,\tau) = a_{n-1} + \cfrac{1}{-b_{n-1}z + \cfrac{1}{a_{n-2} + \cfrac{1}{-b_{n-2}z + \cdots + \cfrac{1}{a_1 + \cfrac{1}{-b_1 z + \cfrac{1}{a_0}}}}}} \tag{3.2.11}$$

with $a_k > 0$ and $b_k > 0$ for each k. The coefficients on the right-hand side depend on τ. Recall that $\tau = l_n^{-1}$ corresponds to the Dirichlet condition at the right endpoint and $\tau = 0$ corresponds to the Neumann condition at the right endpoint.

We put $l_0 = (1-c)a_0$, and for $k = 1,\dots,n-1$, we identify a_k with the length l_k of k-th interval and b_k with the k-th mass m_k of a Stieltjes string. Since $\widetilde{R}(\cdot,\tau)$ is independent of c, it is clear that amongst these numbers only l_0 depends on c and that $\frac{l_0}{1-c} = a_0$ is independent of c. Furthermore, (1.2.19) shows that

$$\frac{\alpha^{-1}\widetilde{Q}(z)}{\widetilde{P}(z) + (m_n z - \tau)\alpha^{-1}\widetilde{Q}(z)} = \frac{R_{2n-2}(z,c)}{R_{2n-3}(z,c)}, \tag{3.2.12}$$

where $R_{2n-2}(\cdot,c)$ and $R_{2n-3}(\cdot,c)$ are the Cauer-Fry polynomials for this Stieltjes string as defined in Subsection 1.2.1. Consequently,

$$\alpha^{-1}\widetilde{Q}(z) = CR_{2n-2}(z,c),$$
$$\widetilde{P}(z) + (m_n z - \tau)\alpha^{-1}\widetilde{Q}(z) = CR_{2n-3}(z,c),$$

where

$$C = \frac{\alpha^{-1}\widetilde{Q}(0)}{R_{2n-2}(0,c)} = \left(\frac{1}{\tilde{l} - l_n} + \tau\right)^{-1} \left(1 + \frac{1-c}{l_0}\sum_{k=1}^{n-1} l_k\right)^{-1}$$

in view of (3.2.6) and (1.2.16). Therefore,

$$\begin{aligned}\Phi(\lambda) &= \widetilde{P}(\lambda^2) + i\lambda\widetilde{Q}(\lambda^2)\\ &= C[R_{2n-3}(\lambda^2, c) + (-m_n\lambda^2 + \tau + i\lambda\alpha)R_{2n-2}(\lambda^2, c)].\end{aligned} \tag{3.2.13}$$

Since we require $\alpha_k = 0$ for $k \le n-1$, it follows from (3.1.9)–(3.1.11) by induction that $S_k(\lambda, c) = R_k(\lambda^2, c)$ for $k = -1, \ldots, 2n-2$. Hence, letting $\alpha_n := \alpha$,

$$\begin{aligned}\frac{l_n}{C}\Phi(\lambda) &= l_n S_{2n-3}(\lambda, c) + l_n(-m_n\lambda^2 + \tau + i\lambda\alpha_n)S_{2n-2}(\lambda, c)\\ &= l_n S_{2n-1}(\lambda, c) + \tau l_n S_{2n-2}(\lambda, c).\end{aligned} \tag{3.2.14}$$

When $\tau = l_n^{-1}$, then (3.2.14) shows that

$$\frac{l_n}{C}\Phi(\lambda) = S_{2n}(\lambda, c). \tag{3.2.15}$$

In view of (1.2.16), (3.1.9), (3.1.10), (3.1.11) and $\Phi(0) = 1$ we find

$$\begin{aligned}l_0 + (1-c)\sum_{k=1}^{n} l_k &= l_0 R_{2n-2}(0, c) + (1-c)l_n = l_0 S_{2n-2}(0, c) + l_n l_0 S_{2n-1}(0, c)\\ &= l_0 S_{2n}(0, 0) = \frac{l_0 l_n}{C} = l_0 l_n \left(\frac{1}{\widetilde{l} - l_n} + \frac{1}{l_n}\right)\left(1 + \frac{1-c}{l_0}\sum_{k=1}^{n-1} l_k\right)\\ &= \left(\frac{l_n}{\widetilde{l} - l_n} + 1\right)\left(l_0 + (1-c)\sum_{k=1}^{n} l_k\right) - \frac{\widetilde{l} l_n(1-c)}{\widetilde{l} - l_n}.\end{aligned}$$

It follows that $\frac{l_0}{1-c} + \sum_{k=1}^{n} l_k = \widetilde{l}$. Since the eigenvalues of the problem (3.1.4), (3.2.1) are the zeros of $S_{2n}(\cdot, c)$ by (3.2.15) and (3.2.3), the existence proof is complete when $\tau = l_n^{-1}$.

When $\tau = 0$, then (3.2.13) shows that

$$\frac{l_n}{C}\Phi(\lambda) = l_n S_{2n-1}(\lambda, c). \tag{3.2.16}$$

In view of (3.2.11), (3.2.10) and (3.2.6) we find

$$\frac{l_0}{1-c} + \sum_{k=1}^{n-1} l_k = R_{2n-2}(0, c) = \widetilde{R}(0, 0) = \alpha^{-1}\widetilde{Q}(0) = \widetilde{l} - l_n.$$

Since the eigenvalues of the problem (3.1.4), (3.2.2) are the zeros of $S_{2n-1}(\cdot, c)$ by (3.2.16) and (3.2.3), the existence proof is also complete when $\tau = 0$.

To prove uniqueness consider any two strings, indexed with 1 and 2, which solve this inverse problem. We denote the corresponding data and functions with upper indices (j) for $j = 1$ and $j = 2$, respectively. That is, these Stieltjes strings satisfy $\frac{l_0^{(j)}}{1-c} + \sum_{k=1}^{n-1} l_k^{(j)} + l_n = \widetilde{l}$, the length of the interval between the right end and the first bead from the right is l_n, the remaining intervals have lengths $l_k^{(j)}$ $(k = 0, \dots, n-1)$, the masses of the beads are $m_k^{(j)}$ $(k = 1, \dots, n)$, the damping coefficient of the damped bead is $\alpha^{(j)} > 0$, and the sequence of the eigenvalues of the two strings is $(\nu_k)_{k=-n,k\neq 0}^{n}$. Let $\tau = l_n^{-1}$ and $\widehat{S}^{(j)} = S_{2n}^{(j)}(\cdot, c)$ for the problem (3.1.4), (3.2.1) and let $\tau = 0$ and $\widehat{S}^{(j)} = l_n S_{2n-1}^{(j)}(\cdot, c)$ for the problem (3.1.4), (3.2.2). Then the polynomials $\widehat{S}^{(j)}$ have the same sequences of zeros $(\nu_k)_{k=-n,k\neq 0}^{n}$ for $j = 1$ and $j = 2$. We know from (3.2.15), (3.2.16) and (3.2.13) that

$$\widehat{S}^{(j)}(\lambda) = l_n R_{2n-3}^{(j)}(\lambda^2, c) + l_n(-m_n^{(j)}\lambda^2 + \tau + i\lambda\alpha^{(j)}) R_{2n-2}^{(j)}(\lambda^2, c). \tag{3.2.17}$$

In particular,

$$\begin{aligned}\widehat{S}^{(j)}(0) &= l_n R_{2n-3}^{(j)}(0, c) + \tau l_n R_{2n-2}^{(j)}(0, c)\\ &= (1-c)\frac{l_n}{l_0^{(j)}} + \tau l_n \left(1 + \frac{1-c}{l_0^{(j)}} \sum_{k=1}^{n-1} l_k^{(j)}\right) = (1-c)\frac{\widehat{l}}{l_0^{(j)}}\end{aligned}$$

by (1.2.15) and (1.2.16), where $\widehat{l} = l_n$ for the problem (3.1.4), (3.2.2) and

$$\widehat{l} = l_n + \frac{l_0^{(j)}}{1-c} + \sum_{k=1}^{n-1} l_k^{(j)} = \widetilde{l}$$

for the problem (3.1.4), (3.2.1) is independent of j. It follows that

$$l_0^{(1)} \widehat{S}^{(1)} = l_0^{(2)} \widehat{S}^{(2)}. \tag{3.2.18}$$

Taking the real parts and the imaginary parts of the polynomials in (3.2.18) we arrive at

$$\begin{aligned}&l_0^{(1)}[R_{2n-3}^{(1)}(z, c) + (-m_n^{(1)} z + \tau) R_{2n-2}^{(1)}(z, c)]\\ &\qquad = l_0^{(2)}[R_{2n-3}^{(2)}(z, c) + (-m_n^{(2)} z + \tau) R_{2n-2}^{(2)}(z, c)], \qquad (3.2.19)\\ &l_0^{(1)}\alpha^{(1)} R_{2n-2}^{(1)}(z, c) = l_0^{(2)}\alpha^{(2)} R_{2n-2}^{(2)}(z, c). \qquad (3.2.20)\end{aligned}$$

Then (3.2.19) and (3.2.20) imply

$$\frac{R_{2n-3}^{(1)}(z, c)}{\alpha^{(1)} R_{2n-2}^{(1)}(z, c)} - \frac{m_n^{(1)} z}{\alpha^{(1)}} + \frac{\tau}{\alpha^{(1)}} = \frac{R_{2n-3}^{(2)}(z, c)}{\alpha^{(2)} R_{2n-2}^{(1)}(z, c)} - \frac{m_n^{(2)} z}{\alpha^{(2)}} + \frac{\tau}{\alpha^{(2)}}. \tag{3.2.21}$$

The polynomials $R_{2n-2}^{(j)}(\cdot, c)$ and $R_{2n-3}^{(j)}(\cdot, c)$ are of equal degree $n-1$, and therefore (3.2.21) implies

$$\frac{m_n^{(1)}}{\alpha^{(1)}} = \frac{m_n^{(2)}}{\alpha^{(2)}} \tag{3.2.22}$$

and

$$\frac{R^{(1)}_{2n-3}(z,c)}{\alpha^{(1)}R^{(1)}_{2n-2}(z,c)}+\frac{\tau}{\alpha^{(1)}}=\frac{R^{(2)}_{2n-3}(z,c)}{\alpha^{(2)}R^{(2)}_{2n-2}(z,c)}+\frac{\tau}{\alpha^{(2)}} \tag{3.2.23}$$

and, consequently,

$$\frac{1}{\alpha^{(1)}}\left(\frac{R^{(1)}_{2n-3}(0,c)}{R^{(1)}_{2n-2}(0,c)}+\tau\right)=\frac{1}{\alpha^{(2)}}\left(\frac{R^{(2)}_{2n-3}(0,c)}{R^{(2)}_{2n-2}(0,c)}+\tau\right). \tag{3.2.24}$$

From (1.2.15) and (1.2.16) it follows that

$$\frac{R^{(j)}_{2n-2}(0,c)}{R^{(j)}_{2n-3}(0,c)}=\frac{l^{(j)}_0}{1-c}+\sum_{k=1}^{n-1}l^{(j)}_k=\widetilde{l}-l_n>0$$

is independent of j, and therefore (3.2.24) and (3.2.22) imply that $\alpha^{(1)}=\alpha^{(2)}$ and $m^{(1)}_n=m^{(2)}_n$. Finally, (3.2.23) gives

$$\frac{R^{(1)}_{2n-2}(\cdot,c)}{R^{(1)}_{2n-3}(\cdot,c)}=\frac{R^{(2)}_{2n-2}(\cdot,c)}{R^{(2)}_{2n-3}(\cdot,c)},$$

and since the continued fraction expansion (1.2.19), with n replaced by $n-1$, is unique by Lemma A.3.5, it follows that $l^{(1)}_k=l^{(2)}_k$ $(k=0,\dots,n-1)$ and $m^{(1)}_k=m^{(2)}_k$ $(k=1,\dots,n-1)$. □

Corollary 3.2.3. *Let $n\in\mathbb{N}$, $\widetilde{l}>0$, $\alpha>0$ and $c\in[0,1)$ be given together with the sequence $(\nu_k)_{k=-n,\,k\neq0}^{n}$ of complex numbers which satisfy the conditions:*

(i) *$\operatorname{Im}\nu_k>0$ for $k=\pm1,\dots,\pm n$;*

(ii) *$\nu_{-k}=-\overline{\nu_k}$ for not pure imaginary ν_{-k}.*

Let DP *be either the problem (3.1.4), (3.2.1) or the problem (3.1.4), (3.2.2).*

(iii) *If* DP *is the problem (3.1.4), (3.2.1), then it is additionally assumed that $\alpha\widetilde{l}\geq4\sum_{k=-n,k\neq0}^{n}\frac{i}{\nu_k}$.*

(iv) *If* DP *is the problem (3.1.4), (3.2.2), then it is additionally assumed that $\alpha\widetilde{l}>\sum_{k=-n,k\neq0}^{n}\frac{i}{\nu_k}$.*

Then there exists a Stieltjes string, i.e., a sequence of positive numbers $(l_k)_{k=0}^{n}$ with $\frac{l_0}{1-c}+\sum\limits_{k=1}^{n}l_k=\widetilde{l}$ and a sequence of positive numbers $(m_k)_{k=1}^{n}$ which generate DP *with $\alpha_k=0$ for $k\leq n-1$ and $\alpha_n=\alpha$ such that its eigenvalues coincide with the sequence $(\nu_k)_{k=-n,\,k\neq0}^{n}$. The string data are uniquely determined by the given data in the case of problem (3.1.4), (3.2.1) when $\alpha\widetilde{l}=4\sum_{k=-n,k\neq0}^{n}\frac{i}{\nu_k}$ and in the case of problem (3.1.4), (3.2.2). However, in the case of problem (3.1.4), (3.2.1)* *for* $\alpha\widetilde{l} > 4\sum_{k=-n,k\neq 0}^{n}\frac{i}{\nu_k}$, *there are two solutions, one with* $l_n < \frac{1}{2}\widetilde{l}$ *and one with* $l_n > \frac{1}{2}\widetilde{l}$. *In each case,* l_0 *is the only string data which depends on* c, *and* $\frac{l_0}{1-c}$ *is independent of* c.

Proof. We introduce the notation $\beta = \sum_{k=-n,k\neq 0}^{n}\frac{i}{\nu_k}$ and recall that we have shown in the proof of Theorem 3.2.2 that the assumptions (i) and (ii) imply that $\beta > 0$. In view of Theorem 3.2.2 and its proof it suffices to prove that the equation (3.2.7) has exactly one or two solutions $l_n \in (0, \widetilde{l})$ in the respective cases. Then the existence and uniqueness statements of Theorem 3.2.3 would complete the proof of this corollary.

First consider the condition (3.2.1), i.e., $\tau = l_n^{-1}$. Then (3.2.7) becomes

$$\alpha\beta^{-1} = \frac{1}{\widetilde{l} - l_n} + \frac{1}{l_n},$$

which can be rewritten as

$$\left(l_n - \frac{1}{2}\widetilde{l}\right)^2 = \frac{\widetilde{l}}{4}(\widetilde{l} - 4\alpha^{-1}\beta).$$

This equation has a solution if and only if $\alpha\widetilde{l} \geq 4\beta$, in which case each solution l_n is in $(0, \widetilde{l})$. When $\alpha\widetilde{l} = 4\beta$, then the solution $l_n = \frac{\widetilde{l}}{2}$ is unique, whereas for $\alpha\widetilde{l} > 4\beta$, there is exactly one solution l_n in each of the intervals $(0, \frac{\widetilde{l}}{2})$ and $(\frac{\widetilde{l}}{2}, \widetilde{l})$.

For the condition (3.2.2), i.e., $\tau = 0$, the equation (3.2.7) becomes $\alpha(\widetilde{l} - l_n) = \beta$, which has a solution $l_n \in (0, \widetilde{l})$ if and only if $\alpha\widetilde{l} > \beta$, and then this solution is unique. □

Corollary 3.2.4. *Let* $n \in \mathbb{N}$, $l_n > 0$, $\alpha > 0$ *and* $c \in [0, 1)$ *be given together with the sequence* $(\nu_k)_{k=-n,\,k\neq 0}^{n}$ *of complex numbers which satisfy the conditions:*

(i) $\operatorname{Im}\nu_k > 0$ *for* $k = \pm 1, \dots, \pm n$;

(ii) $\nu_{-k} = -\overline{\nu_k}$ *for not pure imaginary* ν_{-k}.

Let BP *be either the problem* (3.1.4), (3.2.1) *or the problem* (3.1.4), (3.2.2).

(iii) *If* BP *is the problem* (3.1.4), (3.2.1), *we additionally assume that* $\alpha l_n > \sum_{k=-n,k\neq 0}^{n}\frac{i}{\nu_k}$.

Then there exists a unique Stieltjes string, i.e., a unique sequence of positive numbers $(l_k)_{k=0}^{n-1}$ *and a sequence of positive numbers* $(m_k)_{k=1}^{n}$ *which generate* BP *with* $\alpha_k = 0$ *for* $k \leq n-1$ *and* $\alpha_n = \alpha$ *such that its eigenvalues coincide with the sequence* $(\nu_k)_{k=-n,\,k\neq 0}^{n}$. *All generated string data are independent of* l_n, l_0 *is the only string data which depends on* c, *and* $\frac{l_0}{1-c}$ *is independent of* c.

Proof. Using a reasoning similar to that in the proof of Corollary 3.2.3 it follows that (3.2.7) gives a unique

$$\widetilde{l} = l_n + \frac{1}{\alpha\beta^{-1} - \tau},$$

and the existence and uniqueness statement of Theorem 3.2.3 immediately implies the statement of this corollary if we further observe that the function $\widetilde{R}$ given by (3.2.10), which generates the data, does not depend on l_n. □

Remark 3.2.5. Observing Corollary A.6.7 it is easy to see that the distribution of the beads depends continuously on the given data in Theorem 3.2.2 and in Corollaries 3.2.3 and 3.2.4.

Example 3.2.6. For the sequence $(-2+i, -1+i, i, 2i, 1+i, 2+i)$ find strings of total length $l = 4.2$ with three beads and $l_3 = 1.3$ which are damped at the third bead with mass m_3 such that the given sequence is the sequence of the eigenvalues of the Dirichlet-Dirichlet and Dirichlet-Neumann problem, respectively.

Solution. The functions $\widetilde{P}$ and $\widetilde{Q}$ defined by (3.2.4) are

$$\widetilde{P}(z) = -\frac{1}{20}z^3 + \frac{5}{4}z^2 - \frac{37}{10}z + 1, \quad \widetilde{Q}(z) = \frac{7}{20}z^2 - \frac{11}{4}z + \frac{29}{10},$$

and therefore

$$\lim_{|z|\to\infty} \frac{\widetilde{P}(z)}{z\widetilde{Q}(z)} = -\frac{1}{7}.$$

We first consider the Dirichlet-Dirichlet problem. We find

$$\alpha = \widetilde{Q}(0)\left(\frac{1}{l-l_n} + \frac{1}{l_n}\right) = \frac{29}{10}\left(\frac{10}{29} + \frac{10}{13}\right) = \frac{42}{13}, \quad m_3 = \frac{6}{13}.$$

Then

$$\widetilde{R}\left(z, \frac{10}{13}\right) = \frac{\alpha^{-1}\widetilde{Q}(z)}{\widetilde{P}(z) + (m_3 z - 10)\alpha^{-1}\widetilde{Q}(z)} = \frac{7z^2 - 105z + 58}{50z^2 - 170z + 20}$$
$$= \frac{7}{50} + \cfrac{1}{-\frac{125}{78}z + \cfrac{1}{\frac{507}{1325} + \cfrac{1}{-\frac{2809}{1638}z + \cfrac{1}{\frac{126}{53}}}}}$$

shows that the string data for the Dirichlet-Dirichlet problem with the given data are

$$l_0 = \frac{126}{53}, \; l_1 = \frac{507}{1325}, \; l_2 = \frac{7}{50}, \; m_1 = \frac{2809}{1638}, \; m_2 = \frac{125}{78}, \; m_3 = \frac{6}{13}, \; \alpha = \frac{42}{13}.$$

For the Dirichlet-Neumann problem we have

$$\alpha = \widetilde{Q}(0)\frac{1}{l-l_n} = 1, \quad m_3 = \frac{1}{7}.$$

Then

$$\begin{aligned}\widetilde{R}(z,0) &= \frac{\alpha^{-1}\widetilde{Q}(z)}{\widetilde{P}(z)+m_3 z\alpha^{-1}\widetilde{Q}(z)} = \frac{49z^2-385z+406}{120z^2-460z+140}\\ &= \tfrac{49}{120} + \cfrac{1}{-\tfrac{720}{1183}z + \cfrac{1}{\tfrac{2197}{2760} + \cfrac{1}{-\tfrac{529}{507}z + \cfrac{1}{\tfrac{39}{23}}}}}\end{aligned}$$

shows that the string data for the Dirichlet-Neumann problem with the given data are

$$l_0 = \frac{39}{23},\ l_1 = \frac{2197}{2760},\ l_2 = \frac{49}{120},\ m_1 = \frac{529}{507},\ m_2 = \frac{720}{1183},\ m_3 = \frac{1}{7},\ \alpha = 1. \qquad \square$$

Example 3.2.7. For the sequence (i, i, i, i) find strings of total length $l = 3$ with two beads and $l_2 = 1$ which are damped at the second bead with mass m_2 such that the given sequence is the sequence of the eigenvalues of the Dirichlet-Dirichlet and Dirichlet-Neumann problem, respectively.

Solution. The functions $\widetilde{P}$ and $\widetilde{Q}$ defined by (3.2.4) are

$$\widetilde{P}(z) = z^2 - 6z + 1, \quad \widetilde{Q}(z) = -4z + 4,$$

and therefore

$$\lim_{|z|\to\infty} \frac{\widetilde{P}(z)}{z\widetilde{Q}(z)} = -\frac{1}{4}.$$

We first consider the Dirichlet-Dirichlet problem. We find

$$\alpha = \widetilde{Q}(0)\left(\frac{1}{l-l_n} + \frac{1}{l_n}\right) = 4\left(\frac{1}{2}+1\right) = 6, \quad m_2 = \frac{3}{2}.$$

Then

$$\begin{aligned}\widetilde{R}(z,1) &= \frac{\alpha^{-1}\widetilde{Q}(z)}{\widetilde{P}(z)+(m_2 z-1)\alpha^{-1}\widetilde{Q}(z)} = \frac{2z-2}{13z-1}\\ &= \tfrac{2}{13} + \cfrac{1}{-\tfrac{169}{24}z + \cfrac{1}{\tfrac{24}{13}}}\end{aligned}$$

shows that the string data for the Dirichlet-Dirichlet problem with the given data are

$$l_0 = \frac{24}{13},\ l_1 = \frac{2}{13},\ m_1 = \frac{169}{24},\ m_2 = \frac{3}{2},\ \alpha = 6.$$

For the Dirichlet-Neumann problem we have

$$\alpha = \widetilde{Q}(0)\frac{1}{l - l_n} = 4 \cdot \frac{1}{2} = 2, \quad m_2 = \frac{1}{2}.$$

Then

$$\begin{aligned}\widetilde{R}(z,0) &= \frac{\alpha^{-1}\widetilde{Q}(z)}{\widetilde{P}(z) + m_2 z\alpha^{-1}\widetilde{Q}(z)} = \frac{2z-2}{5z-1} \\ &= \tfrac{2}{5} + \cfrac{1}{-\tfrac{25}{8}z + \cfrac{1}{\tfrac{8}{5}}}\end{aligned}$$

shows that the string data for the Dirichlet-Neumann problem with the given data are

$$l_0 = \frac{8}{5},\ l_1 = \frac{2}{5},\ m_1 = \frac{25}{8},\ m_2 = \frac{1}{2},\ \alpha = 2.$$ □

3.3 Comparison of characteristic functions

In this section we will give necessary and sufficient conditions for two polynomials of even degree $2n$ to be the characteristic functions $S_{2n}(\cdot,0)$ and $S_{2n-1}(\cdot,0)$ of the problem (3.1.4) with Dirichlet-Dirichlet boundary conditions (3.1.5) and Dirichlet-Neumann boundary conditions (3.1.6), respectively, for the same string which is damped only at the n-th bead.

Theorem 3.3.1. *Let $n \in \mathbb{N}$ and let ϕ_1 and ϕ_2 be two polynomials of degree $2n$. Then there is a string as considered in Section 3.1 with n beads, where only the n-th bead is damped, such that $\phi_1 = \upsilon S_{2n-1}(\cdot,0)$ and $\phi_2 = \upsilon S_{2n}(\cdot,0)$ for some $\upsilon > 0$ if and only if the following properties are satisfied:*

(i) *ϕ_1 and ϕ_2 are real on the imaginary axis, i.e., there are real polynomials $\widetilde{P}_1$, $\widetilde{Q}_1$, $\widetilde{P}_2$, $\widetilde{Q}_2$ such that*

$$\phi_j(\lambda) = \widetilde{P}_j(\lambda^2) + i\lambda\widetilde{Q}_j(\lambda^2), \quad j = 1,2,\ \lambda \in \mathbb{C};$$

(ii) *The polynomials ϕ_1 and ϕ_2 do not have common zeros;*

(iii) *$\phi_1(0) > 0$ and $\phi_2(0) > 0$;*

(iv) *$\widetilde{Q}_1 \neq 0$ and there is a positive number β such that*

$$\widetilde{P}_2\widetilde{Q}_1 - \widetilde{P}_1\widetilde{Q}_2 = \beta\widetilde{Q}_1^2 \quad \text{and} \quad \widetilde{Q}_2 = \lim_{|z|\to\infty} \frac{\widetilde{P}_2(z)}{\widetilde{P}_1(z)}\widetilde{Q}_1;$$

(v) *$\frac{\widetilde{P}_2}{\widetilde{P}_1}$ is a rational $\mathcal{S}_0$-function.*

Proof. To prove necessity we assume that there is a string with n beads such that only the n-th bead is damped and that $\phi_1 = \upsilon S_{2n-1}(\cdot, 0)$ and $\phi_2 = \upsilon S_{2n}(\cdot, 0)$ for some $\upsilon > 0$. Since the properties (i)–(iv) are independent of the choice of υ, we may assume that $\upsilon = 1$. From (3.1.9)–(3.1.11) it follows that $S_{2n-1}(\cdot, 0)$ and $S_{2n}(\cdot, 0)$ are real on the imaginary axis, so that property (i) is satisfied.

Property (ii) is satisfied because if $S_{2n}(\lambda, 0) = 0 = S_{2n-1}(\lambda, 0)$, then (3.1.9)–(3.1.10) would lead to $S_0(\lambda, 0) = 0$, which is impossible since $S_0(\lambda, 0) = 1$ by (3.1.11). Property (iii) is clear since $S_{2n-1}(0, 0) = l_0^{-1}$ and $S_{2n}(0, 0) = l l_0^{-1}$.

The first $n - 1$ beads are undamped and therefore the recurrence relations (1.2.10)–(1.2.14) and (3.1.9)–(3.1.11) show that

$$\begin{aligned} S_{2n-1}(\lambda, 0) &= (-\lambda^2 m_n + i\alpha\lambda) R_{2n-2}(\lambda^2, 0) + R_{2n-3}(\lambda^2, 0), \\ S_{2n}(\lambda, 0) &= l_n S_{2n-1}(\lambda, 0) + R_{2n-2}(\lambda^2, 0), \end{aligned}$$

whence

$$\begin{aligned} \widetilde{P}_1(z) &= -m_n z R_{2n-2}(z, 0) + R_{2n-3}(z, 0), \quad \widetilde{Q}_1(z) = \alpha R_{2n-2}(z, 0), \\ \widetilde{P}_2(z) &= l_n \widetilde{P}_1(z) + R_{2n-2}(z, 0), \quad \widetilde{Q}_2(z) = l_n \widetilde{Q}_1(z). \end{aligned}$$

It follows that

$$\begin{aligned} \widetilde{P}_2\widetilde{Q}_1 - \widetilde{P}_1\widetilde{Q}_2 &= (l_n \widetilde{P}_1 + R_{2n-2}(\cdot, 0))\widetilde{Q}_1 - \widetilde{P}_1 l_n \widetilde{Q}_1 \\ &= R_{2n-2}(\cdot, 0)\widetilde{Q}_1 = \alpha^{-1}\widetilde{Q}_1^2, \end{aligned}$$

which proves the first identity in (iv) with $\beta = \alpha^{-1}$. The second identity holds because $\lim\limits_{|z|\to\infty} \frac{\widetilde{P}_2(z)}{\widetilde{P}_1(z)} = l_n$ since $\widetilde{P}_1$ has degree n whereas $R_{2n-2}(\cdot, 0)$ has degree $n - 1$.

From

$$R_{2n-1}(z, 0) = -m_n z R_{2n-2}(z, 0) + R_{2n-3}(z, 0) = \widetilde{P}_1(z)$$

and

$$R_{2n}(z, 0) = l_n R_{2n-1}(z, 0) + R_{2n-2}(z, 0) = l_n \widetilde{P}_1(z) + R_{2n-2}(z, 0) = \widetilde{P}_2(z)$$

it follows that

$$\frac{R_{2n}(\cdot, 0)}{R_{2n-1}(\cdot, 0)} = \frac{\widetilde{P}_2}{\widetilde{P}_1}.$$

The representation (1.2.19) and Lemma A.3.6 show that $\frac{\widetilde{P}_2}{\widetilde{P}_1}$ is a rational $\mathcal{S}$-function. In view of $\widetilde{P}_1(0) \neq 0$ we conclude that also property (v) holds.

For the proof of the sufficiency let ϕ_1 and ϕ_2 be given which satisfy properties (i)–(v). Since ϕ_1 and ϕ_2 are polynomials of degree $2n$, it follows from the representation in (i) that the polynomials $\widetilde{P}_1$ and $\widetilde{P}_2$ have degree n. Assume that $\widetilde{P}_1$ and $\widetilde{P}_2$ have a common zero. In view of property (iv) this would also be a zero of $\widetilde{Q}_1$ and $\widetilde{Q}_2$. Therefore ϕ_1 and ϕ_2 would have common zeros by property (i); but this is impossible by property (ii). Therefore $\frac{\widetilde{P}_2}{\widetilde{P}_1}$ is a rational $\mathcal{S}_0$-function with n poles and n zeros by assumption (iv). By Lemma A.3.5 the function $\frac{\widetilde{P}_2}{\widetilde{P}_1}$ has the representation (A.3.8) with $c_0 > 0$ and $a_n > 0$. With the notation from (A.3.8) we define a string with n beads whose masses are $m_k = b_k$ $(k = 1, \dots, n)$ and whose threads have lengths $l_k = a_k$ $(k = 1, \dots, n)$ and $l_0 = \frac{1}{c_0}$. By (1.2.19) the Cauer-Fry polynomials for this string satisfy

$$\frac{R_{2n}(\cdot, 0)}{R_{2n-1}(\cdot, 0)} = \frac{\widetilde{P}_2}{\widetilde{P}_1}.$$

Therefore there is a nonzero complex number υ such that

$$\widetilde{P}_2 = \upsilon R_{2n}(\cdot, 0), \quad \widetilde{P}_1 = \upsilon R_{2n-1}(\cdot, 0),$$

and property (iii) implies $\upsilon > 0$. Furthermore, the representation (1.2.19) gives

$$\lim_{|z|\to\infty} \frac{\widetilde{P}_2(z)}{\widetilde{P}_1(z)} = \lim_{|z|\to\infty} \frac{R_{2n}(z, 0)}{R_{2n-1}(z, 0)} = l_n,$$

and therefore

$$\widetilde{Q}_2 = l_n \widetilde{Q}_1$$

by property (iv). Putting $\alpha = \beta^{-1}$, property (iv) implies

$$\begin{aligned}
\alpha^{-1}\widetilde{Q}_1 = \widetilde{P}_2 - \widetilde{P}_1 \frac{\widetilde{Q}_2}{\widetilde{Q}_1} &= \widetilde{P}_2 - l_n \widetilde{P}_1 \\
&= \upsilon\left[R_{2n}(\cdot, 0) - l_n R_{2n-1}(\cdot, 0)\right] = \upsilon R_{2n-2}(\cdot, 0).
\end{aligned}$$

Therefore

$$\begin{aligned}
\phi_1(\lambda) &= \widetilde{P}_1(\lambda^2) + i\lambda \widetilde{Q}_1(\lambda^2) \\
&= \upsilon\left[R_{2n-1}(\lambda^2, 0) + i\lambda\alpha R_{2n-2}(\lambda^2, 0)\right] \\
&= \upsilon\left[(-m_n\lambda^2 + i\lambda\alpha)R_{2n-2}(\lambda^2, 0) + R_{2n-3}(\lambda^2, 0)\right] \\
&= \upsilon S_{2n-1}(\lambda, 0)
\end{aligned}$$

and

$$\begin{aligned}
\phi_2(\lambda) &= \widetilde{P}_2(\lambda^2) + i\lambda \widetilde{Q}_2(\lambda^2) \\
&= \upsilon\left[R_{2n}(\lambda^2, 0) + i\lambda\alpha l_n R_{2n-2}(\lambda^2, 0)\right] \\
&= \upsilon\left[l_n\left\{R_{2n-1}(\lambda^2, 0) + i\lambda\alpha R_{2n-2}(\lambda^2, 0)\right\} + R_{2n-2}(\lambda^2, 0)\right] \\
&= \upsilon\left\{l_n S_{2n-1}(\lambda, 0) + S_{2n-2}(\lambda, 0)\right\} \\
&= \upsilon S_{2n}(\lambda, 0).
\end{aligned}$$

□

Remark 3.3.2. 1. The number υ in Theorem 3.3.1 can be determined once the string data have been found, e.g.,

$$\upsilon = \frac{\phi_1(0)}{S_{2n-1}(0,0)} = l_0\phi_1(0).$$

2. Theorem 3.3.1 is most useful when prescribing the eigenvalues of the Dirichlet-Dirichlet and the Dirichlet-Neumann problems, i.e., the zeros of ϕ_1 and ϕ_2. Then one can assume without loss of generality that $\phi_1(0) > 0$ and $\phi_2(0) > 0$, and if for one such pair (ϕ_1, ϕ_2) properties (i)–(v) are satisfied, then these conditions hold for all pairs $(\upsilon_1\phi_1, \upsilon_2\phi_2)$ $(\upsilon_1 > 0,\ \upsilon_2 > 0)$, and we obtain a one-parameter family of strings, where the parameter is the total length of the string $\upsilon_2\phi_2(0)\,(\upsilon_1\phi_1(0))^{-1}$.
3. Given two sequences of complex numbers which are the eigenvalues of the Dirichlet-Dirichlet and Dirichlet-Neumann problem, respectively, one can uniquely determine which of the sequences belongs to which problem. Indeed, if we have that the zeros of ϕ_2 are the eigenvalues of the Dirichlet-Dirichlet problem and the zeros of ϕ_1 are the eigenvalues of the Dirichlet-Neumann problem, then $\frac{\widetilde{P}_2}{\widetilde{P}_1}$ is a rational $\mathcal{S}_0$-function by property (v). But then $\frac{\widetilde{P}_1}{\widetilde{P}_2}$ is a not a rational $\mathcal{S}_0$-function, so that it cannot be true that the zeros of ϕ_1 are the eigenvalues of the Dirichlet-Dirichlet problem and the zeros of ϕ_2 are the eigenvalues of the Dirichlet-Neumann problem.

3.4 Stieltjes strings with a damping condition at the right end

In this section we consider small transverse vibrations of a Stieltjes string with Robin boundary condition (1.2.42) at the left end and with an additional bead of mass $m_{n+1} > 0$ at the right end that can move in the direction orthogonal to the equilibrium position of the string subject to damping with coefficient $\alpha > 0$. In this case the condition at the right end is given by

$$\frac{v_{n+1}(t) - v_n(t)}{l_n} + m_{n+1}v''_{n+1}(t) + \alpha v'_{n+1}(t) = 0.$$

Substituting $v_k(t) = u_k e^{i\lambda t}$ we obtain

$$\frac{u_{n+1} - u_n}{l_n} - m_{n+1}\lambda^2 u_{n+1} + i\alpha\lambda u_{n+1} = 0. \tag{3.4.1}$$

A mathematically equivalent formulation of (3.4.1) is obtained by introducing u_{n+2} and any $l_{n+1} > 0$ and writing (3.4.1) as

$$\frac{u_{n+1} - u_{n+2}}{l_{n+1}} + \frac{u_{n+1} - u_n}{l_n} - m_{n+1}\lambda^2 u_{n+1} + i\alpha\lambda u_{n+1} = 0, \tag{3.4.2}$$

$$u_{n+2} = u_{n+1}. \tag{3.4.3}$$

With the notation γ_0 from (1.2.43) it follows that problem (1.2.4), (3.4.2), (1.2.42), (3.4.3) is a special case of problem (1.2.4), (1.2.42), (3.1.12) with n replaced by $n+1$, with $c = \gamma_0$, and with $d = 1$. Here $\alpha_1 = \cdots = \alpha_n = 0$ and $\alpha_{n+1} = \alpha > 0$. In particular, $S_{2n+1}(\lambda, \gamma_0)$ is defined by (3.1.10) for $k = n+1$. Recalling the notation T_c^d from (3.1.14), we have therefore that the sequence of the $2n+2$ eigenvalues $(\lambda_k)_{k=-(n+1),k\neq 0}^{n+1}$, counted with algebraic multiplicity, is the sequence of the eigenvalues of the problem (1.2.4), (1.2.42), (3.4.1).

Theorem 3.4.1. 1. *The sequence $(\lambda_k)_{k=-(n+1),k\neq 0}^{n+1}$ is the sequence of the zeros of $S_{2n+1}(\cdot, \gamma_0)$, counted with multiplicity.*

2. *The sequence $(\lambda_k)_{k=-(n+1),k\neq 0}^{n+1}$ can be indexed in such a way that for all $k \in \{\pm 1, \dots, \pm(n+1)\}$, $\lambda_{-k} = -\overline{\lambda_k}$ if λ_k is not pure imaginary.*

3. *For all $k \in \{\pm 1, \dots, \pm(n+1)\}$, $\operatorname{Im} \lambda_k > 0$.*

Proof. In the proof of Theorem 3.1.5 we may replace T_0^1 with $T_{\gamma_0}^1$. □

Theorem 3.4.2. *Let $n \in \mathbb{N}$, $l > 0$ and $\alpha > 0$ be given together with a string of length l and the eigenvalue sequence $(\lambda_k)_{k=-(n+1),\, k\neq 0}^{n+1}$ of the corresponding problem (1.2.4), (1.2.42), (3.4.1). Then the sequence $(l_k)_{k=0}^{n}$, the sequence $(m_k)_{k=1}^{n+1}$ and the number γ are uniquely determined by $(\lambda_k)_{k=-(n+1),k\neq 0}^{n+1}$, l and α.*

Theorem 3.4.3. *Let $n \in \mathbb{N}$ and $\alpha > 0$ be given together with the sequence of complex numbers $(\lambda_k)_{k=-(n+1),\, k\neq 0}^{n+1}$ which satisfy the conditions:*

(i) $\operatorname{Im} \lambda_k > 0$ *for* $k = \pm 1, \dots, \pm(n+1)$;

(ii) $\lambda_{-k} = -\overline{\lambda_k}$ *for not pure imaginary* λ_{-k}.

Let $\beta := \sum_{k=-n+1,k\neq 0}^{n+1} \frac{i}{\lambda_k}$. Then there exists a number $\widehat{l} \in (0, \alpha^{-1}\beta)$ such that for all $l \in (\widehat{l}, \alpha^{-1}\beta)$ there exist a unique number $\gamma > 0$, a unique sequence of positive numbers $(l_k)_{k=0}^{n}$ with $\sum\limits_{k=0}^{n} l_k = l$ and a unique sequence of positive numbers $(m_k)_{k=1}^{n+1}$ such that $(\lambda_k)_{k=-(n+1),k\neq 0}^{n+1}$ is the sequence of the eigenvalues of the corresponding problem (1.2.4), (1.2.42), (3.4.1).

Proof of Theorems 3.4.2 *and* 3.4.3. We are going to apply Corollary 3.2.4 with n replaced by $n+1$. To this end we first consider an arbitrary $c \in (0,1)$ and an auxiliary $l_{n+1} > 0$. By Corollary 3.2.4 there are a sequence of positive numbers $(l_k)_{k=0}^{n}$ and a sequence of positive numbers $(m_k)_{k=1}^{n+1}$ which generate (3.1.4), (3.2.2) with $\alpha_k = 0$ for $k \leq n$ and $\alpha_{n+1} = \alpha$ such that its eigenvalues coincide with the sequence $(\lambda_k)_{k=-(n+1),\, k\neq 0}^{n+1}$. Furthermore, $l_1, \dots, l_n, m_1, \dots, m_{n+1}$ are independent of c and l_{n+1}, and also the number $a_0 = \frac{l_0}{1-c}$ is independent of c and l_{n+1}. These numbers are therefore uniquely determined by the given data. When l is given, also l_0 is uniquely determined, and

$$l^{\#} := a_0 + \sum_{k=1}^{n} l_k = \frac{l_0}{1-c} + \sum_{k=1}^{n} l_k = \sum_{k=0}^{n} l_k + ca_0 = l + ca_0 \tag{3.4.4}$$

is independent of c and l_{n+1}. But since we have to satisfy $\gamma_0 = c$ we obtain $1 + \gamma l_0 = \frac{1}{c}$ by (1.2.43), and therefore

$$\gamma = \frac{1-c}{cl_0} = \frac{1}{ca_0} - \frac{1}{l^{\#} - l}$$

is uniquely determined by the given data. This completes the proof of Theorem 3.4.2 and the uniqueness part of Theorem 3.4.3.

For the existence part in Theorem 3.4.3 we require $c \in (0,1)$, and (3.4.4) leads to the condition $l \in (l^{\#} - a_0, l^{\#})$. In the notation of Theorem 3.2.2 we have $\widetilde{l} = l_{n+1} + l^{\#}$, and it follows from (3.2.7) that $l^{\#} = \alpha^{-1}\beta$. Putting

$$\widehat{l} = l^{\#} - a_0 = \sum_{k=1}^{n} l_k$$

completes the proof. □

3.5 The inverse problem by parts of spectra of Robin-Regge problems

In this section we consider two problems for the Stieltjes string equations (1.2.4) with the same damping condition (3.4.1) at the right end, but with possibly different Robin conditions at the left end,

$$\frac{u_1 - u_0}{l_0} - \gamma_1 u_0 = 0, \tag{3.5.1}$$

$$\frac{u_1 - u_0}{l_0} - \gamma_2 u_0 = 0, \tag{3.5.2}$$

where $\gamma_1 > 0$ and $\gamma_2 > 0$. According to Theorem 3.4.1, the characteristic functions of these problems are $S_{2n+1}(\cdot, \gamma_{1,0})$ and $S_{2n+1}(\cdot, \gamma_{2,0})$, where $\gamma_{j,0} = (1 + \gamma_j l_0)^{-1}$ $(j = 1, 2)$ by (1.2.43). Furthermore, the sequence of the zeros of $S_{2n+1}(\cdot, \gamma_{j,0})$, i.e., the eigenvalues of problem (1.2.4), (3.4.1) with (3.5.1) for $j = 1$ and with (3.5.2) for $j = 2$, can be indexed as $(\lambda_k(\gamma_j))_{k=-(n+1),k\neq 0}^{n+1}$ so that the properties 2 and 3 of Theorem 3.4.1 are satisfied. Recalling from Section 3.4 that (3.1.10) also holds for $k = n+1$ here, it follows from $\alpha_1 = \cdots = \alpha_n = 0$ that

$$S_{2n+1}(\lambda, \gamma_{j,0}) = (-m_{n+1}\lambda^2 + i\alpha\lambda) R_{2n}(\lambda^2, \gamma_{j,0}) + R_{2n-1}(\lambda^2, \gamma_{j,0}). \tag{3.5.3}$$

Proposition 3.5.1. *For all $\alpha > 0$, $\gamma_1 > 0$, $\gamma_2 > 0$ and $\lambda \in \mathbb{C}$ it is true that*

$$S_{2n+1}(\lambda,\gamma_{2,0})S_{2n+1}(-\lambda,\gamma_{1,0}) - S_{2n+1}(-\lambda,\gamma_{2,0})S_{2n+1}(\lambda,\gamma_{1,0}) \\ = 2i\alpha\lambda\left(\frac{1}{\gamma_2}-\frac{1}{\gamma_1}\right)S_{2n+1}(0,\gamma_{1,0})S_{2n+1}(0,\gamma_{2,0}). \quad (3.5.4)$$

Proof. Proposition A.6.1 and the Lagrange identity (1.2.47) show that

$$\begin{aligned} &S_{2n+1}(\lambda,\gamma_{2,0})S_{2n+1}(-\lambda,\gamma_{1,0}) - S_{2n+1}(-\lambda,\gamma_{2,0})S_{2n+1}(\lambda,\gamma_{1,0}) \\ &\quad = 2i\alpha\lambda[R_{2n-1}(\lambda^2,\gamma_{1,0})R_{2n}(\lambda^2,\gamma_{2,0}) - R_{2n-1}(\lambda^2,\gamma_{2,0})R_{2n}(\lambda^2,\gamma_{1,0})] \\ &\quad = 2i\alpha\lambda\frac{\gamma_{2,0}-\gamma_{1,0}}{l_0}. \end{aligned}$$

But in view of (3.5.3) and (1.2.44),

$$\begin{aligned} \frac{\gamma_{2,0}-\gamma_{1,0}}{l_0} &= \frac{1-\gamma_{1,0}}{l_0} - \frac{1-\gamma_{2,0}}{l_0} \\ &= \left(\frac{1-\gamma_{1,0}}{l_0}\right)\left(\frac{1-\gamma_{2,0}}{l_0}\right)\left(\frac{l_0}{1-\gamma_{2,0}} - \frac{l_0}{1-\gamma_{1,0}}\right) \\ &= S_{2n+1}(0,\gamma_{1,0})S_{2n+1}(0,\gamma_{2,0})\left(\frac{1+\gamma_2 l_0}{\gamma_2} - \frac{1+\gamma_1 l_0}{\gamma_1}\right) \\ &= S_{2n+1}(0,\gamma_{1,0})S_{2n+1}(0,\gamma_{2,0})\left(\frac{1}{\gamma_2}-\frac{1}{\gamma_1}\right). \end{aligned}$$ □

If $\gamma_1 \neq \gamma_2$, then (3.5.4) implies that

$$\{\lambda_k(\gamma_1) : 0 < |k| \le n+1\} \cap \{\lambda_k(\gamma_2) : 0 < |k| \le n+1\} = \emptyset. \quad (3.5.5)$$

The information given in Theorem 3.4.1, Proposition 3.5.1 and (3.5.5) suffices to recover string data from parts of the eigenvalues of two problems, which are given together with the total length of the string, the damping constant and the difference of the inverses of the Robin parameters.

Theorem 3.5.2. *Let $n \in \mathbb{N}$, $l > 0$ and $\alpha > 0$ be given together with a string of length l. Let $p \in \mathbb{N}_0$ with $0 \le p \le n+1$, let $(\theta_k)_{0<|k|\le p}$ be a subsequence of the sequence of eigenvalues of the problem (1.2.4), (3.4.1), (3.5.1) (Robin condition for some $\gamma_1 > 0$), and let $(\tau_k)_{p<|k|\le n+1}$ be a subsequence of the sequence of eigenvalues of the problem (1.2.4), (3.4.1), (3.5.2) (Robin condition for some $\gamma_2 > 0$). If additionally $\{\theta_k : 0 < |k| \le p\} \cap \{\tau_k : p < |k| \le n+1\} = \emptyset$ and $\gamma_2^{-1} - \gamma_1^{-1}$ is known, then the string data $(m_k)_{k=1}^{n+1}$, $(l_k)_{k=0}^{n}$, γ_1, γ_2 are uniquely determined by l, α, $(\theta_k)_{0<|k|\le p}$ and $(\tau_k)_{p<|k|\le n+1}$.*

Proof. Define

$$\omega_1(\lambda) = \prod_{0<|k|\le p}\left(1-\frac{\lambda}{\theta_k}\right), \quad \omega_2(\lambda) = \prod_{p<|k|\le n+1}\left(1-\frac{\lambda}{\tau_k}\right),$$
$$\omega(\lambda) = \omega_1(-\lambda)\omega_2(\lambda).$$

Since all θ_k $(0 < |k| \le p)$ and τ_k $(p < |k| \le n+1)$ have positive imaginary parts by Theorem 3.4.1, part 3, it follows that $\omega(\lambda) \ne 0$ whenever $\omega(-\lambda) = 0$. By Proposition A.6.2 the functional equation

$$\omega(\lambda)\psi(-\lambda) - \omega(-\lambda)\psi(\lambda) = 2i\alpha\lambda\left(\gamma_2^{-1} - \gamma_1^{-1}\right)$$

has a unique solution ψ in the class of polynomials of degree at most $2n+2$ with $\psi(0) = 1$. Let $(\theta_k)_{0<|k|\le n+1}$ be the sequence of eigenvalues of the problem (1.2.4), (3.4.1), (3.5.1) (Robin condition for $\gamma_1 > 0$), and let $(\tau_k)_{0<|k|\le n+1}$ be the sequence of eigenvalues of the problem (1.2.4), (3.4.1), (3.5.2) (Robin condition for $\gamma_2 > 0$). In view of (3.5.4), the unique ψ must have the representation

$$\psi(\lambda) = \prod_{p<|k|\le n+1}\left(1 + \frac{\lambda}{\theta_k}\right)\prod_{0<|k|\le p}\left(1 - \frac{\lambda}{\tau_k}\right),$$

which means that the full sequences $(\theta_k)_{0<|k|\le n+1}$ and $(\tau_k)_{0<|k|\le n+1}$ are uniquely determined by the given subsequences, α and $\gamma_2^{-1} - \gamma_1^{-1}$, where we have used again Theorem 3.4.1, part 3, to uniquely determine which zeros of ψ are the zeros of $S_{2n+1}(\cdot, \gamma_{1,0})$ and which are the zeros of $S_{2n+1}(\cdot, \gamma_{2,0})$. Since also l is given, it follows from Theorem 3.4.2 that the string data $(m_k)_{k=1}^{n+1}$, $(l_k)_{k=0}^{n}$, γ_1 and γ_2 are uniquely determined by the given data. □

The next theorem shows that for given sequences as in Theorem 3.5.2 which satisfy the conditions (i) and (ii) of Theorem 3.4.3 one can always find a range of total lengths l and a range of $\gamma_2^{-1} - \gamma_1^{-1}$ for which the pair of string problems (1.2.4), (3.4.1), (3.5.1) and (3.5.2) has a solution. The given range of l and $\gamma_2^{-1} - \gamma_1^{-1}$ only shows existence; the maximal range will be quite difficult to determine.

Theorem 3.5.3. *Let $n \in \mathbb{N}$, $\alpha > 0$, $p \in \mathbb{N}_0$ with $0 \le p \le n+1$, and two sequences $(\theta_k)_{0<|k|\le p}$ and $(\tau_k)_{p<|k|\le n+1}$ of complex numbers be given so that*

(i) $\{\theta_k : 0 < |k| \le p\} \cap \{\tau_k : p < |k| \le n+1\} = \emptyset$;

(ii) $\operatorname{Im}\theta_k > 0$ $(0 < |k| < p)$, $\operatorname{Im}\tau_k > 0$ $(p < |k| \le n+1)$;

(iii) $\theta_{-k} = -\overline{\theta_k}$, $\tau_{-k} = -\overline{\tau_k}$ *for not pure imaginary* θ_k, τ_k.

Then there exist a number $\varepsilon > 0$ and a compact interval $\Lambda \subset (0, \infty)$ such that for all $\Delta \in (-\varepsilon, \varepsilon)$ and $l \in \Lambda$ there exist unique positive real numbers γ_1, γ_2 with $\gamma_2^{-1} - \gamma_1^{-1} = \Delta$ and unique sequences of positive numbers $(m_k)_{k=1}^{n+1}$ and $(l_k)_{k=0}^{n}$, with $\sum\limits_{k=0}^{n} l_k = l$ such that $(\tau_k)_{0<|k|\le p}$ are eigenvalues of the problem (1.2.4), (3.4.1), (3.5.1) (Robin condition with γ_1), and $(\theta_k)_{p<|k|\le n+1}$ are eigenvalues of the problem (1.2.4), (3.4.1), (3.5.2) (Robin condition with γ_2).

Proof. As in the proof of Theorem 3.5.2 define

$$\omega_1(\lambda) = \prod_{0<|k|\le p} \left(1 - \frac{\lambda}{\theta_k}\right), \quad \omega_2(\lambda) = \prod_{p<|k|\le n+1} \left(1 - \frac{\lambda}{\tau_k}\right),$$

$$\omega(\lambda) = \omega_1(-\lambda)\omega_2(\lambda).$$

By assumption (iii) the function ω is symmetric here, and then Corollary A.6.3 shows that there is an unbounded open interval $J_{\omega,\alpha}$ with $0 \in J$, depending on the given sequences and α, such that for all $\delta \in J_{\omega,\alpha}$ the functional equation

$$\omega(\lambda)\psi(-\lambda) - \omega(-\lambda)\psi(\lambda) = 2i\lambda\alpha\delta \tag{3.5.6}$$

has a unique solution ψ in the class of polynomials of degree $2n+2$ with $\psi(0) = 1$ and such that the coefficient of $\lambda^{4(n+1)}$ in $\chi(\lambda) := \omega(\lambda)\psi(-\lambda)$ is positive. Furthermore, ψ is symmetric. Writing $\chi(\lambda) = \widetilde{P}(\lambda^2) + i\lambda\widetilde{Q}(\lambda^2)$ with real polynomials $\widetilde{P}$ and $\widetilde{Q}$, it follows from (3.5.6) that

$$2i\lambda\widetilde{Q}(\lambda^2) = \chi(\lambda) - \chi(-\lambda) = 2i\lambda\alpha\delta,$$

which means that $\widetilde{Q}$ is constant. Dividing χ by its leading coefficient, it becomes a monic polynomial which is positive at 0. Then Proposition A.5.6 shows that χ has $2n+2$ zeros, counted with multiplicity, both in the open upper half-plane and in the open lower half-plane. Hence the given sequences $(\theta_k)_{0<|k|\le p}$ and $(\tau_k)_{p<|k|\le n+1}$ can be extended to sequences $(\theta_k)_{0<|k|\le n+1}$ and $(\tau_k)_{0<|k|\le n+1}$ such that (ii) and (iii) hold for all indices k with $0 < |k| \le n+1$ and such that

$$\chi(\lambda) = \prod_{0<|k|\le n+1} \left(1 + \frac{\lambda}{\theta_k}\right) \prod_{0<|k|\le n+1} \left(1 - \frac{\lambda}{\tau_k}\right). \tag{3.5.7}$$

We notice that there is a bounded open interval $\Lambda_\delta \subset (0,\infty)$ such that the sequence $(\tau_k)_{0<|k|\le n+1}$, together with α and $l \in \Lambda_\delta$ satisfy the assumptions of Theorem 3.4.3. Hence there are $(m_k)_{k=1}^{n+1}$, $(l_k)_{k=0}^{n}$ and $\gamma_2 > 0$ with $\sum\limits_{k=0}^{n} l_k = l$ such that $(\tau_k)_{0<|k|\le n+1}$ is the sequence of eigenvalues of the corresponding problem (1.2.4), (3.4.1), (3.5.2). By Corollary A.6.7 and the construction in the proof of Theorem 3.4.3 all quantities depend continuously on δ, including the (positive) endpoints of the interval Λ_δ.

Choosing $\varepsilon_0 > 0$ such that $[-\varepsilon_0, \varepsilon_0] \subset J_{\omega,\alpha}$ and

$$\widetilde{\Lambda} := \bigcap_{\delta=-\varepsilon_0}^{\varepsilon_0} \Lambda_\delta \ne \emptyset,$$

we let Λ be a nonempty compact subinterval of $\widetilde{\Lambda}$. Then it is clear that the generated data depend continuously on $(\delta, l) \in [-\varepsilon_0, \varepsilon_0] \times \Lambda$, and we will indicate this dependence when it is convenient. Defining

$$\varepsilon_1 := \min\{\gamma_2^{-1}(\delta, l) : (\delta, l) \in [-\varepsilon_0, \varepsilon_0] \times \Lambda\}$$

it is clear that $\varepsilon_1 > 0$. We finally put

$$\varepsilon := \min\{\varepsilon_0, \varepsilon_1\}.$$

Now let $l \in \Lambda$, $\Delta \in (-\varepsilon, \varepsilon)$, and set

$$\gamma_1 := \left(\gamma_2^{-1} - \Delta\right)^{-1}.$$

Clearly, $\gamma_1 > 0$ and $\gamma_2^{-1} - \gamma_1^{-1} = \Delta$. Comparing the identities (3.5.6) with $\delta = \Delta$ and (3.5.4), and observing the uniqueness of ψ as a solution of (3.5.6), it follows that

$$\frac{S_{2n+1}(\lambda, \gamma_{1,0})S_{2n+1}(-\lambda, \gamma_{2,0})}{S_{2n+1}(0, \gamma_{1,0})S_{2n+1}(0, \gamma_{2,0})} = \omega(\lambda)\psi(-\lambda).$$

Therefore the problem (1.2.4), (3.4.1), (3.5.1) generated by $(m_k)_{k=1}^{n+1}$, $(l_k)_{k=0}^{n}$ and the given α, l has the given sequence $(\theta_k)_{0<|k|\le n+1}$ as the sequence of the eigenvalues of this problem. This completes the existence proof.

Uniqueness follows from Theorem 3.5.2. □

3.6 Stieltjes strings with damping at an interior point

3.6.1 The direct problem

As in Section 1.4 we consider a Stieltjes string S composed two strings S_1 and S_2 bearing n_1 and n_2 beads, respectively, being damped at the point where S_1 and S_2 meet. The coefficient of damping will be denoted by $\alpha > 0$. With the notation introduced in Subsection 1.4.1 the problem is

$$\frac{u_k^{(j)} - u_{k+1}^{(j)}}{l_k^{(j)}} + \frac{u_k^{(j)} - u_{k-1}^{(j)}}{l_{k-1}^{(j)}} - m_k^{(j)}\lambda^2 u_k^{(j)} = 0, \quad (k = 1, \dots, n_j;\ j = 1, 2), \tag{3.6.1}$$

$$u_{n_1+1}^{(1)} = u_{n_2+1}^{(2)}, \tag{3.6.2}$$

$$\frac{u_{n_1+1}^{(1)} - u_{n_1}^{(1)}}{l_{n_1}^{(1)}} + \frac{u_{n_2+1}^{(2)} - u_{n_2}^{(2)}}{l_{n_2}^{(2)}} = -i\alpha\lambda u_{n_1+1}^{(1)}, \tag{3.6.3}$$

$$u_0^{(j)} = 0, \quad (j = 1, 2). \tag{3.6.4}$$

Due to (3.6.1)–(3.6.4) we have

$$R_{2n_1}^{(1)}(\lambda^2, 0)u_1^{(1)} = R_{2n_2}^{(2)}(\lambda^2, 0)u_1^{(2)},$$
$$R_{2n_1-1}^{(1)}(\lambda^2, 0)u_1^{(1)} + R_{2n_2-1}^{(2)}(\lambda^2, 0)u_1^{(2)} = -i\alpha\lambda R_{2n_1}^{(1)}(\lambda^2, 0)u_1^{(1)}.$$

This linear system has a nontrivial solution $(u_1^{(1)}, u_1^{(2)})$ if and only if the coefficient matrix is singular. Therefore, the eigenvalues of problem (3.6.1)–(3.6.4) coincide with the zeros of the polynomial Φ given by

$$R_{2n_1}^{(1)}(\lambda^2, 0)R_{2n_2-1}^{(2)}(\lambda^2, 0) + R_{2n_1-1}^{(1)}(\lambda^2, 0)R_{2n_2}^{(2)}(\lambda^2, 0) + i\alpha\lambda R_{2n_1}^{(1)}(\lambda^2, 0)R_{2n_2}^{(2)}(\lambda^2, 0).$$

Proposition 3.6.1. *The function Φ is of $\overline{\mathrm{SHB}}$ class. If $R^{(1)}_{2n_1}(\cdot,0)$ and $R^{(2)}_{2n_2}(\cdot,0)$ do not have common zeros, then $\Phi \in \mathrm{SHB}$. The function Φ has at most $2\min\{n_1,n_2\}$ real zeros, and $\Phi(0) \neq 0$. The real zeros of Φ are simple, and $\lambda \in \mathbb{R}$ is a zero of Φ if and only if $R^{(1)}_{2n_1}(\lambda^2,0) = 0$ and $R^{(2)}_{2n_2}(\lambda^2,0) = 0$. More precisely, if λ_0 is a real zero of Φ, then $\Phi'(\lambda_0) \in \mathbb{R} \setminus \{0\}$ and $\operatorname{Im}\Phi''(\lambda_0) \neq 0$.*

Proof. Evidently, $\Phi(-\overline{\lambda}) = \overline{\Phi(\lambda)}$. For the decomposition $\Phi(\lambda) = \widetilde{P}(\lambda^2) + i\lambda\widetilde{Q}(\lambda^2)$ according to (A.5.1) we have

$$\begin{aligned}\widetilde{P} &= R^{(1)}_{2n_1}(\cdot,0)R^{(2)}_{2n_2-1}(\cdot,0) + R^{(1)}_{2n_1-1}(\cdot,0)R^{(2)}_{2n_2}(\cdot,0),\\ \widetilde{Q} &= \alpha R^{(1)}_{2n_1}(\cdot,0)R^{(2)}_{2n_2}(\cdot,0).\end{aligned}$$

The function ϕ defined in (1.4.5) equals $\widetilde{P}$, and therefore Lemma 1.4.1 shows that $\frac{\widetilde{Q}}{\widetilde{P}}$ is an $\mathcal{S}_0$-function. Also, by Proposition 1.2.8, all zeros of $\widetilde{Q}$ are real and positive. Hence common zeros of $\widetilde{P}$ and $\widetilde{Q}$, if any, must be positive real numbers. Therefore $\Phi \in \overline{\mathrm{SHB}}$ by Proposition A.5.8.

If λ is a common zero of $\widetilde{P}$ and $\widetilde{Q}$, we may assume without loss of generality that $R^{(1)}_{2n_1}(\lambda,0) = 0$. We recall from (1.2.36) that $R^{(1)}_{2n_1-1}(\lambda,0) \neq 0$, and therefore $\widetilde{P}(\lambda) = 0$ if and only if $R^{(2)}_{2n_2}(\lambda,0) = 0$. Hence the common zeros of $\widetilde{P}$ and $\widetilde{Q}$, if any, are precisely the common zeros of $R^{(1)}_{2n_1}(\cdot,0)$ and $R^{(2)}_{2n_2}(\cdot,0)$. If there are no such common zeros, then it follows from Proposition A.5.8 that $\Phi \in \mathrm{SHB}$.

From Propositions 1.2.8 and 1.2.9 it follows that $\Phi(0) = \widetilde{P}(0) > 0$. Furthermore, any real zero λ of Φ is a common real zero of $\widetilde{P}$ and $\widetilde{Q}$, which in turn is equivalent to $R^{(1)}_{2n_1}(\lambda^2,0) = 0$ and $R^{(2)}_{2n_2}(\lambda^2,0) = 0$. Hence Φ has at most $2\min\{n_1,n_2\}$ distinct real zeros. We observe that the representation (1.2.19) and Propositions 1.2.8 and 1.2.9 show that $\frac{R^{(j)}_{2n_j}(\cdot,0)}{R^{(j)}_{2n_j-1}(\cdot,0)}$ $(j=1,2)$ are $\mathcal{S}_0$-functions. Then it follows for a common zero z_0 of $R^{(1)}_{2n_1}(\cdot,0)$ and $R^{(2)}_{2n_2}(\cdot,0)$ that

$$\begin{aligned}\frac{\widetilde{P}'(z_0)}{R^{(1)}_{2n_1-1}(z_0^2,0)R^{(2)}_{2n_2-1}(z_0^2,0)} &= \frac{(R^{(1)}_{2n_1}(\cdot,0))'(z_0)}{R^{(1)}_{2n_1-1}(z_0,0)} + \frac{(R^{(2)}_{2n_2}(\cdot,0))'(z_0)}{R^{(2)}_{2n_2-1}(z_0,0)}\\ &= \left(\frac{R^{(1)}_{2n_1}(\cdot,0)}{R^{(1)}_{2n_1-1}(\cdot,0)}\right)'(z_0) + \left(\frac{R^{(2)}_{2n_2}(\cdot,0)}{R^{(2)}_{2n_2-1}(\cdot,0)}\right)'(z_0) > 0\end{aligned}$$

in view of Lemma A.2.5. Also, since z_0 is a simple zero of $R^{(1)}_{2n_1}(\cdot,0)$ and $R^{(2)}_{2n_2}(\cdot,0)$, see Proposition 1.2.8, it follows that the zero z_0 of $\widetilde{Q}$ has multiplicity 2. For a real zero λ_0 of Φ it is therefore immediately clear that $\operatorname{Re}\Phi$ has a simple zero at λ_0 and that $\operatorname{Im}\Phi$ has a zero of multiplicity 2 at λ_0. □

Theorem 3.6.2. *Problem (3.6.1)–(3.6.4) has $2n_1 + 2n_2 + 1$ eigenvalues, counted with multiplicity, and the sequence $(\lambda_k)_{-(n_1+n_2)}^{n_1+n_2}$ of the eigenvalues can be indexed in such a way that:*

1. $\operatorname{Im}\lambda_k \geq 0$ *for* $k = -n_1 - n_2, \ldots, n_1 + n_2$;
2. $\lambda_{-k} = -\overline{\lambda_k}$ *for not pure imaginary* λ_{-k};
3. *all real eigenvalues (if any) are simple and nonzero;*
4. *for each real eigenvalue* λ_k *it is true that* $\operatorname{Im}\Phi'(\lambda_k) = 0$ *and* $\operatorname{Im}\Phi''(\lambda_k) \neq 0$;
5. *the number of real eigenvalues does not exceed* $2\min\{n_1, n_2\}$.

Proof. The degrees of the polynomials $R_{2n_j}^{(j)}(\cdot, 0)$ and $R_{2n_j-1}^{(j)}(\cdot, 0)$ are n_j $(j = 1, 2)$, and therefore the polynomial Φ has degree $2n_1 + 2n_2 + 1$. Hence this theorem is an immediate consequence of Proposition 3.6.1. □

Let us briefly discuss the physical meaning of real eigenvalues for this problem. Consider an undamped Stieltjes string with fixed ends. Then the spectrum of the corresponding boundary problem, i.e., of problem (3.6.1)–(3.6.4) is real. Let $(u_{1,k}^{(1)}, \ldots, u_{n_1,k}^{(1)}, u_{n_2,k}^{(2)}, \ldots, u_{1,k}^{(2)})$ be an eigenvector corresponding to the eigenvalue λ_k with $1 < k \leq n_1 + n_2$. Here $u_{j,k}^{(1)}$ and $u_{j,k}^{(2)}$ are the amplitudes of vibrations of the beads. Then the piecewise linear curve describing the threads between beads, i.e., the amplitude function of vibrations of the k-th frequency possesses $k - 1$ nodes (see Theorem 1.2.22). If now we apply one-dimensional damping at a node, then the real λ_k remains an eigenvalue for the damped string. The number of eigenvalues of the undamped string with the eigenvectors whose piecewise linear curves have nodes at the same point does not exceed $2\min\{n_1, n_2\}$. This is because the piecewise linear curves must be eigenfunctions of the problems on the parts of the interval. Thus we obtain statement 5 of Theorem 3.6.2 using just physical arguments.

3.6.2 The inverse problem

Theorem 3.6.2 states that conditions 1–5 therein are necessary for a sequence of complex numbers to be the eigenvalues of problem (3.6.1)–(3.6.4). Now we prove that they are sufficient.

Theorem 3.6.3. *Let two positive numbers* $l > 0$ *and* $\widehat{l} \in (0, l)$ *and two natural numbers* n_1 *and* n_2 *be given together with the sequence of complex numbers* $(\lambda_k)_{k=-(n_1+n_2)}^{n_1+n_2}$ *which satisfy properties 1–5 of Theorem 3.6.2 with*

$$\Phi(\lambda) = \prod_{k=-(n_1+n_2)}^{n_1+n_2} \left(1 - \frac{\lambda}{\lambda_k}\right). \tag{3.6.5}$$

Then there exist a positive real number α, *sequences of positive numbers* $(m_k^{(j)})_{k=1}^{n_j}$ *and* $(l_k^{(j)})_{j=0}^{n_j}$ $(j = 1, 2)$ *with* $\sum_{k=0}^{n_1} l_k^{(1)} = \widehat{l}$ *and* $\sum_{k=0}^{n_2} l_k^{(2)} = l - \widehat{l}$ *such that the eigenvalues of the problem (3.6.1)–(3.6.4) for these data coincide with* $(\lambda_k)_{k=-(n_1+n_2)}^{n_1+n_2}$.

Proof. By properties 2 and 3, the number of indices k for which λ_k is real is a nonnegative even integer, which we denote by $2m$, with $0 \le m \le n_1 + n_2$ in view of property 5. For convenience we may therefore assume that the given sequence is indexed in such a way that λ_k is real if and only if $|k| > n_1 + n_2 - 2m$. Defining

$$R(z) = \prod_{k=n_1+n_2-m+1}^{n_1+n_2} \left(1 - \frac{z}{\lambda_k^2}\right), \quad \Phi_1(\lambda) = \prod_{k=-(n_1+n_2-m)}^{n_1+n_2-m} \left(1 - \frac{\lambda}{\lambda_k}\right), \tag{3.6.6}$$

it follows that

$$\Phi(\lambda) = R(\lambda^2)\Phi_1(\lambda). \tag{3.6.7}$$

Clearly, R is a real polynomial of degree m with m simple positive real zeros, whereas Φ_1 is a polynomial of degree $2n_1 + 2n_2 + 1 - 2m$ belonging to the SHB class in view of properties 1 and 2. Hence there are real polynomials $\widetilde{P}$ and $\widetilde{Q}$ such that

$$\Phi_1(\lambda) = \widetilde{P}(\lambda^2) + i\lambda\widetilde{Q}(\lambda^2), \tag{3.6.8}$$

see (A.5.1). Proposition A.5.4 gives

$$\widetilde{P}(0) = \Phi_1(0) = 1 \quad \text{and} \quad \widetilde{Q}(0) = \sum_{k=-(n_1+n_2-m)}^{n_1+n_2-m} \frac{i}{\lambda_k} > 0. \tag{3.6.9}$$

We set

$$\alpha = \widetilde{Q}(0)\left(\widehat{l}^{-1} + (l - \widehat{l})^{-1}\right) \tag{3.6.10}$$

and observe that (3.6.9) shows that $\alpha > 0$.

By Theorems A.5.2 and A.5.5 the zeros of $\widetilde{P}$ and $\widetilde{Q}$ are real positive and interlace, and $\frac{\widetilde{Q}}{\widetilde{P}}$ is a rational $\mathcal{S}_0$-function. Since Φ_1 has degree $2n_1+2n_2+1-2m$, the polynomial $\widetilde{Q}$ has degree $n_1 + n_2 - m$ and $\widetilde{P}$ has degree at most $n_1 + n_2 - m$. On the other hand, since $\frac{\widetilde{Q}}{\widetilde{P}}$ is a rational $\mathcal{S}_0$-function, the degree of $\widetilde{Q}$ does not exceed the degree of $\widetilde{P}$. Hence also $\widetilde{P}$ has degree n_1+n_2-m. Denoting the (simple positive real) zeros of $\widetilde{Q}$ by ν_k $(k = 1, \dots, n_1+n_2-m)$, we conclude from Lemmas A.2.4 and A.2.7 that

$$\frac{\widetilde{P}(z)}{\widetilde{Q}(z)} = \sum_{k=1}^{n_1+n_2-m} \frac{A_k}{z - \nu_k} + B$$

with a real number B and positive numbers A_k $(k = 1, \dots, n_1 + n_2 - m)$. Substituting $z = 0$ it follows from (3.6.9) and (3.6.10) that

$$B = \sum_{k=1}^{n_1+n_2-m} \frac{A_k}{\nu_k} + \frac{1}{\alpha}\left(\frac{1}{\widehat{l}} + \frac{1}{l - \widehat{l}}\right). \tag{3.6.11}$$

If $m > 0$, then property 4 implies that each of the real zeros $\lambda_{n_1+n_2-m+k}$ $(k = 1, \dots, m)$ of Φ is a double zero of $\operatorname{Im}\Phi$, which means that $\lambda^2_{n_1+n_2-m+k}$

is a double zero of $R\widetilde{Q}$. Since $\lambda^2_{n_1+n_2-m+k}$ is a simple zero of R, it is therefore also a simple zero of $\widetilde{Q}$. We may therefore assume without loss of generality that $\nu_k = \lambda^2_{n_1+n_2-m+k}$ $(k = 1, \ldots, m)$. Setting $\nu_k^{(1)} := \nu_k^{(2)} = \nu_k$ for $k = 1, \ldots, m$ and writing the set $\{\nu_{m+1}, \ldots, \nu_{n_1+n_2-m}\}$ with $n_1 + n_2 - 2m$ elements in an arbitrary manner as the union of two sets $\{\nu^{(1)}_{m+1}, \ldots, \nu^{(1)}_{n_1}\}$ and $\{\nu^{(2)}_{m+1}, \ldots, \nu^{(2)}_{n_2}\}$ of $n_1 - m$ and $n_2 - m$ elements, respectively, it follows that

$$\frac{\widetilde{P}(z)}{\widetilde{Q}(z)} = \sum_{k=1}^{n_1} \frac{A_k^{(1)}}{z - \nu_k^{(1)}} + \sum_{k=1}^{n_2} \frac{A_k^{(2)}}{z - \nu_k^{(2)}} + B, \tag{3.6.12}$$

where $A_k^{(j)} := A_r$ when $\nu_k^{(j)} = \nu_r$ for $k = m + 1, \ldots, n_j$ and $j = 1, 2$, whereas $A_k^{(j)} > 0$ $(j = 1, 2)$ with $A_k^{(1)} + A_k^{(2)} = A_k$ for $j = 1, \ldots, m$. We define

$$B_1 = \sum_{k=1}^{n_1} \frac{A_k^{(1)}}{\nu_k^{(1)}} + \frac{1}{\alpha\widehat{l}}, \tag{3.6.13}$$

$$B_2 = \sum_{k=1}^{n_2} \frac{A_k^{(2)}}{\nu_k^{(2)}} + \frac{1}{\alpha(l - \widehat{l})}. \tag{3.6.14}$$

Evidently, $B_1 > 0$, $B_2 > 0$ and, by (3.6.11), $B_1 + B_2 = B$.

There are nonzero constants C_1 and C_2 such that $R\widetilde{Q} = Q_1Q_2$, where $Q_j(z) = C_j \prod\limits_{k=1}^{n_j} (z - \nu_k^{(j)})$, $(j = 1, 2)$, and we define

$$P_j(z) = Q_j(z) \left(\sum_{k=1}^{n_j} \frac{A_k^{(j)}}{z - \nu_k^{(j)}} + B_j \right), \quad (j = 1, 2). \tag{3.6.15}$$

Clearly, the representation (3.6.15) shows that $-\frac{P_j}{Q_j}$ is a sum of Nevanlinna functions and hence a Nevanlinna function with poles at $\nu_k^{(j)}$, $(k = 1, \ldots, n_j;\ j = 1, 2)$. Then also $\frac{Q_j}{P_j}$ is a Nevanlinna function. From $\lim\limits_{z\to-\infty} \frac{P_j(z)}{Q_j(z)} = B_j > 0$ it follows that both P_j and Q_j have the same degree n_j. Since $\frac{Q_j}{P_j}$ is increasing by Lemma A.2.5, the smallest pole of this function is smaller than the smallest zero. Furthermore, $\frac{Q_1(0)}{P_1(0)} = \alpha\widehat{l} > 0$ by (3.6.15) and (3.6.13), whereas $\frac{Q_2(0)}{P_2(0)} = \alpha(l - \widehat{l}) > 0$ by (3.6.15) and (3.6.14). Since the smallest zero of Q_j and thus also of $\frac{Q_j}{P_j}$ is positive it follows that the smallest pole of $\frac{Q_j}{P_j}$ must also be positive. Thus we have shown that $\frac{Q_j}{\alpha P_j}$ is an $\mathcal{S}_0$-function for $j = 1, 2$.

By Lemma A.3.5 we obtain the continued fraction expansions

$$\frac{Q_j(z)}{\alpha P_j(z)} = l_{n_j}^{(j)} + \cfrac{1}{-m_{n_j}^{(j)} z + \cfrac{1}{l_{n_j-1}^{(j)} + \cfrac{1}{-m_{n_j-1}^{(j)} z + \dots + \cfrac{1}{l_1^{(j)} + \cfrac{1}{-m_1^{(j)} z + \cfrac{1}{l_0^{(j)}}}}}}} \tag{3.6.16}$$

with $l_k^{(j)} > 0$ ($k = 0, \dots, n_j$, $j = 1, 2$) and $m_k^{(j)} > 0$ ($k = 1, \dots, n_j$, $j = 1, 2$). Denoting the Cauer-Fry polynomials for these string data by $R_k^{(j)}(\cdot, 0)$, we know from (1.2.19) that the right-hand side of (3.6.16) is equal to $\frac{R_{2n_1}^{(j)}(z,0)}{R_{2n_1-1}^{(j)}(z,0)}$, ($j = 1, 2$). Therefore it follows from (3.6.12), (3.6.15) and (3.6.16) that

$$\begin{aligned}\frac{\alpha R\widetilde{P}}{R\widetilde{Q}} = \frac{\alpha\widetilde{P}}{\widetilde{Q}} = \frac{\alpha P_1}{Q_1} + \frac{\alpha P_2}{Q_2} &= \frac{R_{2n_1-1}^{(1)}(\cdot,0)}{R_{2n_1}^{(1)}(\cdot,0)} + \frac{R_{2n_2-1}^{(2)}(\cdot,0)}{R_{2n_2}^{(2)}(\cdot,0)} \\ &= \frac{R_{2n_1-1}^{(1)}(\cdot,0)R_{2n_2}^{(2)}(\cdot,0) + R_{2n_1}^{(1)}(\cdot,0)R_{2n_2-1}^{(2)}(\cdot,0)}{R_{2n_1}^{(1)}(\cdot,0)R_{2n_2}^{(2)}(\cdot,0)}.\end{aligned}$$

Since Q_j and $R_{2n_j}^{(j)}(\cdot, 0)$ have the same degree n_j and the same zeros, there are numbers $T_j \in \mathbb{C}\backslash\{0\}$ such that $Q_j = T_j R_{2n_j}^{(j)}(\cdot, 0)$. With $T = \alpha^{-1}T_1T_2$ we conclude

$$R\widetilde{Q} = Q_1Q_2 = T\alpha R_{2n_1}^{(1)}(\cdot,0)R_{2n_2}^{(2)}(\cdot,0)$$

and then

$$R\widetilde{P} = T[R_{2n_1-1}^{(1)}(\cdot,0)R_{2n_2}^{(2)}(\cdot,0) + R_{2n_1}^{(1)}(\cdot,0)R_{2n_2-1}^{(2)}(\cdot,0)].$$

We conclude that

$$\begin{aligned}\Phi(\lambda) &= R(\lambda^2)\widetilde{P}(\lambda^2) + i\lambda R(\lambda^2)\widetilde{Q}(\lambda^2) \\ &= T[R_{2n_1}^{(1)}(\lambda^2,0)R_{2n_2-1}^{(2)}(\lambda^2,0) + R_{2n_1-1}^{(1)}(\lambda^2,0)R_{2n_2}^{(2)}(\lambda^2,0) \\ &\quad + i\alpha\lambda R_{2n_1}^{(1)}(\lambda^2,0)R_{2n_2}^{(2)}(\lambda^2,0)].\end{aligned}$$

In Subsection 3.6.1 we have seen that the zeros of the function on the right-hand side are the eigenvalues of problem (3.6.1)–(3.6.4) associated with the generated data. Hence these eigenvalues coincide with the sequence $(\lambda_k)_{k=-(n_1+n_2)}^{n_1+n_2}$.

Finally, (3.6.13)–(3.6.15) gives

$$\frac{\alpha P_1(0)}{Q_1(0)} = \frac{1}{\widehat{l}}, \quad \frac{\alpha P_2(0)}{Q_2(0)} = \frac{1}{l-\widehat{l}}.$$

According to (3.6.16) and Propositions 1.2.8 and 1.2.9 this implies that $\sum\limits_{k=0}^{n_1} l_k^{(1)} = \widehat{l}$ and $\sum\limits_{k=0}^{n_2} l_k^{(2)} = l - \widehat{l}$. □

3.7 Star graphs with damped central vertex

3.7.1 The direct problem

We consider a plane star graph of $g \geq 2$ Stieltjes strings. We assume all pendant vertices to be fixed. A bead of mass $M > 0$ may be placed at the central vertex; in case there is no bead at the central vertex we put $M = 0$. In Section 2.1 we considered two problems of small transverse vibrations of the graph: one with the the central vertex fixed, called Dirichlet problem (D1), and one with the central vertex free to move in the direction orthogonal to the equilibrium position of the strings, called Neumann problem (N1). In this section it is assumed that the central vertex can move subject to damping with constant coefficient of damping $\alpha > 0$, and we call this the *damped Neumann problem.* It is clear that the damping only affects the balance of forces condition, and therefore the damped Neumann problem is described by (2.1.4), (2.1.5), (2.1.7) and

$$\sum_{j=1}^{g} \frac{u_{n_j+1}^{(j)} - u_{n_j}^{(j)}}{l_{n_j}^{(j)}} - (M\lambda^2 - i\alpha\lambda)u_{n_1+1}^{(1)} = 0. \tag{3.7.1}$$

If we clamp the central vertex, then the damping has no effect, so that we obtain the Dirichlet problem (D1).

According to (1.2.8) we look for solutions of (2.1.4) and (2.1.7) in the form

$$u_k^{(j)} = R_{2k-2}^{(j)}(\lambda^2)u_1^{(j)} \quad (k = 1, \ldots, n_j),$$

where $R_{2k-2}^{(j)} := R_{2k-2}^{(j)}(\cdot, 0)$ are the Cauer-Fry polynomials of degree $k-1$ corresponding to the j-th string. Since (2.1.5) and (3.7.1) have to be satisfied simultaneously, they are equivalent to (2.1.5) and

$$\sum_{j=1}^{g} \frac{u_{n_j+1}^{(j)} - u_{n_j}^{(j)}}{l_{n_j}^{(j)}} = \sum_{j=1}^{g} \frac{1}{g}(M\lambda^2 - i\alpha\lambda)u_{n_j+1}^{(j)}. \tag{3.7.2}$$

Then (2.1.5) and (3.7.2) can be written as

$$R_{2n_1}^{(1)}(\lambda^2)u_1^{(1)} = \cdots = R_{2n_g}^{(g)}(\lambda^2)u_1^{(g)}, \tag{3.7.3}$$

$$\sum_{j=1}^{g} R_{2n_j-1}^{(j)}(\lambda^2)u_1^{(j)} = \sum_{j=1}^{g} \frac{1}{g}(M\lambda^2 - i\alpha\lambda)R_{2n_j}^{(j)}(\lambda^2)u_1^{(j)}. \tag{3.7.4}$$

As in Subsection 2.1.1 we find that $(u_k^{(j)})_{j=1,k=0}^{g,n_j+1}$ is a nontrivial solution of the damped problem (2.1.4), (2.1.5), (2.1.7), (3.7.1) if and only if the matrix

$$A_\alpha(\lambda) = \begin{pmatrix} R_{2n_1}^{(1)}(\lambda^2) & -R_{2n_2}^{(2)}(\lambda^2) & 0 & \cdots & 0 & 0 \\ 0 & R_{2n_2}^{(2)}(\lambda^2) & -R_{2n_3}^{(3)}(\lambda^2) & \cdots & 0 & 0 \\ \vdots & \ddots & \ddots & \ddots & \vdots & \vdots \\ 0 & 0 & 0 & \cdots & R_{2n_{g-1}}^{(g-1)}(\lambda^2) & -R_{2n_g}^{(g)}(\lambda^2) \\ a_{\alpha,1}(\lambda) & a_{\alpha,2}(\lambda) & a_{\alpha,3}(z) & \cdots & a_{\alpha,g-1}(\lambda) & a_{\alpha,g}(\lambda) \end{pmatrix}$$

is singular, where

$$a_{\alpha,j}(\lambda) = R_{2n_j-1}^{(j)}(\lambda^2) - \frac{1}{g}(M\lambda^2 - i\alpha\lambda)R_{2n_j}^{(j)}(\lambda^2) \quad (j = 1, \ldots, g).$$

Expanding the determinant of $A_\alpha(\lambda)$ with respect to the last row and recalling the definitions of $\phi_{D,g}$ and $\phi_{N,g}$ in (2.1.9) and (2.1.10), it follows that

$$\phi(\lambda) := \det A_\alpha(\lambda) = \phi_{N,g}(\lambda^2) + i\alpha\lambda\phi_{D,g}(\lambda^2). \tag{3.7.5}$$

We have shown

Proposition 3.7.1. *The eigenvalues of* (2.1.4), (2.1.5), (2.1.7), (3.7.1) *are the zeros of the polynomial* ϕ *given by* (3.7.5).

Remark 3.7.2. From Theorem 2.1.5 and its proof we know that the degree of the polynomial $\phi_{N,g}$ is $1 + \sum_{j=1}^g n_j$ if $M > 0$ and $\sum_{j=1}^g n_j$ if $M = 0$, and that the degree of $\phi_{D,g}$ is $\sum_{j=1}^g n_j$ in any case. Therefore the degree of ϕ is $2 + 2\sum_{j=1}^g n_j$ if $M > 0$ and $1 + 2\sum_{j=1}^g n_j$ if $M = 0$.

Next we consider the damped problem in the case that none of the threads has an interior bead.

Example 3.7.3. Assume that $n_j = 0$ for all $j = 1, \ldots, g$.

1. If $M = 0$, then the damped problem has exactly one simple eigenvalue $\frac{i}{\alpha}\sum\limits_{j=1}^{g}\frac{1}{l^{(j)}}$.
2. If $M > 0$, then the damped problem has exactly two eigenvalues, counted with multiplicity, namely

$$\frac{i\alpha}{2M} \pm \left(\frac{1}{M}\sum_{j=1}^{g}\frac{1}{l^{(j)}} - \frac{\alpha^2}{4M^2}\right)^{\frac{1}{2}}.$$

Proof. For $n_j = 0$ we have $R_{2n_j}^{(j)} = R_0^{(j)} = 1$ and $R_{2n_j-1}^{(j)} = R_{-1} = \frac{1}{l^{(j)}}$, see (1.2.12) and (1.2.14). Therefore $\phi_{D,g}(\lambda^2) = 1$ and

$$\phi_{N,g}(\lambda^2) = \sum_{j=1}^{g}\left[\frac{1}{l^{(j)}} - \frac{M}{g}\lambda^2\right] = \sum_{j=1}^{g}\frac{1}{l^{(j)}} - M\lambda^2,$$

which gives

$$\phi(\lambda) = -M\lambda^2 + \sum_{j=1}^{g} \frac{1}{l^{(j)}} + i\alpha\lambda.$$

Solving $\phi(\lambda) = 0$ completes the proof. □

Theorem 3.7.4. *The polynomial ϕ belongs to the class $\overline{\mathrm{SHB}}$.*

Proof. Let ϕ_0 be the monic polynomial whose zeros (if any) are the common zeros of $\phi_{N,g}$ and $\phi_{D,g}$, counted with multiplicity, and let

$$\widetilde{\phi}_N = \frac{\phi_{N,g}}{\phi_0}, \quad \widetilde{\phi}_D = \frac{\phi_{D,g}}{\phi_0}.$$

Then $\widetilde{\phi}_N$ and $\widetilde{\phi}_D$ have no common zeros, and

$$\frac{\widetilde{\phi}_D}{\widetilde{\phi}_N} = \frac{\phi_D}{\phi_N}$$

is an S_0-function by Theorem 2.1.2. Setting $P(\lambda) = \widetilde{\phi}_N(\lambda^2)$ and $Q(\lambda) = \alpha\lambda\widetilde{\phi}_D(\lambda^2)$ it follows that the zeros of P and Q are real and interlace and that

$$Q'(0)P(0) - Q(0)P'(0) = \alpha\widetilde{\phi}_D(0)\widetilde{\phi}_N(0) > 0.$$

Hence Theorem A.5.2 gives that

$$\widetilde{\phi}(\lambda) := \widetilde{\phi}_N(\lambda^2) + i\alpha\lambda\widetilde{\phi}_D(\lambda^2)$$

describes a Hermite-Biehler polynomial $\widetilde{\phi}$. By Theorems 2.1.3, 2.1.4 and 2.1.5, the zeros of $\phi_{N,g}$ and $\phi_{D,g}$ are real and positive, and therefore all zeros of $\lambda \mapsto \phi_0(\lambda^2)$ are real. Since $\phi(\lambda) = \phi_0(\lambda^2)\widetilde{\phi}(\lambda)$, it follows that ϕ is a generalized Hermite-Biehler polynomial. It is clear that the polynomial ϕ is symmetric. □

The following factorization of ϕ immediately follows from Theorem 3.7.4, (2.1.9), (2.1.10), (1.2.15) and (1.2.17).

Corollary 3.7.5. *There are $p_1, p_2, p_3 \in \mathbb{N}_0$ such that the zeros of ϕ, counted with multiplicity, can be represented as the union of (possibly zero length) sequences $(\beta_k)_{k=-p_1,k\neq 0}^{p_1}$, $(\gamma_k)_{k=1}^{p_2}$, and $(\delta_k)_{k=-p_3,k\neq 0}^{p_3}$ with the following properties:*

(i) $\beta_k \in \mathbb{R} \setminus \{0\}$ *and* $\beta_{-k} = \beta_k$ *for* $k = 1, \dots, p_1$;

(ii) γ_k *lies on the positive imaginary semiaxis for* $k = 1, \dots, p_2$;

(iii) $\operatorname{Re}\delta_k \neq 0$, $\operatorname{Im}\delta_k > 0$ *and* $\delta_{-k} = -\overline{\delta_k}$ *for* $k = 1, \dots, p_3$.

Then

$$\phi(\lambda) = \phi(0) \prod_{k=-p_1, k\neq 0}^{p_1} \left(1 - \frac{\lambda}{\beta_k}\right) \prod_{k=-p_3, k\neq 0}^{p_3} \left(1 - \frac{\lambda}{\delta_k}\right) \prod_{k=1}^{p_2} \left(1 - \frac{\lambda}{\gamma_k}\right),$$

where

$$\phi(0) = \left(\prod_{j=1}^{g} \frac{l^{(j)}}{l_0^{(j)}}\right)\left(\sum_{j=1}^{g} \frac{1}{l^{(j)}}\right).$$

We recall from Theorem 2.1.5 that the zeros $(\mu_k)_{k=1}^{n'}$ of $\phi_{N,g}$ and the zeros $(\nu_k)_{k=1}^{n}$ of $\phi_{D,g}$ are positive real numbers, where $n = \sum_{j=1}^{g} n_j$ and $n' = n$ if $M = 0$ or $n' = n + 1$ if $M > 0$. For any of those zeros λ, $m(\lambda)$ denotes their multiplicity as a zero of the corresponding function.

Theorem 3.7.6. 1. $(\beta_k^2)_{k=0}^{p_1} = (\mu_k)_{k=1}^{n'} \cap (\nu_k)_{k=1}^{n}$.

2. *If* $\beta_k^2 = \mu_p = \nu_p$, *then* $m(\mu_p) = m(\beta_k)$, $m(\nu_p) = m(\beta_k)+1$ *and* $m(\beta_k) \le g-1$.
3. *Let* $\widetilde{p}_1$ *be the number of distinct positive real zeros of* ϕ. *Then* $2\widetilde{p}_1 \le p_2 + 2p_3 - 1$.
4. *The number* p_2 *is odd if and only if* $M = 0$.
5. $\operatorname{Im} \phi^{(s)}(\beta_k) = 0$ *for* $s = 0, \ldots, m(\beta_k)$.

Proof. 1, 2. Since $\phi_{N,g}$ and $\phi_{D,g}$ are real polynomials, it follows for real λ that $\phi(\lambda) = 0$ if and only if $\phi_{N,g}(\lambda^2) = 0$ and $\phi_{D,g}(\lambda^2) = 0$. In this case, the multiplicity of λ as a zero of ϕ is clearly the minimum of the multiplicities of λ^2 as a zero of $\phi_{N,g}$ and $\phi_{D,g}$. This proves statement 1. If $\beta_k^2 = \mu_p$ is such a common zero, we may assume without loss of generality that p is the smallest index such that $\beta_k^2 = \mu_p = \nu_p$, and we let r be the largest index such that $\beta_k^2 = \mu_r = \nu_r$. In view of Theorem 2.1.5, parts 1 and 2, we have $p \ge 2$ and

$$\mu_{p-1} < \nu_{p-1} = \mu_p = \nu_p = \cdots = \mu_r = \nu_r .$$

Hence $m(\nu_p) = m(\mu_p) + 1$, which proves the first part of statement 2. Since $m(\nu_k) \le g$ by Theorem 2.1.5, part 3, also $m(\beta_k) \le g - 1$ has been shown.

3. By part 2, every real zero of ϕ is the square root of a zero of multiplicity at least 2 of $\phi_{D,g}$. In view of Remark 3.7.2 and $n = \sum_{j=1}^{g} n_j$ it follows that

$$2\widetilde{p}_1 \le n < \frac{1}{2}(2p_1 + p_2 + 2p_3).$$

Therefore

$$\widetilde{p}_1 < p_1 - \widetilde{p}_1 + \frac{1}{2}p_2 + p_3 \le \frac{1}{2}p_2 + p_3,$$

which implies $2\widetilde{p}_1 \le p_2 + 2p_3 - 1$.

4. Since the degree $2p_1+p_2+2p_3$ of ϕ is odd if and only if p_2 is odd, it follows from Remark 3.7.2 that p_2 is odd if and only if $M=0$.

5. By (3.7.5), $\operatorname{Im}\phi(\lambda)=\alpha\lambda\phi_{D,g}(\lambda^2)$ for real λ. By part 2, every β_k is a zero of multiplicity $m(\beta_k)+1$ of $\phi_{D,g}$, and therefore $\operatorname{Im}\phi^{(s)}(\beta_k)=0$ for $s=0,\dots,m(\beta_k)$. The representation of ϕ is clear from $\phi(0)=\phi_{N,g}(0)$. □

3.7.2 The inverse problem

In this subsection it will be shown that the necessary properties of the eigenvalues which were derived in Subsection 3.7.1 are also sufficient to find a star graph which has those eigenvalues. Furthermore, apart from the eigenvalues, also the number of edges and their total lengths can be prescribed.

Theorem 3.7.7. *Let $g\ge 2$ be an integer and let positive numbers $l^{(j)}$, $j=1,\dots,g$, be given. Let the sequences of numbers $(\beta_k)_{k=-p_1,k\neq 0}^{p_1}$, $(\gamma_k)_{k=1}^{p_2}$ and $(\delta_k)_{k=-p_3,k\neq 0}^{p_3}$ be given, where $p_1,p_2,p_3\in\mathbb{N}_0$. Assume that the following conditions are satisfied:*

(i) *For all indices k, $\beta_k\in\mathbb{R}\setminus\{0\}$, $\beta_{-k}=-\beta_k$, $2\widetilde{p}_1\le p_2+2p_3-1$, where $\widetilde{p}_1$ is the number of distinct elements of the sequence $(\beta_k)_{k=0}^{p_1}$.*

(ii) *For each k, $m(\beta_k)$ does not exceed $g-1$, where $m(\beta_k)$ is the number of times β_k occurs in the sequence $(\beta_j)_{j=-p_1,k\neq 0}^{p_1}$.*

(iii) *$\gamma_k=i|\gamma_k|$ and $\gamma_k\neq 0$ for all k.*

(iv) *$\operatorname{Im}\delta_k>0$, $\operatorname{Re}\delta_k\neq 0$ and $\delta_{-k}=-\overline{\delta_k}$ for all k.*

(v) *$\operatorname{Im}\Phi^{(s)}(\beta_k)=0$ for all k and for $s=0,\dots,m(\beta_k)$, where*

$$\Phi(\lambda)=\prod_{k=-p_1,k\neq 0}^{p_1}\left(1-\frac{\lambda}{\beta_k}\right)\prod_{k=-p_3,k\neq 0}^{p_3}\left(1-\frac{\lambda}{\delta_k}\right)\prod_{k=1}^{p_2}\left(1-\frac{\lambda}{\gamma_k}\right). \tag{3.7.6}$$

Then there exists a collection of sequences of positive real numbers $(m_k^{(j)})_{k=1}^{n_j}$ and $(l_k^{(j)})_{k=0}^{n_j}$ with $\sum_{k=0}^{n_j}l_k^{(j)}=l^{(j)}$ for $j=1,\dots,g$, numbers $M\ge 0$ and $\alpha>0$ such that problem (2.1.4), (2.1.5), (2.1.7), (3.7.1) with these data has eigenvalues $(\beta_k)_{k=-p_1,k\neq 0}^{p_1}\cup(\gamma_k)_{k=1}^{p_2}\cup(\delta_k)_{k=-p_3,k\neq 0}^{p_3}$, counted with multiplicity. Here $M=0$ if and only if p_2 is odd.

Proof. For $\lambda\in\mathbb{C}$ define

$$R(\lambda)=\prod_{k=1}^{p_1}\left(1-\frac{\lambda}{\beta_k^2}\right),\quad \omega(\lambda)=\prod_{k=-p_3,k\neq 0}^{p_3}\left(1-\frac{\lambda}{\delta_k}\right)\prod_{k=1}^{p_2}\left(1-\frac{\lambda}{\gamma_k}\right).$$

By definition, $\Phi(\lambda)=R(\lambda^2)\omega(\lambda)$ for $\lambda\in\mathbb{C}$, and it is clear that ω is a symmetric Hermite-Biehler polynomial. Hence there are real polynomials $\widetilde{P}$ and $\widetilde{Q}$ such that

$\omega(\lambda) = \widetilde{P}(\lambda^2) + i\lambda\widetilde{Q}(\lambda^2)$ for $\lambda \in \mathbb{C}$. By Theorems A.5.2 and A.5.5 and Proposition A.5.4, $\widetilde{P}$ and $\widetilde{Q}$ do not have common zeros and $\frac{\widetilde{Q}}{\widetilde{P}}$ is an $\mathcal{S}_0$-function with $\widetilde{P}(z) > 0$ and $\widetilde{Q}(z) > 0$ for $z \in (-\infty, 0]$.

Case I: p_2 is even. The numbers $n = p_1 + \frac{1}{2}p_2 + p_3 - 1$ and $n - p_1$ are nonnegative integers because $\frac{1}{2}p_2 + p_3 > 0$ by condition (i).

First we consider the case that $n - p_1 = 0$. Then $p_2 + 2p_3 = 2(n - p_1) + 2 = 2$ and $2\widetilde{p}_1 \le p_2 + 2p_3 - 1 = 1$ shows that $\widetilde{p}_1 = 0$ and therefore $p_1 = 0$. Hence

$$\Phi(\lambda) = \left(1 - \frac{\lambda}{\lambda_1}\right)\left(1 - \frac{\lambda}{\lambda_2}\right) = 1 - \lambda\left(\frac{1}{\lambda_1} + \frac{1}{\lambda_2}\right) + \frac{\lambda^2}{\lambda_1\lambda_2},$$

where λ_1 and λ_2 are nonzero complex numbers satisfying $\lambda_1 = i|\lambda_1|$ and $\lambda_2 = i|\lambda_2|$ or $\lambda_2 = -\overline{\lambda_1}$ and $\operatorname{Im}\lambda_1 > 0$. Then problem (2.1.4), (2.1.5), (2.1.7), (3.7.1) with

$$M = -\frac{1}{\lambda_1\lambda_2}\sum_{j=1}^{g}\frac{1}{l(j)} > 0, \quad \alpha = i\left(\frac{1}{\lambda_1} + \frac{1}{\lambda_2}\right)\sum_{j=1}^{g}\frac{1}{l(j)} > 0,$$

and $n_j = 0$ for $j = 0, \dots, g$ has the two eigenvalues (λ_1, λ_2) by Example 3.7.3.

Now let $n - p_1 > 0$. Since ω is a polynomial of even degree $p_2 + 2p_3$, the degree of the polynomial $\widetilde{P}$ is $\frac{1}{2}p_2 + p_3 = n + 1 - p_1$ and the degree of $\widetilde{Q}$ is smaller than the degree of $\widetilde{P}$. Recalling that $\widetilde{Q}$ and $\widetilde{P}$ do not have common zeros and that $\frac{\widetilde{Q}}{\widetilde{P}}$ is an $\mathcal{S}_0$-function, we conclude from Proposition A.3.2 and Theorem A.2.6 that there are positive numbers $\widetilde{\mu}_k$ $(k = 1, \dots, n + 1 - p_1)$ and $\widetilde{\nu}_k$ $(k = 1, \dots, n - p_1)$ such that $(\widetilde{\mu}_k)_{k=1}^{n+1-p_1}$ is the sequence of the zeros of $\widetilde{P}$, such that $(\widetilde{\nu}_k)_{k=1}^{n-p_1}$ is the sequence of the zeros of $\widetilde{Q}$, and such that

$$0 < \widetilde{\mu}_1 < \widetilde{\nu}_1 < \widetilde{\mu}_2 < \widetilde{\nu}_2 < \dots < \widetilde{\nu}_{n-p_1} < \widetilde{\mu}_{n+1-p_1}.$$

In particular, the degree of $\widetilde{Q}$ is $n - p_1 > 0$. Due to condition (v) we see that each β_k^2 is a zero of multiplicity at least $m(\beta_k) + 1$ of $R\widetilde{Q}$, whereas β_k^2 is a zero of R of multiplicity $m(\beta_k)$. Therefore $\widetilde{Q}(\beta_k^2) = 0$, and each β_k^2 coincides with some $\widetilde{\nu}_r$. Let

$$(\mu_k)_{k=1}^{n+1} := (\widetilde{\mu}_k)_{k=1}^{n+1-p_1} \cup (\beta_k^2)_{k=1}^{p_1},$$
$$(\nu_k)_{k=1}^{n} := (\widetilde{\nu}_k)_{k=1}^{n-p_1} \cup (\beta_k^2)_{k=1}^{p_1},$$

with $\mu_k \le \mu_{k+1}$ and $\nu_k \le \nu_{k+1}$.

We claim that these two sequences $(\mu_k)_{k=1}^{n+1}$ and $(\nu_k)_{k=1}^{n}$ satisfy the assumptions of Theorem 2.1.8 with $n' = n + 1$. For $k \in \{1, \dots, p_1\}$ we have $\widetilde{\nu}_1 \le \beta_k^2 \le \widetilde{\nu}_{n-p_1}$ and therefore $\mu_1 = \widetilde{\mu}_1 < \widetilde{\nu}_1 = \nu_1$ and $\nu_n = \widetilde{\nu}_{n-p_1} < \widetilde{\mu}_{n+1-p_1} = \mu_{n+1}$. Hence condition (i) of Theorem 2.1.8 holds (see Lemma A.4.4). Condition (ii) then also holds if we observe that $\mu_r = \nu_r$ holds if and only if there is k such that $\mu_r = \nu_r = \beta_k^2$, and in this case $m(\mu_r) = m(\beta_k) = m(\nu_r) - 1$. Finally, condition (iii) of Theorem 2.1.8 is satisfied because of assumption (ii).

Thus, in view of Theorem 2.1.8, there exists a star graph of g Stieltjes strings with all quantities as indicated in the statement of that theorem, and in particular $M > 0$, except for α, such that $(\mu_k)_{k=1}^{n+1}$ and $(\nu_k)_{k=1}^{n}$ are the sequences of the eigenvalues of the Neumann and Dirichlet problems, respectively. Let $\phi_{N,g}$ and $\phi_{D,g}$ be the corresponding characteristic functions, see (2.1.9) and (2.1.10). The polynomials $R\widetilde{P}$ and $\phi_{N,g}$ have the same zeros, counted with multiplicity, and the polynomials $R\widetilde{Q}$ and $\phi_{D,g}$ have the same zeros, counted with multiplicity. Observing that $R(0) = 1$, $\widetilde{P}(0) = 1$, $\widetilde{Q}(0) > 0$ by (A.5.2), $\phi_{N,g}(0) > 0$ and $\phi_{D,g}(0) > 0$, we obtain positive real numbers α_N and α_D such that $R\widetilde{P} = \alpha_N \phi_{N,g}$ and $R\widetilde{Q} = \alpha_D \phi_{D,g}$. Putting $\alpha = \alpha_D \alpha_N^{-1}$, it follows that

$$\Phi(\lambda) = R(\lambda^2)\widetilde{P}(\lambda^2) + i\lambda R(\lambda^2)\widetilde{Q}(\lambda^2) = \alpha_N[\phi_{N,g}(\lambda^2) + i\alpha\lambda\phi_{D,g}(\lambda^2)], \tag{3.7.7}$$

and hence Proposition 3.7.1 shows that the eigenvalues of the damped problem for this star graph and this α coincide with the zeros of Φ, that is, the given sequences.

Case II: p_2 is odd. The numbers $n = p_1 + \frac{1}{2}p_2 + p_3 - \frac{1}{2}$ and $n - p_1$ are nonnegative integers because $\frac{1}{2}p_2 + p_3 - \frac{1}{2} \geq 0$ by condition (i).

First we consider the case that $n - p_1 = 0$. Then $p_2 + 2p_3 = 2(n - p_1) + 1 = 1$ and $2\widetilde{p}_1 \leq p_2 + 2p_3 - 1 = 0$ shows that $p_2 = 1$, $p_3 = 0$ and $\widetilde{p}_1 = 0$, and therefore $p_1 = 0$. Hence

$$\Phi(\lambda) = 1 - \frac{\lambda}{\gamma_1},$$

where $\delta_1 = i|\delta_1|$ and $\delta_1 \neq 0$. Then problem (2.1.4), (2.1.5), (2.1.7), (3.7.1) with

$$M = 0, \quad \alpha = \frac{i}{\gamma_1}\sum_{j=1}^{q}\frac{1}{l^{(j)}} > 0,$$

and $n_j = 0$ for $j = 0, \dots, g$ has the single eigenvalue γ_1 by Example 3.7.3.

Now let $n - p_1 > 0$. Since ω is a polynomial of odd degree $p_2 + 2p_3$, the degree of $\widetilde{Q}$ is $\frac{1}{2}p_2 + p_3 - \frac{1}{2} = n - p_1$ and the degree of $\widetilde{P}$ does not exceed the degree of $\widetilde{Q}$. Recalling that $\widetilde{Q}$ and $\widetilde{P}$ do not have common zeros and that $\frac{\widetilde{Q}}{\widetilde{P}}$ is an S_0-function, we conclude from Proposition A.3.2 that there are positive numbers $\widetilde{\mu}_k$ $(k = 1, \dots, n-p_1)$ and $\widetilde{\nu}_k$ $(k = 1, \dots, n-p_1)$ such that $(\widetilde{\mu}_k)_{k=1}^{n-p_1}$ is the sequence of the zeros of $\widetilde{P}$, such that $(\widetilde{\nu}_k)_{k=1}^{n-p_1}$ is the sequence of the zeros of $\widetilde{Q}$, and such that

$$0 < \widetilde{\mu}_1 < \widetilde{\nu}_1 < \widetilde{\mu}_2 < \widetilde{\nu}_2 < \cdots < \widetilde{\nu}_{n-p_1}.$$

Proceeding as in Case I we obtain a star graph of strings, but now with $M = 0$, and the constant $\alpha > 0$, such that the corresponding damped problem has the zeros of Φ, that is, the given sequences of numbers as eigenvalues. □

Remark 3.7.8. The star graphs and the damping coefficient in Theorem 3.7.7 can be constructed explicitly. A construction of star graphs is given in the proof of [114, Theorem 2.9] for $n - p_1 > 0$ and in the proof of Theorem 3.7.7 for $n - p_1 = 0$.

To find α, we deduce from the proof of Theorem 3.6.3 and from (A.5.2), (2.1.13), (1.2.15), (1.2.17) that

$$\begin{aligned}\alpha = \alpha_D \alpha_N^{-1} &= \frac{R(0)\widetilde{Q}(0)}{\phi_{D,g}(0)} \frac{\phi_{N,g}(0)}{R(0)\widetilde{P}(0)} \\ &= i \left(\sum_{k=-p_3, k \neq 0}^{p_3} \frac{1}{\delta_k} + \sum_{k=1}^{p_2} \frac{1}{\gamma_k} \right) \sum_{j=1}^{q} \frac{1}{l^{(j)}}.\end{aligned}$$

Remark 3.7.9. We have seen in Remark 3.7.2 that the number of eigenvalues uniquely determines the total number of beads in the interiors of the strings and whether or not there is a bead at the joint vertex. We have also seen in Remark 3.7.8 that the damping coefficient α is uniquely determined by the spectrum. Finally, we know from (3.7.7) that

$$\frac{\phi_{N,g}(\lambda^2)}{\phi_{D,g}(\lambda^2)} = i\alpha\lambda \frac{\Phi(\lambda) + \Phi(-\lambda)}{\Phi(\lambda) - \Phi(-\lambda)}.$$

Hence we conclude from (2.1.13) that

$$\begin{aligned}M &= -\lim_{\lambda\to\infty} \frac{\phi_{N,g}(\lambda^2)}{\lambda^2 \phi_{D,g}(\lambda^2)} \\ &= -i\alpha \lim_{\lambda\to\infty} \frac{\Phi(\lambda) + \Phi(-\lambda)}{\lambda(\Phi(\lambda) - \Phi(-\lambda))},\end{aligned}$$

which means that M is uniquely determined by the eigenvalues of the damped problem.

3.8 Notes

The investigation of spectral problems for damped systems was started in [47]. As far as we know first results on inverse problems for damped strings were obtain in [4] and [80], [81]. In these papers the left end of the string was supposed to be free and the right end damped or in other words the right end of the string could move with viscous friction in the direction orthogonal to the equilibrium position of the string. Very wide classes of strings were considered. In [4] the class of the so-called regular strings was used, i.e., strings of finite mass and length, while in [80], [81] the class of so-called S-strings, i.e., strings of finite lengths and finite first momentum of mass distribution. Necessary and sufficient conditions for a sequence of complex numbers to be the spectrum of a damped string were given in [4] in implicit form and in [80] and [81] explicitly. If the string is so smooth that its density ρ satisfies $\rho \in W_2^2(0,l)$ and $\rho(s) \geq \epsilon > 0$ where l is the length of the string, then one can apply the Liouville transformation (see [40, p. 202]) to transform the equation of the string into the Sturm-Liouville equation. The corresponding boundary value problem with the spectral parameter in the boundary conditions was considered in many publications (see [35], [125], [104], [105] and references therein). In [65] and [112] necessary and sufficient conditions are given such that a sequence of complex numbers is the spectrum of a smooth inhomogeneous string damped at one end for any value of the damping parameter except of one crucial value.

The inverse problem, i.e., the problem of recovering the parameters of the string from the spectrum of its vibrations and from the total length of the string, for a Stieltjes strings with a finite number of beads with point-wise (i.e., one-dimensional) damping at the right end and the left end free was solved in [4]. Also, this problem can be reduced to the problem of damped oscillators considered in [133], [134]. Another approach to inverse problem for damped finite-dimensional system was developed in [85] where the given data included not only eigenvalues but also the so-called Jordan pairs. Inverse problems for Stieltjes strings damped at an interior point were solved in [16] and [142]. The inverse problem for star graphs of Stieltjes strings with damping at the central vertex was solved in [97].

In Sections 3.4 and 3.5 we have considered inverse problems with damping at one endpoint and one or two Robin conditions at the other endpoint, where in the case of two Robin conditions parts of the spectra for each of the two problems are known. In [141] the inverse spectral problem is solved in the case of one Robin condition, where part of the string data and part of the spectrum are known.

In Section 3.1 we have considered the matrix operator pencil

$$T_c^d(\lambda) = \lambda^2 M - i\lambda K - A_c^d$$

where a symmetric nonnegative matrix K describes damping. The quadratic matrix operator pencil

$$\lambda^2 M - \lambda(B - iK) - A$$

with symmetric B also occurs in physics where the term λB with symmetric B describes gyroscopic forces (see [144], [7]).

Chapter 4

Trees

4.1 Characteristic polynomials

Let T be a plane tree with g straight edges. Then this tree has $g+1$ vertices, and an arbitrary of those vertices will be chosen as the root. For convenience, the root will be denoted by v_0 and the tree will be considered as being directed towards the root (see Lemma C.2.6). The other vertices will be denoted by v_j ($j=1,\dots,g$). For each $j=1,\dots,g$ there is a unique simple path from v_j to v_0, and the first edge in this path, which is the unique outgoing edge from v_j, will be denoted by e_j. In this way we have uniquely labelled the g edges of T from the given labelling of the vertices.

We recall some notation from Appendix C. For $j=0,\dots,g$, $I_j^+ := I^+(v_j)$ is the set of those indices k for which e_k is an incoming edge into the vertex v_j. The set of the indices of the outgoing edges from the vertex v_j is $I^-(v_j)=\{j\}$ for $j=1,\dots,n$ and $I^-(v_0)=\emptyset$. Obviously, v_j ($j=1,\dots,g$) is a pendant vertex if and only if $I_j^+=\emptyset$. For $j=0,\dots,g$, $I_j := I_j^+ \cup I_j^-$ is the set of indices of those edges of T which are incident with v_j, and $d(v_j)=\#I_j$ is the degree of v_j. Clearly, $d(v_0)=\#I_0^+$ and $d(v_j)=1+\#I_j^+$ ($j=1,\dots,g$). Also let $I_0' := \{1,\dots,g\}\setminus I_0^+$ be the index set of all edges which are not incident with the root v_0. For each $j\in\{1,\dots,g\}$, the vertex v_j is the entrance $v_+(e_j)$ of e_j, whereas the exit $v_-(e_j)$ of e_j can be written as $v_{\sigma(j)}$ for some $\sigma(j)\in\{0,\dots,g\}\setminus\{j\}$. Note that e_j is an incoming edge at $v_{\sigma(j)}$, and therefore a pair (r,j) satisfies $r\in\{0,\dots,g\}$ and $j\in I_r^+$ if and only if $j\in\{1,\dots,g\}$ and $r=\sigma(j)$.

Each edge e_j ($j=1,\dots,g$) is divided into n_j+1 segments of the lengths $l_0^{(j)}$, …, $l_{n_j}^{(j)}$ by beads with masses $m_1^{(j)}$, …, $m_{n_j}^{(j)}$ ($l_k^{(j)}>0$, $m_k^{(j)}>0$). Here $n_j\ge 0$ and, clearly, $l^{(j)} := \sum\limits_{k=0}^{n_j} l_k^{(j)}$ is the length of e_j. The indexing of the beads and the segments are such that the lower indices increase in the direction of the edge. This means that for $j=1,\dots,g$, each vertex v_j is located at the beginning

M. Möller, V. Pivovarchik, *Direct and Inverse Finite-Dimensional Spectral Problems on Graphs*, Operator Theory: Advances and Applications 283, https://doi.org/10.1007/978-3-030-60484-4_4

of the segment of length $l_0^{(j)}$. For $j = 0, \ldots, g$ with $I_j^+ \neq \emptyset$, each incoming edge e_r ($r \in I_j^+$) ends at v_j with a segment of length $l_{n_r}^{(r)}$. It is assumed that the tree is stretched. The edges of the tree are strings which can vibrate in the direction orthogonal to the equilibrium position of the strings. We denote the transverse displacement of the j-th bead with mass $m_k^{(j)}$ by $v_k^{(j)}(t)$. For $j = 1, \ldots, g$, the displacement of the entrance v_j of the edge e_j is denoted by $v_0^{(j)}(t)$, whereas the displacement of the exit of the edge e_j is denoted by $v_{n_j+1}^{(j)}(t)$. With this notation, vibrations of the graph can be described by the system of equations

$$\frac{v_k^{(j)}(t) - v_{k-1}^{(j)}(t)}{l_{k-1}^{(j)}} - \frac{v_{k+1}^{(j)}(t) - v_k^{(j)}(t)}{l_k^{(j)}} + m_k^{(j)} \frac{\partial^2 v_k^{(j)}}{\partial t^2}(t) = 0 \tag{4.1.1}$$

$$(j = 1, \ldots, g,\ n_j \geq 1,\ k = 1, \ldots, n_j).$$

The vibrating tree is still assumed to be connected. For the interior vertices we therefore impose the continuity conditions

$$v_0^{(r)}(t) = v_{n_j+1}^{(j)}(t)\ (j \in I_r^+) \text{ for } r = 1, \ldots, g \text{ with } I_r^+ \neq \emptyset, \tag{4.1.2}$$

$$v_{n_r+1}^{(r)}(t) = v_{n_s+1}^{(s)}(t)\ (r, s \in I_0^+) \text{ if } d(v_0) > 1. \tag{4.1.3}$$

Balance of forces at an interior vertex which is not the root implies

$$\frac{v_1^{(j)}(t) - v_0^{(j)}(t)}{l_0^{(j)}} = \sum_{r \in I_j^+} \frac{v_{n_r+1}^{(r)}(t) - v_{n_r}^{(r)}(t)}{l_{n_r}^{(r)}} \text{ for } j = 1, \ldots, g \text{ with } I_j^+ \neq \emptyset. \tag{4.1.4}$$

For a pendant vertex (except for the root if it is pendant) we impose a mixed boundary condition

$$v_0^{(j)}(t) = c_j v_1^{(j)}(t) \text{ for } j = 1, \ldots, g \text{ with } I_j^+ = \emptyset, \tag{4.1.5}$$

with $c_j \in \mathbb{R}$. We need to impose one more condition at the root. We consider two cases: the Dirichlet case with

$$v_{n_r+1}^{(r)}(t) = 0 \text{ for some } r \in I_0^+ \tag{4.1.6}$$

and the Neumann case

$$\sum_{r \in I_0^+} \frac{v_{n_r+1}^{(r)}(t) - v_{n_r}^{(r)}(t)}{l_{n_r}^{(r)}} = 0. \tag{4.1.7}$$

In (4.1.2) we observe that due to the discussion at the beginning of this section we may write (r, j) with $r \in \{1, \ldots, g\}$ and $j \in I_r^+$ as $j \in I_0'$ and $r = \sigma(j)$. Furthermore, we observe that there is ambiguity in (4.1.3) since there are exactly

$d(v_0) - 1$ independent conditions. Also, we want to make the formulation of the conditions independent of any reference to the details of the indexing of the vertices and edges. Clearly, in the Dirichlet case, (4.1.3) and (4.1.6) are equivalent to

$$v_{n_j+1}^{(j)}(t) = 0 \ (j \in I_0). \tag{4.1.8}$$

In the Neumann case, (4.1.3) and (4.1.7) imply

$$d(v_0)v_{n_j+1}^{(j)}(t) = \sum_{r\in I_0}\left(v_{n_r+1}^{(r)}(t) - \frac{v_{n_r+1}^{(r)}(t) - v_{n_r}^{(r)}(t)}{l_{n_r}^{(r)}}\right) \quad (j \in I_0). \tag{4.1.9}$$

Conversely, since the right-hand sides in (4.1.9) are independent of j, (4.1.3) immediately follows from (4.1.9), whereas (4.1.7) follows from adding up the $d(v_0)$ equations in (4.1.9). Later in this section we will discuss the reason for this choice and possible simplifications.

Substituting $v_k^{(j)}(t) = e^{i\lambda t}u_k^{(j)}$ into (4.1.1)–(4.1.7) we obtain the Dirichlet problem

$$\frac{u_k^{(j)} - u_{k-1}^{(j)}}{l_{k-1}^{(j)}} - \frac{u_{k+1}^{(j)} - u_k^{(j)}}{l_k^{(j)}} - m_k^{(j)}\lambda^2 u_k^{(j)} = 0 \tag{4.1.10}$$

$$(j = 1, \dots, g, \ n_j \geq 1, \ k = 1, \dots, n_j),$$

$$u_0^{(\sigma(j))} = u_{n_j+1}^{(j)} \ (j \in I_0'), \tag{4.1.11}$$

$$\frac{u_1^{(j)} - u_0^{(j)}}{l_0^{(j)}} = \sum_{r\in I_j^+}\frac{u_{n_r+1}^{(r)} - u_{n_r}^{(r)}}{l_{n_r}^{(r)}} \ \text{for } j = 1, \dots, g \text{ with } I_j^+ \neq \emptyset, \tag{4.1.12}$$

$$u_0^{(j)} = c_j u_1^{(j)} \ \text{for } j = 1, \dots, g \text{ with } I_j^+ = \emptyset, \tag{4.1.13}$$

$$u_{n_j+1}^{(j)} = 0 \ (j \in I_0), \tag{4.1.14}$$

and the Neumann problem (4.1.10)–(4.1.13),

$$d(v_0)u_{n_j+1}^{(j)} = \sum_{r\in I_0}\left(u_{n_r+1}^{(r)} - \frac{u_{n_r+1}^{(r)} - u_{n_r}^{(r)}}{l_{n_r}^{(r)}}\right) \ (j \in I_0). \tag{4.1.15}$$

The conditions (4.1.14) are called Dirichlet conditions at the root, whereas the conditions (4.1.15) are called Neumann conditions at the root. If the root is a pendant vertex, i.e., $d(v_0) = 1$, then (4.1.15) is equivalent to the usual Neumann condition. In what follows, problem (4.1.10)–(4.1.14) is called the Dirichlet problem for the tree T and problem (4.1.10)–(4.1.13), (4.1.15) is called the Neumann problem for the tree T.

We observe that both the Dirichlet and the Neumann problem are well-posed. Indeed, letting $n := \sum_{j=1}^{g} n_j$, there are $n + 2g$ unknown $u_k^{(j)}$ $(j = 1, \dots, g,$

$k = 0, \dots, n_j + 1$), (4.1.10) consists of n equations, (4.1.11) consists of $g - d(v_0)$ equations, (4.1.12) and (4.1.13) together consist of g equations, and both (4.1.14) and (4.1.15) consist of $d(v_0)$ equations. Therefore the Dirichlet problem as well as the Neumann problem consists of a system of $n + 2g$ linear equations for $n + 2g$ unknown variables.

However, we can immediately eliminate the n unknowns $u_k^{(j)}$ for $k \geq 2$ by using the n equations in (4.1.10). Indeed, for fixed $j \in \{1, \dots, g\}$, the n_j equation in (4.1.10) are of the form (1.2.4). With the functions $R_k^{(j)}(\cdot, c)$ for $k = -2, \dots, 2n_j$ defined according to the recurrence relations (1.2.10) and (1.2.11) with the initial conditions (1.2.12) and (1.2.13), also the linear combinations

$$\widetilde{R}_k^{(j)} := R_k^{(j)}(\cdot, 0)(u_1^{(j)} - u_0^{(j)}) + R_k^{(j)}(\cdot, 1)u_0^{(j)}$$

satisfy these recurrence relations (1.2.10) and (1.2.11), whereas the initial conditions are $\widetilde{R}_0^{(j)}(z) = u_1^{(j)}$ and $\widetilde{R}_{-2}^{(j)}(z) = u_0^{(j)}$. It follows that $u_k^{(j)} = \widetilde{R}_{2k-2}^{(j)}$ for $k = 0, \dots, n_j + 1$, that is,

$$u_k^{(j)} = R_{2k-2}^{(j)}(\cdot, 0)(u_1^{(j)} - u_0^{(j)}) + R_{2k-2}^{(j)}(\cdot, 1)u_0^{(j)} \quad (j = 1, \dots, g;\ k = 0, \dots, n_j + 1). \tag{4.1.16}$$

Therefore, we are left with $2g$ unknown variables

$$u_{2j-1} := u_0^{(j)} \text{ and } u_{2j} := u_1^{(j)} - u_0^{(j)} \quad (j = 1, \dots, g), \tag{4.1.17}$$

and in view of (4.1.16) the conditions (4.1.11)–(4.1.13) can be expressed in terms of linear functionals operating on $U := (u_1, \dots, u_{2g})$ which are defined as follows:

$$\mathcal{L}_{2j}(z)U := R_{2n_j}^{(j)}(z, 1)u_{2j-1} + R_{2n_j}^{(j)}(z, 0)u_{2j} - u_{2\sigma(j)-1} \quad (j \in I_0'), \tag{4.1.18}$$

$$\mathcal{L}_{2j-1}(z)U := -\frac{u_{2j}}{l_0^{(j)}} + \sum_{r \in I_j^+} (R_{2n_r-1}^{(r)}(z, 1)u_{2r-1} + R_{2n_r-1}^{(r)}(z, 0)u_{2r}) \quad \text{for } j = 1, \dots, g \text{ with } I_j^+ \neq \emptyset, \tag{4.1.19}$$

$$\mathcal{L}_{2j-1}(z)U := (1 - c_j)u_{2j-1} - c_j u_{2j} \text{ for } j = 1, \dots, g \text{ with } I_j^+ = \emptyset. \tag{4.1.20}$$

For the Dirichlet problem, the conditions at the root lead to the functionals

$$\mathcal{L}_{2j}^D(z)U := R_{2n_j}^{(j)}(z, 1)u_{2j-1} + R_{2n_j}^{(j)}(z, 0)u_{2j} \ (j \in I_0), \tag{4.1.21}$$

and for the Neumann problem, the conditions at the root lead to the functionals

$$\mathcal{L}_{2j}^N(z)U := \mathcal{L}_{2j}^D(z)U - \frac{1}{d(v_0)} \sum_{r \in I_0} \mathcal{L}_{2r}^D(z)U + \sum_{r \in I_0} \left(R_{2n_r-1}^{(r)}(z, 1)u_{2r-1} + R_{2n_r-1}^{(r)}(z, 0)u_{2r} \right) \quad (j \in I_0). \tag{4.1.22}$$

For convenience, we set $\mathcal{L}_j^N := \mathcal{L}_j^D := \mathcal{L}_j$ for $j = 1, \dots, 2g$ with $j \notin 2I_0$. Defining the $(2g) \times (2g)$ matrices $\Phi_D(z)$ and $\Phi_N(z)$ by

$$\Phi_D(z)U = \begin{pmatrix} \mathcal{L}_1^D(z)U \\ \vdots \\ \mathcal{L}_{2g}^D(z)U \end{pmatrix}, \quad \Phi_N(z)U = \begin{pmatrix} \mathcal{L}_1^N(z)U \\ \vdots \\ \mathcal{L}_{2g}^N(z)U \end{pmatrix}, \tag{4.1.23}$$

it follows that the Dirichlet problem (4.1.10)–(4.1.14) is equivalent to $\Phi_D(z)U = 0$, and the Neumann problem (4.1.10)–(4.1.13), (4.1.15) is equivalent to $\Phi_N(z)U = 0$.

We call the function ϕ_D defined by

$$\phi_D(z) = \det(\Phi_D(z)) \tag{4.1.24}$$

the *characteristic polynomial* of the Dirichlet problem (4.1.10)–(4.1.14). Similarly, the function ϕ_N defined by

$$\phi_N(z) = \det(\Phi_N(z)) \tag{4.1.25}$$

is called the *characteristic polynomial* of the Neumann problem (4.1.10)–(4.1.13), (4.1.15). The zeros of the characteristic polynomials will be called the *characteristic values* of the Dirichlet and the Neumann problem, respectively.

Proposition 4.1.1. *ϕ_D and ϕ_N are real polynomials which are independent of the indexing of the vertices and edges of the tree T.*

Proof. Since all entries of the matrices Φ_D and Φ_N are real polynomials, it is clear that also their determinants ϕ_D and ϕ_N are real polynomials. Reindexing the edges and vertices would result in simultaneous permutations of rows and columns of the matrices $\Phi_D(z)$ and $\Phi_N(z)$, which leaves their determinants unchanged. □

The linear functionals (4.1.22) for the Neumann conditions at the root have complicated right-hand sides, but with this definition we obtain symmetry, that is, the conditions are invariant under permutation of indices in I_0 and therefore lead to an unambiguously defined characteristic polynomial. However, it may be more convenient to consider functionals which are directly derived from (4.1.3) and (4.1.7). But for $d(v_0) > 1$ such choice is ambiguous, and the characteristic polynomial obtained in this way may be the negative of the one we have defined in (4.1.25). In the proposition below we list a class of such choices which always give the correct characteristic polynomial.

Proposition 4.1.2. *Let $\tau : \{1, \dots, d(v_0)\} \to I_0$ be bijective and define*

$$\widetilde{\mathcal{L}}_{2\tau(j)}^N(z)U := \mathcal{L}_{2\tau(j)}^D(z)U - \mathcal{L}_{2\tau(j+1)}^D(z)U \quad (j = 1, \dots, d(v_0) - 1), \tag{4.1.26}$$

$$\widetilde{\mathcal{L}}_{2\tau(d(v_0))}^N(z)U := \sum_{r \in I_0} \left(R_{2n_r-1}^{(r)}(z,1)u_{2r-1} + R_{2n_r-1}^{(r)}(z,0)u_{2r} \right). \tag{4.1.27}$$

Let Φ_N^τ be derived from Φ_N by replacing the linear functionals $\mathcal{L}_{2\tau(j)}^N(z)$ in (4.1.23) with the linear functionals $\widetilde{\mathcal{L}}_{2\tau(j)}^N(z)$ $(j = 1, \dots, d(v_0))$. Then $\det(\Phi_N^\tau(z)) = \phi_N(z)$.

Proof. We will show that Φ_N^τ is derived from Φ_N by elementary row operations which do not change the determinant. For $j = 1, \dots, d(v_0) - 1$ it immediately follows from (4.1.22) that

$$\widetilde{\mathcal{L}}^N_{2\tau(j)} = \mathcal{L}^N_{2\tau(j)} - \mathcal{L}^N_{2\tau(j+1)}.$$

Together with

$$\begin{aligned}\sum_{j=1}^{d(v_0)-1} j\widetilde{\mathcal{L}}^N_{2\tau(j)} &= \sum_{j=1}^{d(v_0)-1} j\mathcal{L}^D_{2\tau(j)} - \sum_{j=1}^{d(v_0)} (j-1)\mathcal{L}^D_{2\tau(j)} \\ &= -d(v_0)\mathcal{L}^D_{2\tau(d(v_0))} + \sum_{j=1}^{d(v_0)} \mathcal{L}^D_{2\tau(j)}\end{aligned}$$

and (4.1.22) we obtain

$$\begin{aligned}\widetilde{\mathcal{L}}^N_{2\tau(d(v_0))}(z) &= \sum_{r\in I_0} \left(R^{(r)}_{2n_r-1}(z,1)u_{2r-1} + R^{(r)}_{2n_r-1}(z,0)u_{2r}\right) \\ &= \mathcal{L}^N_{2\tau(d(v_0))}(z) + \frac{1}{d(v_0)} \sum_{j=1}^{d(v_0)-1} j\widetilde{\mathcal{L}}^N_{2\tau(j)}(z).\end{aligned}$$

□

4.2 Relations between characteristic polynomials for a tree and for complementary subtrees

Assume that the root v_0 of the tree T is an interior vertex. We cut the tree T at v_0 into two subtrees T_1 and T_2 (see Figure 4.1). We say that T_1 and T_2 are

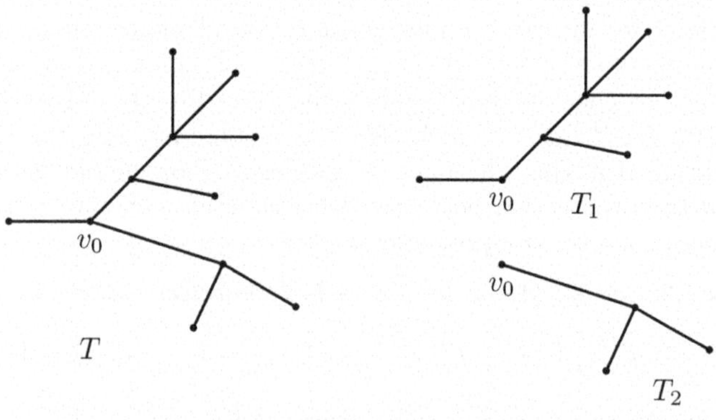

Figure 4.1: Complementary subtrees for an interior root

complementary subtrees of T. In the notation of Section 4.1 let $I^{(\iota)}$ be the set of all indices j for which e_j is an edge of the subtree T_ι ($\iota = 1, 2$). With $g_1 := \#I^{(1)}$ we have $1 \le g_1 < g$ and $\#I^{(2)} = g - g_1$. For convenience, the indexing of the edges of T will be chosen in such a way that $I^{(1)} = \{1, \dots, g_1\}$. We also let $I_0^{(\iota)} = I_0 \cap I^{(\iota)}$ ($\iota = 1, 2$).

Next we are going to introduce notation for the Dirichlet and Neumann problems on the subtrees T_1 and T_2. Let $\mathcal{L}_j^{D,1}(z)$, $\mathcal{L}_j^{N,1}(z)$ ($j = 1, \dots, 2g_1$) and $\mathcal{L}_j^{D,2}(z)$, $\mathcal{L}_j^{N,2}(z)$ ($j = 2g_1+1, \dots, 2g$) be the linear functionals as defined in Section 4.1 associated with the trees T_1 and T_2, respectively. With u_j ($j = 1, \dots, 2g$) defined in Section 4.1 we let $U^{(1)} = (u_j)_{j=1}^{2g_1}$ and $U^{(2)} = (u_j)_{j=2g_1+1}^{2g}$.

It is clear that

$$\mathcal{L}_j^D(z)U = \mathcal{L}_j^{D,1}(z)U^{(1)} \quad (j = 1, \dots, 2g_1), \tag{4.2.1}$$

$$\mathcal{L}_j^D(z)U = \mathcal{L}_j^{D,2}(z)U^{(2)} \quad (j = 2g_1 + 1, \dots, 2g). \tag{4.2.2}$$

Hence the Dirichlet problem for T splits into two Dirichlet problems for T_1 and T_2, and the corresponding coefficient matrices $\Phi_D^{(1)}(z)$ and $\Phi_D^{(2)}(z)$ are given by

$$\Phi_D^{(1)}(z)U^{(1)} = \begin{pmatrix} \mathcal{L}_1^{D,1}(z)U^{(1)} \\ \vdots \\ \mathcal{L}_{2g_1}^{D,1}(z)U^{(1)} \end{pmatrix}, \quad \Phi_D^{(2)}(z)U^{(2)} = \begin{pmatrix} \mathcal{L}_{2g_1+1}^{D,2}(z)U^{(2)} \\ \vdots \\ \mathcal{L}_{2g}^{D,2}(z)U^{(2)} \end{pmatrix}. \tag{4.2.3}$$

Then the characteristic polynomials $\phi_D^{(\iota)}$ of the Dirichlet problems on the subtrees T_ι ($\iota = 1, 2$) are

$$\phi_D^{(\iota)}(z) = \det(\Phi_D^{(\iota)}(z)) \quad (\iota = 1, 2) \tag{4.2.4}$$

The coefficient matrices $\Phi_N^{(1)}(z)$ and $\Phi_N^{(2)}(z)$ for the Neumann problems on T_1 and T_2 are given by

$$\Phi_N^{(1)}(z)U^{(1)} = \begin{pmatrix} \mathcal{L}_1^{N,1}(z)U^{(1)} \\ \vdots \\ \mathcal{L}_{2g_1}^{N,1}(z)U^{(1)} \end{pmatrix}, \quad \Phi_N^{(2)}(z)U^{(2)} = \begin{pmatrix} \mathcal{L}_{2g_1+1}^{N,2}(z)U^{(2)} \\ \vdots \\ \mathcal{L}_{2g}^{N,2}(z)U^{(2)} \end{pmatrix}, \tag{4.2.5}$$

where $\mathcal{L}_j^{N,\iota}(z) = \mathcal{L}_j^{D,\iota}(z)$ is as in (4.2.1), (4.2.2) when $j \notin 2I_0$ since the conditions (4.1.18) and (4.1.19) are the same for the Dirichlet and the Neumann case. For $j \in 2I_0$, the functionals $\mathcal{L}_{2j}^{N,\iota}(z)$ ($j \in I_0^{(\iota)}$, $\iota = 1, 2$) are obtained from (4.1.22) by replacing I_0 in the sums by $I_0^{(\iota)}$ and $d(v_0)$ by $d^{(\iota)}(v_0) := \#I_0^{(\iota)}$. Then the the characteristic polynomials $\phi_N^{(\iota)}$ of the Neumann problems on the subtrees T_ι ($\iota = 1, 2$) are

$$\phi_N^{(\iota)}(z) = \det(\Phi_N^{(\iota)}(z)) \quad (\iota = 1, 2). \tag{4.2.6}$$

Theorem 4.2.1. *Let the root v_0 of a tree T be an interior vertex. Let T_1 and T_2 be complementary subtrees of T having v_0 as the only common vertex and being directed towards the root. Then*

$$\phi_D(z) = \phi_D^{(1)}(z)\phi_D^{(2)}(z), \tag{4.2.7}$$

$$\phi_N(z) = \phi_N^{(1)}(z)\phi_D^{(2)}(z) + \phi_D^{(1)}(z)\phi_N^{(2)}(z). \tag{4.2.8}$$

Proof. It immediately follows from (4.2.3) and (4.1.23) that

$$\Phi_D(z) = \begin{pmatrix} \Phi_D^{(1)}(z) & 0 \\ 0 & \Phi_D^{(2)}(z) \end{pmatrix},$$

which proves (4.2.7).

For the proof of (4.2.8) we observe that according to Proposition 4.1.1 we may choose any arbitrary indexing of the edges. It is convenient to choose the indices in such a way that $j \in I_0$ if and only if $g_1 - d^{(1)}(v_0) < j \le g_1 + d^{(2)}(v_0)$. The function τ defined by

$$\tau(j) = \begin{cases} g_1 - d^{(1)}(v_0) + j & \text{for } j = 1, \dots, d^{(1)}(v_0), \\ g_1 + d(v_0) + 1 - j & \text{for } j = d^{(1)}(v_0) + 1, \dots, d(v_0), \end{cases}$$

is a bijection from $\{1, \dots, d(v_0)\}$ to I_0, and therefore Proposition 4.1.2 shows that

$$\phi_N(z) = \det(\Phi_N^\tau(z)). \tag{4.2.9}$$

Writing $\Phi_N^\tau(z)$ as a block matrix

$$\Phi_N^\tau(z) = \begin{pmatrix} \widetilde{\Phi}^{1,1}(z) & \widetilde{\Phi}^{1,2}(z) \\ \widetilde{\Phi}^{2,1}(z) & \widetilde{\Phi}^{2,2}(z) \end{pmatrix}$$

with respect to the decomposition $\mathbb{C}^{2g} = \mathbb{C}^{2g_1} \oplus \mathbb{C}^{2(g-g_1)}$, it is clear that the only nonzero elements of $\widetilde{\Phi}^{1,2}$ and $\widetilde{\Phi}^{2,1}$ are in the last row of $\widetilde{\Phi}^{1,2}$ and in the second row of $\widetilde{\Phi}^{2,1}$. Observing (4.2.9), an expansion of $\det(\Phi_N^\tau(z))$ with respect to the $2g_1$-th row gives

$$\phi_N(z) = \det(\widetilde{\Phi}_1(z)) + \det(\widetilde{\Phi}_2(z)),$$

where

$$\widetilde{\Phi}_1(z) = \begin{pmatrix} \widetilde{\Phi}^{1,1}(z) & 0 \\ \widetilde{\Phi}^{2,1}(z) & \widetilde{\Phi}^{2,2}(z) \end{pmatrix}$$

and where the matrix $\widetilde{\Phi}_2(z)$ is obtained from $\Phi_N^\tau(z)$ by setting the first $2g_1$ elements in its $2g_1$-th row equal 0, that is, by replacing the last row of $\widetilde{\Phi}^{1,1}(z)$ by the zero row.

Clearly, $\det(\widetilde{\Phi}_1(z)) = \det(\widetilde{\Phi}^{1,1}(z))\det(\widetilde{\Phi}^{2,2}(z))$. In view of Proposition 4.1.2, the last row of $\widetilde{\Phi}^{1,1}$ is given by

$$\widetilde{\mathcal{L}}^N_{2g_1}(z)\begin{pmatrix}U^{(1)}\\0\end{pmatrix} = \mathcal{L}^D_{2g_1}(z)\begin{pmatrix}U^{(1)}\\0\end{pmatrix} = \mathcal{L}^{D,1}_{2g_1}(z)U^{(1)},$$

and, in case $d^{(1)}(v_0) > 1$, recursively adding the new $2(j+1)$-th row in $\widetilde{\Phi}^{1,1}(z)$ to its $(2j)$-th row for $j = 2(g_1-1),\dots,2(g_1-d^{(1)}(v_0)+1)$ one arrives at $\Phi^{(1)}_D(z)$. This shows that $\det(\widetilde{\Phi}^{1,1}(z)) = \det(\Phi^{(1)}_D(z))$.

With $\tau_2(j) = g_1 + d^{(2)}(v_0)+1-j$ for $j = 1,\dots,d^{(2)}(v_0)$ it is easy to verify that $\widetilde{\Phi}^{2,2}(z) = \Phi^{2,\tau_2}_N(z)$, and $\det(\widetilde{\Phi}^{2,2}(z)) = \phi^{(2)}_N(z)$ follows in view of Proposition 4.1.2. We have shown that

$$\det(\widetilde{\Phi}_1(z)) = \phi^{(1)}_D(z)\phi^{(2)}_N(z).$$

The $(2g_1)$-th row of $\widetilde{\Phi}_2(z)$ is given by

$$\widetilde{\mathcal{L}}^N_{2g_1}(z)\begin{pmatrix}0\\U^{(2)}\end{pmatrix} = -\mathcal{L}^D_{2(g_1+1)}(z)U.$$

Replacing the $(2g_1)$-th row of $\widetilde{\Phi}_2(z)$ with its negative and then interchanging the $2g_1$-th row with the $2(g_1+1)$-th row leads to a matrix

$$\widehat{\Phi}(z) = \begin{pmatrix}\widehat{\Phi}^{1,1}(z) & \widehat{\Phi}^{1,2}(z)\\ 0 & \widehat{\Phi}^{2,2}(z)\end{pmatrix}$$

with $\det(\widehat{\Phi}(z)) = \det(\widetilde{\Phi}_2(z))$. With $\tau_1(j) = g_1 - d^{(1)}(v_0)+j$ for $j = 1,\dots,d^{(1)}(v_0)$ it follows from Proposition 4.1.2 that $\widehat{\Phi}^{1,1}(z) = \Phi^{1,\tau_1}_N(z)$, and hence $\det(\widehat{\Phi}^{1,1}(z)) = \det(\Phi^{(1)}_N(z)) = \phi^{(1)}_N(z)$. The second row of $\widehat{\Phi}^{2,2}(z)$ is given by the functional $\mathcal{L}^{D,2}_{2(g_1+1)}(z)U^{(2)}$, and if $d^{(2)}(v_0) > 1$, its $2j$-th row for $j = 2,\dots,d^{(2)}(v_0)$ is given by the functional $(\mathcal{L}^{D,2}_{2j}(z) - \mathcal{L}^{D,2}_{2(j-1)}(z))U^{(2)}$. Recursively adding the $2(j-1)$-th row of this matrix to its $2j$-th for $j = 2,\dots,d^{(2)}(v_0)$, the resulting matrix is $\Phi^{(2)}_D(z)$. Hence $\det(\widehat{\Phi}^{2,2}(z)) = \phi^{(2)}_D(z)$. This completes the proof of (4.2.8). □

Corollary 4.2.2. *Let the root v_0 of a tree T be an interior vertex, let $2 \le p \le d(v_0)$ and let T_j $(j = 1,\dots,p)$ be complementary subtrees of T having v_0 as the only common vertex and being directed towards the root. For $j = 1,\dots,p$ let $\phi^{(j)}_D$ and $\phi^{(j)}_N$ be the characteristic polynomials of the Dirichlet and Neumann problem, respectively, on the subtree T_j. Then*

$$\phi_D(z) = \prod_{j=1}^{p}\phi^{(j)}_D(z), \tag{4.2.10}$$

$$\phi_N(z) = \sum_{j=1}^{p}\phi^{(j)}_N(z)\prod_{k=1,\ k\ne j}^{p}\phi^{(k)}_D(z). \tag{4.2.11}$$

Remark 4.2.3. 1. If one takes $p = d(v_0)$ in Corollary 4.2.2, then the root v_0 is a pendant vertex for all subtrees T_j ($j = 1, \dots, d(v_0)$).

2. Let e_j be an edge of the tree T which is incident with v_0 such that v_j is a pendant vertex of T. We consider e_j as a subtree T_j of T. Then $\Phi_D^{(j)}(z)$ and $\Phi_N^{(j)}(z)$ are 2×2 matrices, and the first functional is given by (4.1.20) for both $\Phi_D^{(j)}(z)$ and $\Phi_N^{(j)}(z)$, whereas the second functional is given by (4.1.21) for $\Phi_D^{(j)}(z)$ and by (4.1.22) for $\Phi_N^{(j)}(z)$. Therefore, for $j = 1, \dots, g$,

$$\Phi_D^{(j)}(z) = \begin{pmatrix} 1 - c_j & -c_j \\ R_{2n_j}^{(j)}(z,1) & R_{2n_j}^{(j)}(z,0) \end{pmatrix}, \quad \Phi_N^{(j)}(z) = \begin{pmatrix} 1 - c_j & -c_j \\ R_{2n_j-1}^{(j)}(z,1) & R_{2n_j-1}^{(j)}(z,0) \end{pmatrix},$$

and taking their determinants gives

$$\phi_D^{(j)}(z) = R_{2n_j}^{(j)}(z, c_j), \quad \phi_N^{(j)}(z) = R_{2n_j-1}^{(j)}(z, c_j)$$

in view of part 4 of Remark 1.2.7.

3. Let the tree T be a star graph with the centre being the root and consider Corollary 4.2.2 with $p = d(v_0) = g$. Then for each subtree, the only edge is incident with the root, and $\phi_D^{(j)}$ and $\phi_N^{(j)}$ are as found in part 2 of this remark. Therefore the characteristic polynomials (4.2.10) and (4.2.11) with $c_j = 0$ ($j = 1, \dots, g$) are identical to the characteristic polynomials (2.1.9) and (2.1.10) for the star graph with central mass $M = 0$. That is, for a star graph without central mass and with Dirichlet conditions at the pendant vertices, the characteristic polynomials obtained in this section agree with those obtained in Section 2.1.

Next we consider trees with the root being a pendant vertex. Removing the edge which is incident with v_0, a new tree T' is obtained (see Figure 4.2).

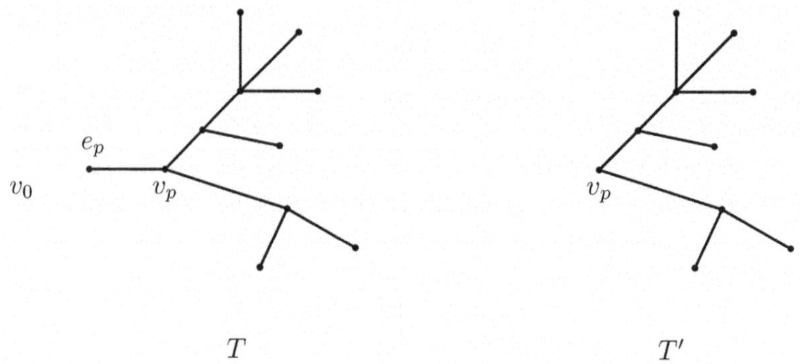

Figure 4.2: Subtree for a pendant root

Proposition 4.2.4. *Let T be a tree with $g \geq 2$ edges and assume that v_0 is a pendant vertex. Removing the edge e_p which is incident with v_0, a subtree T' is obtained, where v_p becomes the root of T'. Let $\phi_D^{(\prime)}$ and $\phi_N^{(\prime)}$ be the characteristic functions of the Dirichlet and the Neumann problem for T'. Then*

$$\phi_D = \frac{1}{l_0^{(p)}} R_{2n_p}^{(p)}(\cdot, 1)\phi_D^{(\prime)} + R_{2n_p}^{(p)}(\cdot, 0)\phi_N^{(\prime)}, \tag{4.2.12}$$

$$\phi_N = \frac{1}{l_0^{(p)}} R_{2n_p-1}^{(p)}(\cdot, 1)\phi_D^{(\prime)} + R_{2n_p-1}^{(p)}(\cdot, 0)\phi_N^{(\prime)}. \tag{4.2.13}$$

Proof. Without loss of generality we may assume that $p = 1$. We first introduce a tree $\widehat{T}$ which is obtained from the tree T by choosing v_1 as the root of $\widehat{T}$. To fix notation, we put $\widehat{e}_j = e_j$ for $j = 1, \ldots, g$ but with $\widehat{e}_1$ directed from v_0 to v_1. Then, in the notation of Section 4.1, $\widehat{v}_0 = v_1$, $\widehat{v}_1 = v_0$, $\widehat{v}_j = v_j$ for $j = 1, \ldots, g$. Furthermore, $\widehat{l}_0^{(1)} = l_{n_0}^{(1)}$ and $\widehat{l}_{n_0}^{(1)} = l_0^{(1)}$. The amplitudes of the displacements are $\widehat{u}_k^{(j)} = u_k^{(j)}$ ($j = 2, \ldots, g$; $k = 0, \ldots, n_j + 1$), whereas $\widehat{u}_k^{(1)} = u_{n_1+1-k}^{(1)}$ ($k = 0, \ldots, n_1 + 1$) due to the reversal of the orientation of e_1. Clearly, $\widehat{I}_j = I_j$ for $j = 2, \ldots, g$, $\widehat{I}_1 = I_0 = \{1\}$, $\widehat{I}_0 = I_1$, $\widehat{I}_0' = I_0' \setminus I_1$, $d(\widehat{v}_0) = d(v_1)$.

Observe that the Dirichlet and Neumann boundary conditions for T are given by $u_{n_1+1}^{(1)} = c_0 u_{n_1}^{(1)}$ with $c_0 = 0$ for the Dirichlet problem and $c_0 = 1$ for the Neumann problem. Hence, most of the following reasoning will be the same for the Dirichlet and the Neumann case, and we briefly write Φ for the matrix functions Φ_D, Φ_N given by (4.1.23) and ϕ for the characteristic functions ϕ_D and ϕ_N given by (4.1.24), (4.1.25) with a generic value for c_0. In particular, we will write $\mathcal{L}_j(z)$ for the functionals $\mathcal{L}_j^D(z)$ and $\mathcal{L}_j^N(z)$ ($j = 1, \ldots, 2g$). Since this notation is only used in this proof, no confusion with the definition of $\mathcal{L}_j(z)$ in Section 4.1 should arise.

Introducing the notation $\widehat{u}_{2j-1} := \widehat{u}_0^{(j)}$, $\widehat{u}_{2j} := \widehat{u}_1^{(j)} - \widehat{u}_0^{(j)}$ ($j = 1, \ldots, g$) (see (4.1.17)), and $\widehat{U} = (\widehat{u}_j)_{j=1}^{2g}$, we are going to express the linear functionals associated with the tree T in terms of the linear functionals associated with the tree $\widehat{T}$. Since the indexing of the edges does not influence the characteristic functions (see Proposition 4.1.1), we will assume now that $I_1 = \{1, \ldots, d(v_1)\}$. Clearly, (4.1.11) for $j \in \widehat{I}_0'$ as well as (4.1.12), (4.1.13) for $j = 2, \ldots, g$ are identical for both trees, so that

$$\mathcal{L}_{2j}(x)U = \widehat{\mathcal{L}}_{2j}(z)\widehat{U} \quad (j = d(\widehat{v}_0) + 1, \ldots, g), \tag{4.2.14}$$

$$\mathcal{L}_{2j-1}(x)U = \widehat{\mathcal{L}}_{2j-1}(z)\widehat{U} \quad (j = 2, \ldots, g). \tag{4.2.15}$$

The conditions (4.1.11) for T and $j \in I_1$ can be written in the form $\widehat{u}_{n_1+1}^{(1)} = \widehat{u}_{n_j+1}^{(j)}$ for $j = 2, \ldots, d(\widehat{v}_0)$, whereas (4.1.12) for $j = 1$ gives

$$\sum_{r \in \widehat{I}_0} \frac{\widehat{u}_{n_r+1}^{(r)} - \widehat{u}_{n_r}^{(r)}}{\widehat{l}_{n_r}^{(r)}} = 0,$$

which means that the condition at the root of $\widehat{T}$ is a Neumann condition. We define $\tau : \{1, \dots, d(\widehat{v}_0)\} \to \widehat{I}_0 = \{1, \dots, d(\widehat{v}_0)\}$ by $\tau(j) = d(\widehat{v}_0) + 1 - j$. Therefore it follows by Proposition 4.1.2, (4.1.21), (4.1.16) and (4.1.17) that

$$\widehat{\mathcal{L}}_{2j}^N(z)\widehat{U} = \widehat{u}_{n_j+1}^{(j)} - \widehat{u}_{n_{j-1}+1}^{(j-1)} \quad (j = 2, \dots, d(\widehat{v}_0)),$$
$$\widehat{\mathcal{L}}_2^N(z)\widehat{U} = \sum_{r \in \widehat{I}_0} \frac{\widehat{u}_{n_r+1}^r - \widehat{u}_{n_r}^r}{\widehat{l}_{n_r}^{(r)}}.$$

On the other hand, by (4.1.18), (4.1.16) and (4.1.17),

$$\mathcal{L}_{2j}(z)U = u_{n_j+1}^{(j)} - u_0^{(1)} \quad (j = 2, \dots, d(\widehat{v}_0)),$$

which gives

$$(\mathcal{L}_{2j}(z) - \mathcal{L}_{2(j-1)})U = u_{n_j+1}^{(j)} - u_{n_{j-1}+1}^{(j-1)} = \widehat{L}_{2j}^N(z)\widehat{U} \ (j = 3, \dots, d(\widehat{v}_0)), \tag{4.2.16}$$

whereas

$$\mathcal{L}_4(z)U = u_{n_2+1}^{(2)} - u_0^{(1)} = \widehat{u}_{n_2+1}^{(2)} - \widehat{u}_{n_1+1}^{(1)} = \widehat{L}_4(z)\widehat{U}. \tag{4.2.17}$$

Next we find in view of (4.1.19), (4.1.16) and (4.1.17) that

$$\begin{aligned}
\mathcal{L}_1(z)U &= -\frac{u_2}{l_0^{(1)}} + \sum_{r \in I_1^+} \frac{u_{n_r+1}^{(r)} - u_{n_r}^{(r)}}{l_{n_r}^{(r)}} \\
&= \sum_{r \in \widehat{I}_0} \frac{\widehat{u}_{n_r+1}^{(r)} - \widehat{u}_{n_r}^{(r)}}{\widehat{l}_{n_r}^{(r)}} \\
&= \widehat{\mathcal{L}}_2^N(z)\widehat{U}.
\end{aligned} \tag{4.2.18}$$

The functional (4.1.20) for $\widehat{T}$ and $j = 1$ gives

$$\widehat{\mathcal{L}}_1(z)\widehat{U} = (1 - c_0)\widehat{u}_1 - c_0\widehat{u}_2.$$

Then, observing (4.1.16) and (4.1.17), the Dirichlet condition (4.1.21) or the Neumann condition (4.1.22) at the root of T give

$$\begin{aligned}
\mathcal{L}_2(z)U &= (1 - c_0)u_{n_1+1}^{(1)} + c_0\frac{u_{n_1+1}^{(1)} - u_{n_1}^{(1)}}{l_{n_1}^{(1)}} \\
&= (1 - c_0)\widehat{u}_1 - \frac{c_0}{l_{n_1}^{(1)}}\widehat{u}_2 \\
&= \left(1 - c_0 + \frac{c_0}{l_{n_1}^{(1)}}\right)\widehat{\mathcal{L}}_1(z)\widehat{U}.
\end{aligned} \tag{4.2.19}$$

The linear combinations of the functionals on the left-hand side of (4.2.16) are described by a $2g \times 2g$ matrix Ψ with $\det \Psi = 1$, whereas the transformations (4.2.18) and (4.2.19) are described by a $2g \times 2g$ matrix $\widehat{\Psi}$ with

$$-\det \widehat{\Psi} = 1 - c_0 + \frac{c_0}{l_{n_1}^{(1)}}.$$

From (4.2.14)–(4.2.19) it follows that

$$\Psi \Phi(z) U = \widehat{\Psi} \widehat{\Phi}_N^{\mathcal{T}}(z) \widehat{U}. \tag{4.2.20}$$

Furthermore,

$$\widehat{U} = \begin{pmatrix} \Psi_1(z) & 0 \\ 0 & I_{2g-2} \end{pmatrix} U, \tag{4.2.21}$$

where I_{2g-2} is the $(2g-2) \times (2g-2)$ identity matrix. Therefore,

$$\det(\Phi(z)) = (\det \widehat{\Psi})(\det(\widehat{\Phi}_N^{\mathcal{T}}(z)))(\det(\Psi_1(z))).$$

From

$$\begin{pmatrix} \widehat{u}_1 \\ \widehat{u}_2 \end{pmatrix} = \Psi_1(z) \begin{pmatrix} u_1 \\ u_2 \end{pmatrix},$$

$\widehat{u}_1 = u_{n_1+1}^{(1)}$, $\widehat{u}_2 = u_{n_1}^{(1)} - u_{n_1+1}^{(1)}$, (4.1.16) and (4.1.17) it follows that

$$\Psi_1(z) = \begin{pmatrix} R_{2n_1}^{(1)}(z,1) & R_{2n_1}^{(1)}(z,0) \\ -l_{n_1}^{(1)} R_{2n_1-1}^{(1)}(z,1) & -l_{n_1}^{(1)} R_{2n_1-1}^{(1)}(z,0) \end{pmatrix}.$$

Hence

$$\det \Psi_1(z) = -\frac{l_{n_1}^{(1)}}{l_0^{(1)}}$$

by the Lagrange identity (1.2.47). Therefore,

$$\phi(z) = \det(\Phi(z)) = \frac{(1-c_0) l_{n_1}^{(1)} + c_0}{l_0^{(1)}} \det(\widehat{\Phi}_N^{\mathcal{T}}(z)) = \frac{(1-c_0) l_{n_1}^{(1)} + c_0}{l_0^{(1)}} \widehat{\phi}_N(z). \tag{4.2.22}$$

Denoting the characteristic functions for the string $\widehat{e}_1$ by $\widehat{\phi}_D^{(1)}(\cdot, c_0)$ and $\widehat{\phi}_N^{(1)}(\cdot, c_0)$ and observing that $\widehat{e}_1$ and $T^{(\prime)}$ are complementary subtrees of $\widehat{T}$ with common vertex $\widehat{v}_0$, it follows from Theorem 4.2.1 and Remark 4.2.3 that

$$\begin{aligned} \widehat{\phi}_N &= \widehat{\phi}_N^{(1)}(\cdot, c_0) \phi_D^{(\prime)} + \widehat{\phi}_D^{(1)}(\cdot, c_0) \phi_N^{(\prime)} \\ &= \widehat{R}_{2n_1-1}^{(1)}(\cdot, c_0) \phi_D^{(\prime)} + \widehat{R}_{2n_1}^{(1)}(\cdot, c_0) \phi_N^{(\prime)}, \end{aligned} \tag{4.2.23}$$

where the $\widehat{R}_k^{(1)}$ are the Cauer-Fry polynomials for the string $\widehat{e}_1$. The relations between the Cauer-Fry polynomials of a string for opposite orientations stated in Proposition 1.2.12 immediately leads to (4.2.12) and (4.2.13) for $p = 1$. □

Remark 4.2.5. The star graph considered in Section 2.3 is a particular case of the tree $\widehat{T}$ introduced in the proof of Proposition 4.2.4. With the obvious identification of notation in this section and in Section 2.3 it is clear that (2.3.8) is a particular case of (4.2.23).

Lemma 4.2.6. *Let the assumptions and notation be as in Proposition* 4.2.4. *Then*

$$\frac{\phi_D(z)}{\phi_N(z)} = l^{(p)}_{n_p} + \cfrac{1}{-m^{(p)}_{n_p} z + \cfrac{1}{l^{(p)}_{n_p-1} + \cfrac{1}{-m^{(p)}_{n_p-1} z + \cdots + \cfrac{1}{-m^{(p)}_1 z + \cfrac{1}{l^{(p)}_0 + \cfrac{\phi^{(\prime)}_D(z)}{\phi^{(\prime)}_N(z)}}}}}}. \tag{4.2.24}$$

Proof. This proof is similar to the proof of (2.3.12). Defining

$$\Psi_k = \begin{pmatrix} \widehat{R}^{(p)}_{2k-1}(\cdot,0) & \widehat{R}^{(p)}_{2k}(\cdot,0) \\ \widehat{R}^{(p)}_{2k-1}(\cdot,1) & \widehat{R}^{(p)}_{2k}(\cdot,1) \end{pmatrix} \quad (k = 0,\dots,n_p),$$

it follows in view of Lemma 1.2.4 that

$$\Psi_k = \Psi_{k-1}\Theta_k \quad (k = 1,\dots,n_p),$$

where

$$\Theta_k(z) = \begin{pmatrix} 1 & \widehat{l}^{(p)}_k \\ -\widehat{m}^{(p)}_k z & 1 - \widehat{l}^{(p)}_k \widehat{m}^{(p)}_k z \end{pmatrix} \quad (k = 1,\dots,n_p,\ z \in \mathbb{C}).$$

Hence

$$l^{(p)}_0 \begin{pmatrix} \phi_D \\ \phi_N \end{pmatrix} = \begin{pmatrix} l^{(p)}_{n_p} & 0 \\ 0 & 1 \end{pmatrix} \Psi_{n_p} \begin{pmatrix} \phi^{(\prime)}_D \\ \phi^{(\prime)}_N \end{pmatrix} = \begin{pmatrix} l^{(p)}_{n_p} & 0 \\ 0 & 1 \end{pmatrix} \Psi_0\Theta_1\cdots\Theta_{n_g} \begin{pmatrix} \phi^{(\prime)}_D \\ \phi^{(\prime)}_N \end{pmatrix} \tag{4.2.25}$$

by (4.2.22) and (4.2.23) with edge index p. Observing that

$$\begin{pmatrix} l^{(p)}_{n_p} & 0 \\ 0 & 1 \end{pmatrix} \Psi_0 = \begin{pmatrix} 1 & l^{(p)}_{n_p} \\ 0 & 1 \end{pmatrix},$$

the representation (4.2.24) now follows by a recursive application of (A.4.1), observing that $\widehat{m}_k = m_{n_p+1-k}$ and $\widehat{l}_k = l_{n_p-k}$. □

Theorem 4.2.7. *Let T be a tree with $g \geq 2$ edges such that T is directed towards its root. Let n be the number of beads on T and assume that the numbers c_j in the boundary conditions (4.1.13) at the pendant vertices v_j ($j = 1, \dots, g$ with $I_j^+ = \emptyset$) satisfy $0 \leq c_j \leq 1$. Then the characteristic functions ϕ_D and ϕ_N of the tree defined in (4.1.24) and (4.1.25) are polynomials of degree n and satisfy $\phi_D(z) > 0$ for $z \in (-\infty, 0]$ and $\phi_N(z) > 0$ for $z \in (-\infty, 0)$. Furthermore, $\phi_N(0) = 0$ if and only if $c_j = 1$ for all pendant vertices v_j. The rational function $\frac{\phi_D}{\phi_N}$ is an $\mathcal{S}$-function.*

Proof. We are going to prove the statement by induction on the height h of T, that is, the maximal number of edges traversed from any vertex towards the root. If $h = 1$, then T is a star graph, and Corollary 4.2.2 and Remark 4.2.3 show that

$$\phi_D = \prod_{j=1}^{g} R_{2n_j}^{(j)}(\cdot, c_j),$$

$$\phi_N = \sum_{j=1}^{g} R_{2n_j-1}^{(j)}(\cdot, c_j) \prod_{k=1,\, k\neq j}^{g} R_{2n_k}^{(k)}(\cdot, c_k).$$

The statements of this theorem, except for the last one, now immediately follow in this case $h = 1$ from Remark 1.2.7, part 4, and Propositions 1.2.8–1.2.11. Observing that

$$-\frac{\phi_N}{\phi_D} = \sum_{j=1}^{g} \left(-\frac{R_{2n_j-1}^{(j)}(\cdot, c_j)}{R_{2n_j}^{(j)}(\cdot, c_j)} \right),$$

it follows from (1.2.19) and Lemma A.2.4 that each summand on the right-hand side is a rational Nevanlinna function. Hence the right-hand side and therefore also the left-hand side is a rational Nevanlinna function, and by Lemma A.2.4 and what we already have shown the proof in the case $h = 1$ is complete.

Now assume that $h > 1$ and that the statement of this theorem is true for trees of height less than h. First let v_0 be a pendant vertex. In the notation of Proposition 4.2.4 with $p = 1$, T' is a tree of height $h - 1$ since each path from v_j ($j = 2, \dots, g$) to v_0 has to pass through $v_1 = \widehat{v}_0$ and is therefore shorter by one edge than the path from v_j to v_0. In view of (4.2.12) and (4.2.13) the statement of this theorem follows with a reasoning similar to that for the case $h = 1$ if we take note of the following. The pendant vertices which are not the root are the same for T and T'. Hence $\phi_N(0) = R_{2n_1-1}^{(1)}(0, 0)\phi_N^{(\prime)}(0)$ is 0 by induction hypothesis if and only if all $c_j = 1$ for the pendant vertices v_j of T', and hence of T, which are not the root. By induction hypothesis, the rational function $\frac{\phi_D^{(\prime)}}{\phi_N^{(\prime)}}$ is an $\mathcal{S}$-function and has therefore a representation of the form (A.3.8). Then, by Lemma 4.2.6, also $\frac{\phi_D}{\phi_N}$ has the form (A.3.8), and Lemma A.3.6 shows that $\frac{\phi_D}{\phi_N}$ is a rational $\mathcal{S}$-function.

Finally, if the root of T is not a pendant vertex, then the statements of this theorem hold for T in view of Corollary 4.2.2, Remark 4.2.3, part 1, and the fact that the theorem has already been proved for trees of height h whose roots are pendant vertices. In particular,

$$-\frac{\phi_N}{\phi_D} = \sum_{j=1}^{g} \left(-\frac{\phi_N^{(j)}(\cdot, c_j)}{\phi_D^{(j)}(\cdot, c_j)} \right)$$

is a rational $\mathcal{S}$-function. □

4.3 The direct spectral problem on a tree

In this section, let T be a rooted tree with $g \geq 2$ edges as considered in Section 4.1 where each of its pendant vertices which is not the root is considered with Dirichlet conditions. We will use the notation introduced in Sections 4.1 and 4.2. Recall that n is the number of beads on the tree.

Theorem 4.3.1. *All zeros of ϕ_D and ϕ_N are real, and the nondecreasing sequence $(\nu_k)_{k=1}^n$ of zeros of ϕ_D interlaces with the nondecreasing sequence $(\mu_k)_{k=1}^n$ of zeros of ϕ_N:*

$$0 < \mu_1 \leq \nu_1 \leq \mu_2 \leq \nu_2 \leq \cdots \leq \nu_n.$$

Proof. By Theorem 4.2.7, ϕ_D and ϕ_N are polynomials of degree n, $\phi_N(0) \neq 0$ and $\frac{\Phi_D}{\Phi_N}$ is an S-function. Hence the poles and zeros of $\frac{\Phi_D}{\Phi_N}$ are positive real numbers and interlace strictly by Corollary A.3.3. Observing that ϕ_D and ϕ_N may have common zeros, an application of Lemma A.4.4 completes the proof. □

Theorem 4.3.2. *Let $z \in \mathbb{C}$ and denote by $n_D(z)$ the multiplicity of z as a zero of ϕ_D and by $n_N(z)$ the multiplicity of z as a zero of ϕ_N. Let $\varkappa$ be the number of interior vertices of the tree T which are not the root. Then*

1. $|n_N(z) - n_D(z)| \leq 1$;
2. $n_D(z) \leq g - d(v_0) - \varkappa$ *if* $n_D(z) \leq n_N(z)$;
3. $n_D(z) \leq g - \varkappa$ *if* $n_D(z) > n_N(z)$.

Proof. The estimate in part 1 immediately follows from Theorem 4.3.1.

The statements 2 and 3 will be proved by induction on g. Since we might have to apply the induction hypothesis to the case $g = 1$ and since the case $g = 1$ is easy, we take the induction base to be $g = 1$. In this case, $\varkappa = 0$, $d(v_0) = 1$, and Remark 4.2.3 gives $\phi_D = R_{2n}(\cdot, 0)$ and $\phi_N = R_{2n-1}(\cdot, 0)$. But the zeros of these two polynomials are simple and distinct (see (1.2.36)), so that $n_N(z) \geq n_D(z)$ is only possible if $n_D(z) = 0$, and in this case $n_D(z) = 0 = g - d(v_0) - \varkappa$, whereas $n_D(z) = 1 = g - \varkappa$ if $n_D(z) > n_N(z)$.

Now let $g \geq 2$. Our induction hypothesis is that statements 2 and 3 are true for any tree with at most $g - 1$ edges. First assume that the root is an interior vertex. Then T can be cut into two complementary subtrees T_1 and T_2, as considered in Theorem 4.2.1. Denoting by g_j the number of edges of T_j, by $d^{(j)}(v_0)$

the number of edges of T_j incident with v_0 and by $\varkappa_j$ the number of interior edges of T_j distinct from the root v_0 $(j = 1, 2)$, it is clear that $g = g_1 + g_2$, $\varkappa = \varkappa_1 + \varkappa_2$, and $d(v_0) = d_1(v_0) + d_2(v_0)$. Also, let $n_D^{(j)}(z)$ be the multiplicity of z as a zero of $\phi_D^{(j)}$ and let $n_N^{(j)}(z)$ be the multiplicity of z as a zero of $\phi_N^{(j)}$. Without loss of generality we may assume that $n_N^{(1)}(z) - n_D^{(1)}(z) \geq n_N^{(2)}(z) - n_D^{(2)}(z)$. Clearly, (4.2.7) shows that $n_D(z) = n_D^{(1)}(z) + n_D^{(2)}(z)$. Furthermore, the right-hand side of (4.2.8) is a sum of two polynomials for which the multiplicity of the zero z is $n_N^{(1)}(z)+n_D^{(2)}(z) = n_D^{(1)}(z)+n_D^{(2)}(z)+n_N^{(1)}(z)-n_D^{(1)}(z) = n_D(z)+n_N^{(1)}(z)-n_D^{(1)}(z)$ and $n_D(z)+n_N^{(2)}(z)-n_D^{(2)}(z)$, respectively. If these two numbers are distinct, then $n_N(z)$ equals the minimum of these two numbers, whereas $n_N(z)$ may be larger otherwise. Taking statement 1 into account, we have to consider the following 9 cases:

Case	I	II	III	IV	V	VI	VII	VIII	IX
$n_N^{(1)}(z) - n_D^{(1)}(z)$	1	1	1	0	0	0	-1	-1	-1
$n_N^{(2)}(z) - n_D^{(2)}(z)$	1	0	-1	0	0	-1	-1	-1	-1
$n_N(z) - n_D(z)$	1	0	-1	1	0	-1	1	0	-1

In cases I, II, IV and V we have

$$\begin{aligned} n_D(z) &= n_D^{(1)}(z) + n_D^{(2)}(z) \\ &\leq (g_1 - d^{(1)}(v_0) - \varkappa_1) + (g_2 - d^{(2)}(v_0) - \varkappa_2) = g - d(v_0) - \varkappa, \end{aligned}$$

which proves statement 2 for these cases. In cases III, VI and IX we have

$$n_D(z) = n_D^{(1)}(z) + n_D^{(2)}(z) \leq (g_1 - \varkappa_1) + (g_2 - \varkappa_2) = g - \varkappa,$$

which proves statement 3.

This part of the proof will be complete if we show that cases VII and VIII cannot occur. Indeed, by (4.2.7) and (4.2.8),

$$-\frac{\phi_N}{\phi_D} = -\frac{\phi_N^{(1)}}{\phi_D^{(1)}} - \frac{\phi_N^{(2)}}{\phi_D^{(2)}},$$

where z is a pole of both summands on the right-hand side, and these summands are rational Nevanlinna functions by Theorem 4.2.7 and Lemma A.2.4. Since rational Nevanlinna functions are locally increasing on the real axis by Lemma A.2.5, these poles cannot cancel, and hence z is a pole of the left-hand side, which means that Cases VII and VIII are impossible.

Finally, let v_0 be a pendant vertex of T. With the notation from Proposition 4.2.4 and from (4.2.25) we know that

$$l_0^{(p)} \begin{pmatrix} \phi_D \\ \phi_N \end{pmatrix} = \begin{pmatrix} l_{n_p}^{(p)} & 0 \\ 0 & 1 \end{pmatrix} \Psi_{n_p} \begin{pmatrix} \phi_D^{(\prime)} \\ \phi_N^{(\prime)} \end{pmatrix}.$$

Therefore, if z is a zero of multiplicity at least ν (we may include $\nu = 0$) of both $\phi_D^{(\prime)}$ and $\phi_N^{(\prime)}$, then z is also a zero of multiplicity at least ν of both ϕ_D and ϕ_N, and vice versa because $w \mapsto (\Psi_{n_p}(w))^{-1}$ is a matrix polynomial as the determinant of Ψ_{n_p} is constant by the Lagrange identity. It follows that

$$\min\{n_D(z), n_N(z)\} = \min\{n_D^{(\prime)}(z), n_N^{(\prime)}(z)\}.$$

Taking statement 1 into account, we have to consider the following 9 cases, with the obvious notation for the numbers associated with the tree T':

Case	I	II	III	IV	V	VI	VII	VIII	IX
$n_N^{(\prime)}(z) - n_D^{(\prime)}(z)$	1	1	1	0	0	0	-1	-1	-1
$n_N(z) - n_D(z)$	1	0	-1	1	0	-1	1	0	-1

We observe that $g^{(\prime)} = g - 1$ since one edge was removed, $\varkappa^{(\prime)} = \varkappa - 1$ since one interior vertex has become the new root, and $d(v_0^{(\prime)}) \ge 1 = d(v_0)$. In cases I, II, IV and V,

$$n_D(z) = \min\{n_D(z), n_N(z)\} = \min\{n_D^{(\prime)}(z), n_N^{(\prime)}(z)\} = n_D^{(\prime)}(z),$$

and therefore

$$n_D(z) = n_D^{(\prime)}(z) \le g^{(\prime)} - d(v_0^{(\prime)}) - \varkappa^{(\prime)} \le g - d(v_0) - \varkappa,$$

which proves statement 2 for these cases. In cases III, VI and IX we have

$$n_D(z) = \min\{n_D(z), n_N(z)\} + 1 = \min\{n_D^{(\prime)}(z), n_N^{(\prime)}(z)\} + 1.$$

For cases III and VI this leads to

$$n_D(z) = n_D^{(\prime)}(z) + 1 \le g^{(\prime)} - d(v_0^{(\prime)}) - \varkappa^{(\prime)} + 1 \le g - \varkappa,$$

whereas in case IX we get

$$n_D(z) = n_D^{(\prime)}(z) \le g^{(\prime)} - \varkappa^{(\prime)} = g - \varkappa.$$

Thus statement 2 in cases III and VI and statement 3 in case IX has been proved. Finally, statement 2 in cases VII and VIII follows from

$$\begin{aligned} n_D(z) &= \min\{n_D(z), n_N(z)\} = \min\{n_D^{(\prime)}(z), n_N^{(\prime)}(z)\} = n_D^{(\prime)}(z) - 1 \\ &\le g^{(\prime)} - \varkappa^{(\prime)} - 1 = g - \varkappa - 1 = g - d(v_0) - 1. \end{aligned}$$

□

Proposition 4.3.3. *Under the assumptions of this section, the number* $\frac{\phi_N(0)}{\phi_D(0)}$ *only depends on the shape of the rooted tree and the lengths of its edges.*

Proof. We are going to prove this result by induction on g, the number of edges of T. Let $l^{(j)}$ be the length of the edge e_j $(j = 1, \dots, g)$. For $g = 1$ it follows from Remark 4.2.3, part 2, (1.2.25) and (1.2.28) that

$$\frac{\phi_D(0)}{\phi_N(0)} = l^{(1)}. \tag{4.3.1}$$

Now let $g > 1$ and assume that the statement is true for trees with fewer than g edges. First consider the case that v_0 is an interior vertex. Then Corollary 4.2.2 gives

$$\frac{\phi_N(0)}{\phi_D(0)} = \sum_{j=1}^{d(v_0)} \frac{\phi_N^{(j)}(0)}{\phi_D^{(j)}(0)}, \tag{4.3.2}$$

where each of the summands on the right-hand side satisfies the statement by induction hypothesis. Hence the statement of this proposition holds in this case. Finally, if v_0 is a pendant vertex, then Proposition 4.2.4 and (1.2.25), (1.2.28), (1.2.30), (1.2.31) show that

$$\frac{\phi_D(0)}{\phi_N(0)} = l^{(p)} + \frac{\phi_D^{(\prime)}(0)}{\phi_N^{(\prime)}(0)}, \tag{4.3.3}$$

which completes the proof since the tree T' satisfies the induction hypothesis. □

Remark 4.3.4. Proposition 4.3.3 shows that $\frac{\phi_N(0)}{\phi_D(0)}$ is completely determined by the shape of the tree, the lengths of its edges and the location of the root v. We call this number the shape characteristic of the rooted tree T and denote it by $\chi_{T,v}$.

The recursions (4.3.1), (4.3.2) and (4.3.3) give a constructive way to find the shape characteristic of a rooted tree T. Cutting T at the root into $d(v_0)$ subtrees T_j $(j \in I_0)$ such that e_j is an edge of T_j, v_0 becomes a pendant vertex of T_j. In case $I_j^+ \neq \emptyset$, let T_j' be the tree with root v_j obtained by removing the vertex v_0 and the edge e_j from T_j. Then it follows from (4.3.2), (4.3.3) and (4.3.1) that

$$\chi_{T,v_0} = \sum_{r\in I_0} \chi_{T_j,v_0} = \sum_{\substack{r\in I_0\\ I^+(v_r)=\emptyset}} \frac{1}{l^{(r)}} + \sum_{\substack{r\in I_0\\ I^+(v_r)\neq\emptyset}} \frac{1}{l^{(r)} + \cfrac{1}{\chi_{T_r',v_r}}}. \tag{4.3.4}$$

As an example, consider a metric star graph with central vertex v_0. If v_0 is the root, then

$$\chi_{T,v_0} = \sum_{r=1}^{g} \frac{1}{l^{(r)}}.$$

If, however, the pendant vertex v_1 is the root, then

$$\chi_{T,v_1} = \frac{1}{l^{(1)} + \cfrac{1}{\chi_{T_1',v_0}}} = \frac{1}{l^{(1)} + \cfrac{1}{\sum\limits_{r=2}^{g} \cfrac{1}{l^{(r)}}}}.$$

4.4 The inverse problem for trees of Stieltjes strings: strict interlacing of the characteristic values

Here we give a positive answer to the following question: given two sequences of interlacing real positive numbers, a rooted metric tree of a prescribed shape with prescribed lengths of edges, does there exist a distribution of beads on the tree such that the two sequences are Neumann and Dirichlet characteristic values for this tree?

Theorem 4.4.1. *Let n be a nonnegative integer, let $(\mu_k)_{k=1}^n$ and $(\nu_k)_{k=1}^n$ be sequences of real numbers which interlace:*

$$0 < \mu_1 < \nu_1 < \cdots < \mu_n < \nu_n.$$

Let T be a tree with $g \geq 2$ edges rooted at a vertex v_0 and let positive numbers $l^{(j)}$ $(j = 1, \ldots g)$ be given. Then there exist nonnegative integers n_j $(j = 1, \ldots, g)$ with $\sum\limits_{j=1}^{g} n_j = n$, sequences $(m_k^{(j)})_{k=1}^{n_j}$ and $(l_k^{(j)})_{k=0}^{n_j}$ $(j = 1, \ldots, g)$ of positive numbers with $\sum\limits_{k=0}^{n_j} l_k^{(j)} = l^{(j)}$, and for each $j = 1, \ldots, g$ with $I_j^+ \neq \emptyset$ there is a set Λ_j of positive numbers with $\#\Lambda_j \leq n$ such that the following holds. If $l^{(j)} \notin \Lambda_j$ for $j = 1, \ldots, g$ with $I_j^+ \neq \emptyset$, then the sequence of the characteristic values of the Dirichlet problem (4.1.10)–(4.1.14) for T with Dirichlet conditions, i.e., $c_j = 0$, at the pendant vertices v_j which are not the root coincides with $(\nu_k)_{k=1}^n$, and the sequence of the characteristic values of the Neumann problem (4.1.10)–(4.1.13), (4.1.15) for T with Dirichlet conditions, i.e., $c_j = 0$, at the pendant vertices v_j which are not the root coincides with $(\mu_k)_{k=1}^n$.

Proof. We are going to prove the statement by induction on g. It is convenient to include the case $g = 1$ since trees with one edge may occur in the induction step of the proof. But for $g = 1$ the statement of this theorem is the existence statement of Theorem 1.3.1. We note that here we do not need any restrictions on $l^{(1)}$, which implies that in the induction step no restrictions on $l^{(j)}$ are needed when $I_j^+ = \emptyset$.

Now assume that $g \geq 2$ and that the statement of this theorem is true for every tree with fewer than g edges. Consider the rational function ϕ defined by

$$\phi(z) = \chi_{T,v_0} \prod_{k=1}^{n} \frac{1 - \dfrac{z}{\mu_k}}{1 - \dfrac{z}{\nu_k}}, \tag{4.4.1}$$

where the shape characteristic $\chi_{T,v_0} > 0$ of the tree is known because the shape of T as well as the lengths of its edges are given (see Proposition 4.3.3 and Remark 4.3.4). We will construct a distribution of beads on the interior of the edges of the tree T such that the function ϕ defined in (4.4.1) equals $\frac{\phi_N}{\phi_D}$ for our given graph and the constructed distribution of beads. By Corollary A.3.3, the function

$\frac{1}{\phi}$ belongs to S_0 and therefore $-\phi$ is a Nevanlinna function by Lemma A.2.4. In view of Lemma A.2.7 it follows that

$$\phi(z) = \sum_{k=1}^{n} \frac{A_k}{z - \nu_k} + B, \tag{4.4.2}$$

where $A_k > 0$ and $B = \chi_{T,v_0} + \sum\limits_{k=1}^{n} \frac{A_k}{\nu_k}$.

First assume that $d(v_0) > 1$. Then we cut the tree into $d(v_0)$ complementary subtrees T_ι which are rooted at v_0 ($\iota = 1, \dots, d(v_0)$). Denote by $I^{(\iota)}$ the set of those indices $j \in \{1, \dots, g\}$ for which e_j is an edge of T_ι ($\iota = 1, \dots, d(v_0)$) and choose nonnegative integers N_ι ($\iota = 1, \dots, d(v_0)$) such that $\sum\limits_{\iota=1}^{d(v_0)} N_\iota = n$. We arrange the set $\{1, \dots, n\}$ in an arbitrary manner as the union of disjoint sets $\{k_j^{(\iota)} : j = 1, \dots, N_\iota\}$ ($\iota = 1, \dots, d(v_0)$). Put

$$\phi_\iota(z) := \sum_{j=1}^{N_\iota} \frac{A_{k_j^{(\iota)}}}{z - \nu_{k_j^{(\iota)}}} + B_\iota \quad (\iota = 1, \dots, d(v_0)), \tag{4.4.3}$$

where

$$B_\iota := \chi_{T_\iota,v_0} + \sum_{j=1}^{N_\iota} \frac{A_{k_j^{(\iota)}}}{\nu_{k_j^{(\iota)}}} \quad (\iota = 1, \dots, d(v_0)).$$

Recall from (4.3.2) that $\chi_{T,v_0} = \sum\limits_{\iota=1}^{d(v_0)} \chi_{T_\iota,v_0}$. Each summand on the right-hand side of (4.4.3) is the negative of a Nevanlinna function (see Theorem A.2.6), so that for each $\iota = 1, \dots, d(v_0)$, $-\phi_\iota$ is a Nevanlinna function with N_ι simple positive real poles. Furthermore, $B_\iota > 0$, and therefore ϕ_ι has also N_ι zeros, which are real and interlace with its poles. It follows that $\frac{1}{\phi_\iota} \in \mathcal{S}_0$. In view of $\phi_\iota(0) = \chi_{T_\iota,v_0} > 0$ and $\lim\limits_{z\to-\infty} \phi_\iota(z) = B_\iota > 0$, all zeros of ϕ_ι are positive. Hence, by induction hypothesis there exist sequences of positive numbers $(m_k^{(j)})_{k=1}^{n_j}$ ($j = 1, \dots, g$) and $(l_k^{(j)})_{k=0}^{n_j}$ such that $\sum\limits_{k=0}^{n_j} l_k^{(j)} = l^{(j)}$, the sequence of the characteristic values of the Dirichlet problem (4.1.10)–(4.1.14) for T_ι coincides with the poles of ϕ_ι, and the sequence of the characteristic values of the Neumann problem (4.1.10)–(4.1.13), (4.1.15) for T_ι coincides with the zeros of ϕ_ι. The functions ϕ_ι and $\frac{\phi_N^{(\iota)}}{\phi_D^{(\iota)}}$ have the same (simple) zeros and poles, and together with $\phi_\iota(0) = \chi_{T_\iota,v_0} = \frac{\phi_N^{(\iota)}(0)}{\phi_D^{(\iota)}(0)}$ this means that $\phi_\iota = \frac{\phi_N^{(\iota)}}{\phi_D^{(\iota)}}$ ($\iota = 1, \dots, d(v_0)$). Hence $\phi = \frac{\phi_N}{\phi_D}$ by Corollary 4.2.2, and therefore the generated problem for T has the requested characteristic values.

Now consider the case that v_0 is a pendant vertex of T. Since $\frac{1}{\phi}$ is an S_0-function, Lemma A.3.5 gives a continued fraction expansion

$$\frac{1}{\phi(z)} = a_0 + \cfrac{1}{-b_1 z + \cfrac{1}{a_1 + \cfrac{1}{-b_2 z + \dots + \cfrac{1}{-b_n z + \cfrac{1}{a_n}}}}}. \tag{4.4.4}$$

Without loss of generality we may assume that e_1 is incident with v_0. With the notation from Proposition 4.2.4 and in view of (4.3.3) we have

$$0 < l^{(1)} < l^{(1)} + \frac{1}{\chi_{T',v_1}} = \chi_{T,v_0} = \frac{1}{\phi(0)} = \sum_{k=0}^{n} a_k.$$

Letting Λ_1 be the set of the numbers $\sum\limits_{k=0}^{j} a_k$ $(j = 0, \dots, n-1)$, it follows that there is a number $n_1 \in \{0, \dots, n\}$ such that

$$\sum_{k=0}^{n_1-1} a_k < l^{(1)} < \sum_{k=0}^{n_1} a_k. \tag{4.4.5}$$

Define $l_k^{(1)} = a_{n_1-k}$ and $m_k^{(1)} = b_{n_1+1-k}$ for $k = 1, \dots, n_1$, $l_0^{(1)} = l^{(1)} - \sum\limits_{k=0}^{n_1-1} a_k$, $a_{n_1}^{(\prime)} = a_{n_1} - l_0^{(1)}$, $a_k^{(\prime)} = a_k$ for $k = n_1+1, \dots, n$. Then

$$\frac{1}{\phi(z)} = l_{n_1}^{(1)} + \cfrac{1}{-m_{n_1}^{(1)} z + \cfrac{1}{l_{n_1-1}^{(1)} + \cfrac{1}{-m_{n_1-1}^{(1)} z + \dots + \cfrac{1}{-m_1^{(1)} z + \cfrac{1}{l_0^{(1)} + \cfrac{1}{\phi^{(\prime)}(z)}}}}}},$$

where

$$\frac{1}{\phi^{(\prime)}(z)} = a_{n_1}^{(\prime)} + \cfrac{1}{-b_{n_1+1}^{(\prime)} z + \cfrac{1}{a_{n_1+1}^{(1)} + \cfrac{1}{-b_{n_1+2}^{(\prime)} z + \dots + \cfrac{1}{-b_n^{(\prime)} z + \cfrac{1}{a_n^{(\prime)}}}}}}.$$

By Lemma A.3.6, $\frac{1}{\phi^{(\prime)}}$ has $n-n_1$ positive real poles and zeros which interlace, and $\phi^{(\prime)}(0)=\chi_{T',v_1}$. By induction hypothesis, there are sequences of positive numbers $(m_k^{(j)})_{k=1}^{n_j}$ and $(l_k^{(j)})_{k=0}^{n_j}$ $(j=2,\ldots,g)$ such that $\sum\limits_{k=0}^{n_j} l_k^{(j)} = l^{(j)}$, the sequence of the characteristic values of the Dirichlet problem (4.1.10)–(4.1.14) for T' coincides with the poles of $\phi^{(\prime)}$, and the sequence of the characteristic values of the Neumann problem (4.1.10)–(4.1.13), (4.1.15) for T' coincides with the zeros of $\phi^{(\prime)}$. Therefore $\phi^{(\prime)} = \frac{\phi_N^{(\prime)}}{\phi_D^{(\prime)}}$, and the two above formulas for $\frac{1}{\phi}$ and $\frac{1}{\phi^{(\prime)}}$ and (4.2.24) show that $\phi = \frac{\phi_N}{\phi_D}$. This implies that the generated problem for T has the requested characteristic values. □

Remark 4.4.2.

1. The restrictions in Theorem 4.4.1 are quite implicit and cannot be checked a priori. However, the proof of the theorem is constructive. If the root is a pendant vertex, then the set Λ_1 is determined by the given data, and in case $l^{(1)} \in \Lambda_1$ the procedure will stop, and there is no solution to the inverse problem. If however, the root is an interior vertex, then firstly there is no restriction on the lengths of the strings in the first step and secondly there is some choice in allocating the Stieltjes functions for the subgraphs. It may be possible, if some of the subgraphs down the line have a pendant vertex as the root, to avoid the termination of the procedure by a suitable choice of the Stieltjes functions.
2. As a simple example let us consider a star graph where one of the pendant vertices is the root. We know from Theorem 2.3.8 that the inverse problem has a solution if one allows a bead at the central vertex.
3. It seems to be an open problem if the inverse problem considered in Theorem 4.4.1 has a solution for each metric tree if one allows beads at interior vertices.

4.5 Notes

Theorem 4.3.1 and Proposition 4.3.3 were proved in [109]. Theorem 4.4.1 can also be found in [109], but the version in [109] contains a mistake: the condition $l^{(j)} \in \Lambda_j$ is missing there.

Branching continued fractions which appear in Lemma 4.2.6 and in the proof of Theorem 4.4.1 is a well-known subject (see, e.g., [126], [13]).

Chapter 5

Spectral problems on graphs

5.1 The problem and its spectrum

Let G be a metric digraph with g edges; G may be connected or not connected. We refer to Appendix C for definitions and notation. The edges of G are denoted by e_j and the lengths of the edges by $l^{(j)} > 0$ $(j = 1, \ldots, g)$. In our constructions below we may encounter degenerate cases, and therefore we only require that G has at least one vertex. In particular, it will be allowed that the graph has no edges.

Each edge e_j is a Stieltjes string of the total length $l^{(j)}$ bearing n_j beads. The beads on e_j have the masses $m_k^{(j)} > 0$ $(k = 1, \ldots, n_j)$ and divide the edge into threads of lengths $l_k^{(j)} > 0$ $(k = 0, \ldots, n_j)$, where the lower index is chosen in such a way that it increases in the direction of the edge.

The amplitudes of the vibrations of the beads and the ends of the strings are subject to the equations (1.2.4), and with the indexing as in (4.1.10) these amplitudes will be denoted by $u_0^{(j)}, \ldots, u_{n_j+1}^{(j)}$ on the edges e_j $(j = 1, \ldots, g)$.

We recall from Appendix C that $\widehat{V}$ denotes the set of all vertices of G which are not isolated. Each vertex $v \in \widehat{V}$ may be clamped or freely moving. The set of clamped vertices is denoted by V^{c} and the set of freely moving vertices is denoted by V^{f}. For each interior vertex v we require the continuity condition that all values $u_0^{(j)}$ with $j \in I^-(v)$ and $u_{n_j+1}^{(j)}$ with $j \in I^+(v)$ are equal. Defining

$$U(v) := \{u_0^{(j)} : j \in I^-(v)\} \cup \{u_{n_j+1}^{(j)} : j \in I^+(v)\},$$

this condition can be written as

$$\#U(v) = 1 \quad (v \in V^{\mathrm{int}}). \tag{5.1.1}$$

Hence each vertex $v \in \widehat{V}$ has a unique amplitude, which will be denoted by $u(v)$.

M. Möller, V. Pivovarchik, *Direct and Inverse Finite-Dimensional Spectral Problems on Graphs*, Operator Theory: Advances and Applications 283, https://doi.org/10.1007/978-3-030-60484-4_5

If $v \in V^c$, then $u(v) = 0$, and we arrive at

$$u_0^{(j)} = 0 \ (j \in I^-(v)), \quad u_{n_j+1}^{(j)} = 0 \ (j \in I^+(v)). \tag{5.1.2}$$

For an interior vertex v, the condition (5.1.2) is called the *generalized Dirichlet condition* at v. If, however, $v \in V^f$, then

$$\sum_{j\in I^-(v)} \frac{u_1^{(j)} - u_0^{(j)}}{l_0^{(j)}} = \sum_{j\in I^+(v)} \frac{u_{n_j+1}^{(j)} - u_{n_j}^{(j)}}{l_{n_j}^{(j)}} \tag{5.1.3}$$

is imposed, which is the balance of forces condition when the vertex v is an interior vertex.

We will call the pairs of conditions (5.1.1), (5.1.3) *generalized Neumann conditions* at the interior vertex v. For pendant vertices the conditions (5.1.2) and (5.1.3) are the usual Dirichlet and Neumann conditions, respectively. In this chapter we will consider the problem (4.1.10), (5.1.1), (5.1.2), (5.1.3), i.e., (4.1.10) with (generalized) Dirichlet conditions at each $v \in V^c$ and with (generalized) Neumann conditions at each $v \in V^f$.

We are now going to write the problem (4.1.10), (5.1.1) (5.1.2), (5.1.3) as a matrix eigenvalue problem

$$(A - zM)Y = 0. \tag{5.1.4}$$

For convenience, v_k $(k = 1, \ldots, n_{g+1})$ is an arbitrary indexing of the freely moving vertices in V with $n_{g+1} := \#V^f$ and $u_k^{(g+1)} := u(v_k)$ $(k = 1, \ldots, n_{g+1})$ is the amplitude of the vibration at v_k. Then Y will be written as $Y = (u_k^{(j)})_{\substack{j=1,\ldots,g+1 \\ k=1,\ldots,n_j}}$. Although the indices (j, k) may be arranged in an arbitrary manner, for definiteness we will use lexicographic order, that is, $(j, k) < (r, s)$ if and only if $j < r$ or $j = r$ and $k < s$.

Denoting by

$$n = \sum_{j=1}^{g} n_j$$

the total number of beads on the graph G, we formally allow that $n = 0$. Of course, also $n_{g+1} = 0$ is possible. In order for Y to have at least one component we will require that $n + n_{g+1} > 0$.

We observe that (5.1.1) and (5.1.2) mean that the values of $u_0^{(j)}$ and $u_{n_j+1}^{(j)}$ are either 0 or some $u_k^{(g+1)}$. Therefore the solutions of (4.1.10), (5.1.1), (5.1.2), (5.1.3) are uniquely determined by (4.1.10) and (5.1.3) for Y. Hence A and M will be $(n + n_{g+1}) \times (n + n_{g+1})$ matrices. The component of the vector $(A - zM)Y$

with index (j,k) $(j=1,\dots,g+1,\ k=1,\dots,n_j)$ will be

$$\frac{u_k^{(j)}-u_{k-1}^{(j)}}{l_{k-1}^{(j)}}-\frac{u_{k+1}^{(j)}-u_k^{(j)}}{l_k^{(j)}}-m_k^{(j)}zu_k^{(j)}\quad (j=1,\dots,g,\ n_j\geq 1,\ k=1,\dots,n_j), \tag{5.1.5}$$

$$\sum_{\iota\in I^+(v_k)}\frac{u_k^{(g+1)}-u_{n_\iota}^{(\iota)}}{l_{n_\iota}^{(\iota)}}-\sum_{\iota\in I^-(v_k)}\frac{u_1^{(\iota)}-u_k^{(g+1)}}{l_0^{(\iota)}}\quad (j=g+1, k=1,\ \dots, n_{g+1}). \tag{5.1.6}$$

Of course, for $u_0^{(j)}$ and $u_{n_j+1}^{(j)}$ the appropriate value of $u_s^{(g+1)}$ for freely moving ends and the value 0 for clamped ends has to be substituted in (5.1.5) and also in (5.1.6) if $n_r=0$ for some $r=1,\dots,g$.

Definition 5.1.1. The problem (4.1.10), (5.1.1) (5.1.2), (5.1.3) on the graph G is called degenerate if G has a component G_1 with at least one edge such that no edge of G_1 carries a bead and such that all vertices of G_1 are freely moving.

Theorem 5.1.2. *Let $n+n_{g+1}>0$. Then the matrix A is symmetric and positive semidefinite and the matrix M is a positive semidefinite diagonal matrix. The matrix $A-zM$ is singular for all $z\in\mathbb{C}$ if and only if the problem is degenerate. If the problem is not degenerate, then the eigenvalue problem $z\mapsto(A-zM)$ has n eigenvalues, counted with multiplicity, all eigenvalues are real and nonnegative, and the geometric multiplicity of each eigenvalue coincides with its algebraic multiplicity. The number 0 is an eigenvalue if and only if there is a component of G with at least one edge such that each vertex of this component is freely moving. The multiplicity of the eigenvalue 0 is the number of components with at least one edge in which each vertex is freely moving.*

Proof. We will first show that the matrix A is symmetric. We will denote it by

$$A=:(a_{j,k,r,s}),$$

where (j,k) is the row index and (r,s) is the column index. To show the symmetry of A we have to verify that $a_{j,k,r,s}=a_{r,s,j,k}$ for all $(j,k)<(r,s)$. We will consider several cases:

Case 1: $j<r\leq g$. In this case it is clear that (5.1.5) with index (j,k) does not contain $u_s^{(r)}$ and (5.1.5) with index (r,s) does not contain $u_k^{(j)}$. Therefore $a_{j,k,r,s}=0=a_{r,s,j,k}$.

Case 2: $j=r\leq g$ and $s\geq k+2$. Here it is obvious from (5.1.5) that $a_{j,k,j,s}=0=a_{j,s,j,k}$.

Case 3: $j=r\leq g$ and $s=k+1$. It is clear by (5.1.5) that

$$a_{j,k,j,k+1}=-\frac{1}{l_k^{(j)}}=a_{j,k+1,j,k}.$$

Case 4: $j \le g$, $r = g+1$, $k = 1$, $j \in I^-(v_s)$. Then $u_s^{(g+1)} = u_0^{(j)}$ and therefore it follows in view of (5.1.5) and (5.1.6) that

$$a_{j,k,g+1,s} = -\frac{1}{l_0^{(j)}} = a_{g+1,s,j,k}.$$

Case 5: $j \le g$, $r = g+1$, $k = n_j$, $j \in I^+(v_s)$. Then $u_s^{(g+1)} = u_{n_j+1}^{(j)}$ and therefore it follows in view of (5.1.5) and (5.1.6) that

$$a_{j,k,g+1,s} = -\frac{1}{l_{n_j}^{(j)}} = a_{g+1,s,j,k}.$$

Case 6: $j \le g$, $r = g+1$, and k and s are different from the cases 4 and 5. Then it follows in view of (5.1.5) and (5.1.6) that $a_{j,k,g+1,s} = 0 = a_{g+1,s,j,k}$.

Case 7: $j = r = g+1$. For $l, m \in \{1, \dots, n_{g+1}\}$ let

$$\widehat{I}(l,m) = \{\iota \in \{1, \dots, g\} : n_\iota = 0,\ \iota \in I^+(v_l) \cap I^-(v_m)\}.$$

Then

$$a_{g+1,k,g+1,s} = -\sum_{\iota \in \widehat{I}(k,s) \cup \widehat{I}(s,k)} \frac{1}{l_0^{(\iota)}} = a_{g+1,s,g+1,k}.$$

Thus we have shown that A is a symmetric matrix, and we have also explicitly calculated the off-diagonal entries of A. The diagonal elements of A are easily obtained from (5.1.5) and (5.1.6), namely,

$$a_{j,k,j,k} = \frac{1}{l_{k-1}^{(j)}} + \frac{1}{l_k^{(j)}} \quad (j = 1, \dots, g,\ k = 1, \dots, n_j),$$

$$a_{g+1,k,g+1,k} = \sum_{\iota \in I^+(v_k) \setminus \widehat{I}(k,k)} \frac{1}{l_{n_\iota}^{(\iota)}} + \sum_{\iota \in I^-(v_k) \setminus \widehat{I}(k,k)} \frac{1}{l_0^{(\iota)}} \quad (k = 1, \dots, n_{g+1}).$$

To show that A is positive semidefinite we observe that the definition of AY in (5.1.5) and (5.1.6) gives

$$Y^* AY = S_1 + S_2,$$

where

$$S_1 = \sum_{j=1}^{g} \sum_{k=1}^{n_j} \overline{u_k^{(j)}} \left(\frac{u_k^{(j)} - u_{k-1}^{(j)}}{l_{k-1}^{(j)}} - \frac{u_{k+1}^{(j)} - u_k^{(j)}}{l_k^{(j)}} \right),$$

$$S_2 = \sum_{k=1}^{n_{g+1}} \overline{u_k^{(g+1)}} \left(\sum_{\iota \in I^+(v_k)} \frac{u_k^{(g+1)} - u_{n_\iota}^{(\iota)}}{l_{n_\iota}^{(\iota)}} - \sum_{\iota \in I^-(v_k)} \frac{u_1^{(\iota)} - u_k^{(g+1)}}{l_0^{(\iota)}} \right).$$

We split $S_1 =: S_{11} + S_{12} + S_{13}$ with

$$S_{11} := \sum_{\substack{j=1\\ n_j>0}}^{g} \frac{1}{l_0^{(j)}} \left(|u_1^{(j)}|^2 - \overline{u_1^{(j)}} u_0^{(j)} \right)$$

$$= \sum_{v\in V^c} \sum_{\substack{j\in I^-(v)\\ n_j>0}} \frac{1}{l_0^{(j)}} |u_1^{(j)}|^2 + \sum_{k=1}^{n_{g+1}} \sum_{\substack{j\in I^-(v_k)\\ n_j>0}} \frac{1}{l_0^{(j)}} \left(|u_1^{(j)}|^2 - \overline{u_1^{(j)}} u_k^{(g+1)} \right),$$

$$S_{12} := \sum_{\substack{j=1\\ n_j>0}}^{g} \frac{1}{l_{n_j}^{(j)}} \left(|u_{n_j}^{(j)}|^2 - \overline{u_{n_j}^{(j)}} u_{n_j+1}^{(j)} \right)$$

$$= \sum_{v\in V^c} \sum_{\substack{j\in I^+(v)\\ n_j>0}} \frac{1}{l_{n_j}^{(j)}} |u_{n_j}^{(j)}|^2 + \sum_{k=1}^{n_{g+1}} \sum_{\substack{j\in I^+(v_k)\\ n_j>0}} \frac{1}{l_{n_j}^{(j)}} \left(|u_{n_j}^{(j)}|^2 - \overline{u_{n_j}^{(j)}} u_k^{(g+1)} \right),$$

and

$$S_{13} := \sum_{j=1}^{g} \sum_{k=1}^{n_j-1} \frac{1}{l_k^{(j)}} |u_{k+1}^{(j)} - u_k^{(j)}|^2.$$

We can now write $S_{11} + S_{12} + S_2 = S_{21} + S_{22}$, where

$$S_{21} = \sum_{v\in V^c} \sum_{\substack{j\in I^-(v)\\ n_j>0}} \frac{1}{l_0^{(j)}} |u_1^{(j)}|^2 + \sum_{v\in V^c} \sum_{\substack{j\in I^+(v)\\ n_j>0}} \frac{1}{l_{n_j}^{(j)}} |u_{n_j}^{(j)}|^2,$$

$$S_{22} = \sum_{k=1}^{n_{g+1}} \sum_{j\in I^-(v_k)} \frac{\alpha_j}{l_0^{(j)}} |u_1^{(j)} - u_k^{(g+1)}|^2 + \sum_{k=1}^{n_{g+1}} \sum_{j\in I^+(v_k)} \frac{\alpha_j}{l_{n_j}^{(j)}} |u_{n_j}^{(j)} - u_k^{(g+1)}|^2,$$

where $\alpha_j = \frac{1}{2}$ if $n_j = 0$, $v_-(e_j) \in V^{\mathrm{f}}$ and $v_+(e_j) \in V^{\mathrm{f}}$, whereas $\alpha_j = 1$ otherwise. This identity is clear for all terms with $n_j > 0$. Now let $n_j = 0$. We have to compare terms in S_2 and S_{22}. If both $v_-(e_j)$ and $v_+(e_j)$ are clamped vertices, then this edge e_j does not contribute to S_2 and S_{22}. If $v_+(e_j)$ belongs to V^{f}, say $v_+(e_j) = v_k$, then $j \in I^+(v_k)$ gives the term

$$\overline{u_k^{(g+1)}} \frac{u_k^{(g+1)} - u_0^j}{l_0^{(j)}}$$

in S_2. Similarly, if $v_-(e_j)$ belongs to V^{f}, say $v_-(e_j) = v_l$, then $j \in I^-(v_l)$ gives the term

$$-\overline{u_l^{(g+1)}} \frac{u_1^j - u_l^{(g+1)}}{l_0^{(j)}}$$

in S_2. We are going to consider three cases.

If $v_+(e_j) \in V^{\rm f}$ and $v_-(e_j) \in V^{\rm c}$, then $u_0^{(j)} = 0$, and the contribution from j in S_2 is $\frac{1}{l_0^{(j)}}|u_k^{(g+1)}|^2$, whereas the contribution in S_{22} is $\frac{1}{l_0^{(j)}}|u_0^{(j)} - u_k^{(g+1)}|^2 = \frac{1}{l_0^{(j)}}|u_k^{(g+1)}|^2$. Therefore, the terms in S_2 and S_{22} for such j are the same. Similarly, if $v_+(e_j) \in V^{\rm c}$ and $v_-(e_j) \in V^{\rm f}$, then $u_1^{(j)} = 0$, and j gives the contribution $\frac{1}{l_0^{(j)}}|u_l^{(g+1)}|^2$ in S_2 as well as S_{22}.

Finally, if $v_+(e_j) \in V^{\rm f}$ and $v_-(e_j) \in V^{\rm f}$, then the contribution of j in S_2 is

$$\frac{1}{l_0^{(j)}}\left[\overline{u_k^{(g+1)}}(u_k^{(g+1)} - u_l^{(g+1)}) - \overline{u_l^{(g+1)}}(u_k^{(g+1)} - u_l^{(g+1)})\right] = \frac{1}{l_0^{(j)}}\left|u_k^{(g+1)} - u_l^{(g+1)}\right|^2,$$

which coincides with the contribution in S_{22}.

From $S_{13} \geq 0$, $S_{21} \geq 0$, $S_{22} \geq 0$ and $Y^*AY = S_{13} + S_{21} + S_{22}$ it follows that A is positive semidefinite.

Clearly, M is a diagonal matrix with positive diagonal entries $m_k^{(j)}$ in the row and column with index (j,k) $(j = 1,\dots,g,\ k = 1,\dots,n_j)$, whereas all other diagonal entries are 0.

Now assume that the problem is degenerate and let G_1 be as in Definition 5.1.1. Define the vector Y by setting $u_k^{(g+1)} = 1$ for all $k = 1,\dots,n_{g+1}$ with $v_k \in V^{G_1}$ and by setting $u_k^{(j)} = 0$ for all other components. Then clearly $S_{13} = 0$ and $S_{21} = 0$. For the terms in S_{22} we observe the following. If $u_k^{(g+1)} = 0$ in the term $|u_1^{(j)} - u_k^{(g+1)}|^2$ and if $n_j > 0$, then also $u_1^{(j)} = 0$. If $u_k^{(g+1)} = 0$ in the term $|u_1^{(j)} - u_k^{(g+1)}|^2$ and if $n_j = 0$, then $u_1^{(j)} = u_{n_j+1}^{(j)}$ is the amplitude at $v_+(e_j)$. But since v_k is not a vertex of the component G_1, also $v_+(e_j)$ is not a vertex of G_1 since v_k and $v_+(e_j)$ are ends of the edge e_j. Hence $u_1^{(j)} = 0$. A corresponding reasoning holds for $|u_{n_j}^{(j)} - u_k^{(g+1)}|^2$. If $u_k^{(g+1)} = 1$ then the term $|u_1^{(j)} - u_k^{(g+1)}|^2$ and the term $|u_{n_j}^{(j)} - u_k^{(g+1)}|^2$ are of the form $|u_l^{(g+1)} - u_k^{(g+1)}|^2$ with $v_l \in V^{G_1}$ since $v_k \in V^{G_1}$ implies that e_j is an edge in G_1. Altogether, it follows that all summands in S_{22} are 0. Therefore we have shown that $Y^*AY = 0$, which implies $AY = 0$ since A is positive semidefinite. Obviously, $MY = 0$, and hence $(A - zM)Y = 0$. Thus $A - zM$ is a singular matrix for all $z \in \mathbb{C}$.

For the remainder of this proof let the problem be nondegenerate.

If $n > 0$ and $n_{g+1} = 0$, then M is a positive definite diagonal matrix, and by Proposition B.2.1, all eigenvalues are real and nonnegative, and the geometric multiplicity of each eigenvalue coincides with its algebraic multiplicity.

Now let $n_{g+1} > 0$ and let A_{22} be the lower right $n_{g+1} \times n_{g+1}$ submatrix of A. We are going to show that A_{22} is invertible. To this end let $Y_2 = (u_k^{(g+1)})_{k=1}^{n_{g+1}}$ such that $Y_2^*A_{22}Y_2 = 0$. Putting $Y = \begin{pmatrix} 0 \\ Y_2 \end{pmatrix}$, it follows that $Y^*AY = Y_2^*A_{22}Y_2 = 0$ and in particular $S_{22} = 0$. Hence each summand in S_{22} is zero.

Let $k \in \{1, \dots, n_{g+1}\}$ and let e_j be an edge of G which is incident with v_k. Then at least one of $\frac{\alpha_j}{l_0^{(j)}}|u_1^{(j)} - u_k^{(g+1)}|^2$, $\frac{\alpha_j}{l_{n_j}^{(j)}}|u_{n_j}^{(j)} - u_k^{(g+1)}|^2$ is a summand in A_{22}. Let G_0 be the component of G with $v_k \in V^{G_0}$ and let V_0 be the set of those vertices of G_0 which are either clamped vertices or freely moving vertices having at least one incident edge e_j with $n_j > 0$. Since G is nondegenerate, $V_0 \neq \emptyset$. If $v_k \in V_0$, then there is an incident edge e_j with $n_j > 0$, and in view of the definition of Y, $u_1^{(j)} = 0$ and $u_{n_j}^{(j)} = 0$, so that also $u_k^{(g+1)} = 0$.

If, however, $v_k \notin V_0$, then we choose a path $(v_k = w_0, e_{j_1}, w_1, \dots, e_{j_l}, w_l)$ in G_0 with $w_l \in V_0$. Without loss of generality we may assume that $w_i \notin V_0$ for $i < l$. By definition of V_0, $n_{j_i} = 0$ for $i = 1, \dots, l$ and w_i is a freely moving vertex for $i = 1, \dots, l-1$. Therefore $\frac{1}{l_0^{(j_i)}}|u(w_{i-1}) - u(w_i)|^2$ occurs in S_{22} for $i = 1, \dots, l$, which shows that $u_k^{(g+1)} = u(w_0) = \cdots = u(w_l)$. If w_l is a clamped vertex, then $u(w_l) = 0$ by definition of clamped vertices, and if w_l is a freely moving vertex, then $u(w_l) = 0$ has already been shown for freely moving vertices in V_0. Therefore $u_k^{(g+1)} = 0$ also follows for $v_k \notin V_0$.

We have shown that $Y_2 = 0$, which means that A_{22} is invertible. In case $n = 0$, the matrix $A - zM$ becomes A_{22}. Therefore the problem has no eigenvalues, formally in accordance with the statement of this theorem.

Now let $n > 0$ and $n_{g+1} > 0$. Writing A and M as $(n + n_{g+1}) \times (n + n_{g+1})$ block matrices we get

$$A - zM = \begin{pmatrix} A_{11} - zM_1 & A_{12} \\ A_{21} & A_{22} \end{pmatrix}.$$

The Schur factorization (see (B.4.2))

$$A - zM = \begin{pmatrix} I & A_{12}A_{22}^{-1} \\ 0 & I \end{pmatrix} \begin{pmatrix} A_{11} - A_{12}A_{22}^{-1}A_{21} - zM_1 & 0 \\ 0 & A_{22} \end{pmatrix} \begin{pmatrix} I & 0 \\ A_{22}^{-1}A_{21} & I \end{pmatrix}$$

shows that all spectral data are given by $A_{11} - A_{12}A_{22}^{-1}A_{21} - zM_1$. But since A is symmetric and positive semidefinite, also A_{11} and A_{22} are symmetric and positive semidefinite, and $A_{21} = A_{12}^*$. Therefore $A_{11} - A_{12}A_{22}^{-1}A_{21}$ is symmetric and positive semidefinite in view of the Schur factorization, and an application of Proposition B.2.1 shows that all eigenvalues are real and that the geometric multiplicity of the eigenvalues coincides with their algebraic multiplicity also in case $n > 0$ and $n_g > 0$.

Since M has exactly n nonzero entries it is clear in all cases that there are exactly n eigenvalues, counted with multiplicity.

If the graph has a component without clamped vertices, let $\widetilde{I}$ be the index set of the edges in that component. For $j = 1, \dots, g$ and $k = 1, \dots, n_j$ put $u_k^{(j)} = 1$ if $j \in \tilde{I}$ and $u_k^{(j)} = 0$ if $j \notin \tilde{I}$. We put $u_k^{(g+1)} = 1$ if there is $j \in \widetilde{I}$ such that $j \in I(v_k)$ and $u_k^{(g+1)} = 0$ otherwise. Clearly, $S_{21} = 0$. On each edge of the graph, either all amplitudes are 0 or all amplitudes are 1. Hence $S_{13} = 0$ and $S_{22} = 0$. Therefore 0 is an eigenvalue in this case.

Let Y be such that $Y^*AY = 0$. Note that this implies $S_{13} = 0$, $S_{21} = 0$ and $S_{22} = 0$. Let $j \in \{1, \dots, g\}$. Then $S_{13} = 0$ gives $u_{k+1}^{(j)} = u_k^{(j)}$ ($j = 1, \dots, g$, $k = 1, \dots, n_j - 1$). If $v_-(e_j) \in V^c$, then $u_0^{(j)} = 0$, and $u_1^{(j)} = 0$ if $n_j > 0$ in view of $S_{21} = 0$. If $v_-(e_j) \in V^f$, then $u_0^{(j)} = u_1^{(j)}$ due to $S_{22} = 0$. Similarly, if $v_+(e_j) \in V^c$, then $u_{n_j+1}^{(j)} = 0$, and $u_{n_j}^{(j)} = 0$ if $n_j > 0$ in view of $S_{21} = 0$. If $v_+(e_j) \in V^f$, then $u_{n_j+1}^{(j)} = u_{n_j}^{(j)}$ due to $S_{22} = 0$. It follows that $u_0^{(j)} = u_1^{(j)} = \dots = u_{n_j+1}^{(j)}$ for all $j = 1, \dots, g$.

For any two edges (including loops) in a component of G there is a walk G which contains these two edges. Hence the identities from the previous paragraph and the continuity condition (5.1.1) show that $Y^*AY = 0$ implies that all $u_k^{(j)}$ have the same value for all j for which e_j is in the same component and all $k = 0, \dots, n_j+1$. If a component has a clamped vertex, then this value must be 0 in view of (5.1.2). This proves the statement on the multiplicity of the eigenvalue 0. □

Remark 5.1.3. Reversing the direction of an edge, say e_j, results in reversing the order of the rows and columns with index (j, k) ($k = 1, \dots, n_j$). Hence the eigenvalue problem is invariant under changing the direction of the edges, and in particular $\det(A - zM)$ does not depend on the direction of the edges. When $n > 0$ and $n_{g+1} > 0$, i.e., when the graph carries at least one bead and has freely moving vertices, then we set

$$A_1 = A_{11} - \widetilde{A}_1, \quad \widetilde{A}_1 = A_{12}A_{22}^{-1}A_{21},$$

recalling the notation from the proof of Theorem 5.1.2. Then it is clear that also $\det(A_1 - zM_1)$ does not depend on the direction of the edges. For convenience we may also use this notation when $n_{g+1} = 0$, in which case $A_1 = A$ and $M_1 = M$.

Proposition 5.1.4. *Assume that the graph G carries at least one bead. Let v be a freely moving vertex of G with $d(v) = 2$ and two incident edges. Let G' be the graph obtained by concatenating the two edges incident with v to one edge. Then, with appropriate indexing, $A_1 - zM_1$ is the same for G as for G'.*

Proof. Observe that the assumption implies that there are exactly two edges which are incident with v; that is, v is not the vertex of a loop. We choose an indexing and orientation of the edges and freely moving vertices in such a way that $v = v_{n_{g+1}}$, $I^+(v) = \{g - 1\}$ and $I^-(v) = \{g\}$. Therefore the edges of G' are $e_1, \dots, e_{g-2}$ and the edge e'_{g-1} obtained by concatenating e_{g-1} and e_g. Then the amplitudes of the ends and beads of e'_{g-1} are $u_0^{(g-1)}, \dots, u_{n_{g-1}}^{(g-1)}, u_1^{(g)}, \dots, u_{n_g+1}^{(g)}$. Writing A as an $([n + n_{g+1} - 1] + 1) \times ([n + n_{g+1} - 1] + 1)$ block matrix,

$$A =: \begin{pmatrix} \widehat{A}_{11} & \widehat{A}_{12} \\ \widehat{A}_{21} & \widehat{A}_{22} \end{pmatrix},$$

we define

$$A^\dagger = \widehat{A}_{11} - \widehat{A}_{12}\widehat{A}_{22}^{-1}\widehat{A}_{21}$$

and claim that $A' = A^\dagger$.

To prove this claim, we first evaluate $\widehat{A}_{12}$, that is, the elements $a_{j,k,g+1,n_{g+1}}$ of A with $(j,k) < (g+1, n_{g+1})$. According to the cases 4 to 7 in the proof of Theorem 5.1.2, the nonzero elements of $\widehat{A}_{1,2}$ are

$$a_{g,1,g+1,n_{g+1}} = -\frac{1}{l_0^{(g)}} \text{ if } n_g > 0,$$

$$a_{g-1,n_{g-1},g+1,n_{g+1}} = -\frac{1}{l_{n_{g-1}}^{(g-1)}} \text{ if } n_{g-1} > 0,$$

$$a_{g+1,k,g+1,n_{g+1}} = -\frac{1}{l_0^{(g)}} \text{ if } n_g = 0,\ v_+(e_g) = v_k \text{ and } (n_{g-1} > 0 \text{ or } v_-(e_{g-1}) \neq v_k),$$

$$a_{g+1,k,g+1,n_{g+1}} = -\frac{1}{l_{n_{g-1}}^{(g-1)}} \text{ if } n_{g-1} = 0,\ v_-(e_{g-1}) = v_k \text{ and } (n_g > 0 \text{ or } v_+(e_g) \neq v_k),$$

$$a_{g+1,k,g+1,n_{g+1}} = -\frac{1}{l_{n_{g-1}}^{(g-1)}} - \frac{1}{l_0^{(g)}} \text{ if } n_{g-1} = n_g = 0 \text{ and } v_-(e_{g-1}) = v_+(e_g) = v_k.$$

Furthermore,

$$a_{g+1,n_{g+1},g+1,n_{g+1}} = \frac{1}{l_{n_{g-1}}^{(g-1)}} + \frac{1}{l_0^{(g)}}.$$

The variables in the linear functionals defining A' are the same as the variables for $\widehat{A}_{11}$, and the corresponding coefficients in the linear functionals are the same for $\widehat{A}_{11}$ and A' if they do not contain $l_0^{(g)}$ or $l_{n_{g-1}}^{g-1}$. We may assume without loss of generality that $n_{g-1} \geq n_g$. Due to the symmetry of A' and $A^\dagger$ we only have to consider indices (j,k,r,s) with $(j,k) \leq (r,s)$. The first row to consider is the row with index $(g-1, n_{g-1})$, which is present when $n_{g-1} > 0$. In $\widehat{A}_{11}$ it is represented by the linear functional

$$\frac{u_{n_{g-1}}^{(g-1)} - u_{n_{g-1}-1}^{(g-1)}}{l_{n_{g-1}-1}^{(g-1)}} + \frac{u_{n_{g-1}}^{(g-1)}}{l_{n_{g-1}}^{(g-1)}},$$

whereas its representation in A' is

$$\frac{u_{n_{g-1}}^{(g-1)} - u_{n_{g-1}-1}^{(g-1)}}{l_{n_{g-1}-1}^{(g-1)}} - \frac{u_1^{(g)} - u_{n_{g-1}}^{(g-1)}}{l_{n_{g-1}}^{(g-1)} + l_0^{(g)}}.$$

Then

$$a'_{g-1,n_{g-1},g-1,n_{g-1}} = \frac{1}{l_{n_{g-1}-1}^{(g-1)}} + \frac{1}{l_{n_{g-1}}^{(g-1)} + l_0^{(g)}},$$

whereas

$$\begin{aligned}a^{\dagger}_{g-1,n_{g-1},g-1,n_{g-1}} &= a_{g-1,n_{g-1},g-1,n_{g-1}} - \frac{a_{g-1,n_{g-1},g+1,n_{g+1}}a_{g+1,n_{g+1},g-1,n_{g-1}}}{a_{g+1,n_{g+1},g+1,n_{g+1}}}\\ &= \frac{1}{l^{(g-1)}_{n_{g-1}-1}} + \frac{1}{l^{(g-1)}_{n_{g-1}}} - \frac{1}{(l^{(g-1)}_{n_{g-1}})^2}\frac{l^{(g-1)}_{n_{g-1}}l^{(g)}_0}{l^{(g-1)}_{n_g-1}+l^{(g)}_0}\\ &= \frac{1}{l^{(g-1)}_{n_{g-1}-1}} + \frac{1}{l^{(g-1)}_{n_g-1}+l^{(g)}_0},\end{aligned}$$

which proves $a'_{g-1,n_{g-1},g-1,n_{g-1}} = a^{\dagger}_{g-1,n_{g-1},g-1,n_{g-1}}$. These are all relevant entries in this row if $n_g = 0$ and $v_+(e_g) \in V^{\mathrm{c}}$. Otherwise, the coefficient of $u^{(g)}_1$ needs to be considered. If $n_g > 0$ then we find

$$a'_{g-1,n_{g-1},g,1} = -\frac{1}{l^{(g-1)}_{n_g-1}+l^{(g)}_0}$$

and

$$\begin{aligned}a^{\dagger}_{g-1,n_{g-1},g,1} &= -\frac{a_{g-1,n_{g-1},g+1,n_{g+1}}a_{g+1,n_{g+1},g,1}}{a_{g+1,n_{g+1},g+1,n_{g+1}}}\\ &\quad -\frac{1}{l^{(g-1)}_{n_{g-1}}l^{(g)}_0}\frac{l^{(g-1)}_{n_{g-1}}l^{(g)}_0}{l^{(g-1)}_{n_g-1}+l^{(g)}_0},\end{aligned}$$

which proves $a'_{g-1,n_{g-1},g,1} = a^{\dagger}_{g-1,n_{g-1},g,1}$. When $n_g = 0$ and $v_+(e_g) \in V^{\mathrm{f}}$, say $v_+(e_g) = v_s$, then

$$a'_{g-1,n_{g-1},g+1,s} = -\frac{1}{l^{(g-1)}_{n_g-1}+l^{(g)}_0}$$

and

$$\begin{aligned}a^{\dagger}_{g-1,n_{g-1},g+1,s} &= -\frac{a_{g-1,n_{g-1},g+1,n_{g+1}}a_{g+1,n_{g+1},g+1,s}}{a_{g+1,n_{g+1},g+1,n_{g+1}}}\\ &= -\frac{1}{l^{(g-1)}_{n_{g-1}}l^{(g)}_0}\frac{l^{(g-1)}_{n_{g-1}}l^{(g)}_0}{l^{(g-1)}_{n_g-1}+l^{(g)}_0},\end{aligned}$$

which proves $a'_{g-1,n_{g-1},g+1,s} = a^{\dagger}_{g-1,n_{g-1},g+1,s}$.

The next row to consider is the row with index $(g,1)$, which is present when $n_g > 0$. This can be formally obtained from the previous case by reorienting and reindexing. Hence also this row is the same for A' and $A^{\dagger}$.

We still have to consider rows with index $(g+1,k)$ for which $n_{g-1} = 0$ and $v_-(e_{g-1}) = v_k$ or $n_g = 0$ and $v_+(e_g) = v_k$. First let $n_g = 0$ and $v_+(e_g) = v_k$ but

$n_{g-1} > 0$ or $n_{g-1} = 0$ and $g-1 \notin I^-(v_k)$. In view of (5.1.6), the difference of the linear functionals for A_{11} and A' is

$$\frac{u_k^{(g+1)}}{l_0^{(g)}} - \frac{u_k^{(g+1)} - u_{n_{g-1}}^{(g-1)}}{l_{n_{g-1}}^{(g-1)} + l_0^{(g)}}.$$

The only relevant component with index $(g+1,k,r,s)$ and $(g+1,k) \le (r,s)$ is when $(r,s) = (g+1,k)$. From

$$\frac{a_{g+1,k,g+1,n_{g+1}} a_{g+1,n_{g+1},g+1,k}}{a_{g+1,n_{g+1},g+1,n_{g+1}}} = \frac{1}{(l_0^{(g)})^2} \frac{l_{n_{g-1}}^{(g-1)} l_0^{(g)}}{l_{n_g-1}^{(g-1)} + l_0^{(g)}}$$

it follows that

$$a^\dagger_{g+1,k,g+1,k} - a'_{g+1,k,g+1,k} = \frac{1}{l_0^{(g)}} - \frac{1}{l_{n_{g-1}}^{(g-1)} + l_0^{(g)}} - \frac{1}{(l_0^{(g)})^2} \frac{l_{n_{g-1}}^{(g-1)} l_0^{(g)}}{l_{n_g-1}^{(g-1)} + l_0^{(g)}} = 0.$$

The case that $n_{g-1} = n_g = 0$, $v_-(e_{g-1}) = v_k$ and $g \notin I^+(v_k)$ is similar; in the two formulas above, the numbers $l_{n_{g-1}}^{(g-1)}$ and $l_0^{(g)}$ just have to be interchanged.

Finally, let $n_{g-1} = n_g = 0$, $v_-(e_{g-1}) = v_+(e_g) = v_k$. In view of (5.1.6), the difference of the linear functionals for A_{11} and A' is

$$\frac{u_k^{(g+1)}}{l_0^{(g)}} + \frac{u_k^{(g+1)}}{l_{n_{g-1}}^{(g-1)}} - 2\frac{u_k^{(g+1)} - u_k^{(g+1)}}{l_{n_{g-1}}^{(g-1)} + l_0^{(g)}}.$$

From

$$\frac{a_{g+1,k,g+1,n_{g+1}} a_{g+1,n_{g+1},g+1,k}}{a_{g+1,n_{g+1},g+1,n_{g+1}}} = \left(\frac{1}{l_{n_{g-1}}^{(g-1)}} + \frac{1}{l_0^{(g)}} \right)^2 \frac{l_{n_{g-1}}^{(g-1)} l_0^{(g)}}{l_{n_g-1}^{(g-1)} + l_0^{(g)}}$$

it follows that

$$a^\dagger_{g+1,k,g+1,k} - a'_{g+1,k,g+1,k} = \frac{1}{l_0^{(g)}} + \frac{1}{l_{n_{g-1}}^{(g-1)}} - \frac{l_{n_g-1}^{(g-1)} + l_0^{(g)}}{l_{n_{g-1}}^{(g-1)} l_0^{(g)}} = 0.$$

Thus we have shown that $A' = A^\dagger$. Since $A_1^\dagger = A_1$ in view of Lemma B.4.1, it follows that $A_1' = A_1$. □

5.2 Characteristic polynomials

Unless otherwise stated we are going to assume in this section that $g > 0$.

The vector of all amplitudes of problem (4.1.10) has the components $u_k^{(j)}$ ($j = 1, \ldots, g$, $k = 0, \ldots, n_j+1$). In Section 5.1 we have eliminated some redundant amplitudes at the vertices to arrive at a matrix representation $A - zM$ of the problem. This procedure has been appropriate to prove that the characteristic values of the problem are real and nonnegative. For further results, we will use the Cauer-Fry polynomials for (4.1.10), which are described in (4.1.16) with

$$u_{2j-1} := u_0^{(j)} \text{ and } u_{2j} := u_1^{(j)} - u_0^{(j)} \quad (j = 1, \dots, g), \quad U = (u_j)_{j=1}^{2g}. \tag{5.2.1}$$

For an interior vertex, there is in general no canonical way to write the boundary conditions, and we therefore observe that similar to (4.1.9) it follows that for the freely moving interior vertices v of G the identities (5.1.1) and (5.1.3) imply that

$$u_0^{(j)} = \sum_{r\in I^+(v)} \left(\frac{u_{n_r+1}^{(r)}}{d(v)} - \frac{u_{n_r+1}^{(r)} - u_{n_r}^{(r)}}{l_{n_r}^{(r)}} \right) + \sum_{r\in I^-(v)} \left(\frac{u_0^{(r)}}{d(v)} + \frac{u_1^{(r)} - u_0^{(r)}}{l_0^{(r)}} \right) \quad (j \in I^-(v)), \tag{5.2.2}$$

$$u_{n_j+1}^{(j)} = \sum_{r\in I^+(v)} \left(\frac{u_{n_r+1}^{(r)}}{d(v)} - \frac{u_{n_r+1}^{(r)} - u_{n_r}^{(r)}}{l_{n_r}^{(r)}} \right) + \sum_{r\in I^-(v)} \left(\frac{u_0^{(r)}}{d(v)} + \frac{u_1^{(r)} - u_0^{(r)}}{l_0^{(r)}} \right) \quad (j \in I^+(v)). \tag{5.2.3}$$

Conversely, since the right-hand sides in (5.2.2) and (5.2.3) are independent of j, (5.1.1) immediately follows from (5.2.2) and (5.2.3), whereas (5.1.3) follows from adding up the $d(v)$ equations in (5.2.2) and (5.2.3).

The left-hand sides in (5.2.2) and (5.2.3) are the same as in (5.1.2), whereas the right-hand sides only depend on v. Hence it is convenient to introduce the following linear functionals:

$$\mathcal{L}_{2j-1}^{\mathrm{c}}(z)U := u_{2j-1} \quad (j = 1, \dots, g), \tag{5.2.4}$$

$$\mathcal{L}_{2j}^{\mathrm{c}}(z)U := R_{2n_j}^{(j)}(z,1)u_{2j-1} + R_{2n_j}^{(j)}(z,0)u_{2j} \quad (j = 1, \dots, g), \tag{5.2.5}$$

$$\begin{aligned} \mathcal{L}_v^{\mathrm{f}}(z)U := &-\frac{1}{d(v)} \sum_{r\in I^+(v)} \left(R_{2n_r}^{(r)}(z,1)u_{2r-1} + R_{2n_r}^{(r)}(z,0)u_{2r} \right) \\ &+ \sum_{r\in I^+j(v)} \left(R_{2n_r-1}^{(r)}(z,1)u_{2r-1} + R_{2n_r-1}^{(r)}(z,0)u_{2r} \right) \\ &- \sum_{r\in I^-(v)} \left(\frac{u_{2r-1}}{d(v)} + \frac{u_{2r}}{l_0^{(r)}} \right) \quad (v \in V^{\mathrm{f}}). \end{aligned} \tag{5.2.6}$$

It is clear that being imposed at a pendant clamped vertex, (5.1.2) just becomes the usual Dirichlet condition (condition (1.2.23) or (1.2.24)), whereas imposed at a freely moving pendant vertex, (5.2.2), (5.2.3) reduces to exactly one condition, which is the usual Neumann condition (condition (1.2.27) or (1.2.29)).

As in Section 4.1, we have $2g$ unknown variables, and the conditions (5.1.2), (5.2.2), (5.2.3) can be expressed in terms of the linear functionals

$$\mathcal{L}_{2j-1}(z)U := \mathcal{L}^{\mathrm{c}}_{2j-1}(z)U \quad (j \in I^-(v),\ v \in V^{\mathrm{c}}), \tag{5.2.7}$$

$$\mathcal{L}_{2j}(z)U := \mathcal{L}^{\mathrm{c}}_{2j}(z)U \quad (j \in I^+(v),\ v \in V^{\mathrm{c}}), \tag{5.2.8}$$

$$\mathcal{L}_{2j-1}(z)U := \mathcal{L}^{\mathrm{c}}_{2j-1}(z)U + \mathcal{L}^{\mathrm{f}}_{v}(z)U \quad (j \in I^-(v),\ v \in V^{\mathrm{f}}), \tag{5.2.9}$$

$$\mathcal{L}_{2j}(z)U := \mathcal{L}^{\mathrm{c}}_{2j}(z)U + \mathcal{L}^{\mathrm{f}}_{v}(z)U \quad (j \in I^+(v),\ v \in V^{\mathrm{f}}). \tag{5.2.10}$$

Defining the $(2g) \times (2g)$ matrix $\Phi(z)$ by

$$\Phi(z)U = \begin{pmatrix} \mathcal{L}_1(z)U \\ \vdots \\ \mathcal{L}_{2g}(z)U \end{pmatrix}, \tag{5.2.11}$$

it follows that the system of equations $(\mathcal{L}_j(z)U)_{j=1}^{2g} = 0$ can be written in the form $\Phi(z)U = 0$. The matrix function Φ is called the *characteristic matrix function* of (4.1.10), (5.1.2), (5.2.2), (5.2.3), and the function ϕ defined by

$$\phi(z) := \det(\Phi(z)) \tag{5.2.12}$$

is called the *characteristic polynomial* of problem (4.1.10), (5.1.2), (5.2.2), (5.2.3). For convenience we let $\phi = 1$ when $g = 0$.

It is clear that the characteristic polynomials are independent of the indexing of the edges since a permutation if the edges would result in simultaneous permutations of columns and rows of Φ, which would leave the corresponding determinants unchanged.

Although it appears by the above reasoning that the spectral data of $A - zM$ and $\Phi(z)$ are the same, we will now formally state and prove it.

Proposition 5.2.1. *There is a constant $C \neq 0$ such that* $\det(A - zM) = C\Phi(z)$ *for all $z \in \mathbb{C}$. The multiplicity of each zero z of ϕ equals the defect of the matrix $\Phi(z)$.*

Proof. Let $Y_0 = (u_k^{(j)})_{\substack{j=1,\dots,g \\ k=0,\dots,n_j+1}}$ and let $A_0 - zM_0$ be the $(n+2g) \times (n+2g)$ matrix for which the components of $(A_0 - zM_0)Y_0 = 0$ are given by (4.1.10), (5.1.1), (5.1.2), (5.1.3). Here the order of the components will be arbitrary. Indeed, we will not compare the spectral data of $A - zM$ and $\Phi(z)$ directly, but instead we will compare them with conveniently chosen matrices $A_0 - zM_0$ in the class of matrices obtained from one particular representation by permutation of rows and columns, and in different parts of this proof we may use different matrices from this class. In order to simplify the notation, we will denote any such matrix with $A_0 - zM_0$. Since the determinants of the matrices in this class are unique up to a multiplicative factor -1 and since solutions Y_0 of $(A_0 - zM_0)Y_0 = 0$ would only differ by a permutation of components, it is clear that we can take any $A_0 - zM_0$ for this comparison.

For each vertex $v \in V$ choose an enumeration $I(v) = \{j_1(v), \dots, j_{d(v)}(v)\}$ and put $s_r(v) = 0$ if $j_r(v) \in I^-(v)$ and $s_r(v) = n_{j_r(v)} + 1$ if $j_r(v) \in I^+(v)$. Note that for each $v \in V^{\text{int}}$, (5.1.1) consists of $d(v) - 1$ conditions. Then we can write (5.1.1) as

$$u_{s_r(v)}^{(j_r(v))} - u_{s_1(v)}^{(j_1(v))} = 0 \quad (v \in V^{\text{int}},\ r = 2, \dots, d(v)), \tag{5.2.13}$$

whereas (5.1.2) becomes

$$u_{s_r(v)}^{(j_r(v))} = 0 \quad (v \in V^{\text{c}},\ r = 1, \dots, d(v)). \tag{5.2.14}$$

First we will compare the spectral data of $A - zM$ and $A_0 - zM_0$. For each $v \in V^{\text{c}}$ consecutive components of $(A_0 - zM_0)Y_0 = 0$ will be defined by (5.2.14), and for each $v \in V^{\text{int}} \cap V^{\text{f}}$ consecutive components of $(A_0 - zM_0)Y_0 = 0$ will be defined by (5.2.13). The remaining components of $(A_0 - zM_0)Y_0$ are defined by (4.1.10) and (5.1.3) in the same order as in $(A - zM)Y = 0$. We now arrange the columns of $A_0 - zM_0$ such that the column index of $u_{s_r(v)}^{(j_r(v))}$ equal the row index of $u_{s_r(v)}^{(j_r(v))} = 0$ for $v \in V^{\text{c}}$ and $r = 1, \dots, d(v)$, and that the column index of $u_{s_r(v)}^{(j_r(v))}$ for $v \in V^{\text{int}} \cap V^{\text{f}}$ and $r = 2, \dots, d(v)$ coincides with the row index given the equation from (5.2.13) which contains $u_{s_r(v)}^{(j_r(v))}$. For $k = 1, \dots, n_{g+1}$ we may identify $u_{s_1(v_k)}^{(j_1(v_k))}$ with $u_k^{(g+1)}$. That is, we can write $Y_0^\mathsf{T} = (Y_{00}^\mathsf{T}, Y^\mathsf{T})$, which leads to a block representation

$$(A_0 - zM_0)Y_0 = \begin{pmatrix} I & B_1 \\ B_2 & \widetilde{A}_0 - zM \end{pmatrix} \begin{pmatrix} Y_{00} \\ Y \end{pmatrix},$$

where the matrix I is an identity matrix. The matrix B_1 has entries -1 when the column corresponds to some v_k $(k = 1, \dots, g)$ and the row index corresponds to an equation (5.2.13) for v_k, and all other entries of B_1 are 0. By the Schur factorization,

$$A_0 - zM_0 = \begin{pmatrix} I & 0 \\ B_2 & I \end{pmatrix} \begin{pmatrix} I & 0 \\ 0 & \widetilde{A}_0 - B_2B_1 - zM \end{pmatrix} \begin{pmatrix} I & B_1 \\ 0 & I \end{pmatrix}.$$

By definition of the matrices $\widetilde{A}_0$ and A their columns corresponding to amplitudes at the beads are equal, whereas those columns of B_1 and therefore also of B_2B_1 are zero. It remains to compare the columns corresponding to $u_k^{(g+1)}$. But in A these columns are obtained by identifying all $u_l^{(j)}$ which are amplitudes at v_k with $u_k^{(g+1)}$, whereas the corresponding columns of $\widetilde{A}_0$ are the columns corresponding to $u_{s_1(v_k)}^{(j_1(v_k))}$, and the corresponding columns of $-B_2B_1$ are the sums of the columns corresponding to $u_{s_2(v)}^{(j_2(v))}, \dots, u_{s_{d(v)}(v)}^{(j_{d(v)}(v))}$. Therefore $\widetilde{A}_0 - B_2B_1 = A$ has been shown.

The above Schur factorization implies $\det(A_0 - zM_0) = \det(A - zM)$, and $(A_0 - zM_0)Y_0 = 0$ is equivalent to $(A - zM)Y = 0$ with $Y_0^\mathsf{T} = (-(B_1Y)^\mathsf{T}, Y^\mathsf{T})$.

We are going to use (4.1.16) to relate $A_0 - zM_0$ to $\Phi(z)$. To this end, let $j \in \{1, \dots, g\}$ with $n_j > 0$, let $u^{(j)} = (u_0^{(j)}, \dots, u_{n_j+1}^{(j)})^\mathsf{T}$, define $\widehat{\mathcal{L}}_k^{(j)}(z)u^{(j)}$ by the left-hand side of (4.1.10), with $\lambda^2 = z$, that is,

$$l_k^{(j)}\widehat{\mathcal{L}}_k^{(j)}(z)u^{(j)} = -u_{k+1}^{(j)} + \left(1 + \frac{l_k^{(j)}}{l_{k-1}^{(j)}} - l_k^{(j)}m_k^{(j)}z\right)u_k^{(j)} - \frac{l_k^{(j)}}{l_{k-1}^{(j)}}u_{k-1}^{(j)}$$

and let

$$\mathcal{L}_k^{(j)}(z)u^{(j)} := -u_{k+1}^{(j)} + R_{2k}^{(j)}(z,0)(u_1^{(j)} - u_0^{(j)}) + R_{2k}^{(j)}(z,1)u_0^{(j)} \quad (k = 1, \dots, n_j)$$

be the linear functionals corresponding to (4.1.16).

We are going to prove that for $j = 1, \dots, g$ with $n_j > 0$ and for $k = 1, \dots, n_j$ and $r = 1, \dots, k$ there are polynomials $\alpha_{k,r}^{(j)}$ with $\alpha_{k,k}^{(j)}(z) = \frac{1}{l_k^{(j)}}$ such that the identities

$$\widehat{\mathcal{L}}_k^{(j)}(z)u^{(j)} = \sum_{r=1}^{k} \alpha_{k,r}^{(j)}(z)\mathcal{L}_r^{(j)}(z)u^{(j)} \quad (k = 1, \dots, n_j) \tag{5.2.15}$$

hold.

To prove (5.2.15) we observe that the recursion (1.2.9) for R_{2k} gives

$$R_{2k}^{(j)}(z,c) = \left(1 + \frac{l_k^{(j)}}{l_{k-1}^{(j)}} - l_k^{(j)}m_k^{(j)}z\right)R_{2k-2}^{(j)}(z,c) - \frac{l_k^{(j)}}{l_{k-1}^{(j)}}R_{2k-4}^{(j)}(z,c).$$

In view of (1.2.12) and (1.2.13) it follows that

$$R_2^{(j)}(z,c) = 1 + \frac{l_1^{(j)}}{l_0^{(j)}} - l_1^{(j)}m_1^{(j)}z - \frac{l_1^{(j)}}{l_0^{(j)}}c.$$

Hence

$$\begin{aligned} l_1^{(j)}\widehat{\mathcal{L}}_1^{(j)}(z)u^{(j)} &= -u_2^{(j)} + \left(1 + \frac{l_1^{(j)}}{l_0^{(j)}} - l_1^{(j)}m_1^{(j)}z\right)(u_1^{(j)} - u_0^{(j)}) \\ &\quad + \left(1 - l_1^{(j)}m_1^{(j)}z\right)u_0^{(j)} \\ &= -u_2^{(j)} + R_2^{(j)}(z,0)(u_1^{(j)} - u_0^{(j)}) + R_2^{(j)}(z,1)u_0^{(j)} \\ &= \mathcal{L}_1^{(j)}(z)u^{(j)}, \end{aligned}$$

which shows that (5.2.15) holds for $k = 1$. For $k \geq 2$ we calculate

$$\begin{aligned}
l_k^{(j)} \widehat{\mathcal{L}}_k^{(j)}(z)u^{(j)} &= -u_{k+1}^{(j)} + \left(1 + \frac{l_k^{(j)}}{l_{k-1}^{(j)}} - l_k^{(j)} m_k^{(j)} z\right) \\
&\quad \times \left(-\mathcal{L}_{k-1}^{(j)}(z)u^{(j)} + R_{2k-2}^{(j)}(z,0)(u_1^{(j)} - u_0^{(j)}) + R_{2k-2}^{(j)}(z,1)u_0^{(j)}\right) \\
&\quad - \frac{l_k^{(j)}}{l_{k-1}^{(j)}} u_{k-1}^{(j)} \\
&= -u_{k+1}^{(j)} - \left(1 + \frac{l_k^{(j)}}{l_{k-1}^{(j)}} - l_k^{(j)} m_k^{(j)} z\right) \mathcal{L}_{k-1}^{(j)}(z)u^{(j)} \\
&\quad + \left(R_{2k}^{(j)}(z,0) + \frac{l_k^{(j)}}{l_{k-1}^{(j)}} R_{2k-4}^{(j)}(z,0)\right)(u_1^{(j)} - u_0^{(j)}) \\
&\quad + \left(R_{2k}^{(j)}(z,1) + \frac{l_k^{(j)}}{l_{k-1}^{(j)}} R_{2k-4}^{(j)}(z,1)\right) u_0^{(j)} \\
&\quad - \frac{l_k^{(j)}}{l_{k-1}^{(j)}} u_{k-1}^{(j)} \\
&= \mathcal{L}_k^{(j)}(z)u^{(j)} - \left(1 + \frac{l_k^{(j)}}{l_{k-1}^{(j)}} - l_k^{(j)} m_k^{(j)} z\right) \mathcal{L}_{k-1}^{(j)}(z)u^{(j)} \\
&\quad + \frac{l_k^{(j)}}{l_{k-1}^{(j)}} \left(-u_{k-1}^{(j)} + R_{2k-4}^{(j)}(z,0)(u_1^{(j)} - u_0^{(j)}) + R_{2k-4}^{(j)}(z,1)u_0^{(j)}\right).
\end{aligned}$$

Hence

$$l_2^{(j)} \widehat{\mathcal{L}}_2^{(j)}(z)u^{(j)} = \mathcal{L}_2^{(j)}(z)u^{(j)} - \left(1 + \frac{l_2^{(j)}}{l_1^{(j)}} - l_2^{(j)} m_2^{(j)} z\right) \mathcal{L}_1^{(j)}(z)u^{(j)}$$

and

$$\begin{aligned}
l_k^{(j)} \widehat{\mathcal{L}}_k^{(j)}(z)u^{(j)} &= \mathcal{L}_k^{(j)}(z)u^{(j)} - \left(1 + \frac{l_k^{(j)}}{l_{k-1}^{(j)}} - l_k^{(j)} m_k^{(j)} z\right) \mathcal{L}_{k-1}^{(j)}(z)u^{(j)} \\
&\quad + \frac{l_k^{(j)}}{l_{k-1}^{(j)}} \mathcal{L}_{k-2}^{(j)}(z)u^{(j)}
\end{aligned}$$

when $k > 2$. Thus we have shown (5.2.15).

Now the rows of $A_0 - zM_0$ will be arranged in such a way that we first take the conditions (4.1.10), that is, the rows corresponding to the linear functionals $\widehat{\mathcal{L}}_k^{(j)}(z)$ ($j = 1, \ldots, g$, $k = 1, \ldots, n_j$) with lexicographic order in (j, k)

and then the rows corresponding to (5.1.1), (5.1.2) and (5.1.3). The columns of $A_0 - zM_0$ are arranged according to the permutation $(Y_{01}^\mathsf{T}, Y_{02}^\mathsf{T})^\mathsf{T}$ of Y_0, where $Y_{01}^\mathsf{T} = (u_2^{(1)}, \dots, u_{n_1+1}^{(1)}, \dots, u_2^{(g)}, \dots, u_{n_g+1}^{(g)})^\mathsf{T}$ and $Y_{02}^\mathsf{T} = (u_0^{(1)}, u_1^{(1)}, \dots, u_0^{(g)}, u_1^{(g)})^\mathsf{T}$. In accordance with the indexing of the rows, the components $u_k^{(j)}$ ($j = 1, \dots, g$, $k = 2, \dots, n_{j+1}$) of Y_{01} will be given the index $(j, k-1)$. Then this representation of $A_0 - zM_0$ can be written as

$$A_0 - zM_0 = \begin{pmatrix} \Psi_{11}(z) & 0 \\ 0 & \Psi_{12} \end{pmatrix} \Phi_0(z) \begin{pmatrix} I & 0 \\ 0 & \Psi_2 \end{pmatrix}$$

in the following way. The matrix Ψ_2 is a block diagonal matrix with g diagonal blocks $\begin{pmatrix} 1 & 0 \\ -1 & 1 \end{pmatrix}$, and therefore $\Psi_2 Y_{02} = U$ with U as defined in (5.2.1). The matrix $\Psi_{11}(z)$ consists of the coefficients of (5.2.15). In particular, $\Psi_{11}(z)$ is a triangular matrix with diagonal elements $\frac{1}{l_k^{(j)}} > 0$. At the beginning of this section we have seen that the conditions (5.1.1) and (5.1.3) can be written as (5.2.2) and (5.2.3) via an invertible linear transformation. Hence we arrive at an invertible matrix Ψ_{12}, where we may assume that the rows corresponding to (5.1.2), (5.2.2) and (5.2.3) in $\Phi_0(z)$ are ordered as in $\Phi(z)$. This shows that there is a constant C_0 such that $\det(A_0 - zM_0) = C_0 \det \Phi_0(z)$ for all $z \in \mathbb{C}$ and that the defect of $A_0 - zM_0$ equals the defect of $\Phi_0(z)$ for all $z \in \mathbb{C}$.

The first n rows of $\Phi_0(z)$ are given by the linear functionals $\mathcal{L}_k^{(j)}$, and therefore

$$\Phi_0(z) = \begin{pmatrix} -I & \Phi_{01}(z) \\ \Phi_{02}(z) & \Phi_{03}(z) \end{pmatrix},$$

where the entries of $\Phi_{01}(z)$ with row index (j, k) and column index $2j - r$ ($r = 0, 1$) are $R_{2k}^{(j)}(z, r)$, and all other entries of $\Phi_{01}(z)$ are 0. With the above indexing of the first n rows of $\Phi_0(z)$ and the components of Y_{01} we let $e_{j,k}$ be the corresponding unit vectors in $\mathbb{C}^n$. Hence we can write the above description of $\Phi_{01}(z)$ as

$$e_{j,k}^\mathsf{T} \Phi_{01}(z) U = R_{2k}^{(j)}(z,1) u_{2j-1} + R_{2k}^{(j)}(z,0) u_{2j} \quad (j = 1, \dots, g,\ k = 1, \dots, n_j).$$

We will have to express the conditions (5.1.2), (5.2.2) and (5.2.3) in terms of the components of Y_{01} and U. Let $j \in \{1, \dots, g\}$. Clearly, $u_0^{(j)} = u_{2j-1}$ and $u_1^{(j)} - u_0^{(j)} = u_{2j}$ by definition of U. When $n_j = 0$, then

$$u_{n_j+1}^{(j)} = u_1^{(j)} = u_{2j} + u_{2j-1} = R_{2n_j}^{(j)}(z,1) u_{2j-1} + R_{2n_j}^{(j)}(z,0) u_{2j}$$

and

$$u_{n_j+1}^{(j)} - u_{n_j}^{(j)} = u_1^{(j)} - u_0^{(j)} = u_{2j}.$$

When $n_j = 1$, then

$$u_{n_j+1}^{(j)} - u_{n_j}^{(j)} = u_2^{(j)} - u_1^{(j)} = u_{n_j+1}^{(j)} - R_{2n_j}^{(j)}(z,1) u_{2j-1} - R_{2n_j}^{(j)}(z,0) u_{2j}.$$

In all other cases, $u^{(j)}_{n_j+1}$ and $u^{(j)}_{n_j+1} - u^{(j)}_{n_j}$ are written in terms of components of Y_{01}.

With $Y_1 := (Y_{01}^\mathsf{T}, U^\mathsf{T})^\mathsf{T}$ define

$$\widetilde{\mathcal{L}}^{\mathrm{c}}_{2j-1}(z)Y_1 := u^{(j)}_0 \quad (j = 1, \dots, g),$$
$$\widetilde{\mathcal{L}}^{\mathrm{c}}_{2j}(z)Y_1 := u^{(j)}_{n_j+1} \quad (j = 1, \dots, g),$$
$$\widetilde{\mathcal{L}}^{\mathrm{f}}_v(z)Y_1 := -\sum_{r\in I^+(v)} \left(\frac{u^{(r)}_{n_r+1}}{d(v)} - \frac{u^{(r)}_{n_r+1} - u^{(r)}_{n_r}}{l^{(r)}_{n_r}} \right) - \sum_{r\in I^-(v)} \left(\frac{u^{(r)}_0}{d(v)} + \frac{u^{(r)}_1 - u^{(r)}_0}{l^{(r)}_0} \right)$$
$$(v \in V^{\mathrm{f}}),$$

where the $u^{(j)}_k$ have to be expressed in terms of the components of Y_1 as discussed above. Putting

$$\widetilde{\mathcal{L}}_{2j-1}(z)Y_1 := \widetilde{\mathcal{L}}^{\mathrm{c}}_{2j-1}(z)Y_1 \quad (j \in I^-(v),\ v \in V^{\mathrm{c}}),$$
$$\widetilde{\mathcal{L}}_{2j}(z)Y_1 := \widetilde{\mathcal{L}}^{\mathrm{c}}_{2j}(z)Y_1 \quad (j \in I^+(v),\ v \in V^{\mathrm{c}}),$$
$$\widetilde{\mathcal{L}}_{2j-1}(z)Y_1 := \widetilde{\mathcal{L}}^{\mathrm{c}}_{2j-1}(z)Y_1 + \widetilde{\mathcal{L}}^{\mathrm{f}}_v(z)Y_1 \quad (j \in I^-(v),\ v \in V^{\mathrm{f}}),$$
$$\widetilde{\mathcal{L}}_{2j}(z)Y_1 := \widetilde{\mathcal{L}}^{\mathrm{c}}_{2j}(z)Y_1 + \widetilde{\mathcal{L}}^{\mathrm{f}}_v(z)Y_1 \quad (j \in I^+(v),\ v \in V^{\mathrm{f}})$$

it follows that

$$e_l^\mathsf{T} \left(\Phi_{02}(z) \quad \Phi_{03}(z)\right) Y_1 = \widetilde{\mathcal{L}}_l(z)Y_1 \quad (l = 1, \dots, 2g),$$

where e_l is the l-th unit vector in $\mathbb{C}^{2g}$.

Now we are going to compare the matrix $\Phi(z)$ with the Schur complement $\Phi_{04}(z) := \Phi_{02}(z)\Phi_{01}(z) + \Phi_{03}(z)$ of $\Phi_0(z)$. From above we conclude that

$$e_l^\mathsf{T}\Phi_{04}(z)U = \widetilde{\mathcal{L}}_l(z) \begin{pmatrix} \Phi_{01}(z)U \\ U \end{pmatrix} \quad (l = 1, \dots, 2g).$$

Let $j \in \{1, \dots, g\}$. Then

$$\widetilde{\mathcal{L}}^{\mathrm{c}}_{2j-1}(z) \begin{pmatrix} \Phi_{01}(z)U \\ U \end{pmatrix} = u_{2j-1} = \mathcal{L}^{\mathrm{c}}_{2j-1}(z)U.$$

When $n_j = 0$, then

$$\widetilde{\mathcal{L}}^{\mathrm{c}}_{2j}(z) \begin{pmatrix} \Phi_{01}(z)U \\ U \end{pmatrix} = R^{(j)}_{2n_j}(z,1)u_{2j-1} + R^{(j)}_{2n_j}(z,0)u_{2j} = \mathcal{L}^{\mathrm{c}}_{2j}(z)U,$$

and when $n_j > 0$, then

$$\widetilde{\mathcal{L}}^{\mathrm{c}}_{2j}(z) \begin{pmatrix} \Phi_{01}(z)U \\ U \end{pmatrix} = e^\mathsf{T}_{j,n_j}\Phi_{01}(z)U = \mathcal{L}^{\mathrm{c}}_{2j}(z)U.$$

For $v \in V^{\mathrm{f}}$ we have

$$\begin{aligned}
\widetilde{\mathcal{L}}_v^{\mathrm{f}}(z)\begin{pmatrix}\Phi_{01}(z)U\\ U\end{pmatrix} &= -\sum_{\substack{r\in I^+(v)\\ n_r=0}}\frac{u_{2r}+u_{2r-1}}{d(v)} - \sum_{\substack{r\in I^+(v)\\ n_r>0}}\frac{e_{r,n_r}^{\mathsf{T}}\Phi_{01}(z)U}{d(v)} + \sum_{\substack{r\in I^+(v)\\ n_r=0}}\frac{u_{2r}}{l_{n_r}^{(r)}}\\
&\quad + \sum_{\substack{r\in I^+(v)\\ n_r=1}}\frac{e_{r,n_r}^{\mathsf{T}}\Phi_{01}(z)U - u_{2r} - u_{2r-1}}{l_{n_r}^{(r)}}\\
&\quad + \sum_{\substack{r\in I^+(v)\\ n_r>1}}\frac{e_{r,n_r}^{\mathsf{T}}\Phi_{01}(z)U - e_{r,n_r-1}^{\mathsf{T}}\Phi_{01}(z)U}{l_{n_r}^{(r)}}\\
&\quad - \sum_{r\in I^-(v)}\left(\frac{u_{2r-1}}{d(v)}+\frac{u_{2r}}{l_0^{(r)}}\right)\\
&= -\sum_{r\in I^+(v)}\frac{R_{2n_r}^{(r)}(z,1)u_{2r-1}+R_{2n_r}^{(r)}(z,0)u_{2r}}{d(v)}\\
&\quad + \sum_{r\in I^+(v)}\left(R_{2n_r-1}^{(r)}(z,1)u_{2r-1}+R_{2n_r-1}^{(r)}(z,0)u_{2r}\right)\\
&\quad - \sum_{r\in I^-(v)}\left(\frac{u_{2r-1}}{d(v)}+\frac{u_{2r}}{l_0^{(r)}}\right)\\
&= \mathcal{L}_v^{\mathrm{f}}(z)U.
\end{aligned}$$

Hence

$$e_l^{\mathsf{T}}\Phi_{04}(z)U = \widetilde{\mathcal{L}}_l(z)\begin{pmatrix}\Phi_{01}(z)U\\ U\end{pmatrix} = \mathcal{L}_{2j-1}^{\mathrm{c}}(z)U = e_l^{\mathsf{T}}\Phi(z) \quad (l=1,\ldots,2g),$$

which shows that $\Phi_{04}(z) = \Phi(z)$ for all $z \in \mathbb{C}$. Since $\det\Phi_0(z) = (-1)^n\det\Phi_{04}(z)$ and since the defects of $\Phi_0(z)$ and $\Phi_{04}(z)$ are equal, the proof is complete. □

In Chapter 1 we have found characteristic polynomials for string problems in terms of Cauer-Fry polynomials. Vibrations on strings are special cases of the vibrations considered in this chapter. However, since characteristic polynomials are only unique up to multiplication by a constant factor and since in this chapter some constructions require to take sums of (products of) characteristic polynomials, these multiplicative constants are not redundant, in general. Considering the metric edge e_j with the distribution of beads as defined in Section (5.1), let $\varphi_{AB}^{(j)}$ with $A, B \in \{D, N\}$ be the characteristic polynomial of the string problem on e_j as defined by (5.2.12), where A is the condition at the entrance, B is the condition at the exit, D stands for the Dirichlet condition, and N stands for the Neumann condition.

The following proposition gives the explicit representation of $\varphi^{(j)}_{AB}$ in terms of Cauer-Fry polynomials.

Proposition 5.2.2. *The characteristic polynomials for the string problems on* e_j $(j = 1, \ldots, g)$ *have the representations*

$$\varphi^{(j)}_{DD}(z) = R^{(j)}_{2n_j}(z,0), \quad \varphi^{(j)}_{DN}(z) = R^{(j)}_{2n_j-1}(z,0),$$
$$\varphi^{(j)}_{ND}(z) = \frac{1}{l^{(j)}_0} R^{(j)}_{2n_j}(z,1), \quad \varphi^{(j)}_{NN}(z) = \frac{1}{l^{(j)}_0} R^{(j)}_{2n_j-1}(z,1).$$

Proof. Without loss of generality we let $j = 1$. Then the characteristic matrix function is a 2×2 matrix function whose entries are determined by $\mathcal{L}_1$ and $\mathcal{L}_2$ as given by (5.2.7)–(5.2.10). Denoting these functionals by $\mathcal{L}^A_k$ for $k = 1, 2$ and $A \in \{D, N\}$, it follows that $\varphi^{(1)}_{AB}$ is the determinant of the matrix whose rows are given by $\mathcal{L}^A_1$ and $\mathcal{L}^B_2$. We calculate

$$\begin{aligned}
\mathcal{L}^D_1(z)(u_1,u_2)^\mathsf{T} &= u_1,\\
\mathcal{L}^N_1(z)(u_1,u_2)^\mathsf{T} &= -\frac{1}{l^{(1)}_0} u_2,\\
\mathcal{L}^D_2(z)(u_1,u_2)^\mathsf{T} &= R^{(1)}_{2n_1}(z,1)u_1 + R^{(1)}_{2n_1}(z,0)u_2,\\
\mathcal{L}^N_2(z)(u_1,u_2)^\mathsf{T} &= R^{(1)}_{2n_1-1}(z,1)u_1 + R^{(1)}_{2n_1-1}(z,0)u_2.
\end{aligned}$$

Hence

$$\begin{aligned}
\varphi^{(1)}_{DD}(z) &= \begin{vmatrix} 1 & 0 \\ R^{(1)}_{2n_1}(z,1) & R^{(1)}_{2n_1}(z,1) \end{vmatrix} = R^{(1)}_{2n_1}(z,0),\\
\varphi^{(1)}_{DN}(z) &= \begin{vmatrix} 1 & 0 \\ R^{(1)}_{2n_1-1}(z,1) & R^{(1)}_{2n_1-1}(z,1) \end{vmatrix} = R^{(1)}_{2n_1-1}(z,0),\\
\varphi^{(1)}_{ND}(z) &= \begin{vmatrix} 0 & -\frac{1}{l^{(1)}_0} \\ R^{(1)}_{2n_1}(z,1) & R^{(1)}_{2n_1}(z,1) \end{vmatrix} = \frac{1}{l^{(1)}_0} R^{(1)}_{2n_1}(z,1),\\
\varphi^{(1)}_{NN}(z) &= \begin{vmatrix} 0 & -\frac{1}{l^{(1)}_0} \\ R^{(1)}_{2n_1-1}(z,1) & R^{(1)}_{2n_1-1}(z,1) \end{vmatrix} = \frac{1}{l^{(1)}_0} R^{(1)}_{2n_1-1}(z,1).
\end{aligned}$$
□

5.3 Characteristic polynomials for related graphs

In this section we will represent characteristic polynomials of graphs via characteristic polynomials of subgraphs.

Lemma 5.3.1. *Let* $G = G_1 \dot{\cup} G_2$, *and let* ϕ, ϕ_1 *and* ϕ_2 *be the corresponding characteristic polynomials. Then* $\phi = \phi_1\phi_2$.

Proof. When G_1 or G_2 has no edges, then $\phi_1 = 1$ or $\phi_2 = 1$, and $\phi = \phi_1\phi_2$ is trivially true. Now let both G_1 and G_2 have at least one edge. With a suitable indexing of the edges, the corresponding characteristic matrices can be written in the block diagonal form

$$\Phi(z) = \begin{pmatrix} \Phi_1(z) & 0 \\ 0 & \Phi_2(z) \end{pmatrix}.$$

Taking determinants completes the proof. □

With some abuse of notation, an arbitrarily chosen vertex $v_0 \in V$ will be called the root of G. At the root we will consider both the (generalized) Dirichlet and the (generalized) Neumann problem. Hence it is convenient for (generalized) Neumann conditions at the root to use the notation $V_N^{\mathrm{c}} := V^{\mathrm{c}} \setminus \{v_0\}$ as well as $V_N^{\mathrm{f}} := V^{\mathrm{f}} \cup \{v_0\}$ for the clamped and freely moving vertices, respectively. For (generalized) Dirichlet conditions at the root the notation $V_D^{\mathrm{c}} := V^{\mathrm{c}} \cup \{v_0\}$ and $V_D^{\mathrm{f}} := V^{\mathrm{f}} \setminus \{v_0\}$ for the clamped and freely moving edges, respectively, will be used. For $A \in \{D, N\}$, the corresponding characteristic matrix $\Phi_A(z)$ and the characteristic polynomial ϕ_A are obtained by replacing V^{c} and V^{f} in (5.2.7)–(5.2.10) with V_A^{c} and V_A^{f}. For $A = D$, the function ϕ_D is called the characteristic polynomial for the Dirichlet problem at the root, and for $A = N$, the function ϕ_N is called the characteristic polynomial for the Neumann problem at the root.

If v_0 is an interior vertex, we can cut the graph G in the following way. The root v_0 is replaced by two vertices $v_0^{(1)}$ and $v_0^{(2)}$, and each edge incident with v_0 is attached to exactly one of $v_0^{(1)}$ or $v_0^{(2)}$, preserving the orientation of the edges, i.e., $I^-(v_0) = I^-(v_0^{(1)}) \,\dot\cup\, I^-(v_0^{(2)})$ and $I^+(v_0) = I^+(v_0^{(1)}) \,\dot\cup\, I^+(v_0^{(2)})$. It is assumed that at least one edge is attached to each of the new vertices. Denote this new graph by G_0, where the orientation of the edges is the same as for the graph G. For $A, B \in \{D, N\}$ we denote by $\Phi_{AB}(z)$ and ϕ_{AB} the characteristic matrix and the characteristic polynomial, respectively, where A corresponds to the vertex $v_0^{(1)}$ and B corresponds to the vertex $v_0^{(2)}$. For example, ϕ_{ND} is the characteristic polynomial of the graph G_0 with (generalized) Neumann condition at the vertex $v_0^{(1)}$ and with (generalized) Dirichlet condition at the vertex $v_0^{(2)}$.

Lemma 5.3.2. *Let v_0 be an interior vertex and let G_0 be as defined above. Then*

$$\phi_D = \phi_{DD}. \tag{5.3.1}$$

Proof. The edges and linear functionals $\mathcal{L}_k$ $(k = 1, \dots, 2g)$ are identical for the problems on G and G_0. Therefore $\Phi_D = \Phi_{DD}$, and taking the determinants completes the proof. □

For $v \in G$ (or $v \in G_0$) let $\widetilde{I}(v) = \{2j - 1 : j \in I^-(v)\} \cup \{2j : j \in I^+(v)\}$. Let $s = 1, 2$ and choose bijective maps $\tau_s : \{1, \dots, d(v_0^{(s)})\} \to \widetilde{I}(v_0^{(s)})$. We define the matrix function $\Phi_{v_0}^{(s)}$ by replacing the rows of the characteristic matrix

$\Phi_N(z)$ with index $\tau_s(j)$ for $j = 1, \dots, d(v_0^{(s)})$ with the corresponding rows for the (generalized) Neumann conditions at $v_0^{(s)}$ for G_0, whereas the rows with index $\tau_{3-s}(j)$ for $j = 1, \dots, d(v_0^{(3-s)})$ are replaced with the rows given by the linear functionals $\mathcal{L}^{\mathrm{c}}_{\tau_1(d(v_0^{(1)}))}$ for $j = 1$ and $s = 1$, by $\mathcal{L}^{\mathrm{c}}_{\tau_2(1)}$ for $j = 1$ and $s = 2$, and by $\mathcal{L}^{\mathrm{c}}_{\tau_{3-s}(j)} - \mathcal{L}^{\mathrm{c}}_{\tau_{3-s}(j-1)}$ for $j = 2, \dots, d(v_0^{(3-s)})$. We put $\phi^{(s)}_{v_0} = \det \Phi^{(s)}_{v_0}$. See the proof of the next lemma for more details.

Remark 5.3.3. The construction of the matrix functions $\Phi^{(s)}_{v_0}$ $(s = 1, 2)$ is quite technical, but becomes easier when $d(v_0) = 2$, i.e., $d(v_0^{(1)}) = d(v_0^{(2)}) = 1$. In this case, $\widetilde{I}(v_0^{(s)})$ consists of a single index, say k_s. Then $\Phi^{(s)}_{v_0}$ is obtained from Φ_N by replacing the two rows with indices k_1 and k_2, which describe the generalized Neumann conditions at v_0 for G, as follows.

(i) The row with index k_s becomes a row which describes the Neumann condition at $v_0^{(s)}$ in G_0.

(ii) The row with index k_{3-s} becomes a row which describes the Dirichlet condition at $v_0^{(s)}$ in G_0.

Lemma 5.3.4. *Let v_0 be an interior vertex of G and let G_0 be as defined above. Then*

$$\phi_N = \phi_{ND} + \phi_{DN} - \phi^{(1)}_{v_0} - \phi^{(2)}_{v_0}. \tag{5.3.2}$$

Proof. Define $\tau : \{1, \dots, d(v_0)\} \to \widetilde{I}(v_0)$ by $\tau(j) = \tau_1(j)$ for $j = 1, \dots, d(v_0^{(1)})$ and by $\tau(d(v_0^{(1)}) + j) = \tau_2(j)$ for $j = 1, \dots, d(v_0^{(2)})$. Then let

$$\begin{aligned}
\widehat{\mathcal{L}}_{\tau(1)} &= \frac{1}{d(v_0)} \sum_{r=1}^{d(v_0)} \mathcal{L}_{\tau(r)}, \\
\widehat{\mathcal{L}}_{\tau(j)} &= \mathcal{L}_{\tau(j)} - \mathcal{L}_{\tau(j-1)} \quad (j = 2, \dots, d(v_0)), \\
\widehat{\mathcal{L}}_j &= \mathcal{L}_j \quad (j \in \{1, \dots, 2g\} \setminus \widetilde{I}(v_0)).
\end{aligned}$$

Let $\widehat{\Phi}(z)$ be the matrix whose rows are represented by $\widehat{\mathcal{L}}_j(z)$ $(j = 1, \dots, 2g)$. Clearly, $\widehat{\Phi}(z) = \widehat{T}\Phi(z)$ with a $2g \times 2g$ matrix $\widehat{T}$. Recursively expanding the determinant of $\widehat{T}$ with respect to the rows with indices $j \notin \widetilde{I}(v_0)$ and simultaneous permutation of the remaining rows and columns show that $\det \widehat{T} = \det T$, where T is the $d(v_0) \times d(v_0)$ matrix

$$T = \begin{pmatrix} \frac{1}{d(v_0)} & \frac{1}{d(v_0)} & \cdots & \cdots & \frac{1}{d(v_0)} \\ -1 & 1 & 0 & \cdots & 0 \\ 0 & -1 & 1 & \cdots & 0 \\ \vdots & \vdots & \vdots & & \vdots \\ 0 & \cdots & 0 & -1 & 1 \end{pmatrix}.$$

Expanding the determinant of T with respect to its first row it follows that

$$\det T = \frac{1}{d(v_0)} \sum_{j=1}^{d(v_0)} (-1)^{j-1} \det T_j\,,$$

where T_j is a 2×2 block diagonal matrix with the first block of size $(j-1)\times(j-1)$ being upper triangular with diagonal elements -1 and the second block of size $(d(v_0)-j)\times(d(v_0)-j)$ being lower triangular with diagonal elements 1. Hence $\det T = 1$, so that $\det T = 1$ and therefore $\det(\widehat{\Phi}(z)) = \det(\Phi_N(z))$. It is clear from (5.2.9), (5.2.10) and (5.2.6) that

$$\widehat{\mathcal{L}}_{\tau(j)} = \mathcal{L}^{\mathrm{c}}_{\tau(j)} - \mathcal{L}^{\mathrm{c}}_{\tau(j-1)} \quad (j = 2, \dots, d(v_0))$$

and

$$\widehat{\mathcal{L}}_{\tau(1)} = \widehat{\mathcal{L}}^{(1)}_{\tau(1)} + \widehat{\mathcal{L}}^{(2)}_{\tau(1)},$$

where

$$\begin{aligned}\widehat{\mathcal{L}}^{(s)}_{\tau(1)}(z)U = &\sum_{r\in I^+(v_0^{(s)})} \left(R^{(r)}_{2n_r-1}(z,1)u_{2r-1} + R^{(r)}_{2n_r-1}(z,0)u_{2r}\right)\\ &- \sum_{r\in I^-(v_0^{(s)})} \frac{u_{2r}}{l_0^{(r)}} \quad (s = 1,2).\end{aligned}$$

By expansion rules for determinants, it follows that

$$\phi_N = \det\widehat{\Phi}_{11} + \det\widehat{\Phi}_{12} + \det\widehat{\Phi}_{21} + \det\widehat{\Phi}_{22}, \tag{5.3.3}$$

where the matrices $\widehat{\Phi}_{ik}$ $(i,k = 1,2)$ are obtained from $\widehat{\Phi}$ by substituting $\widehat{\mathcal{L}}_{\tau(1)}$ with $\widehat{\mathcal{L}}^{(i)}_{\tau(1)}$ and $\widehat{\mathcal{L}}_{\tau(d(v_0^{(1)})+1)}$ with $(-1)^k\mathcal{L}^{\mathrm{c}}_{\tau(d(v_0^{(1)})+k-1)}$.

In $\widehat{\Phi}_{1k}$ $(k = 1,2)$ we may perform the inverse of the above row operations, now corresponding to τ_1, and therefore the rows with indices $\tau(1), \dots, \tau(d(v_0^{(1)}))$ represent (generalized) Neumann conditions at $v_0^{(1)}$. In particular, replacing the row with index $\tau_k(1)$ with its negative for $k = 1$, we arrive at $\Phi_{v_0^{(1)}}$, and therefore

$$\det\widehat{\Phi}_{11} = -\phi^{(1)}_{v_0}. \tag{5.3.4}$$

For $k = 2$ and $j = d(v_0^{(2)}) + 2, \dots, d(v_0)$ we replace $\widehat{\mathcal{L}}_{\tau(j)}$ with $\mathcal{L}^{\mathrm{c}}_{\tau(j)}$, which is similar to a matrix multiplication of $\widehat{\Phi}_{12}$ by a triangular matrix whose diagonal elements are 1. Thus we have arrived at Φ_{ND} and therefore

$$\det\widehat{\Phi}_{12} = \phi_{ND}. \tag{5.3.5}$$

In $\widehat{\Phi}_{2k}$ $(k = 1,2)$ we take a cyclic permutation of the rows with indices $\tau(j)$ for $j = 1, \dots, d(v_0^{(1)}) + 1$ such that the row with index $\tau(1)$ becomes the row with index $\tau(d(v_0^{(1)})+1)$. Note that this cyclic permutation of $d(v_0^{(1)})+1$ rows is realized by a permutation matrix with determinant $(-1)^{d(v_0^{(1)})}$. With the same reasoning as above for $\widehat{\Phi}_{1k}$ the rows with indices $\tau_2(j)$ $(j=1,\dots,d(v_0^{(2)}))$ correspond to (generalized) Neumann conditions at $v_0^{(2)}$. For $j=1,\dots,d(v_0^{(1)})$, the new rows with index $\tau_1(j)$ are given by $\mathcal{L}^{\mathrm{c}}_{\tau_1(j+1)}-\mathcal{L}^{\mathrm{c}}_{\tau_1(j)}$ $(j=1,\dots,d(v_0^{(1)})-1)$ and $(-1)^k\mathcal{L}^{\mathrm{c}}_{\tau(d(v_0^{(1)})+k-1)}$ $(j=d(v_0^{(1)}))$. For $k=1$ we replace these linear functionals with $\mathcal{L}^{\mathrm{c}}_{\tau(j)}$ for $j=1,\dots,d(v_0^{(1)})$, which is achieved by a transformation which is similar to an upper triangular matrix with $d(v_0^{(1)})$ diagonal entries being -1, whereas the remaining diagonal entries are 1. Therefore this transformation matrix has determinant $(-1)^{d(v_0^{(1)})}$. We have arrived at Φ_{DN}, and hence

$$\det\widehat{\Phi}_{21}=\det\Phi_{DN}. \tag{5.3.6}$$

For $k=2$ we take the cyclic permutation of the rows with indices $\tau_1(j)$ for $j=1,\dots,d(v_0^{(1)})$ such that the row with index $\tau_1(j)$ is given by $\mathcal{L}^{\mathrm{c}}_{\tau_2(1)}$, whereas the rows with indices $\tau_1(j)$ for $j=2,\dots,d(v_0^{(1)})$ are given by $\mathcal{L}^{\mathrm{c}}_{\tau_1(j)}-\mathcal{L}^{\mathrm{c}}_{\tau_1(j-1)}$. Hence we have arrived at $\Phi^{(2)}_{v_0}$. Since this permutation is achieved by a transformation matrix with determinant $(-1)^{d(v_0^{(1)})-1}$, it follows that

$$\det\widehat{\Phi}_{22}=-\phi^{(2)}_{v_0}. \tag{5.3.7}$$

Substituting (5.3.4)–(5.3.7) into (5.3.3) completes the proof. □

Definition 5.3.5. Let $s\in\{1,2\}$. Then $v_0^{(s)}$ is called locally clamped if there is a set V_s of vertices of G_0 such that

(i) $v_0^{(3-s)}\notin V_s$;

(ii) if e is an edge of G_0 which is incident with $v_0^{(s)}$, then both ends of e belong to V_s;

(iii) if $v\in V_s$ has an incident edge whose other end does not belong to V_s, then the vertex v carries the (generalized) Dirichlet condition.

Remark 5.3.6. 1. Let $s\in\{0,1\}$ and let $V^{(s)}$ be the set of vertices v of G_0 such that $v\neq v_0^{(s)}$ and such that there is an edge e in G_0 with the ends of e being v and $v_0^{(s)}$. If $v_0^{(3-s)}\notin V^{(s)}$ and each $v\in V^{(s)}$ is a pendant vertex or carries generalized Dirichlet conditions, then $v_0^{(s)}$ is locally clamped.

2. If $v_0^{(1)}$ and $v_0^{(2)}$ belong to different components of G_0, then $v_0^{(1)}$ and $v_0^{(2)}$ are locally clamped.

Proof. In statement 1 put $V_s=V^{(s)}\cup\{v_0^{(s)}\}$, and in statement 2 let V_s be the component of G_0 which contains $v_0^{(s)}$ $(s=1,2)$. Then it is clear that V_s satisfies the conditions of Definition 5.3.5. □

Proposition 5.3.7. *Let $s \in \{1,2\}$. If $v_0^{(s)}$ is locally clamped, then $\phi_{v_0}^{(s)} = 0$.*

Proof. Let I_s be the set of all $j \in \{1,\dots,g\}$ such that both ends of e_j belong to V_s. In view of condition (ii) in Definition 5.3.5 we have $I(v_0^{(s)}) \subset I_s$. The rows of $\Phi_{v_0}^{(s)}$ with indices $2j-1$ and $2j$ for $j \in I_s$ only depend on u_{2k-1} and u_{2k} for $k \in I_s$. Indeed, this is clear when $v_+(e_j)$ or $v_-(e_j)$ carries the (generalized) Dirichlet condition, and it follows from condition (iii) in Definition 5.3.5 when $v_+(e_j)$ or $v_-(e_j)$ carries the (generalized) Neumann condition. By condition (i) in Definition 5.3.5, the row with index $\tau_{3-s}(1)$ is not one of those rows listed above, and by definition of $\Phi_{v_0}^{(s)}$ this row only depends on u_{2k-1} and u_{2k} for some $k \in I(v_0^{(s)})$. Therefore $\Phi_{v_0}^{(s)}$ has a $(2\#I_s+1)\times(2g)$ submatrix of rank at most $2\#I_s$, which proves that $\phi_{v_0}^{(s)} = \det(\Phi_{v_0}^{(s)}) = 0$. □

5.4 Series connection of graphs

Let us consider two connected graphs G^j $(j=1,2)$ with at least two vertices in each of them. We choose two distinct vertices v_{in}^j and v_{out}^j in each of them as the entrance and exit vertices. In this section we investigate the series connection of these graphs. For $A, B \in \{D, N\}$ and $j = 1,2$ we denote by ϕ_{AB}^j the characteristic polynomial of the boundary value problem on the graph G^j, where A refers to the boundary condition at v_{in}^j, B refers to the boundary condition at v_{out}^j, D stands for (generalized) Dirichlet boundary conditions and N stands for (generalized) Neumann boundary conditions.

If we connect v_{out}^1 with v_{in}^2, then we obtain a new graph $G = G^1 \cup G^2$ which is assumed to have the generalized Neumann condition at the cut vertex $v_{\text{out}}^1 = v_{\text{in}}^2$ with all other conditions inherited from G^1 and G^2. For $A, B \in \{D, N\}$ the functions ϕ_{NN}, ϕ_{ND}, ϕ_{DN} and ϕ_{DD} denote the characteristic polynomials on G, where A corresponds to v_{in}^1, B corresponds to v_{out}^2 and D and N denote the (generalized) Dirichlet and Neumann conditions, respectively.

Theorem 5.4.1. *For $A, B \in \{D, N\}$ the following identities are true:*

$$\phi_{AB} = \phi_{AN}^1\phi_{DB}^2 + \phi_{AD}^1\phi_{NB}^2. \tag{5.4.1}$$

Proof. In the notation of Section 5.3 we have $G_0 = G^1 \dot\cup G^2$ with $v_0^{(1)} = v_{\text{out}}^1$ and $v_0^{(2)} = v_{\text{in}}^2$. Observing part 2 of Remark 5.3.6 and Proposition 5.3.7, (5.4.1) immediately follows from Lemmas 5.3.4 and 5.3.2. □

Using these identities we immediately obtain an analogue of the Lagrange identity.

Corollary 5.4.2. *It is true that*

$$\phi_{ND}\phi_{DN} - \phi_{NN}\phi_{DD} = (\phi_{ND}^1\phi_{DN}^1 - \phi_{NN}^1\phi_{DD}^1)(\phi_{ND}^2\phi_{DN}^2 - \phi_{NN}^2\phi_{DD}^2). \tag{5.4.2}$$

Proof. Defining

$$\Theta = \begin{pmatrix} \phi_{ND} & \phi_{NN} \\ \phi_{DD} & \phi_{DN} \end{pmatrix}, \quad \Theta^j = \begin{pmatrix} \phi^j_{ND} & \phi^j_{NN} \\ \phi^j_{DD} & \phi^j_{DN} \end{pmatrix} \quad (j = 1, 2),$$

the identities (5.4.1) are equivalent to $\Theta = \Theta^1 \Theta^2$. Taking determinants gives (5.4.2). □

Above we have considered the series connection of two graphs, which is obtained by joining the graphs at two vertices. An alternative method would be to add an edge which joins these two vertices. The reverse of this operation is to remove an edge of the graph so that the two ends of this edge belong to different components of the resulting graph. Below we are going to consider the case of a tree with repeated removal of edges, which gives a particularly easy Lagrange type formula.

Let T be a directed tree, choose any two distinct vertices of T and denote them by v_{in} and v_{out} such that the unique path P in T connecting the two vertices is directed from v_{in} to v_{out}. Let $p+1$ ($p \geq 1$) be the number of vertices of the path and denote these vertices by $v_{\text{in}} = v_0, v_1, \dots, v_p = v_{\text{out}}$ in the order they are traversed. Denote the edge from v_{j-1} to v_j by e_j. After removing the edges e_j ($j = 1, \dots, p$) from T the remaining graph has $p+1$ components (see Lemma C.2.3), which can be uniquely indexed as T_j by the requirement that $v_j \in T_j$ ($j = 0, \dots, p$). Each T_j is either an isolated vertex or a tree, including the degenerate case of a tree with only one edge. At each vertex v of T which is distinct from $v_0, \dots, v_p$ the tree T is equipped with (generalized) Dirichlet boundary conditions or (generalized) Neumann boundary conditions, whereas the tree T carries generalized Neumann boundary conditions at $v_1, \dots, v_{p-1}$ for $p > 1$. For $A, B \in \{D, N\}$ we denote by ϕ_{AB} the characteristic polynomial of the boundary value problem on the tree T, where A refers to the boundary condition at v_{in}, B refers to the boundary condition at v_{out}, D stands for the (generalized) Dirichlet boundary conditions and N stands for the (generalized) Neumann boundary conditions. Let $l_0^{(j)}$ have the usual meaning for the string e_j and let $\phi_D^{(j)}$ be the characteristic polynomial for T_j with root v_j ($j = 0, \dots, p$) and (generalized) Dirichlet boundary conditions at v_j whereas the boundary conditions at all other vertices of T_j are inherited from the boundary conditions for T.

Proposition 5.4.3. *For the directed tree T as considered above, the following Lagrange type formula holds:*

$$\phi_{ND}\phi_{DN} - \phi_{NN}\phi_{DD} = \left(\prod_{j=1}^{p} \frac{1}{l_0^{(j)}} \prod_{j=0}^{p} \phi_D^{(j)} \right)^2. \tag{5.4.3}$$

Proof. Recall the definition of the characteristic polynomials $\varphi_{AB}^{(j)}$ from Section 5.2 (see Proposition 5.2.2) and recall that $\phi_D^{(j)} = 1$ if T_j is an isolated vertex. We will prove the statement by induction on p. Hence first consider the case $p = 1$. Applying Lemmas 5.3.2 and 5.3.1 twice if v_0 and v_1 are interior vertices it follows that

$$\phi_{DD} = \phi_D^{(0)} \varphi_{DD}^{(1)} \phi_D^{(1)}.$$

Clearly, the above reasoning also applies, mutatis mutandis, when v_0 or v_1 are pendant vertices. Therefore the above representation of ϕ_{DD} holds in all cases.

For $j = 0, 1$ let $\widetilde{T}_j$ consist of T_j augmented by the vertex v_{1-j} and the edge e_1, with the orientation inherited from the orientation of T. For $A, B \in \{D, N\}$ we denote by $\widetilde{\phi}_{AB}^{(j)}$ the characteristic polynomial of the boundary value problem on $\widetilde{T}_j$, where A refers to the boundary condition at v_j, B refers to the boundary condition at v_{1-j}, D stands for the (generalized) Dirichlet boundary conditions and N stands for the (generalized) Neumann boundary conditions. Another application of Lemma 5.3.2, part 2 of Remark 5.3.6 and Proposition 5.3.7 shows that

$$\phi_{DN} = \phi_D^{(0)} \widetilde{\phi}_{ND}^{(1)}, \quad \phi_{ND} = \phi_D^{(1)} \widetilde{\phi}_{ND}^{(0)}.$$

If v_0 is an interior vertex of T, then T can be considered as a series connection of $G_1 = T_0$ and $G_2 = \widetilde{T}_1$ by letting $v_{\text{out}}^1 = v_0$, $v_{\text{in}}^2 = v_0$ and choosing vertices $v_{\text{in}}^1 \neq v_0$ and v_{out}^2 of T_0 and T_1, respectively. Then Theorem 5.4.1 gives

$$\phi_{NN} = \phi_D^{(0)} \widetilde{\phi}_{NN}^{(1)} + \phi_N^{(0)} \widetilde{\phi}_{ND}^{(1)}.$$

Putting $\phi_N^{(j)} := 0$ if T_j is an isolated vertex, this representation also holds when v_0 is a pendant vertex of T. Similarly, if T_0 or T_1 is not an isolated vertex, then Theorem 5.4.1 shows that

$$\begin{aligned}
\widetilde{\phi}_{ND}^{(0)} &= \varphi_{DD}^{(1)} \phi_N^{(0)} + \varphi_{ND}^{(1)} \phi_D^{(0)}, \\
\widetilde{\phi}_{ND}^{(1)} &= \varphi_{DD}^{(1)} \phi_N^{(1)} + \varphi_{DN}^{(1)} \phi_D^{(1)}, \\
\widetilde{\phi}_{NN}^{(1)} &= \varphi_{ND}^{(1)} \phi_N^{(1)} + \varphi_{NN}^{(1)} \phi_D^{(1)},
\end{aligned}$$

which clearly extends to the case when T_0 or T_1 is an isolated vertex. Therefore the left-hand side of (5.4.3) can be written as

$$\phi_{ND}\phi_{DN} - \phi_{NN}\phi_{DD} = \phi_D^{(0)} \phi_D^{(1)} \psi,$$

where

$$\begin{aligned}
\psi &= \widetilde{\phi}_{ND}^{(0)} \widetilde{\phi}_{ND}^{(1)} - \varphi_{DD}^{(1)} \left[\phi_D^{(0)} \widetilde{\phi}_{NN}^{(1)} + \phi_N^{(0)} \widetilde{\phi}_{ND}^{(1)} \right] \\
&= \left[\widetilde{\phi}_{ND}^{(0)} - \varphi_{DD}^{(1)} \phi_N^{(0)} \right] \widetilde{\phi}_{ND}^{(1)} - \varphi_{DD}^{(1)} \phi_D^{(0)} \widetilde{\phi}_{NN}^{(1)} \\
&= \varphi_{ND}^{(1)} \phi_D^{(0)} \widetilde{\phi}_{ND}^{(1)} - \varphi_{DD}^{(1)} \phi_D^{(0)} \widetilde{\phi}_{NN}^{(1)} \\
&= \phi_D^{(0)} \left[\varphi_{ND}^{(1)} \varphi_{DD}^{(1)} \phi_N^{(1)} + \varphi_{ND}^{(1)} \varphi_{DN}^{(1)} \phi_D^{(1)} - \varphi_{DD}^{(1)} \varphi_{ND}^{(1)} \phi_N^{(1)} - \varphi_{DD}^{(1)} \varphi_{NN}^{(1)} \phi_D^{(1)} \right] \\
&= \phi_D^{(0)} \phi_D^{(1)} \left[\varphi_{ND}^{(1)} \varphi_{DN}^{(1)} - \varphi_{DD}^{(1)} \varphi_{NN}^{(1)} \right].
\end{aligned}$$

In view of Proposition 5.2.2 and the Lagrange identity, Theorem 1.2.16, it follows that

$$\begin{aligned}\psi &= \frac{1}{l_0^{(1)}}\phi_D^{(0)}\phi_D^{(1)}\left[R_{2n_1}^{(1)}(\cdot,1)R_{2n_1-1}^{(1)}(\cdot,0) - R_{2n_1}^{(1)}(\cdot,0)R_{2n_1-1}^{(1)}(\cdot,1)\right] \\ &= \left(\frac{1}{l_0^{(1)}}\right)^2 \phi_D^{(0)}\phi_D^{(1)}.\end{aligned}$$

This complete the proof in case $p = 1$.

For $p > 1$, let T^1 be the tree obtained from T by removing all vertices of T_p and all edges which are incident with a vertex of of T_p. The tree T^1 with path from v_0 to v_{p-1} satisfies the induction hypothesis with $p-1$. Therefore, in the notation of Corollary 5.4.2,

$$\phi_{ND}^1\phi_{DN}^1 - \phi_{NN}^1\phi_{DD}^1 = \left(\prod_{j=1}^{p-1}\frac{1}{l_0^{(j)}}\prod_{j=0}^{p-1}\phi_D^{(j)}\right)^2.$$

Let T^2 be the tree T_p, augmented by the vertex v_{p-1} and the edge e_p. Applying the case $p = 1$ to the tree T^2 and the path from v_{p-1} to v_p it follows that

$$\phi_{ND}^2\phi_{DN}^2 - \phi_{NN}^2\phi_{DD}^2 = \left(\frac{1}{l_0^{(p)}}\phi_D^{(p)}\right)^2,$$

where we have used that removing the edge e_p from T^2 produces an isolated vertex and the tree T_p. Setting $v_{\text{in}}^1 = v_0$, $v_{\text{out}}^1 = v_{p-1}$, $v_{\text{in}}^2 = v_{p-1}$, $v_{\text{out}}^2 = v_p$ and observing that T has generalized Neumann conditions at v_{p-1}, T can be considered as the series connection of T^1 and T^2. An application of Corollary 5.4.2 completes the proof. □

5.5 Parallel connection of graphs

Now we consider parallel connection of two graphs G^1 and G^2, i.e., with the notation from the previous section, a graph G obtained by connecting v_{in}^1 with v_{in}^2 and v_{out}^1 with v_{out}^2. This construction is illustrated in Figure 5.1, where v_{in}^j and v_{out}^j are pendant vertices of G^j for $j = 1, 2$, respectively, where all edges are outgoing from v_{in}^j and incoming into v_{out}^j. However, this is not required in general.

For $A, B \in \{D, N\}$ let ϕ_{AB} be the characteristic polynomial of the problem on G, where A corresponds to the vertex $v_{\text{in}} := v_{\text{in}}^1 = v_{\text{in}}^2$, B corresponds to the vertex $v_{\text{out}} := v_{\text{out}}^1 = v_{\text{out}}^2$, D stands for the generalized Dirichlet conditions and N stands for the generalized Neumann conditions. Corresponding notation is used for G^1 and G^2.

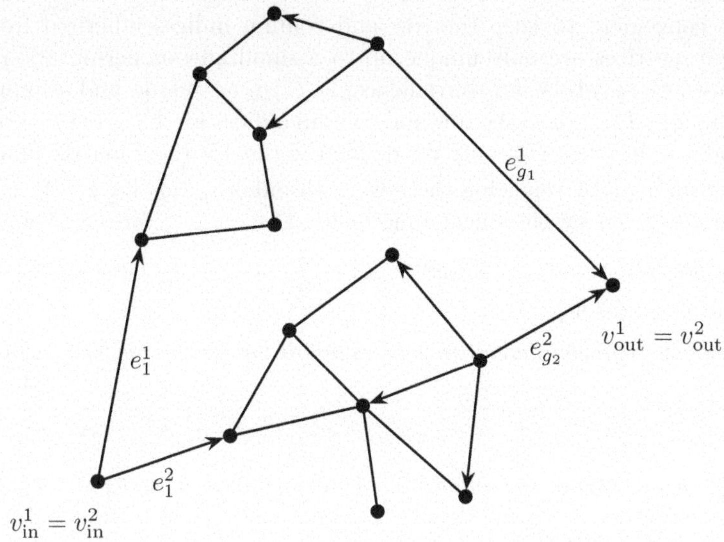

Figure 5.1

Theorem 5.5.1. *For parallel connected graphs with the notation introduced above,*

$$\phi_{DD} = \phi^1_{DD}\phi^2_{DD}, \tag{5.5.1}$$

$$\phi_{DN} = \phi^1_{DN}\phi^2_{DD} + \phi^1_{DD}\phi^2_{DN}, \tag{5.5.2}$$

$$\phi_{ND} = \phi^1_{ND}\phi^2_{DD} + \phi^1_{DD}\phi^2_{ND}. \tag{5.5.3}$$

Proof. The identity (5.5.1) follows by applying Lemma 5.3.2 twice and by then applying Lemma 5.3.1. Let G_0 be the graph obtained by only joining $v^1_{\rm out}$ and $v^2_{\rm out}$. In (5.5.2) we have the generalized Dirichlet condition at $v^1_{\rm in} = v^2_{\rm in}$, and therefore Lemma 5.3.2 shows that ϕ_{DN} is the characteristic polynomial of G_0 with generalized Neumann condition at $v^1_{\rm out} = v^2_{\rm out}$ and Dirichlet conditions at $v^1_{\rm in}$ and $v^2_{\rm in}$. An application of Lemma 5.3.4, Proposition 5.3.7 and Lemma 5.3.1 gives (5.5.2). The same reasoning with incoming and outgoing vertices interchanged proves (5.5.3). □

The formula for ϕ_{NN} is much more involved, and therefore we are going to state and prove it separately. But first we will introduce some notation. Let Φ_{NN}, Φ^1_{NN} and Φ^2_{NN} be characteristic matrix functions of G, G^1 and G^2 with (generalized) Neumann conditions at the entrance and exit vertices, respectively.

Adapting the construction in the paragraph before Lemma 5.3.4 to parallel connected graphs, let $s \in \{1, 2\}$ and $\kappa \in \{\text{in}, \text{out}\}$, let g_s be the number of edges in G^s, and choose bijective maps $\tau_{\kappa,s} : \{1, \dots, d(v^{(s)}_\kappa)\} \to \widetilde{I}(v^{(s)}_\kappa)$. For the matrices

Φ^s_{AB} it is convenient to keep the row and column indices inherited from Φ_{AB}. Since these matrices are only unique up to a simultaneous permutation of rows and columns, we may take, for convenience, $1,\ldots,g_1$ as the row and column indices for Φ^1_{AB} and $g_1+1,\ldots,g$ as the row and column indices for Φ^2_{AB}. Let $\tilde\kappa=$ out when $\kappa=$ in and $\tilde\kappa=$ in when $\kappa=$ out. We define the $(2g_s)\times(2g_s)$ matrix function Φ^s_κ obtained from Φ^s_{NN} by replacing the rows with index $\tau_{\tilde\kappa,s}(j)$ for $j=1,\ldots,d(v^{(s)}_{\tilde\kappa})$ with the rows given by the linear functionals $\mathcal{L}^{\mathrm{c}}_{\tau_{\kappa,s}(d(v^{(s)}_\kappa))}$ for $j=1$ and $s=1$, by $\mathcal{L}^{\mathrm{c}}_{\tau_{\kappa,s}(1)}$ for $j=1$ and $s=2$, and by $\mathcal{L}^{\mathrm{c}}_{\tau_{\tilde\kappa,s}(j)}-\mathcal{L}^{\mathrm{c}}_{\tau_{\tilde\kappa,s}(j-1)}$ for $j=2,\ldots,d(v^{(s)}_{\tilde\kappa})$. Finally put $\phi^s_\kappa=\det\Phi^s_\kappa$.

Theorem 5.5.2. *The characteristic polynomial of the Neumann-Neumann problem on G is*

$$\phi_{NN}=\phi^1_{NN}\phi^2_{DD}+\phi^1_{ND}\phi^2_{DN}+\phi^1_{DN}\phi^2_{ND}+\phi^1_{DD}\phi^2_{NN}-\phi^1_{\mathrm{in}}\phi^2_{\mathrm{out}}-\phi^1_{\mathrm{out}}\phi^2_{\mathrm{in}}. \tag{5.5.4}$$

Proof. The assumptions of Lemma 5.3.4 are satisfied with $v_0^{(j)}=v^j_{\mathrm{in}}$. Due to some clash of notation, we will denote the functions ϕ_{AB} in Lemma 5.3.4 by ϕ^0_{AB}. Applying Lemma 5.3.4 first to G and v_0 being $v^1_{\mathrm{in}}=v^2_{\mathrm{in}}$ and then to G_0 and v_0 being $v^1_{\mathrm{out}}=v^2_{\mathrm{out}}$ and observing part 2 of Remark 5.3.6 and Proposition 5.3.7 shows that

$$\begin{aligned}\phi_{NN}&=\phi^0_{ND}+\phi^0_{DN}-\phi^{(1)}_{v_{\mathrm{in}}}-\phi^{(2)}_{v_{\mathrm{in}}}\\&=\phi^1_{NN}\phi^2_{DD}+\phi^1_{ND}\phi^2_{DN}+\phi^1_{DN}\phi^2_{ND}+\phi^1_{DD}\phi^2_{NN}-\phi^{(1)}_{v_{\mathrm{in}}}-\phi^{(2)}_{v_{\mathrm{in}}}.\end{aligned}$$

We still have to find the representations of $\phi^{(1)}_{v_{\mathrm{in}}}$ and $\phi^{(2)}_{v_{\mathrm{in}}}$. First consider $\phi^{(1)}_{v_{\mathrm{in}}}$. Recalling the construction of $\phi^{(1)}_{v_0}=-\det\widehat\Phi_{11}$ from the proof of Lemma 5.3.4 with $v_0=v_{\mathrm{in}}$, we let Φ^0 be the matrix function $\widehat\Phi_{11}$ as defined by (5.3.3) with respect to $v_0=v_{\mathrm{in}}$. Applying now (5.3.3) with respect to $v_0=v_{\mathrm{out}}$ results in a representation

$$\det\Phi^0=\det\Phi^0_{11}+\det\Phi^0_{12}+\det\Phi^0_{21}+\det\Phi^0_{22}.$$

Since $v_{\mathrm{in}}\neq v_{\mathrm{out}}$ implies $\widetilde I(v_{\mathrm{in}})\cap\widetilde I(v_{\mathrm{out}})=\emptyset$, it is clear that the transformations from Φ_{NN} to $\widehat\Phi_{11}$ and then to the above representation of Φ^0 are independent of each other. For indices $k=1,\ldots,2g_1$ with $k\notin\widetilde I(v_{\mathrm{in}})\cup\widetilde I(v_{\mathrm{out}})$, the rows with index k are the same for Φ_{NN}, Φ^0, Φ^0_{11}, Φ^0_{12}, Φ^0_{21}, and Φ^0_{22} and only depend on the variables u_r with $r=1,\ldots,2g_1$. By construction this extend to $k=1,\ldots,2g_1$ with $k\in\widetilde I(v^1_{\mathrm{in}})$ for the matrix function Φ^0 and then additionally to $\#\widetilde I(v^1_{\mathrm{out}})$ rows with indices in $\widetilde I(v_{\mathrm{out}})$ for the matrix functions Φ^0_{11}, Φ^0_{12} and Φ^0_{21}. Since also one row of $\widehat\Phi_{11}$, and therefore also of Φ^0_{11}, Φ^0_{12} and Φ^0_{21}, corresponding to the boundary conditions at v^2_{in} only depends on the variables corresponding to the edges incident with v^1_{in}, each of the three matrices Φ^0_{11}, Φ^0_{12} and Φ^0_{21} has a $(2g_1+1)\times(2g)$ submatrix of rank at most $2g_1$. Hence their determinants are 0.

From (5.3.7) we therefore see that

$$\phi^{(1)}_{v_{\mathrm{in}}}=\det\left(\left(\Phi^{(1)}_{\mathrm{in}}\right)^{(2)}_{\mathrm{out}}\right).$$

By definition, $\left(\Phi_{\text{in}}^{(1)}\right)_{\text{out}}^{(2)}$ is obtained from Φ_{NN} by replacing the rows with indices in $\widetilde{I}(v_{\text{in}}^{(1)})$ and $\widetilde{I}(v_{\text{out}}^{(2)})$ with rows corresponding to (generalized) Neumann conditions at $v_{\text{in}}^{(1)}$ for G^1 and $v_{\text{out}}^{(2)}$ for G^2, respectively, whereas the remaining changes result in rows corresponding to the conditions at $v_{\text{out}}^{(1)}$ and $v_{\text{in}}^{(2)}$ as in Φ_{in}^1 and Φ_{out}^2, respectively. The unchanged rows correspond to boundary conditions at vertices of G which are distinct from v_{in} and v_{out} and can therefore be identified with rows of Φ_{in}^1 and Φ_{out}^2, respectively. The new row with index $\tau_{\text{in},2}(1)$ is given by the linear functional $\mathcal{L}^{\text{c}}_{\tau_{\text{in},1}}(d(v_{\text{in}}^{(1)}))$, whereas the new row with index $\tau_{\text{out},1}(1)$ is given by the linear functional $\mathcal{L}^{\text{c}}_{\tau_{\text{in},1}}(1)$. Interchanging these two rows, the resulting matrix may be written as a block diagonal matrix, with one block being Φ_{in}^1 and the other block is Φ_{out}^2. This shows that $\phi_{v_0}^{(1)} = -\phi_{\text{in}}^1\phi_{\text{out}}^2$. Interchanging the indices 1 and 2 gives the representation for $\phi_{v_0}^{(2)}$. □

Remark 5.5.3. 1. The construction of the matrix functions Φ_κ^s ($\kappa \in \{\text{in}, \text{out}\}$, $s \in \{1,2\}$) is quite technical, but becomes easier when $d(v_\kappa) = 2$, i.e., $d(v_\kappa^{(1)}) = d(v_\kappa^{(2)}) = 1$. In this case, $\widetilde{I}(v_\kappa^{(s)})$ consists of a single index, say $k_{\kappa,s}$. Then Φ_κ^s is obtained from Φ_{NN}^s by replacing the row which describes the Neumann condition at $v_{\tilde{\kappa}}^{(s)}$ by a row which describes the Dirichlet condition at $v_\kappa^{(s)}$ in G^s.
2. The parallel connection G of $m \geq 2$ connected graphs G^i ($i = 1, \dots, m$) with at least two vertices is obtained by choosing distinct vertices v_{in}^i and v_{out}^i of each graph G^i and identifying all vertices v_{in}^i ($i = 1, \dots, m$) and all vertices v_{out}^i ($i = 1, \dots, m$). A recursive application of Theorems 5.5.1 and 5.5.2 to the case $m \geq 3$ gives formulas which express the characteristic functions of G in terms of characteristic functions of G^i and functions of the form ϕ_{in}^i and ϕ_{out}^i.

Theorems 5.5.1 and 5.5.2 immediately lead to

Corollary 5.5.4. *The following Lagrange type formula holds:*

$$\begin{aligned}\phi_{ND}\phi_{DN} - \phi_{NN}\phi_{DD} &= \left(\phi_{ND}^1\phi_{DN}^1 - \phi_{NN}^1\phi_{DD}^1\right)\left(\phi_{DD}^2\right)^2 \\ &\quad + \left(\phi_{ND}^2\phi_{DN}^2 - \phi_{NN}^2\phi_{DD}^2\right)\left(\phi_{DD}^1\right)^2 \\ &\quad + \phi_{DD}^1\phi_{\text{in}}^1\phi_{DD}^2\phi_{\text{out}}^2 + \phi_{DD}^1\phi_{\text{out}}^1\phi_{DD}^2\phi_{\text{in}}^2 . \end{aligned} \tag{5.5.5}$$

5.6 Multiplicities of characteristic values of spectral problems on graphs

In this section we consider connected graphs with at least one edge. Whenever needed the graphs will be assumed to be digraphs. We recall from Remark 5.1.3 that the spectral data of the problem (4.1.10), (5.1.1), (5.1.2), (5.1.3) do not depend on the orientation, and therefore a convenient orientation of the edges may be chosen when considering spectral data of this problem. We start this section with a lemma, for which we need the following notation.

Notation 5.6.1. Let G be a digraph with a given indexing of the edges.
1. The set $V^\star$ is the set of all $v \in V^{\mathrm{int}} \cap V^{\mathrm{f}}$ satisfying $I^-(v) \neq \emptyset$, $I^+(v) \neq \emptyset$, $\min I^-(v) < \min I^+(v)$ and $\max I^-(v) < \max I^+(v)$. Set $p^\star = \# V^\star$.
2. The set $V^\dagger$ is the set of all $v \in V^{\mathrm{int}} \cap V^{\mathrm{f}}$ satisfying $I^-(v) \neq \emptyset$, $I^+(v) \neq \emptyset$ and $\max I^-(v) \geq \max I^+(v)$. Set $p^\dagger = \# V^\dagger$.

Lemma 5.6.2. *Let G be a connected digraph. Then the maximal multiplicity of a characteristic value of the problem* (4.1.10), (5.1.1), (5.1.2), (5.1.3) *on G does not exceed $g + p^\dagger - p^\star$.*

Proof. First we observe that for each vertex $v \in V^{\mathrm{int}} \cap V^{\mathrm{f}}$ we can choose a bijective map $\tau_v : \{1, \dots, d(v)\} \to \widetilde{I}(v)$ with $\tau_v(\{1, \dots, d^-(v)\}) = \{2j - 1 : j \in I^-(v)\}$. We may further assume that τ_v is increasing on $\{1, \dots, d^-(v)\}$ when $d^-(v) > 1$ and decreasing on $\{d^-(v) + 1, \dots, d(v)\}$ when $d^+(v) > 1$. Let $z \in \mathbb{C}$. For each $v \in V^{\mathrm{int}} \cap V^{\mathrm{f}}$ define the linear functionals $\widehat{\mathcal{L}}_{\tau_v(j)}(z) := \mathcal{L}_{\tau_v(j)}(z) - \mathcal{L}_{\tau_v(j+1)}(z)$ ($j \in \{1, \dots, d(v) - 1\}$) and $\widehat{\mathcal{L}}_{\tau_v(d(v))}(z) := \frac{1}{d(v)} \sum\limits_{r=1}^{d(v)} \mathcal{L}_{\tau_v(r)}(z)$. For all remaining indices $1 \leq k \leq 2g$ let $\widehat{\mathcal{L}}_k(z) := \mathcal{L}_k(z)$.

Then define the $(g + p^\star - p^\dagger) \times (g + p^\star - p^\dagger)$ matrix $\widehat{\Phi}(z)$ as follows. For convenience, we take a subset Λ of the integers from 1 to $2g$ as the index set, defined as follows. For $v \in V^{\mathrm{c}}$ take all indices $2j - 1$ with $j \in I^-(v)$. For $v \in V^\star$ take all indices $2j - 1$ with $j \in I^-(v)$ and the index $\tau_v(1) + 1$. For $v \in V^\dagger$ take all indices $\tau_v(1), \dots, \tau_v(d^-(v) - 1)$. For all $v \in V^{\mathrm{f}} \setminus (V^\star \cup V^\dagger)$ take all indices $\tau_v(1), \dots, \tau_v(\min\{d^-(v), d(v) - 1\})$, and the index $\tau_v(1) + 1$ if $I^+(v) = \emptyset$.

We are now going to verify that all chosen indices are mutually distinct. Indeed, indices of the form $2j - 1$ are odd and are mutually distinct indices from $\widetilde{I}(v)$. On the other hand, indices of the form $\tau_v(1) + 1$ are all even and the numbers $\tau_v(1) + 1$ are mutually distinct. We conclude that the number of chosen indices is $d^-(v) + 1$ for all $v \in V^\star$, $d^-(v) - 1$ for all $v \in V^\dagger$, and $d^-(v)$ for all other $v \in V$. Since $\sum\limits_{v \in V} d^-(v) = g$, it follows that the set Λ of these indices has $g + p^\star - p^\dagger$ elements, which will be the ordered set of row and column indices of $\widehat{\Phi}(z)$.

The rows of $\widehat{\Phi}(z)$ will now be determined by the coefficients of the u_k ($k \in \Lambda$) of the following linear functionals. For an odd index k, the row with index k is determined by $\widehat{\mathcal{L}}_k(z)$. For an even index k, which is of the form $k = \tau_v(1) + 1$ for some $v \in V^{\mathrm{f}}$, the row with index k is determined by $\widehat{\mathcal{L}}_{\tau_v(d(v))}(z)$. In this way, $\widehat{\Phi}(z)$ has been obtained as a submatrix of a linear transformation of $\Phi(z)$. Therefore the rank of $\widehat{\Phi}(z)$ does not exceed the rank of $\Phi(z)$.

Now we are going to prove that $\widehat{\Phi}(z)$ has rank $g + p^\star - p^\dagger$ by showing that it is an upper triangular matrix with nonzero diagonal entries. This will complete the proof in view of Proposition 5.2.1. For $v \in V^{\mathrm{c}}$ and $j \in I^-(v)$ we have $\widehat{\mathcal{L}}_{2j-1}(z)U = \mathcal{L}^{\mathrm{c}}_{2j-1}(z)U = u_{2j-1}$, that is, this row with index $2j - 1$ of $\widehat{\Phi}(z)$ has entry 1 in the diagonal and all other entries are 0. For $v \in V^{\mathrm{f}}$ and $j = 1, \dots, \min\{d^-(v), d(v) - 1\}$

we have

$$\widehat{\mathcal{L}}_{\tau_v(j)}(z)U = \mathcal{L}^{c}_{\tau_v(j)}(z)U - \mathcal{L}^{c}_{\tau_v(j+1)}(z)U = u_{\tau_v(j)} - \mathcal{L}^{c}_{\tau_v(j+1)}(z)U.$$

When $j < d^-(v)$, then $\mathcal{L}^{c}_{\tau_v(j+1)}(z)U = u_{\tau_v(j+1)}$ with $\tau_v(j+1) > \tau_v(j)$, and therefore the row with index $\tau_v(j)$ has entry 1 in the diagonal, an entry -1 in the column with index $\tau_v(j+1)$ if $\tau_v(j+1) \in \Lambda$, and all other entries 0. However, when $j = d^-(v)$, then $\mathcal{L}^{c}_{\tau_v(j+1)}(z)U = R^{(r)}_{2n_r}(z,1)u_{2r-1} + R^{(r)}_{2n_r}(z,0)u_{2r}$ with $2r = \tau_v(d^-(v)+1))$. By definition of Λ we have $v \notin V^{\dagger}$, $I^-(v) \neq \emptyset$ and $I^+(v) \neq \emptyset$. Hence $\tau_v(d^-(v)) = \max I^-(v) < \max I^+(v) = \tau_v(d^-(v)+1)$. Therefore the row with index $\tau_v(d^-(v))$ has entry 1 in the diagonal, possibly two nonzero entries in two columns to the right of the diagonal, and all other entries 0. The remaining rows, which have row index $\tau_v(1)+1$, are given by

$$\begin{aligned}\widehat{\mathcal{L}}_{\tau_v(d(v))}(z)U &= \frac{1}{d(v)}\sum_{r=1}^{d(v)} \mathcal{L}_{\tau_v(r)}(z)U \\ &= \sum_{r\in I^+(v)} \left(R^{(r)}_{2n_r-1}(z,1)u_{2r-1} + R^{(r)}_{2n_r-1}(z,0)u_{2r}\right) - \sum_{r\in I^-(v)} \frac{u_{2r}}{l_0^{(r)}}\end{aligned}$$

for $v \in V^{\star}$ and $v \in V^{\mathrm{f}}$ with $I^+(v) = \emptyset$. This gives in either case that this row has the entry $-\frac{1}{l_0^{(r)}}$ in the columns with indices $2r$ and $r \in I^-(v)$. In particular, for $2r - \tau_v(1) + 1$ this is a nonzero entry in the diagonal, whereas all other of these nonzero entries are to the right of the diagonal. Further nonzero entries can only occur for the case $v \in V^{\star}$, namely possibly in columns with indices $2r-1$ and $2r$ for $r \in I^+(v)$. But these indices are larger than $\tau_v(1)+1$ by definition of V^*. Hence we have indeed shown that $\widehat{\Phi}(z)$ is an upper triangular matrix with nonzero diagonal elements.

Since the rank of $\widehat{\Phi}(z)$ does not exceed the rank of $\Phi(z)$, it follows that the rank of $\Phi(z)$ is at least $g + p^* - p^{\dagger}$. Therefore the multiplicity of a characteristic value z of the problem (4.1.10), (5.1.1), (5.1.2), (5.1.3) on G, which equals the defect of $\Phi(z)$, does not exceed $2g - (g + p^* - p^{\dagger}) = g + p^{\dagger} - p^*$. □

The estimate for the maximal multiplicity of characteristic values in Lemma 5.6.2 depends on the choice of the directions of the edges and will in general be larger than the maximal possible multiplicity. However, below we will apply Lemma 5.6.2 to some graphs where a suitably chosen orientation gives indeed this maximal value for the multiplicity of some characteristic values.

For a vibration problem on a graph G with an indexing e_j $(j = 1, \dots, g)$ of the edges of G let $I_N^{\mathrm{pen}} := \emptyset$ if G has at most one pendant vertex or if G is a string, and if all vertices of G carry (generalized) Neumann conditions. Otherwise, I_N^{pen} denotes the set of indices $j \in \{1, \dots, g\}$ for which one end of e_j is a pendant vertex of G with a Neumann condition. Note that if G has an edge where both ends are pendant vertices, then the connected graph G is a string. Therefore $j \in I_N^{\mathrm{pen}}$ implies that exactly one end of e_j is a pendant vertex with a Neumann condition.

Definition 5.6.3. Let G be a connected digraph with edges e_j $(j = 1, \ldots, g)$. Then a problem (4.1.10), (5.1.1), (5.1.2), (5.1.3) on G is called *rigged* if

$$R^{(j)}_{2n_j}(2,0) = 0,\ R^{(j)}_{2n_j-1}(2,0) = 1,\ R^{(j)}_{2n_j}(2,1) = 1,\ R^{(j)}_{2n_j-1}(2,1) = 0\ (j \notin I_N^{\text{pen}}), \tag{5.6.1}$$

$$R^{(j)}_{2n_j}(2,0) = 1,\ R^{(j)}_{2n_j-1}(2,0) = 0,\ R^{(j)}_{2n_j}(2,1) = 0,\ R^{(j)}_{2n_j-1}(2,1) = -1\ (j \in I_N^{\text{pen}}), \tag{5.6.2}$$

and $l_0^{(j)} = 1$ holds for $j = 1, \ldots, g$. For $j \in I_N^{\text{pen}}$ it will always be assumed that the edge e_j is directed in such a way that $v^+(e_j)$ is the pendant vertex with Neumann condition.

We recall from Examples 1.2.17 and 1.2.18 that every connected digraph with at least 3 beads on each edge has a rigged problem. The identities (5.6.1) and (5.6.2) immediately lead to

Lemma 5.6.4. *Let G be a rigged digraph and $U \in \mathbb{C}^{2g}$. Then*

$$\begin{aligned} \mathcal{L}^{\text{c}}_{2j-1}(2)U &= u_{2j-1} \quad (j = 1, \ldots, g), \\ \mathcal{L}^{\text{c}}_{2j}(2)U &= u_{2j-1} \quad (j \in \{1, \ldots, g\} \setminus I_N^{\text{pen}}), \\ \mathcal{L}^{\text{c}}_{2j}(2)U &= u_{2j} \quad (j \in I_N^{\text{pen}}), \end{aligned}$$

and

$$\mathcal{L}^{\text{f}}_v(2)U = -\frac{1}{d(v)}\left(\sum_{r\in I^+(v)} u_{2r-1} + \sum_{r\in I^-(v)} u_{2r-1}\right) + \sum_{r\in I^+(v)} u_{2r} - \sum_{r\in I^-(v)} u_{2r}$$

for $v \in V^{\text{f}}$ when $v \notin \{v^+(e_j) : j \in I_N^{\text{pen}}\}$, whereas

$$\mathcal{L}^{\text{f}}_{v^+(e_j)}(2)U = -u_{2j-1} - u_{2j} \quad (j \in I_N^{\text{pen}}).$$

In Appendix C we have introduced the matrix of cycles, which will now be augmented to the matrix of paths.

Definition 5.6.5. Let G be a digraph. The matrix $\widehat{M} = (\widehat{M}_{P,j})$, where the row index P runs through all oriented paths in G and $j = 1, \ldots, g$, called the matrix of paths of G, is defined as follows.

1. For an edge e_j which does not belong to the path P, set $\widehat{M}_{P,j} = 0$.
2. For an edge e_j which belongs to the path P and whose direction coincides with the orientation of the path, set $\widehat{M}_{P,j} = 1$.
3. For an edge e_j which belongs to the path P and whose direction is opposite to the orientation of the path, set $\widehat{M}_{P,j} = -1$.

Notation 5.6.6. For a rigged problem on G and a directed path P in G, the vector $U^P = (u_r^P)_{r=1}^{2g}$ is defined by $u_{2j}^P := \widehat{M}_{P,j}$ and $u_{2j-1}^P := 0$ $(j = 1, \dots, g)$. Also, let $U^{(0)} = (u_r^{(0)})_{r=1}^{2g}$ be defined by $u_{2j-1}^{(0)} = 1$ and $u_{2j}^{(0)} = 0$ $(j = 1, \dots, g)$.

Lemma 5.6.7. *Let G be a rigged digraph and let P be a path in the graph G. Then $\mathcal{L}_v^{\mathrm{f}}(2)U^P = 0$ for all vertices $v \in V^{\mathrm{f}}$ which are not ends of P.*

Proof. When the vertex v of G is not a vertex of P, then its incident edges are not edges of P, and therefore $\mathcal{L}_v^{\mathrm{f}}(2)U^P = 0$. However, for a vertex v of P which is not an endpoint of P, there are, in the direction of P, one incoming edge e_{j_1} and one outgoing edge e_{j_2} of P at v. We have to note that $j_1 = j_2$ when P is a loop; but what follows also holds in this case. Only these two edges may give a contribution to $\mathcal{L}_v^{\mathrm{f}}(2)U^P$. By Lemma 5.6.4, for $s \in \{1, 2\}$ the contribution of the edge e_{j_s} is $u_{2j_s}^P$ if $j_s \in I^+(v)$ and $-u_{2j_s}^P$ if $j_s \in I^-(v)$. Due to the definition of U^P it follows that the edge e_{j_1} gives the contribution 1 and the edge e_{j_2} gives the contribution -1 to $\mathcal{L}_{v_j}^{\mathrm{f}}(2)U^P$, and therefore $\mathcal{L}_v^{\mathrm{f}}(2)U^P = 0$. □

Corollary 5.6.8. *Let G be a rigged digraph.*

1. *When P is a cycle in G, then $\Phi(2)U^P = 0$.*
2. *Let P be a chain in G with each of its ends being a pendant vertex of G. If P is not a string with Neumann conditions at both ends, then $\Phi(2)U^P = 0$.*
3. *If all vertices of G carry (generalized) Neumann conditions, then $\Phi(2)U^{(0)} = 0$.*

Proof. Recalling the definition of Φ, statement 1 is an immediate consequence of Lemmas 5.6.4 and 5.6.7 if we observe that $\mathcal{L}_v^{\mathrm{f}}$ occurs in $\mathcal{L}_r$ only if $v \in V^{\mathrm{f}}$.

Statement 2 follows in a similar manner, but we still have to investigate the ends of P. Therefore, let v be an end of P. There is a unique index j such that v is incident with e_j. If $j \in I_N^{\mathrm{pen}}$, then $v = v^+(e_j)$ carries the Neumann condition and $v^-(e_j)$ is an interior vertex or a pendant vertex with Dirichlet boundary condition. Then Lemmas 5.6.4 and 5.6.7 show that $\mathcal{L}_{2j-1}(2)U^P = 0$ and

$$\mathcal{L}_{2j}(2)U^P = u_{2j} + (-u_{2j-1} - u_{2j}) = -u_{2j-1} = 0.$$

If $j \notin I_N^{\mathrm{pen}}$, then $\mathcal{L}_{2j-1}(2)U^P = 0$ and $\mathcal{L}_{2j}(2)U^P = 0$ immediately follow from Lemmas 5.6.4 and 5.6.7.

For the proof of statement 3 observe that $I_N^{\mathrm{pen}} = \emptyset$. Then $\mathcal{L}_r^{\mathrm{c}}(2)U^{(0)} = 1$ for $r = 1, \dots, 2g$ and Lemma 5.6.4 shows that $\mathcal{L}_v^{\mathrm{f}}(2)U^{(0)} = -1$ for all vertices v of G. Hence $\Phi(2)U^{(0)} = 0$ by definition of Φ. □

Let p_{pen} and p_{int} denote the number of pendant vertices and interior vertices of a graph G, respectively.

Corollary 5.6.9. *Let G be a rigged digraph such that all interior vertices carry generalized Neumann conditions. Then 2 is a characteristic value whose multiplicity is at least $\mu + 1$ if all vertices carry Neumann conditions and at least $\mu + p_{\mathrm{pen}} - 1$ otherwise.*

Proof. If $\mu > 0$, then choose a fundamental set $\mathcal{C}$ of μ cycles. If $p_{\rm pen} \geq 2$, fix one pendant vertex and choose a set $\mathcal{P}$ of $p_{\rm pen} - 1$ paths in G which connect this vertex to exactly one of the remaining pendant vertices. Each of these pendant vertices has exactly one incident edge, say e_{j_i} $(i = 1, \dots, p_{\rm pen} - 1)$. Hence the rows of the matrix of paths $\widehat{M}$ with row indices $P \in \mathcal{C} \cup \mathcal{P}$ are linearly independent since exactly one row of this submatrix has a nonzero entry in the column with index j_i $(i = 1, \dots, p_{\rm pen} - 1)$. It follows that the vectors U^P $(P \in \mathcal{C} \cup \mathcal{P})$ and $U^{(0)}$ are linearly independent. If all vertices carry Neumann conditions, then Corollary 5.6.8, parts 1 and 3, shows that the dimension of the null space of $\Phi(2)$ is at least $\mu + 1$. Otherwise, Corollary 5.6.8, parts 1 and 2, shows that the dimension of the null space of $\Phi(2)$ is at least $\mu + p_{\rm pen} - 1$. An application of Proposition 5.2.1 completes the proof. □

Observe that the multiplicity obtained in Corollary 5.6.9 is at least 1. Indeed, if not all vertices carry (generalized) Neumann conditions, then $p_{\rm pen} \geq 1$. If $\mu = 0$, then G is a tree, and property (iii) of Proposition C.2.5 shows that $p_{\rm pen} \geq 2$.

Finally, we consider the case that the vibration problem has generalized Dirichlet conditions at interior vertices.

Lemma 5.6.10. *Let G be a rigged digraph such that the number δ of interior vertices of G with generalized Dirichlet conditions is positive. Then 2 is a characteristic value whose multiplicity is at least $\mu + p_{\rm pen} + \delta - 1$.*

Proof. Cutting the graph G at those interior vertices v which carry generalized Dirichlet conditions generates a graph G_N where each such v is replaced by $d(v)$ pendant vertices with Dirichlet conditions. Then all interior vertices of G_N have generalized Neumann conditions, and G_N has the same characteristic function as G, see Lemma 5.3.2.

Let V_D be the set of interior vertices of G with generalized Dirichlet conditions. Then the number of vertices of G_N is

$$p^{G_N} = p^G + \sum_{v \in V_D} (d(v) - 1)$$

since each of the $d(v)$ edges in G which is incident with $v \in V_D$ gets its own vertex in G_N. The number of edges does not change, i.e., $g^{G_N} = g^G$. However, G_N may no longer be connected, so that κ^{G_N} may be larger than 1. Furthermore, the number of pendant vertices of G_N is clearly

$$p_{\rm pen}^N = p_{\rm pen} + \sum_{v \in V_D} d(v).$$

From Lemma C.2.3 one concludes that

$$\begin{aligned}\mu^{G_N} &= g^{G_N} + \kappa^{G_N} - p^{G_N} \\ &= g^G + \kappa^{G_N} - p^G - \sum_{v\in V_D} (d(v)-1) \\ &= \mu^G - 1 + \kappa^{G_N} - \sum_{v\in V_D} (d(v)-1).\end{aligned}$$

Since G and each component of G_N have at least one vertex with (generalized) Dirichlet condition, each of the components G_j ($j = 1, \dots, \kappa^{G_N}$) of G_N is a rigged digraph and has at least one pendant vertex with Dirichlet conditions. By Corollary 5.6.9 the multiplicity of 2 as a characteristic value of the problem on G_j is $\mu^{G_j} + p_{\text{pen}}^{(j)} - 1$, where $p_{\text{pen}}^{(j)}$ is the number of pendant vertices of G_j. Since the union of fundamental sets of cycles in G_j ($j = 1, \dots, \kappa^{G_N}$) is clearly a fundamental set of cycles in G_N and since the union of the sets of pendant vertices of G_j ($j = 1, \dots, \kappa^{G_N}$) is clearly the set of the pendant vertices of G_N, it follows that

$$\begin{aligned}\sum_{j=1}^{\kappa^{G_N}} (\mu^{G_j} + p_{\text{pen}}^{(j)} - 1) &= \mu^{G_N} + p_{\text{pen}}^{N} - \kappa^{G_N} \\ &= \mu^G - 1 + p_{\text{pen}}^{N} - \sum_{v\in V_D} (d(v)-1) \\ &= \mu^G - 1 + p_{\text{pen}} + \delta.\end{aligned}$$

From Lemmas 5.3.2 and 5.3.1 we can now conclude that 2 is a characteristic value of the problem on G whose multiplicity is at least $\mu + p_{\text{pen}} + \delta - 1$. □

5.6.1 Cyclically connected graphs

The maximum of the multiplicities of characteristic values for cyclically connected graphs is given in the following theorem.

Theorem 5.6.11. *Let G be a cyclically connected graph with generalized Dirichlet conditions at $\delta \le p$ vertices and generalized Neumann condition at the remaining vertices. Then the maximal multiplicity of a characteristic value is $\mu + |\delta - 1|$. The given maximal value is achieved for problems with at least 3 beads on each edge and for a suitable choice of the lengths of the edges and the distribution of the beads.*

Proof. Let m be the maximal multiplicity of a characteristic value. We choose a digraph structure on G according to Proposition C.4.9. First let $\delta = 0$. Then $V^* = V_0^\times$ and hence $p^\star \ge p - 1$ and $p^\dagger \le 1$ by Proposition C.4.9. Observing Lemmas 5.6.2 and C.2.3, it follows that

$$\begin{aligned}m &\le g + p^\dagger - p^\star \le g + 2 - p \\ &= \mu - \kappa + 2 = \mu + 1.\end{aligned}$$

If $\delta > 0$, then we may choose the vertex w in Proposition C.4.9 to be a vertex with generalized Dirichlet conditions. Then $p^\star = p - \delta$ and $p^\dagger = 0$, and therefore

$$\begin{aligned} m &\le g + p^\dagger - p^\star \le g - p + \delta \\ &= \mu - \kappa + \delta = \mu + \delta - 1. \end{aligned}$$

In view of Example 1.2.17, the graph G can be made into a rigged digraph with the given number of beads on each edge. Then Corollary 5.6.8 for $\delta = 0$ and Lemma 5.6.10 for $\delta > 0$ show that the upper bounds are indeed the maximal possible values. □

5.6.2 Trees and generalized quasi-trees

The maximum of the multiplicities of characteristic values for trees and generalized quasi-trees is given in the following theorem.

Theorem 5.6.12. *Let G be a tree or a generalized quasi-tree with generalized Dirichlet conditions at $\delta \le p_{\rm int}$ interior vertices and generalized Neumann condition at the remaining interior vertices. Then the maximal multiplicity of a characteristic value is $\mu + p_{\rm pen} + \delta - 1 = g - p_{\rm int} + \delta$. The maximum is achieved for problems with at least 3 beads on each edge and for a suitable choice of the lengths of the edges and the distribution of the beads.*

Proof. Observing that $p_{\rm pen} + p_{\rm int} = p$, the identity $\mu + p_{\rm pen} = g - p_{\rm int} + 1$ follows immediately from Lemma C.2.3.

Let m be the maximal multiplicity of a characteristic value. Since each maximal cyclically connected subgraph of G has at least two vertices in common with its complement, it is clear that the vertices in the tree $G^{\mathcal{C}}$ representing maximal cyclically connected subgraphs in G are interior vertices of $G^{\mathcal{C}}$. Therefore any pendant vertex of $G^{\mathcal{C}}$ is also a pendant vertex of G. Since $G^{\mathcal{C}}$ is a tree, it follows that $p_{\rm pen} > 0$. Any pendant vertex of G will be chosen as the root. With an indexing and orientation of the edges according to Lemma C.4.8 it follows that $V_0^\times = V^{\rm int}$, which implies $p^\star = p_{\rm int} - \delta$ and $p^\dagger = 0$. Here we have to observe that the proof of Lemma C.4.6 provides the required indexing if G is a tree. Therefore Lemma 5.6.2 gives

$$m \le g + p^\dagger - p^\star \le g - p_{\rm int} + \delta.$$

It follows from Corollary 5.6.8 for $\delta = 0$ and Lemma 5.6.10 for $\delta > 0$ that the given upper bound can be achieved for a suitable digraph structure on G. □

5.6.3 Connected graphs

In this subsection we consider connected graphs G which are not cyclically connected. We exclude the case $g = 0$, i.e., the graphs have at least one edge. Fixing now such a graph G and denoting by $\mathcal{C}$ the set of all maximal cyclically connected subgraphs of G, we know from Proposition C.4.4 and Corollary C.4.5 that $T := G^{\mathcal{C}}$ is a tree.

Theorem 5.6.13. *Let G be a connected but not cyclically connected graph with at least one edge. It is supposed that each interior vertex carries a generalized Neumann condition. Then the maximal possible multiplicity of a characteristic value is $\mu + p_{\rm pen}^T - 1$.*

Proof. Again we are going to apply Lemma 5.6.2. To this end we have to find a lower bound for $p^\star$ and an upper bound for $p^\dagger$. Let $\mathcal{C}_1$ be the set of all $\Gamma \in \mathcal{C}$ which have exactly one common vertex v_Γ with their complement in G and such that v_Γ has exactly one incident edge in the complement of Γ. Let $\mathcal{C}_2$ be the set of all $\Gamma \in \mathcal{C}$ which have exactly one common vertex v_Γ with their complement in G and such that v_Γ has at least two incident edges in the complement of Γ. Let $\mathcal{C}_3$ be the set of all $\Gamma \in \mathcal{C}$ which have at least two common vertices with their complement in G. Clearly, each $\Gamma \in \mathcal{C}$ belongs to exactly one of $\mathcal{C}_1$, $\mathcal{C}_2$ or $\mathcal{C}_3$.

By Proposition C.4.4, the graph $G_3 := G^{\mathcal{C}_1 \cup \mathcal{C}_2}$ (see (C.4.1)) has $\mathcal{C}_3$ as the set of its maximal cyclically connected subgraphs. Since each $\Gamma \in \mathcal{C}_1 \cup \mathcal{C}_2$ has exactly one vertex in common with its complement, it is clear that $G^{\mathcal{C}_1 \cup \mathcal{C}_2}$ is the subgraph of G which consists of all edges in G which are not edges of any $\Gamma \in \mathcal{C}_1 \cup \mathcal{C}_2$ and all vertices which are incident with these edges. In particular, any vertex of $\Gamma \in \mathcal{C}_3$ which is a vertex of the complement of Γ in G is also a vertex of the complement of Γ in G_3. But by definition of $\mathcal{C}_3$, each $\Gamma \in \mathcal{C}_3$ has at least two vertices in common with its complement in G and therefore also in G_3. By definition C.4.7 this means that G_3 is a generalized quasi-tree if $\mathcal{C}_3 \neq \emptyset$ or a tree if $\mathcal{C}_3 = \emptyset$. Furthermore, the pendant vertices of G_3 are exactly the pendant vertices of T.

Let any pendant vertex of G_3 be its root. From the proof of Lemma C.4.6 and from Lemma C.4.8 we know that there is an orientation and indexing of the edges of G_3 satisfying properties 1, 2 and 3 of Lemma C.4.8. For convenience, we allow the indices of the edges of G_3 and of G to be arbitrary real numbers. We now reverse the orientation and the indexing of all edges of G_3; for example, if an edge is originally indexed by e_r, we could now write e_{-r}. Then it is clear that property 3 of Lemma C.4.8 still holds, whereas $I_3^+(v) = \emptyset$ for the root v of G_3 and $I_3^-(v) = \emptyset$ for all other pendant vertices of G_3. Here and for other notation a lower index 3 indicates that this quantity is taken with respect to the graph G_3, whereas a lower index Γ will indicate that it is taken with respect to a corresponding $\Gamma \in \mathcal{C}$.

The maximal cyclically connected subgraphs $\Gamma \in \mathcal{C}_2$ are attached at interior vertices v_Γ of G_3, and therefore $v_\Gamma \in V_3^\star$, which implies that

$$j_\Gamma := \min I_3^-(v_\Gamma) < \min(I_3 \setminus \{j_\Gamma\}) =: j'_\Gamma .$$

Taking v_Γ as the vertex w from Lemma C.4.9 with respect to Γ, the edges of the graph Γ will be oriented and indexed according to Lemma C.4.9, where we may assume that all indices of the edges of Γ lie between j_Γ and j'_Γ. Then it follows from Lemma C.4.9 that $v \in V^\star$ for $v \in \Gamma \setminus \{v_\Gamma\}$. Furthermore, it is not hard to see that $v_\Gamma \in V_3^\star$ gives $v_\Gamma \in V^\star$.

On the other hand, the maximal cyclically connected subgraphs $\Gamma \in \mathcal{C}_1$ are attached at pendant vertices v_Γ of G_3. Then there is a unique index j_Γ such that

$I_3(v_\Gamma) = \{j_\Gamma\}$. Taking v_Γ as the vertex w from Lemma C.4.9 with respect to Γ, the edges of the graph Γ will be oriented and indexed according to Lemma C.4.9, where we may assume that all indices of the edges of Γ lie below j_Γ. Then it follows from Lemma C.4.9 that $v \in V^\star$ for $v \in \Gamma \setminus \{v_\Gamma\}$. Furthermore, it is not hard to see that $v_\Gamma \notin V^\dagger$ when v_Γ is not the root of G.

We first assume that $V^\dagger = \emptyset$. In the last two paragraphs it was shown that all vertices which are not pendant vertices of T belong to $V^\star$. Therefore, $p^\dagger = 0$ and $p^\star \geq p^G - p^T_{\text{pen}}$, and hence Lemma 5.6.2 shows that the multiplicity of a characteristic value of the problem does not exceed $g^G - p^G + p^T_{\text{pen}} = \mu - 1 + p^T_{\text{pen}}$.

The second case is that $V^\dagger \neq \emptyset$. By the above considerations this can only happen when the root of G_3 is some v_Γ for $\Gamma \in \mathcal{C}_1$. We now attach an edge e_{g+1} to G at v_Γ, oriented in such a way that v_Γ is the exit of e_{g+1}. The entrance of e_{g+1} will be denoted by v_+. Denoting this augmented graph by G_+ it is clear that the cyclomatic numbers of G and G_+ are the same. The corresponding tree T_+ is obtained by attaching e_{g+1} at v_Γ, so that v_Γ becomes an interior vertex of T_+, whereas v_+ becomes a pendant vertex of T_+, which we will choose as the root. Therefore $V^\dagger_+ = \emptyset$, and by what has already been shown it follows that the multiplicity of each characteristic value for the problem on G_+ does not exceed $\mu + p^T_{\text{pen}} - 1$.

Now let z be a characteristic value of the problem on G with multiplicity m. By Theorem 1.3.1 there is a distribution of beads on e_{g+1} such that z is a characteristic value of the Dirichlet-Neumann problem on e_{g+1}. If now U is a solution corresponding to the problem on G, then there is u_{2g} such that $R^{(g+1)}_{2n_{g+1}}(z,0)u_{2g}$ is the unique value given by the continuity condition (5.1.1) at v_Γ for the problem on G. Here we recall that the Neumann condition $R^{(g+1)}_{2n_{g+1}-1}(z,0) = 0$ implies that $R^{(g+1)}_{2n_{g+1}}(z,0)u_{2g} \neq 0$. Hence augmenting U by $(0, u_{2g})$ leads to a solution of the problem on G_+. Therefore the multiplicity of the characteristic value z for the problem on G_+ is at least m. Then the discussion in the previous paragraph shows that $m \leq \mu + p^T_{\text{pen}} - 1$.

To show that the bound can be reached we start with a rigged problem on G_3 with Neumann conditions at all v_Γ for $\Gamma \in \mathcal{C}_1$ and consider data according to Example 1.2.17 on all the remaining edges. Note that the indexing and orientation may be different from that chosen in the first part of this proof. In particular, each $\Gamma \in \mathcal{C}$ is a rigged digraph, and Corollary 5.6.8 shows that $\Phi(2)U^P = 0$ for every cycle P in such Γ with U^P considered in Γ. But then a reasoning as in the proof of Lemma 5.6.7 implies that also $\Phi(2)U^P = 0$ when U^P is considered in G. Hence a fundamental set of cycles in G contributes μ linearly independent vectors in the null space of $\Phi(2)$.

If G_3 is a string, then $G \neq G_3$ since G is not a tree. Hence at least one end of G_3 carries a Neumann condition. If both ends carry Neumann conditions, then all vertices of G carry Neumann conditions, and we already know from Corollary 5.6.8 that $\Phi(2)U^{(0)} = 0$. Therefore the null space of $\Phi(2)$ has dimension at least

$\mu + p^T_{\rm pen} - 1$.

Finally we consider the case that G_3 is either a string with one end carrying a Dirichlet condition and the other end carrying a Neumann condition, or that G_3 has at least 3 vertices. Let P be a chain in G_3 connecting a pendant vertex in G_3 to the root of G_3. Then we know that the vector corresponding to this chain in G_3 is in the null space of the characteristic matrix function at 2. For j such that e_j is an edge of G_3, the values u^P_{2j-1} and u^P_{2j} are as defined for G_3. We are going to define values corresponding to all other edges of G. If an end of P is v_Γ with $\Gamma \in \mathcal{C}_1$ and with incident edge e_{j_Γ} in G_3 and hence in P, then let $u^P_{2r-1} := u^P_{2j_\Gamma}$ for all r for which e_r is an edge of Γ. For all remaining indices $s = 1, \dots, 2g$ we put $u^P_s := 0$. With a reasoning as in the proofs of Lemma 5.6.7 and Corollary 5.6.8 we conclude that to show $\Phi(z)U^P = 0$, only $\mathcal{L}_s(2)U^P = 0$ has still to be shown for indices $s = 2k-1$ or $s = 2k$ when e_k is incident with an end v_Γ of P with $\Gamma \in \mathcal{C}_1$ and $k \in I^-(v_\Gamma)$ or $k \in I^+(v_\Gamma)$, respectively. For such an end let $v = v_\Gamma$ and $j = j_\Gamma$. The vibration problem satisfies (5.6.2) on the edge e_j and (5.6.1) on all other edges e_k which are incident with v. Since the edge e_j is an incoming edge into v and the only edge incident with v which is not an edge of Γ, it follows that

$$\mathcal{L}^{\rm f}_v(2)U = -\frac{1}{d(v)}\left(\sum_{\substack{r\in I^+(v)\\ r\neq j}} u^P_{2r-1} + \sum_{r\in I^-(v)} u^P_{2r-1}\right) - \frac{u^P_{2j}}{d(v)} = -u^P_{2j}.$$

Hence $\mathcal{L}_{2j}(2)U^P = \mathcal{L}^{\rm c}_{2j}(2)U^P + \mathcal{L}^{\rm f}_v(2)U^P = u^P_{2j} - u^P_{2j} = 0$, whereas for the other edges e_k which are incident with v we have $\mathcal{L}_s(2)U^P = \mathcal{L}^{\rm c}_s(2)U^P + \mathcal{L}^{\rm f}_v(2)U^P = u^P_{2k-1} - u^P_{2j} = 0$. Summarizing the above construction, we have found for $q = p^T_{\rm pen} - 1$ pendant vertices $v_1, \dots, v_q$ vectors $U^{(1)}, \dots U^{(q)}$ in the null space of $\Phi(2)$ such that $u^{(k)}_{2j_i} \neq 0$ if and only if $k = i$ ($i, k = 1, \dots, q$), where j_i is the unique index of the edge in G_3 which is incident with the vertex v_i ($i = 1, \dots, q$). Therefore the μ vectors U^P corresponding to a fundamental set of cycles and the vectors $U^{(1)}, \dots, U^{(q)}$ are linearly independent.

We have shown in each of the above cases that the matrix $\Phi(2)$ has a null space of dimension at least $\mu - 1 + p^T_{\rm pen}$. In view of Proposition 5.2.1 this means that 2 is a characteristic value of multiplicity at least $\mu - 1 + p^T_{\rm pen}$. □

5.7 Graphs of symmetric Stieltjes strings

In this section let G be a simple connected metric graph with at least two edges. It is assumed that all edges have the same length and the same distribution of beads and that this distribution is symmetric. This means that if $m^{(j)}_k$ ($k = 1, \dots, n_j$) denotes the masses of the beads and $l^{(j)}_k$ ($k = 0, \dots, n_j$) the lengths of the threads for the j-th edge ($j = 1, \dots, g$), then these numbers, as well as n_j, are independent of j, $m_k^{(j)} = m_{n_j-k+1}^{(j)}$ $(k = 1, \dots, n_j)$ and $l_k^{(j)} = l_{n_j-k}^{(j)}$ $(k = 0, \dots, n_j)$. We may therefore remove the upper index for the masses and lengths of the threads, as well as the upper index for the corresponding Cauer-Fry polynomials. However, we will use n_1 for the number of beads on any of the edges. Then the total number of beads on the graph is $n = gn_1$. Throughout this section, we will consider the Neumann problem, i.e., all vertices of G are freely moving.

5.7.1 The direct problem

Because of the special structure of the Stieltjes strings on the edges, the characteristic polynomial can be described in terms of Cauer-Fry polynomials for the common Stieltjes string and the (weighted) adjacency matrix of the graph. We are going to consider first generic $z \in \mathbb{C}$ with $R_{2n_1}(z,0) \neq 0$. Then in (5.2.11) we replace U with the vector $\widetilde{U}$ whose component are defined by $\widetilde{u}_{2j-1} = u_{2j-1}$ and $\widetilde{u}_{2j} = \mathcal{L}^{\mathrm{c}}_{2j}(z)U$ for $j = 1, \dots, g$. In this notation we suppress the dependence of $\widetilde{U}$ on z since it is not important for what follows. We further define $\widetilde{\Phi}(z)$ by

$$\widetilde{\Phi}(z)\widetilde{U} = \Phi(z)U.$$

For $j = 1, \dots, g$ it follows from the definition of $\mathcal{L}^{\mathrm{c}}_{2j}$ in (5.2.5) that

$$\begin{pmatrix} \widetilde{u}_{2j-1} \\ \widetilde{u}_{2j} \end{pmatrix} = \begin{pmatrix} 1 & 0 \\ R_{2n_1}(z,1) & R_{2n_1}(z,0) \end{pmatrix} \begin{pmatrix} u_{2j-1} \\ u_{2j} \end{pmatrix},$$

which shows that $\widetilde{\Phi}(z)$ is well defined and that

$$\phi(z) = \det \Phi(z) = R^g_{2n_1}(z,0) \det \widetilde{\Phi}(z). \tag{5.7.1}$$

Next we will associate with $\widetilde{\Phi}(z)$ a matrix of smaller size but with the same determinant. To this end we introduce for each vertex v of the graph G the unique increasing surjective function $\tau_v : \{1, \dots, d(v)\} \to \widetilde{I}(v)$. For any interior vertex v of G and $j = 2, \dots, d(v)$ we subtract the row with index $\tau_v(1)$ from the row with index $\tau_v(j)$. In view of (5.2.9) and (5.2.10) the new row is represented by

$$\mathcal{L}_{\tau_v(j)}U - \mathcal{L}_{\tau_v(1)}(z)U = \mathcal{L}^{\mathrm{c}}_{\tau_v(j)}(z)U - \mathcal{L}^{\mathrm{c}}_{\tau_v(1)}(z)U = \widetilde{u}_{\tau_v(j)} - \widetilde{u}_{\tau_v(1)} =: \widehat{u}_{\tau_v(j)}.$$

We also define $\widehat{u}_{\tau_v(1)} := \widetilde{u}_{\tau_v(1)}$ for each $v \in V$. The row operations do not change the determinant of the matrix, and the transformation from $\widetilde{U}$ to $\widehat{U} := (\widehat{u}_j)_{j=1}^{2g}$ is given by a triangular matrix whose diagonal elements are 1. Hence the corresponding coefficient matrix with respect to $\widehat{U}$ has the same determinant as $\widetilde{\Phi}(z)$ and its rows with index $\tau_v(2), \dots, \tau_v(d(v))$ have 1 in the diagonal position and zeros everywhere else. Deleting all rows and columns with indices $\tau_v(j)$ with $j = 2, \dots, d(v)$ for interior vertices v leads to a matrix $\Phi^\dagger(z)$ with $\det \Phi^\dagger(z) = \det \widetilde{\Phi}(z)$.

To describe the entries of the matrix $\Phi^\dagger$, we note that $\Phi^\dagger(z)$ can be indexed by the vertices, with indices $\tau_v(1)$ inherited from the original matrix. It is convenient to write $u_v^\dagger := \widehat{u}_{\tau_v(1)}$ $(v \in V)$ and $U^\dagger = (u_v^\dagger)_{v\in V}$. Since we are interested in the determinant, any arrangement of the rows and columns gives the same result, and therefore the order is irrelevant. For vertices $v, w \in V$ let $\Phi^\dagger(v,w)$ denote the element of $\Phi^\dagger$ in row v and column w. Since the above row operations did not change the rows of $\Phi(z)$ with indices $\tau_v(1)$, we have to express $\mathcal{L}_{\tau_v(1)}(z)U$ in terms of $\widehat{U}$; then $\Phi^\dagger(v,w)$ is its coefficient of $\widehat{u}_{\tau_w(1)} = u_w^\dagger$.

We need some preparation. From

$$u_{2j} = \frac{1}{R_{2n_1}(z,0)}\left(\widetilde{u}_{2j} - R_{2n_1}(z,1)\widetilde{u}_{2j-1}\right) \quad (j = 1,\dots,g)$$

we conclude that

$$\begin{aligned}
&R_{2n_1-1}(z,1)u_{2j-1} + R_{2n_1-1}(z,0)u_{2j}\\
&\quad = \frac{1}{R_{2n_1}(z,0)}\left(R_{2n_1-1}(z,1)R_{2n_1}(z,0) - R_{2n_1-1}(z,0)R_{2n_1}(z,1)\right)\widetilde{u}_{2j-1}\\
&\qquad + \frac{R_{2n_1-1}(z,0)}{R_{2n_1}(z,0)}\widetilde{u}_{2j}\\
&\quad = -\frac{1}{l_0 R_{2n_1}(z,0)}\widetilde{u}_{2j-1} + \frac{R_{2n_1-1}(z,0)}{R_{2n_1}(z,0)}\widetilde{u}_{2j}
\end{aligned}$$

in view of the Lagrange identity (1.2.47). For each $v \in V$ it follows from (5.2.9) and (5.2.10) that

$$\begin{aligned}
\mathcal{L}_{\tau_v(1)}(z)U = \widetilde{u}_{\tau_v(1)} &- \frac{1}{d(v)}\sum_{r\in I^+(v)}\widetilde{u}_{2r} - \frac{1}{l_0 R_{2n_1}(z,0)}\sum_{r\in I^+(v)}\widetilde{u}_{2r-1}\\
&+ \frac{R_{2n_1-1}(z,0)}{R_{2n_1}(z,0)}\sum_{r\in I^+(v)}\widetilde{u}_{2r} - \frac{1}{d(v)}\sum_{r\in I^-(v)}\widetilde{u}_{2r-1}\\
&- \frac{1}{l_0 R_{2n_1}(z,0)}\sum_{r\in I^-(v)}\widetilde{u}_{2r} + \frac{R_{2n_1}(z,1)}{l_0 R_{2n_1}(z,0)}\sum_{r\in I^-(v)}\widetilde{u}_{2r-1}.
\end{aligned}$$

In view of $\widetilde{u}_{\tau_v(1)} = \widehat{u}_{\tau_v(1)}$ and $\widetilde{u}_{\tau_v(j)} = \widehat{u}_{\tau_v(j)} + \widehat{u}_{\tau_v(1)}$ $(j = 2,\dots,d(v))$ we find

$$\sum_{r\in I^+(v)}\widetilde{u}_{2r} + \sum_{r\in I^-(v)}\widetilde{u}_{2r-1} = \sum_{j=1}^{d(v)}\widetilde{u}_{\tau_v(j)} = d(v)u_v^\dagger + \sum_{j=2}^{d(v)}\widehat{u}_{\tau_v(j)}.$$

For $r \in I^+(v)$ we have $2r-1 \in \widetilde{I}(v_-(e_r))$, and therefore the contribution of $\widetilde{u}_{2r-1}$ to $U^\dagger$ is $u^\dagger_{v_-(e_r)}$, whereas for $r \in I^-(v)$ we have $2r \in \widetilde{I}(v_+(e_r))$, and therefore the contribution of $\widetilde{u}_{2r}$ to $U^\dagger$ is $u^\dagger_{v_+(e_r)}$. But the vertices $v_-(e_r)$ for $r \in I^+(v)$ and $v_+(e_r)$ for $r \in I^-(v)$ are exactly the vertices which are adjacent with v, and each adjacent vertex occurs exactly once since the graph is simple. In view of Lemma 2.2.1 it follows for $v, w \in V$ that

$$\Phi^\dagger(v,w) = \begin{cases} d(v)\dfrac{R_{2n_1}(z,1)}{l_0 R_{2n_1}(z,0)} & \text{if } w = v, \\ -\dfrac{1}{l_0 R_{2n_1}(z,0)} & \text{if } w \neq v \text{ and } w \text{ is adjacent to } v, \\ 0 & \text{if } w \neq v \text{ and } w \text{ is not adjacent to } v. \end{cases}$$

Putting $\zeta := R_{2n_1}(z,1)$, $T := \operatorname{diag}\{d(v) : v \in V\}$ and denoting by A the adjacency matrix of G, it follows that

$$\Phi^\dagger(z) = \frac{1}{l_0 R_{2n_1}(z,0)}(\zeta T - A). \tag{5.7.2}$$

With the weighted adjacency matrix $\widetilde{A} = T^{-\frac{1}{2}} A T^{-\frac{1}{2}}$ the following result is an immediate consequence of (5.7.1) and (5.7.2).

Proposition 5.7.1. *The characteristic polynomial of the Neumann problem on the graph G of identical symmetric Stieltjes strings satisfies*

$$\phi(z) = l_0^{-p} \prod_{v\in V} d(v) R_{2n_1}^{g-p}(z,0) \det(R_{2n_1}(z,1)I_p - \widetilde{A}).$$

The following two theorems now immediately follow from Lemmas C.5.4 and C.5.5, where we observe that A is irreducible since G is connected.

Theorem 5.7.2. *The characteristic polynomial of the Neumann problem on the graph G of identical symmetric Stieltjes strings is of the form*

$$\phi(z) = R_{2n_1}^{g-p}(z,0)(R_{2n_1}(z,0)-1)\widetilde{P}_{p-1}(R_{2n_1}(z,1)),$$

where $\widetilde{P}_{p-1}$ is a polynomial of degree $p-1$ with $\tilde{P}_{p-1}(1) \neq 0$.

Theorem 5.7.3. *The characteristic polynomial of the Neumann problem on the graph G of identical symmetric Stieltjes strings is of the form*

$$\phi(z) = R_{2n_1}^{g-p}(z,0)(R_{2n_1}^2(z,1)-1)R_{2n_1}^m(z,1)\widehat{P}_{\frac{p-m}{2}-1}(R_{2n_1}^2(z,1)),$$

where $m \in \mathbb{N}_0$, $p - m \in \mathbb{N}$ is an even number, $\widehat{P}_{\frac{p-m}{2}-1}$ is a polynomial of degree $\frac{p-m}{2} - 1$ with $\widehat{P}_{\frac{p-m}{2}-1}(1) \neq 0$, if and only if the graph is bipartite.

For the following results we recall the notation ν_k and ξ_k $(k = 1, \dots, n_1)$ and the properties of the zeros of $R_{2n_1}(\cdot, 0)$ and $R_{2n_1}(\cdot, 1)$ from Section 1.2.

Proposition 5.7.4. 1. *Under the assumptions of this section, let $a \in [-1,1]$. Then the zeros of the polynomial $R_{2n_1}(\cdot,1) - a$ are real and nonnegative and will be denoted by $\xi_{1,a} \leq \xi_{2,a} \leq \cdots \leq \xi_{n_1,a}$.*

2. *The zeros are simple for* $a \in (-1,1)$, $\xi_{1,a} = 0$ *if and only if* $a = 1$, *and* $\nu_{n_1} = \xi_{n_1,(-1)^{n_1}}$.

3. *Let* $-1 \le a < b \le 1$ *and* $j \in \{1,\dots,n_1\}$. *Then* $\xi_{j,a} < \xi_{j,b}$ *when* j *is even and* $\xi_{j,a} > \xi_{j,b}$ *when* j *is odd.*

4. *One has* $\nu_{2j-1} \in \{\xi_{2j-1,-1}\xi_{2j,-1}\}$ *for* $j=1,\dots,[\frac{n_1+1}{2}]$ *and* $\nu_{2j} \in \{\xi_{2j,1}, \xi_{2j+1,1}\}$ *for* $j = 1,\dots,[\frac{n_1}{2}]$.

Proof. For convenience, we put $\nu_0 := 0$. Then $R_{2n_1}(\nu_0, 1) = 1$ by (1.2.17) and $R^2_{2n_1}(\nu_j, 1) = 1$ $(j = 1,\dots,n_1)$ by (2.2.2). Since the zeros $\xi_j = \xi_{j,0}$ $(j = 1,\dots,n_j)$ are simple, we have that $(-1)^j R_{2n_1}(z, 1) > 0$ for $z \in (\xi_j, \xi_{j+1})$ $(j = 1,\dots,n_1-1)$. The interlacing property (1.2.37) gives that $R_{2n_1}(\nu_j, 1) = (-1)^j$ $(j = 1,\dots,n_1)$ and that $R_{2n_1}(\cdot, 1) - 1$ as well as $R_{2n_1}(\cdot, 1) + 1$ have an even number of zeros in $(\xi_j, \xi_j + 1)$ $(j = 1,\dots,n_1 - 1)$.

For $j = 1,\dots,[\frac{n_1}{2}]$ it follows that $R_{2n_1}(\cdot, 1) + 1$ has at least 2 zeros, counted with multiplicity, in (ξ_{2j-1}, ξ_{2j}). But this accounts for at least $n_1 - 1$ zeros of the polynomial $R_{2n_1}(\cdot, 1)+1$, which means that $R_{2n_1}(\cdot, 1)+1$ has exactly two zeros in (ξ_{2j-1}, ξ_{2j}) $(j = 1,\dots,[\frac{n_1}{2}])$ and one more zero $\nu_{n_1} \in (\xi_{n_1}, \infty)$ if n_1 is odd. This means that all zeros of $R_{2n_1}(\cdot, 1) + 1$ are real, and with the indicated indexing

$$\xi_{2j-1} < \xi_{2j-1,-1} \le \nu_{2j-1} \le \xi_{2j,-1} < \xi_{2j} \quad (j = 1,\dots,[\tfrac{n_j}{2}]),$$

which proves part 1 for $a = -1$ and the first statement of part 4.

Similarly, it follows for $j = 1,\dots,[\frac{n_1-1}{2}]$ that $R_{2n_1}(\cdot, 1) - 1$ has at least 2 zeros, counted with multiplicity, in (ξ_{2j}, ξ_{2j+1}), one zero at 0, and one zero at $\nu_{n_1} \in (\xi_{n_1}, \infty)$ if n_1 is even. But this already accounts for all zeros of the polynomial $R_{2n_1}(\cdot, 1) - 1$, so that $\xi_{1,1} = 0$, $\xi_{n_1,1} = \nu_{n_1}$ if n_1, and

$$\xi_{2j} < \xi_{2j,1} \le \nu_{2j} \le \xi_{2j+1,1} < \xi_{2j+1} \quad (j = 1,\dots,[\tfrac{n_j-1}{2}]).$$

This proves part 1 for $a = 1$ and completes the proof of part 4.

Now let $a \in (-1, 1)$. Then $\xi_{j,(-1)^{j-1}} < \xi_j < \xi_{j,(-1)^j}$ $(j = 1,\dots,n_j)$ shows that $R_{2n_1}(\cdot, 1) - a$ has exactly one zero in each of these n_1 intervals. In particular, $R_{2n_1}(\cdot, 1)$ is monotonic on these intervals, which proves part 3 and completes the proof of parts 1 and 2. □

Theorem 5.7.5. *Let* $(\lambda_k)_{k=1}^{gn_1}$ *be the nondecreasing sequence of the characteristic values of the Neumann problem on a simple connected metric graph G with cyclomatic number* $\mu \ge 2$ *and with the same symmetric distribution of beads on each edge. Then* $\lambda_1 = 0$ *and* $\lambda_{gj} = \nu_j$ *for all* $j = 1,\dots,n_1$.

Proof. Let $p' = p-2$ if G is bipartite and $p' = p-1$ otherwise. In view of Lemmas C.5.4 and C.5.5 the weighted adjacency matrix $\widetilde{A}$ has exactly p' eigenvalues in $(-1, 1)$, counted with multiplicity, and therefore Proposition 5.7.4 shows that ϕ has p' zeros, counted with multiplicity, in each of the intervals (ν_{j-1}, ν_j) $(j = 1,\dots,n_1)$; recall that $\nu_0 = 0$. By Theorems 5.7.2 and 5.7.3, Proposition 5.7.4

and Lemmas C.5.4, C.5.5 and C.2.3 the remaining zeros are ν_j with multiplicity $\mu - 1 > 0$, $\xi_{j,1}$, and $\xi_{j,-1}$ in case of a bipartite graph, $(j = 1, \dots, n_1)$. It is also clear that $\lambda_1 \geq 0$ and that $\xi_{1,1} = 0$ gives $\lambda_1 = 0$. In view of Proposition 5.7.4 we can therefore allocate $\xi_{j,1}$ to the interval $[\nu_{j-1}, \nu_j]$ for $j = 1, \dots, n_j$, and, in case of a bipartite graph, we also allocate $\xi_{j,-1}$ to the interval $[\nu_{j-1}, \nu_j]$ for $j = 1, \dots, n_j$. In this way, each of the intervals $[\nu_{j-1}, \nu_j]$ $(j = 1, \dots, n_1)$ has been allocated p zeros of ϕ. The $g - p$-fold zeros ν_j are allocated to $[\nu_{j-1}, \nu_j]$ $(j = 1, \dots, n_1)$. Thus we have allocated g zeros of ϕ to each of the intervals, and ν_j is the biggest of them for $j = 1, \dots, n_1$; this means that $\lambda_{gj} = \nu_j$ $(j = 1, \dots, n_1)$. □

5.7.2 The inverse problem

Theorem 5.7.6. *Let G be a simple connected metric graph with the same edges and with cyclomatic number $\mu \geq 2$. Assume that each edge has the same distribution of beads. Then the characteristic values of the Neumann problem on G and the length of an edge uniquely determine the distribution of beads.*

Proof. By Theorem 5.7.5, the characteristic values of the Dirichlet-Dirichlet problem on an edge of G are uniquely determined by the characteristic values of the Neumann problem on G. An application of Theorem 2.2.3 completes the proof. □

Remark 5.7.7. 1. In addition to the string data on the edges, also some information on the graph can be recovered from the spectrum of the Neumann problem on the graph G. If it is known that $\mu \geq 1$ but the actual value is not known, and if it is known that the graph is bipartite, then $\lambda_{gn_1} = \nu_{n_1}$ is a characteristic value of multiplicity $g - p + 1 = \mu$, and therefore μ can be recovered from the multiplicity of the largest characteristic value of the Neumann problem on G. The same is true when G is not bipartite but n_1 is even. If G is not bipartite, n_1 is odd and $\mu \geq 2$, then the multiplicity of λ_{gn_1} is $g - p = \mu - 1$, so that also in this case μ can be recovered from the multiplicity of the largest eigenvalue of the Neumann problem on G.

2. As a particular case consider simple connected graphs G with $\mu \geq 2$ and with a minimal number of vertices. Hence G has at least two independent cycles. Since G is simple, any cycle has at least 3 vertices, and any two independent cycles must have at least 4 vertices. Hence the minimal number of vertices is $p = 4$. On the other hand, since G is simple, the maximal number of edges is $\binom{4}{2} = 6$. Therefore the cases $g = 5$ and $g = 6$ are possible, which gives the diamond graph  $(g = 5, \mu = 2)$ or the tetrahedral graph  $(g = 6, \mu = 3)$. Observe that both graphs are nonbipartite since they contain cycles with three edges (see Lemma C.5.3). Hence the characteristic values, $p = 4$ and $\mu \geq 2$ uniquely determine the graph by part 1 of this remark, and then Theorem 5.7.6 shows that the distribution of the beads is also unique.

3. If we know two of the numbers g, p, μ or two of the numbers p, μ, n_1 and if the characteristic values $(\lambda_k)_{k=1}^{n}$ of the Neumann problem on G are known,

then all four of g, p, μ, n_1 can be found from $g - p = \mu - 1$ and $gn_1 = n$. In the case $\mu \geq 2$ the distribution of the beads is uniquely determined by the characteristic values, and therefore $R_{2n_1}(\cdot, 1)$ is known, and so are

$$R_{2n_1}(\alpha_k, 1) \quad (k = 1, \ldots, p).$$

From the results in Subsection 5.7.1 we know that these p values are the eigenvalues of the weighted adjacency matrix $\widetilde{A}$. Now it remains to investigate for which graphs with g edges and p vertices the matrix $\widetilde{A}$ has those eigenvalues.

4. We could even consider as little information as possible. That is, we may even consider an arbitrary simple connected graph (including $g = 1$) and the characteristic values $(\lambda_k)_{k=1}^n$ with $n \in \mathbb{N}$. For $n = 1$ this is the string problem with just one bead, which has as unique solution. For $n > 1$, we consider all pairs (g, n_1) with $gn_1 = n$ and then all simple connected graphs with p vertices, where $\frac{1}{2}(1+\sqrt{8g+1}) \leq p \leq g+1$. Then the results from subsection 5.7.1 may give information about the shape of the graph. As a trivial example we can state that if n is a prime number and not all characteristic values are simple, then the graph has n edges, and each edge carries exactly one bead, positioned at the midpoint of the edge.

5.8 Notes

Inverse spectral and scattering problems on graphs is a rapidly developing branch of analysis. Usually the Sturm-Liouville or the Dirac equation or the string equation is considered on the edges of a graph subject to matching and boundary conditions at the vertices.

The tradition of using Dirichlet and Neumann spectra to solve an inverse problem on a graph (see, e.g., [27], [143]) came from the classical results of Borg [15], Levitan and Gasymov [89] and Marchenko [92] for inverse spectral problems on a finite interval. However, in case of graphs it is possible (see, e.g., [86]) to introduce analogues of Neumann and Dirichlet conditions at an interior vertex chosen as the root. Analogues of Theorem 5.4.1 for Sturm-Liouville problem on a graph were obtained in [129] and [86].

To describe the spectrum of a boundary value problem on a metric graph we use characteristic polynomials, i.e., determinants of characteristic matrix functions defined by (5.2.11). This is similar to characteristic functions of Sturm-Liouville problems on graphs (see, e. g. [118]).

A result similar to Theorem 5.5.2 for Sturm-Liouville problems on graphs was obtained in [110], [111].

The condition that the number of beads on each edge be at least 3 can be explained as follows. If it is less than 3 then in case of a loop there are one or two eigenvalues. Since the first eigenvalue is simple (see [53] for the case of Sturm-Liouville problem), the second one is also simple and maximal multiplicity is 1.

Analogues of Theorems 5.6.12 and 5.6.13 for Sturm-Liouville problems were obtained in [91] and [76].

Absence of beads at the interior vertices of a graph makes the spectral problem a bit more complicated than in [62], [88], [101] because instead of the equation $(\lambda I - L)Y = 0$, where L is the discrete Laplacian, we have the equation $(\lambda M - L)Y = 0$, where $M \geq 0$ is a diagonal matrix which has some entries zero when not all interior vertices carry beads.

Theorem 5.6.13 and a particular case of Theorem 5.6.11 were obtained in [14].

Chapter 6

Sturm-Liouville problems on graphs

6.1 Selfadjoint differential operators

Let $\mathcal{I} = (a, b)$ or $\mathcal{I} = \mathbb{R}$. First we recall the definition of the Lebesgue spaces $L_p(\mathcal{I})$ for $1 \le p \le \infty$. For Lebesgue measurable functions $f, h : \mathcal{I} \to \mathbb{C}$ we write $f = h$ ($f \le h$, etc.) if $f(x) = h(x)$ ($f(x) \le h(x)$, etc.) for almost all $x \in \mathcal{I}$, i.e., if there is a Lebesgue measurable set $X \subset \mathcal{I}$ such that $\mathcal{I} \setminus X$ has Lebesgue measure zero and such that $f(x) = h(x)$ ($f(x) \le h(x)$, etc.) for all $x \in X$. The function f is called essentially bounded if there is such a Lebesgue measurable set X such that f is bounded on X. The relation $f = h$ in the above sense defines equivalence classes of Lebesgue measurable functions, and for $1 \le p < \infty$, $L_p(\mathcal{I})$ is defined as the set of equivalence classes of measurable functions f on $\mathcal{I}$ for which

$$\int_{\mathcal{I}} |f(x)|^p \, dx < \infty,$$

whereas $L_\infty(\mathcal{I})$ is the set of equivalence classes of essentially bounded functions on $\mathcal{I}$. As is customary, we will identify equivalence classes with any of the functions representing it. In particular, when such an equivalence class contains a continuous functions, then this continuous function is unique, and we will identify the equivalence class with its continuous representative. The sesquilinear form

$$(f, g) := \int_{\mathcal{I}} f(x)\overline{g(x)} \, dx, \quad f, g \in L_2(\mathcal{I}),$$

defines an inner product on $L_2(\mathcal{I})$ which makes $L_2(\mathcal{I})$ a Hilbert space.

We recall some basic properties of Sturm-Liouville operators. We consider the Sturm-Liouville expression

$$\ell y = -y'' + qy \tag{6.1.1}$$

M. Möller, V. Pivovarchik, *Direct and Inverse Finite-Dimensional Spectral Problems on Graphs*, Operator Theory: Advances and Applications 283, https://doi.org/10.1007/978-3-030-60484-4_6

on an interval $[a, b]$, where $q \in L_1(a, b)$ is a real valued function. The domain $D(\ell)$ of this differential expression is the set of all $y \in C^1[a, b]$ such that y' is absolutely continuous. Then $y'' \in L_1(a, b)$ and therefore $\ell y \in L_1(a, b)$ is well-defined. Hence ℓ with domain $D(\ell)$ can be considered as an operator from $L_2(a, b)$ to $L_1(a, b)$. Let $L_{\max}$ be the maximal differential operator in $L_2(a, b)$ associated with ℓ. That is,

$$D(L_{\max}) = \{y \in D(\ell) : \ell y \in L_2(a, b)\}, \quad L_{\max} y = \ell y, \quad (y \in D(L_{\max})). \tag{6.1.2}$$

Let $l > 0$, $g \in \mathbb{N}$ and for $j = 1, \dots, g$ let ℓ_j be formally selfadjoint differential expressions on the interval $[0, l]$ as defined in (6.1.1), i.e.,

$$\ell_j y_j = -y_j'' + q_j y_j, \tag{6.1.3}$$

where $q_j \in L_1(0, l)$ $(j = 1, \dots, g)$ are real valued functions. We are going to use the notation

$$y = \begin{pmatrix} y_1 \\ \vdots \\ y_g \end{pmatrix}$$

for elements in the Hilbert space

$$H = \bigoplus_{j=1}^{g} L_2(0, l).$$

Furthermore, when $y_j \in D(\ell_j)$ $(j = 1, \dots, g)$ we use the notation

$$Y = \begin{pmatrix} y \\ y' \end{pmatrix}, \quad Y_R = \begin{pmatrix} Y(0) \\ Y(l) \end{pmatrix}.$$

It is clear that $Y_R \in \mathbb{C}^{4g}$. Finally, let U be a $(2g) \times (4g)$ matrix of rank $2g$. Then the operator $\mathcal{L}_U$ in H is defined by

$$\mathcal{L}_U y = \begin{pmatrix} \ell_1 y_1 \\ \vdots \\ \ell_g y_g \end{pmatrix}, \quad D(\mathcal{L}_U) = \left\{ y \in \bigoplus_{j=1}^{g} D(L_{j,\max}) : U Y_R = 0 \right\}. \tag{6.1.4}$$

From the results in [98, Section 2] it follows that

Proposition 6.1.1. *The operator $\mathcal{L}_U$ is selfadjoint if and only if $U J_1 U^* = 0$, where*

$$J_1 = \begin{pmatrix} J_0 & 0 \\ 0 & -J_0 \end{pmatrix}, \quad J_0 = \begin{pmatrix} 0 & -I_g \\ I_g & 0 \end{pmatrix}.$$

Remark 6.1.2. Writing $U = (U_1\, U_2\, U_3\, U_4)$ with $(2g) \times g$ matrices U_j $(j = 1, 2, 3, 4)$, it is clear that the operator $\mathcal{L}_U$ is self-adjoint if and only if

$$U_1 U_2^* - U_2 U_1^* - U_3 U_4^* + U_4 U_3^* = 0.$$

6.2 Selfadjoint differential operators on graphs

Let G be a metric digraph, where it is assumed that all g edges of G have the same length $l > 0$. Thus each edge e_j $(j = 1, \ldots, g)$ of G will be identified with the interval $[0, l]$, and on each edge, the local coordinate x increases in the direction of the edge. On each edge of the graph we consider a Sturm-Liouville problem (6.1.3).

Now we are ready to define the operator $\mathcal{L}$. With the notation from the previous section, its domain $D(\mathcal{L})$ is the set of vector functions $y = (y_j)_{j=1}^g$ with $y_j \in D(L_{j,\max})$ subject to the following conditions at the vertices of G. At pendant vertices, Dirichlet or Neumann conditions are imposed:

$$y_j(0) = 0 \text{ or } y_j'(0) = 0 \text{ if } I^-(v) = \{j\} \text{ and } v \in V^{\text{pen}}, \tag{6.2.1}$$

$$y_j(l) = 0 \text{ or } y_j'(l) = 0 \text{ if } I^+(v) = \{j\} \text{ and } v \in V^{\text{pen}}. \tag{6.2.2}$$

At each interior vertex, a continuity condition

$$\#(\{y_j(0) : j \in I^-(v)\} \cup \{y_j(l) : j \in I^+(v)\}) = 1 \quad (v \in V^{\text{int}}), \tag{6.2.3}$$

as well as the Kirchhoff condition

$$\sum_{j \in I^+(v)} y_j'(l) = \sum_{j \in I^-(v)} y_j'(0) \quad (v \in V^{\text{int}}) \tag{6.2.4}$$

are imposed. The conditions (6.2.1)–(6.2.4) can be written in the form $UY_R = 0$, where U is defined as follows (the indexing of the rows is arbitrary). We use the notation $Y_R = (y_{R,j})_{j=1}^{4g}$ and observe that

$$y_{R,j} = y_j(0),\ y_{R,g+j} = y_j'(0),\ y_{R,2g+j} = y_j(l),\ y_{R,3g+j} = y_j'(l), \quad (j = 1, \ldots, g).$$

Each of the conditions (6.2.1) and (6.2.2) gives rise to a component of UY_R of the form $Y_{R,\alpha_v g+j}$ with $\alpha_v \in \{0, 1, 2, 3\}$. Condition (6.2.3) for a particular $v \in V^{\text{int}}$ involves the components $Y_{R,j}$ with $j \in I_0^\diamond(v) := I^-(v) \cup \{2g + k : k \in I^+(v)\}$. Choosing any $j_1(v) \in I_0^\diamond(v)$ it is clear that the continuity condition at v gives rises to $d(v) - 1$ components of UY_R of the form $Y_{R,j} - Y_{R,j_1(v)}$ $(j \in I_0^\diamond(v),\ j \neq j_1(v))$. Finally, each condition (6.2.4) gives rise to a component of UY_R of the form

$$\sum_{j \in I^+(v)} Y_{R,3g+j} - \sum_{j \in I^-(v)} Y_{R,g+j} \quad (v \in V^{\text{int}}). \tag{6.2.5}$$

Theorem 6.2.1. *The operator $\mathcal{L}$ is selfadjoint.*

Proof. Since the sum of the degrees $d(v)$ $(v \in V^G)$ is $2g$, the matrix U defined above is a $(2g) \times (4g)$ matrix. Since each of the terms $Y_{R,k}$ related to any of the conditions (6.2.1) and (6.2.2) as well as $Y_{R,j}$ $(j \in I_0^\diamond(v),\ j \neq j_1(v),\ v \in V^{\text{int}})$ and any $Y_{R,k}$ occurring in (6.2.5) is present in exactly one component of UY_R it is clear that the rank of U is $2g$.

According to Remark 6.1.2 the proof will be complete if we have shown that

$$W := U_1U_2^* - U_2U_1^* - U_3U_4^* + U_4U_3^* = 0.$$

Using the notation $W = (w_{r,s})_{r,s=1}^{2g}$, $U_jU_k^* = (w_{j,k,r,s})_{r,s=1}^{2g}$, we have

$$w_{r,s} = w_{1,2,r,s} - w_{1,2,s,r} - w_{3,4,r,s} + w_{3,4,s,r} \quad (r, s = 1, \dots, 2g)$$

and, for $(j, k) = (1, 2)$ or $(j, k) = (3, 4)$,

$$w_{j,k,r,s} = \sum_{t=1}^{g} u_{r,(j-1)g+t}u_{s,(k-1)g+t} \quad (r, s = 1, \dots, 2g).$$

Hence $w_{j,k,r,s} \neq 0$ is only possible for pairs of boundary conditions which contain $y_j(0)$ and $y_j'(0)$ or $y_j(l)$ and $y_j'(l)$. So let r be a row index which corresponds to a component $Y_{R,j} - Y_{R,j_1(v)}$ of UY_R, and let s be a row index which corresponds to a component of the form (6.2.5), with the same $v \in V^{\text{int}}$. Without loss of generality we may assume that $j_1(v) \in I^-(v)$; the case $j_1(v) - 2g \in I^+(v)$ is similar. If also $j \in I^-(v)$, then all nonzero entries in the r-th row of U are entries of the matrix U_1, and it is therefore clear that $w_{r,s} = w_{1,2,r,s}$ in this case. But then $u_{r,j} = 1$, $u_{r,j_1(v)} = -1$, $u_{r,t} = 0$ for $t = 1, \dots, g$ with $t \neq j$ and $t \neq j_1(v)$, $u_{s,g+j} = -1$, $u_{s,g+j_1(v)} = -1$ shows that $w_{r,s} = w_{1,2,r,s} = -1 + 1 = 0$. Similarly, if $j - 2g \in I^+(v)$, then $u_{r,j} = 1$, $u_{r,j_1(v)} = -1$, $u_{r,t} = 0$ for $t = 1, \dots, 4g$ with $t \neq j$ and $t \neq j_1(v)$, $u_{s,g+j} = 1$, $u_{s,g+j_1(v)} = -1$ shows that $w_{1,2,r,s} = 1$ and $w_{3,4,r,s} = 1$. This shows that $w_{r,s} = w_{1,2,r,s} - w_{3,4,r,s} = 1 - 1 = 0$. □

Let $v_1, \dots, v_p$ be an enumeration of the vertices of G and put

$$T = \operatorname{diag}\{d(v_1), \dots, d(v_p)\}.$$

Recall that a graph is called *k-regular* if $d(v) = k$ for all $v \in V^G$. Thus $T = kI_p$ for a k-regular graph G. For a simple graph, the matrix

$$A = (A_{i,j})_{i,j=1}^{p}, \quad A_{i,j} = \begin{cases} 1 & \text{if } v_i \text{ and } v_j \text{ are adjacent,} \\ 0 & \text{otherwise,} \end{cases}$$

is called the *adjacency* matrix, and the matrix

$$\widetilde{A} = T^{-1/2}AT^{-1/2} \tag{6.2.6}$$

is called the *weighted adjacency* matrix, which can be written as

$$\widetilde{A} = \frac{1}{k}A$$

if G is k-regular.

6.3 Graphs with Neumann conditions at pendant vertices

In this section we assume the potentials on the g edges of the graph G are the same, i.e., $q_1 = \cdots = q_g = q$. Furthermore, it will be assumed that q is symmetric with respect to the midpoint of the interval $(0, l)$, i.e.,

$$q(l - x) = q(x) \quad (x \in (0, l)).$$

Denote by $s(\lambda, \cdot)$ and $c(\lambda, \cdot)$ the solutions of

$$-y'' + qy = \lambda^2 y \quad (\lambda \in \mathbb{C}) \tag{6.3.1}$$

on $[0, l]$ which satisfy the initial conditions $s(\lambda, 0) = 0$, $s'(\lambda, 0) = 1$, $c(\lambda, 0) = 1$, $c'(\lambda, 0) = 0$. Note that for $q = 0$, the functions s and c are simply (scaled) sine and cosine functions, respectively.

Proposition 6.3.1. *For all $\lambda \in \mathbb{C}$, the functions s and c satisfy*

$$s(\lambda, l - x) = s(\lambda, l)c(\lambda, x) - s'(\lambda, l)s(\lambda, x) \quad (x \in [0, l]), \tag{6.3.2}$$

$$c(\lambda, l - x) = c(\lambda, l)c(\lambda, x) - c'(\lambda, l)s(\lambda, x) \quad (x \in [0, l]), \tag{6.3.3}$$

$$s'(\lambda, l) = c(\lambda, l), \tag{6.3.4}$$

$$c'(\lambda, l)s(\lambda, l) = c^2(\lambda, l) - 1. \tag{6.3.5}$$

Proof. It is easy to see that due to the symmetry of the potential q also the functions $x \mapsto s(\lambda, l - x)$ and $x \mapsto c(\lambda, l - x)$ are solutions of (6.3.1). Hence these functions can be expressed as linear combinations of $c(\lambda, \cdot)$ and $s(\lambda, \cdot)$. Since in (6.3.2) as well as in (6.3.3) the functions and their derivatives on the left-hand side and on the right-hand side have the same values at $x = 0$, the identities (6.3.2) and (6.3.3) follow.

When $x = l$, (6.3.2) gives $s(\lambda, l)(s'(\lambda, l) - c(\lambda, l)) = 0$. Since the spectrum of the Dirichlet problem (6.3.1), $y(0) = 0$, $y(l) = 0$ has discrete (real) spectrum, the analytic function $s(\cdot, l)$ is not the zero function, and it follows that the analytic function $s'(\cdot, l) - c(\cdot, l)$ is the zero function. Hence (6.3.4) is true. Taking $x = l$ in (6.3.3) gives (6.3.5). □

Theorem 6.3.2. *Let G be a connected simple graph with at least one edge and consider the operator $\mathcal{L}$ with Neumann conditions at all pendant vertices. Then $\lambda \in \mathbb{C}$ with $s(\sqrt{\lambda}, l) \neq 0$ is an eigenvalue of the operator $\mathcal{L}$ if and only if $c(\sqrt{\lambda}, l)$ is an eigenvalue of the weighted adjacency matrix $\widetilde{A}$, and the multiplicity of λ as eigenvalue of $\mathcal{L}$ equals the multiplicity of $c(\sqrt{\lambda}, l)$ as eigenvalue of the matrix $\widetilde{A}$.*

Proof. Since both $\mathcal{L}$ and $\widetilde{A}$ are selfadjoint, geometric multiplicities equal algebraic multiplicities, and we have to show that the null spaces of $\mathcal{L} - \lambda I$ and $\widetilde{A} - c(\sqrt{\lambda}, l)I$ have the same dimension. Given $f_{1,0}, \dots, f_{g,0} \in \mathbb{C}$ and $f_{1,l}, \dots, f_{g,l} \in \mathbb{C}$ it is well known and easy to verify that, for $j = 1, \dots, g$, the boundary value problem $\ell_j y_j = \lambda y_j$,

$$y_j(\lambda, 0) = f_{j,0}, \quad y_j(\lambda, l) = f_{j,l} \quad (\lambda \in \mathbb{C}) \tag{6.3.6}$$

has a unique solution if $s(\sqrt{\lambda}, l) \neq 0$ and that this solution is given by

$$y_j(\lambda, x) = \frac{f_{j,l} - f_{j,0}c(\sqrt{\lambda}, l)}{s(\sqrt{\lambda}, l)} s(\sqrt{\lambda}, x) + f_{j,0}c(\sqrt{\lambda}, x). \tag{6.3.7}$$

Hence y is a solution of $\mathcal{L}y = \lambda y$ if and only if $y = (y_j(\lambda, \cdot))_{j=1}^g$ is given by (6.3.7) and satisfies the conditions (6.2.1)–(6.2.4). These solutions are uniquely given by $f_{1,0}, \dots, f_{g,0}, f_{1,l}, \dots, f_{g,l}$. Taking (6.2.3) and (6.3.6) into account, these values can be uniquely expressed in terms of vectors $F = (f_i)_{i=1}^p$, where $f_{j,0} = f_i$ when $v_i = v_-(e_j)$ and $f_{j,l} = f_i$ when $v_i = v_+(e_j)$. We calculate

$$y_j'(\lambda, 0) = \frac{f_{j,l} - f_{j,0}c(\sqrt{\lambda}, l)}{s(\sqrt{\lambda}, l)}$$

and, observing (6.3.4) and (6.3.5),

$$\begin{aligned} y_j'(\lambda, l) &= \frac{f_{j,l} - f_{j,0}c(\sqrt{\lambda}, l)}{s(\sqrt{\lambda}, l)} s'(\sqrt{\lambda}, l) + f_{j,0}c'(\sqrt{\lambda}, l) \\ &= \frac{f_{j,l}s'(\sqrt{\lambda}, l) - f_{j,0}(c(\sqrt{\lambda}, l)s'(\sqrt{\lambda}, l) - c'(\sqrt{\lambda}, l)s(\sqrt{\lambda}, l))}{s(\sqrt{\lambda}, l)} \\ &= \frac{f_{j,l}c(\sqrt{\lambda}, l) - f_{j,0}}{s(\sqrt{\lambda}, l)}. \end{aligned}$$

Hence the conditions (6.2.1), (6.2.2) and (6.2.4) can be written as

$$f_{j,l} - f_i c(\sqrt{\lambda}, l) = 0 \quad (v_i \in V^{\text{pen}},\ I^-(v_i) = \{j\}), \tag{6.3.8}$$

$$f_{j,0} - f_i c(\sqrt{\lambda}, l) = 0 \quad (v_i \in V^{\text{pen}},\ I^+(v_i) = \{j\}), \tag{6.3.9}$$

$$\sum_{j \in I^+(v_i)} f_{j,0} + \sum_{j \in I^-(v_i)} f_{j,l} - d(v)c(\sqrt{\lambda}, l) f_i = 0 \quad (v \in V^{\text{int}}). \tag{6.3.10}$$

But these are exactly the p components of the vector equation

$$AF - c(\sqrt{\lambda}, l)TF = 0,$$

which is equivalent to

$$(\widetilde{A} - c(\sqrt{\lambda}, l))T^{\frac{1}{2}}F = 0.$$

Hence the theorem is proved. □

Recall that for convenience we call a complex number an eigenvalue of multiplicity 0 if it is no eigenvalue.

Theorem 6.3.3. *Let G be a connected graph with at least two edges and consider the operator $\mathcal{L}$ with (generalized) Neumann conditions at all vertices. Let $\lambda \in \mathbb{C}$ with $s(\sqrt{\lambda}, l) = 0$. Then $(c(\sqrt{\lambda}, l))^2 = 1$, and the following holds:*

1. *If $c'(\sqrt{\lambda}, l) = 0$ and G is bipartite or $c(\sqrt{\lambda}, l) = 1$, then λ is an eigenvalue of the operator $\mathcal{L}$ of multiplicity $g - p + 2$.*
2. *If $c'(\sqrt{\lambda}, l) \neq 0$ and G is bipartite or $c(\sqrt{\lambda}, l) = 1$, then λ is an eigenvalue of the operator $\mathcal{L}$ of multiplicity $g - p + 1$.*
3. *If G is not bipartite and $c(\sqrt{\lambda}, l) = -1$, then λ is an eigenvalue of the operator $\mathcal{L}$ of multiplicity $g - p$.*

Proof. Given $f_{1,0}, \dots, f_{g,0} \in \mathbb{C}$ and $f'_{1,0}, \dots, f'_{g,0} \in \mathbb{C}$ it is well known that, for $j = 1, \dots, g$, the initial value problem $\ell_j \widetilde{y}_j = \lambda \widetilde{y}_j$,

$$\widetilde{y}_j(\lambda, 0) = f_{j,0}, \quad \widetilde{y}'_j(\lambda, 0) = f'_{j,0} \quad (\lambda \in \mathbb{C}), \tag{6.3.11}$$

has the unique solution

$$\widetilde{y}_j(\lambda, x) = f_{j,0} c(\sqrt{\lambda}, x) + f'_{j,0} s(\sqrt{\lambda}, x). \tag{6.3.12}$$

Hence y is a solution of $\mathcal{L}y = \lambda y$ if and only if $y = (\widetilde{y}_j(\lambda, \cdot))_{j=1}^g$ is given by (6.3.12) and satisfies the conditions (6.2.1)–(6.2.4). These solutions are uniquely determined by $f_{1,0}, \dots, f_{g,0}, f'_{1,0}, \dots, f'_{g,0}$. We calculate

$$\widetilde{y}_j(\lambda, l) = f_{j,0} c(\sqrt{\lambda}, l) \tag{6.3.13}$$

and, taking (6.3.4) into account,

$$\widetilde{y}'_j(\lambda, l) = f_{j,0} c'(\sqrt{\lambda}, l) + f'_{j,0} s'(\sqrt{\lambda}, l) = f_{j,0} c'(\sqrt{\lambda}, l) + f'_{j,0} c(\sqrt{\lambda}, l). \tag{6.3.14}$$

The continuity condition (6.2.3) implies that $(f_{j,0})_{j=1}^g$ is uniquely given by $F = (f_i)_{i=1}^p$, where $f_{j,0} = f_i$ with $v_i = v_-(e_j)$. The property $(c(\sqrt{\lambda}, l))^2 = 1$ is obvious in view of (6.3.5). Putting $\varepsilon := -c(\sqrt{\lambda}, l)$ it follows that $\varepsilon = 1$ or $\varepsilon = -1$. Since G is connected and has at least one interior vertex, it follows in case $\varepsilon = -1$ from (6.3.11) and (6.3.13) that the continuity conditions (6.2.3) are satisfied if and only if $(f_i)_{i=1}^p$ is a multiple of $\mathbb{1}_p$, the vector in $\mathbb{C}^p$ with all components being 1. If $\varepsilon = 1$ and G is bipartite with decomposition G_1 and G_2 of the vertices, let $F^{\mathrm{b}} = (f_i^{\mathrm{b}})_{i=1}^p$ be defined by $f_i^{\mathrm{b}} = -1$ if $v_i \in G_1$ and $f_i^{\mathrm{b}} = 1$ if $v_i \in G_2$. In view of (6.3.11) and (6.3.13) the continuity conditions (6.2.3) are satisfied if and only if $(f_i)_{i=1}^g$ is a multiple of F^{b}. Finally, if $\varepsilon = 1$ and G is not bipartite, then the continuity conditions (6.2.3) are satisfied if and only if $(f_i)_{i=1}^g = 0$.

Let $F' = (f'_{j,0})_{j=1}^g$. The conditions (6.2.1), (6.2.2) and (6.2.4) can be written as

$$\sum_{j \in I^+(v_i)} f'_{j,0} + \sum_{j \in I^-(v_i)} \varepsilon f'_{j,0} = \varepsilon c'(\sqrt{\lambda}, l) \sum_{j \in I^+(v_i)} f_{j,0} \quad (i = 1, \dots, p). \tag{6.3.15}$$

In case $\varepsilon = -1$ we have seen that all $f_{j,0}$ ($j = 1, \ldots, g$) are equal. In case $\varepsilon = 1$ and G is bipartite, we have seen that $f_{j,0} = -f_i$ for $j \in I(v_i)$. Finally, when $\varepsilon = 1$ and G is not bipartite, then $f_{j,0} = 0$ for all $j = 1, \ldots, p$. Therefore (6.3.15) can be written as

$$\sum_{j\in I^+(v_i)} f'_{j,0} + \sum_{j\in I^-(v_i)} \varepsilon f'_{j,0} = -c'(\sqrt{\lambda}, l) d^+(v) f_i \quad (i = 1, \ldots, p). \tag{6.3.16}$$

Recalling the definition of the (signed) incidence matrix U_ε, the system of equations (6.3.16) has the matrix form

$$U_\varepsilon F' = -c'(\sqrt{\lambda}, l) T^+ F, \tag{6.3.17}$$

where $T^+ := \operatorname{diag}(d^+(v_1), \ldots, d^+(v_p))$. Hence the multiplicity of the eigenvalue λ of $\mathcal{L}$ is the dimension of all vectors (F, F') which satisfy (6.3.17) with F a multiple of $\mathbb{1}_p$ when $\varepsilon = -1$, with F a multiple of F^{b} when $\varepsilon = 1$ and G is bipartite, and with $F = 0$ when $\varepsilon = 1$ and G is not bipartite.

First consider the case that $c'(\sqrt{\lambda}, l) = 0$ or that $\varepsilon = 1$ and G is not bipartite. In this case, (6.3.17) is a homogeneous equation, and therefore

$$\dim N(\mathcal{L} - \lambda I) = \begin{cases} g - \operatorname{rank} U_\varepsilon + 1 & \text{if } \varepsilon = -1 \text{ or } G \text{ is bipartite}, \\ g - \operatorname{rank} U_\varepsilon & \text{if } \varepsilon = 1 \text{ and } G \text{ is not bipartite}. \end{cases}$$

By Proposition C.5.1 and Lemma C.5.5, the rank of U_ε is $p - 1$ when $\varepsilon = -1$ or G is bipartite, and the rank of U_ε is p when $\varepsilon = 1$ and G is not bipartite. This proves parts 1 and 3.

Now consider the case that $c'(\sqrt{\lambda}, l) \neq 0$. When $\varepsilon = -1$, then each column of U_ε has two nonzero elements, one is -1 and the other one is 1. Therefore $\mathbb{1}_p^{\mathsf{T}} U_\varepsilon = 0$, whereas $\mathbb{1}_p^{\mathsf{T}} T^+ \mathbb{1}_p = g > 0$. This shows that

$$\dim N(\mathcal{L} - \lambda I) = g - \operatorname{rank} U_\varepsilon = g - p + 1,$$

which proves part 2 in this case. Similarly, when $\varepsilon = 1$ and G is not bipartite, then each column of U_ε has two nonzero elements, which are both 1. If the corresponding row indices of these nonzero elements are r and s, then $v_r \in G_1$ and $v_s \in G_2$, or vice versa. Therefore $(F^{\mathsf{b}})^{\mathsf{T}} U_\varepsilon = 0$, whereas $(F^{\mathsf{b}})^{\mathsf{T}} T^+ F^{\mathsf{b}} = g > 0$. This shows that $\dim N(\mathcal{L} - \lambda I) = g - p + 1$ also in this case. □

Remark 6.3.4. In Theorem 6.3.3 we have considered $s(\sqrt{\lambda}, l) = 0$ and distinguished the cases $c'(\sqrt{\lambda}, l) = 0$ and $c'(\sqrt{\lambda}, l) \neq 0$. We will show that each of these cases can occur, both for $c(\sqrt{\lambda}, l) = 1$ and $c(\sqrt{\lambda}, l) = -1$. For simplicity, we will write λ instead of $\sqrt{\lambda}$.

First consider (6.3.1) on $[0, 1]$ with $q = 0$. For $\lambda \neq 0$ it is obvious that $s(\lambda, x) = \lambda^{-1} \sin(\lambda x)$, $s'(\lambda, x) = \cos(\lambda x)$, $c(\lambda, x) = \cos(\lambda x)$, $c'(\lambda, x) = -\lambda \sin(\lambda x)$. Clearly, $s(\lambda, 1) = 0$ implies $c'(\lambda, 1) = 0$.

Now consider the case that $l = 4$ and

$$q(x) = \begin{cases} 0, & x \in (0,1) \cup (3,4), \\ 1, & x \in (1,3). \end{cases}$$

The potential is clearly symmetric, and a straightforward calculation gives for $x \in [1,3]$ that

$$\begin{aligned} s(\lambda, x) = {} & \frac{\sqrt{\lambda^2-1}\sin\lambda\sin\sqrt{\lambda^2-1} + \lambda\cos\lambda\cos\sqrt{\lambda^2-1}}{\lambda\sqrt{\lambda^2-1}} \sin\left(\sqrt{\lambda^2-1}x\right) \\ & + \frac{\sqrt{\lambda^2-1}\sin\lambda\cos\sqrt{\lambda^2-1} - \lambda\cos\lambda\sin\sqrt{\lambda^2-1}}{\lambda\sqrt{\lambda^2-1}} \cos\left(\sqrt{\lambda^2-1}x\right), \\ c(\lambda, x) = {} & \frac{\sqrt{\lambda^2-1}\cos\lambda\cos\sqrt{\lambda^2-1} + \lambda\sin\lambda\sin\sqrt{\lambda^2-1}}{\sqrt{\lambda^2-1}} \cos\left(\sqrt{\lambda^2-1}x\right) \\ & + \frac{\sqrt{\lambda^2-1}\cos\lambda\sin\sqrt{\lambda^2-1} - \lambda\sin\lambda\cos\sqrt{\lambda^2-1}}{\sqrt{\lambda^2-1}} \sin\left(\sqrt{\lambda^2-1}x\right). \end{aligned}$$

This gives

$$\begin{pmatrix} s(\lambda,2) \\ c'(\lambda,2) \end{pmatrix} = \begin{pmatrix} \frac{1}{\sqrt{\lambda^2-1}} & \frac{1}{\lambda} \\ -\sqrt{\lambda^2-1} & -\lambda \end{pmatrix} \begin{pmatrix} \cos\lambda\sin\sqrt{\lambda^2-1} \\ \sin\lambda\cos\sqrt{\lambda^2-1} \end{pmatrix} \tag{6.3.18}$$

and

$$\begin{pmatrix} s'(\lambda,2) \\ c(\lambda,2) \end{pmatrix} = \begin{pmatrix} -\frac{\sqrt{\lambda^2-1}}{\lambda} & 1 \\ -\frac{\lambda}{\sqrt{\lambda^2-1}} & 1 \end{pmatrix} \begin{pmatrix} \sin\lambda\sin\sqrt{\lambda^2-1} \\ \cos\lambda\cos\sqrt{\lambda^2-1} \end{pmatrix}. \tag{6.3.19}$$

It is well known that $s(\cdot, 2)$ has infinitely many real zeros which tend to ∞, see, e.g., [96, Lemma 12.5.1]. Therefore choose any $\lambda \in (1,\infty)$ such that $s(\lambda,2) = 0$. With $x = 2$ in (6.3.2) it follows that

$$0 = s(\lambda,2) = s(\lambda,4)c(\lambda,2) - s'(\lambda,4)s(\lambda,2) = s(\lambda,4)c(\lambda,2).$$

But $c(\lambda,2) \neq 0$ since $s(\lambda,\cdot)$ and $c(\lambda,\cdot)$ are linearly independent solutions of (6.3.1). Hence $s(\lambda,4) = 0$. With $x = 2$ in (6.3.3) it follows that $c(\lambda,2) = c(\lambda,4)c(\lambda,2)$ and therefore $c(\lambda,4) = 1$. The matrix in (6.3.18) is invertible and hence in view of $0 < \lambda - \sqrt{\lambda^2-1} < 1$, at most one of the four numbers $\sin\lambda$, $\cos\lambda$, $\sin\sqrt{\lambda^2-1}$, $\cos\sqrt{\lambda^2-1}$ has the value 0. Therefore (6.3.18) shows that $c'(\lambda,2) \neq 0$. Differentiating (6.3.3) and then putting $x = 2$ in gives

$$-c'(\lambda,2) = c(\lambda,4)c'(\lambda,2) - c'(\lambda,4)s'(\lambda,2) = c'(\lambda,2) - c'(\lambda,4)s'(\lambda,2).$$

This shows that $c'(\lambda,4)s'(\lambda,2) = 2c'(\lambda,2) \neq 0$, and $c'(\lambda,4) \neq 0$ follows.

Between any two zeros of $s(\cdot,2)$ there is one zero of $s'(\cdot,2)$ (see Proposition 6.3.5 below and (6.3.4)), and thus we may choose $\lambda \in (1,\infty)$ such that $s'(\lambda,2)=0$. Differentiating (6.3.2) and putting $x=2$ gives

$$0 = -s'(\lambda,2) = s(\lambda,4)c'(\lambda,2) - s'(\lambda,4)s'(\lambda,2) = s(\lambda,4)c'(\lambda,2).$$

But $c'(\lambda,2) \neq 0$ since $s(\lambda,\cdot)$ and $c(\lambda,\cdot)$ are linearly independent solutions of (6.3.1). Hence $s(\lambda,4)=0$. Differentiating (6.3.3) and putting $x=2$ it follows that $-c'(\lambda,2) = c(\lambda,4)c'(\lambda,2)$ and therefore $c(\lambda,4)=-1$. A similar reasoning as in the previous paragraph shows that (6.3.19) gives $c(\lambda,2)\neq 0$. With $x=2$ in (6.3.3) one obtains

$$c(\lambda,2) = c(\lambda,4)c(\lambda,2) - c'(\lambda,4)s(\lambda,2) = -c(\lambda,2) - c'(\lambda,4)s(\lambda,2).$$

This shows that $c'(\lambda,4)s(\lambda,2) = -2c(\lambda,2) \neq 0$, and $c'(\lambda,4)\neq 0$ follows.

The following result is well known. Its statement and proof can be found in [96, Theorem 12.6.2 and Corollary 12.5.2].

Proposition 6.3.5. *Let $q \in L_2(0,l)$. Then the zeros of the functions $\lambda \mapsto s(\sqrt{\lambda},l)$ and of $\lambda \mapsto c(\sqrt{\lambda},l)$ are real, simple and can be arranged as increasing sequences $(\lambda_k)_{k=1}^\infty$ and $(\zeta_k)_{k=1}^\infty$, respectively, which obey*

$$\zeta_1 < \lambda_1 < \zeta_2 < \lambda_2 < \dots \tag{6.3.20}$$

and the asymptotic formula

$$\lambda_k = \frac{\pi^2 k^2}{l^2} + \frac{Q}{l} + \gamma_k, \tag{6.3.21}$$

where $(\gamma_k)_{k=1}^\infty \in l_2$ and

$$Q = \int_0^l q(x)\,dx. \tag{6.3.22}$$

A formula similar to (6.3.21) holds for the terms of the sequence ζ_k. However, we need the following more general case.

Proposition 6.3.6. *Let $q \in L_2(0,l)$ and $a \in (-1,1)$. Then the zeros of the function $\lambda \mapsto c(\sqrt{\lambda},l) - a$ are real, simple and can be arranged as an increasing sequence $(\zeta_{a,k})_{k=1}^\infty$ which obeys*

$$\zeta_{a,1} < \lambda_1 < \zeta_{a,2} < \lambda_2 < \dots \tag{6.3.23}$$

and the asymptotic formula

$$\zeta_{a,k} = \left(\frac{k\pi - \arccos((-1)^k a)}{l}\right)^2 + \frac{Q}{l} + \beta_{a,k}, \tag{6.3.24}$$

where $(\beta_{a,k})_{k=1}^\infty \in l_2$ and Q is given by (6.3.22).

Proof. We will make use of the representation

$$c(\lambda, l) = \cos \lambda l + Q\frac{\sin \lambda l}{2\lambda} + \int_0^l B(t)\frac{\sin \lambda t}{\lambda}\, dt, \qquad (6.3.25)$$

where $B \in L_2(0, l)$ is a real-valued function, see, e.g., [96, (12.2.24)].

The zeros of $\lambda \mapsto \cos \lambda l - a$ are simple and have the representation

$$\mu_{2k-1} = l^{-1}(-\arccos a + 2k\pi) \text{ and } \mu_{2k} = l^{-1}(\arccos a + 2k\pi) \quad (k \in \mathbb{Z}).$$

Clearly, $\mu_j \in (jl^{-1}\pi, (j+1)l^{-1}\pi)$ for all $j \in \mathbb{Z}$. For $k \in \mathbb{Z}$ and $\zeta \in \mathbb{C}$ the equation

$$\cos((\zeta + \mu_{2k})l) = \cos \mu_{2k} l \cos \zeta l - \sin \mu_{2k} l \sin \zeta l = a \cos \zeta l - \sqrt{1-a^2} \sin \zeta l$$

implies that there is a positive constant C_1 such that

$$|\cos((\zeta + \mu_{2k})l) - a| \geq \sqrt{1-a^2} l|\zeta| - C_1|\zeta|^2 \quad (|\zeta| \leq l^{-1},\ k \in \mathbb{Z}).$$

In view of (6.3.25) there is a positive constant C_2 such that

$$|(c(\zeta + \mu_{2k}, l) - a) - (\cos((\zeta + \mu_{2k})l) - a)| \leq \frac{C_2}{|k|} \quad (|\zeta| \leq l^{-1},\ k \in \mathbb{Z},\ k \neq 0).$$

For $|k| > \max\left\{(2C_1)^{-2}(1-a^2)l^4, (4C_1C_2)^2(1-a^2)^{-2}l^{-4}\right\}$ let

$$r_k := (2C_1)^{-1}\sqrt{1-a^2}\, l|k|^{-\frac{1}{2}}.$$

Then $r_k < 1$, and for $|\zeta| = r_k$ it follows that

$$\begin{aligned}
|\cos((\zeta + \mu_{2k})l) - a| &\geq \sqrt{1-a^2} l r_k - C_1 r_k^2 \\
&= \sqrt{1-a^2} l r_k \left(1 - \frac{1}{2}|k|^{-\frac{1}{2}}\right) \geq \frac{1}{2}\sqrt{1-a^2} l r_k \\
&= \frac{1}{4}C_1^{-1}(1-a^2)l^2|k|^{-\frac{1}{2}} \\
&> \frac{C_2}{|k|} \geq |(c(\zeta + \mu_{2k}, l) - a) - (\cos((\zeta + \mu_{2k})l) - a)|.
\end{aligned}$$

Then Rouché's theorem, applied to the circle with centre μ_{2k} and radius r_k, shows that $c(\cdot, l) - a$ has exactly one simple zero inside this circle, which we are going to denote by $\widetilde{\mu}_{2k}$.

Let $f(\lambda) := c(\lambda, l) - a$. Then

$$\begin{aligned}
f(\mu_{2k}) = f(\mu_{2k}) - f(\widetilde{\mu}_{2k}) &= \int_{\widetilde{\mu}_{2k}}^{\mu_{2k}} f'(\lambda)\, d\lambda \\
&= f'(\mu_{2k})(\mu_{2k} - \widetilde{\mu}_{2k}) + \int_{\widetilde{\mu}_{2k}}^{\mu_{2k}} (f'(\lambda) - f'(\mu_{2k}))\, d\lambda \\
&= f'(\mu_{2k})(\mu_{2k} - \widetilde{\mu}_{2k}) + \int_{\widetilde{\mu}_{2k}}^{\mu_{2k}} \int_{\mu_{2k}}^{\lambda} f''(\mu)\, d\mu\, d\lambda.
\end{aligned}$$

From (6.3.25) we find that

$$f'(\lambda) = -l\sin\lambda l + Ql\frac{\cos\lambda l}{2\lambda} - Q\frac{\sin\lambda l}{2\lambda^2} + \int_0^l tB(t)\frac{\cos\lambda t}{\lambda}\,dt - \int_0^l B(t)\frac{\sin\lambda t}{\lambda^2}\,dt.$$

Therefore

$$f'(\mu_{2k}) = -l\sqrt{1-a^2} + \frac{Qla}{2\mu_{2k}} + \frac{\widehat{\beta}_{2k}^{(1)}}{\mu_{2k}} + \frac{\widehat{\beta}_{2k}^{(2)}}{\mu_{2k}},$$

where

$$\widehat{\beta}_{2k}^{(1)} = \int_0^l tB(t)\cos\mu_{2k}t\,dt,$$

$$\widehat{\beta}_{2k}^{(2)} = \frac{1}{\mu_{2k}}\left(-\frac{1}{2}Q\sqrt{1-a^2} - \int_0^l B(t)\sin\mu_{2k}t\,dt\right).$$

Clearly, the sequence $(\widehat{\beta}_{2k}^{(2)})$ is an l_2 sequence. But also $(\widehat{\beta}_{2k}^{(1)})$ is an l_2 sequence, see, e.g., [96, Lemma 12.2.1]. Hence we have

$$\frac{1}{f'(\mu_{2k})} = -\frac{1}{l\sqrt{1-a^2}}\left(1 + \frac{Qa}{2\mu_{2k}\sqrt{1-a^2}} + \frac{\widehat{\beta}_{2k}^{(3)}}{\mu_{2k}}\right)$$

with an l_2 sequence $(\widehat{\beta}_{2k}^{(3)})$. Furthermore, (6.3.25) gives

$$f(\mu_{2k}) = \frac{Q\sqrt{1-a^2}}{2\mu_{2k}} + \frac{1}{\mu_{2k}}\widehat{\beta}_{2k}^{(4)}$$

with an l_2 sequence $(\widehat{\beta}_{2k}^{(4)})$. Differentiating f' again shows that the entire function f'' is bounded on the strip $\{\lambda \in \mathbb{C} : |\operatorname{Im}\lambda| \le l^{-1}\}$. Hence

$$\int_{\widetilde{\mu}_{2k}}^{\mu_{2k}} \int_{\mu_{2k}}^{\lambda} f''(\mu)\,d\mu\,d\lambda = O((\widetilde{\mu}_{2k} - \mu_{2k})^2) = O(r_k^2) = O(|k|^{-1}).$$

Altogether, it follows that

$$\widetilde{\mu}_{2k} - \mu_{2k} = -\frac{f(\mu_{2k}) + O((\widetilde{\mu}_{2k} - \mu_{2k})^2)}{f'(\mu_{2k})}.$$

The above representations of $f(\mu_{2k})$ and $f'(\mu_{2k})$ show that $|\widetilde{\mu}_{2k} - \mu_{2k}| = O(|k|^{-1})$, and therefore

$$\begin{aligned}\widetilde{\mu}_{2k} &= \mu_{2k} - \frac{f(\mu_{2k})}{f'(\mu_k)} + O(|k|^{-2})\\ &= \mu_{2k} + \frac{1}{2\mu_{2k}}\left(\frac{Q}{l} + \widehat{\beta}_{2k}^{(5)}\right)\\ &= \mu_{2k} + \frac{Q}{4k\pi} + \frac{1}{k}\widehat{\beta}_{2k}^{(6)} \qquad (6.3.26)\end{aligned}$$

with l_2 sequences $(\widehat{\beta}_{2k}^{(5)})$ and $(\widehat{\beta}_{2k}^{(6)})$.

Next we are going to compare the zeros of $\lambda \mapsto \cos \lambda l - a$ and $c(\cdot, l) - a$ globally. For $k \in \mathbb{N}$ let Γ_k be the boundary of the square with vertices $\pm l^{-1}k\pi \pm il^{-1}k\pi$. A standard estimate shows that there is $N_a \in \mathbb{N}$ such that

$$\max\{|\sin \lambda x|, |\cos \lambda x| : x \in [0, l], \lambda \in \Gamma_k\} \leq \frac{2|\cos \lambda l - a|}{1 - |a|} \quad (\lambda \in \Gamma_k, \ k \geq N_a). \tag{6.3.27}$$

Indeed, writing $\lambda = u + iv$ $(u, v \in \mathbb{R})$ for $\lambda \in \mathbb{C}$, the formula

$$\cos \lambda x = \cos ux \cosh vx - i \sin ux \sinh vx$$

shows that

$$|\cos \lambda x| \leq \cosh(\operatorname{Im} \lambda l) \quad (\lambda \in \mathbb{C}, \ x \in [0, l]).$$

Similarly,

$$\sin \lambda x = \sin ux \cosh vx + i \cos ux \sinh vx$$

shows that

$$|\sin \lambda x| \leq \cosh(\operatorname{Im} \lambda l) \quad (\lambda \in \mathbb{C}, \ x \in [0, l]).$$

Putting $u = \pm l^{-1}k\pi$ it follows that $\cos \lambda l = (-1)^k \cosh vl$ and therefore

$$|\cos \lambda l - a| \geq (1 - |a|) \cosh(\operatorname{Im} \lambda l) \quad (|\operatorname{Re} \lambda| = l^{-1}k\pi).$$

Observing

$$|\cos \lambda l - a|^2 \geq \sinh^2 vl - 2|a| \cosh vl \quad (\lambda \in \mathbb{C})$$

gives for $v = \pm l^{-1}k\pi$ that

$$|\cos \lambda l - a|^2 \geq \cosh^2 k\pi \left[\frac{\sinh^2 k\pi}{\cosh^2 k\pi} - \frac{2|a|}{\cosh k\pi}\right],$$

where the term in square brackets is larger than $\frac{1}{2}$ for sufficiently large k. Therefore there is $N_a \in \mathbb{N}$ such that

$$|\cos \lambda l - a| \geq \frac{1}{2}(1 - |a|) \cosh(\operatorname{Im} \lambda l) \quad (\lambda \in \Gamma_k, \ k \geq N_a).$$

Combining the above estimates proves (6.3.27).

From (6.3.25) and (6.3.27) we conclude that $|\cos \lambda l - a| > |c(\lambda, l) - \cos \lambda l|$ for all $\lambda \in \Gamma_k$ and sufficiently large k. By Rouché's theorem, the number of zeros of $\lambda \mapsto \cos \lambda l - a$ and $\lambda \mapsto c(\lambda, l) - a$ inside those Γ_k coincide, counted with multiplicity. The functions $\lambda \mapsto \cos \lambda l - a$ and $c(\cdot, l) - a$ are even functions, and $\mu_{2k-1} = -\mu_{-2k}$ $(k \in \mathbb{Z})$. Due to the asymptotic representation (6.3.25) we can index the zeros of $c(\cdot, l) - a$, counted with multiplicity, as a sequence $(\widetilde{\mu}_j)_{j \in \mathbb{Z}}$, where $\widetilde{\mu}_{2k}$ for large $|k|$ is given by (6.3.26) and $\widetilde{\mu}_{2k-1} = -\widetilde{\mu}_{-2k}$ $(k \in \mathbb{Z})$.

It is now clear that the zeros of $\lambda \mapsto c(\sqrt{\lambda}, l) - a$ are given by the sequence $(\widetilde{\mu}_k^2)_{k=0}^{\infty}$. On the other hand, it is well known that the function $s(\sqrt{\lambda_k}, \cdot)$

has $k-1$ simple zeros in $(0,l)$, see, e.g., [139, Theorem 13.2]. It follows from $s'(\sqrt{\lambda_k},0)=1$ that $(-1)^k s'(\sqrt{\lambda_k},l)>0$. Therefore (6.3.4) and (6.3.5) show that $c(\sqrt{\lambda_k},l)=(-1)^k$. Hence the function $\lambda\mapsto c(\sqrt{\lambda},l)-a$ has a zero $\zeta_{a,k}$ in the interval $(\lambda_{k-1},\lambda_k)$ for $k=2,3,\dots$. For $\lambda<0$ and $\lambda\to-\infty$, the asymptotic representations $\cos(\sqrt{\lambda}t)=\frac{1}{2}e^{\sqrt{|\lambda|}t}+o(1)$ and $|\sin(\sqrt{\lambda}t)|=\frac{1}{2}e^{\sqrt{|\lambda|}t}+o(1)$ show that $c(\sqrt{\lambda},l)\to\infty$. From $c(\sqrt{\lambda_1},l)=-1$ it follows that $\lambda\mapsto c(\sqrt{\lambda},l)-a$ has a zero $\zeta_{a,1}$ in the interval $(-\infty,\lambda_1)$.

The asymptotic representation (6.3.26) shows that

$$\lambda_{2k}<\widetilde{\mu}_{2k}^2<\lambda_{2k+1} \text{ and } \lambda_{2k-1}<\widetilde{\mu}_{-2k}^2=\widetilde{\mu}_{2k-1}^2<\lambda_{2k}$$

for sufficiently large k. Thus the numbers $\zeta_{a,k}$ and $\widetilde{\mu}_{k-1}^2$ can be identified for $k\geq 1$, and hence the sequence $(\zeta_{a,k})_{k=1}^\infty$ is the sequence of all zeros of $\lambda\mapsto c(\sqrt{\lambda},l)-a$. Finally, (6.3.26) proves (6.3.24), applying $-\arccos a=\arccos a-\pi$ for odd k. □

Furthermore, we will need a representation of the zeros of the two functions $\lambda\mapsto c(\sqrt{\lambda},l)-\varepsilon$ with $\varepsilon\in\{-1,1\}$. Due to the identity (6.3.5) it is more convenient to state properties of the zeros of $\lambda\mapsto c'(\sqrt{\lambda},l)$.

Proposition 6.3.7. *Let $q\in L_2(0,l)$. Then the zeros of $\lambda\mapsto c'(\sqrt{\lambda},l)$ are real, simple and can be arranged as an increasing sequence $(\xi_k)_{k=0}^\infty$ which obeys*

$$\xi_0<\zeta_{0,1}<\xi_1<\zeta_{0,2}<\dots \tag{6.3.28}$$

and the asymptotic formula

$$\xi_k=\frac{\pi^2k^2}{l^2}+\frac{Q}{l}+\gamma_{1,k}, \tag{6.3.29}$$

where $(\gamma_{1,k})_{k=0}^\infty\in l_2$ and Q is given by (6.3.22). Furthermore, $c(\sqrt{\lambda_k},l)=(-1)^k$ $(k\in\mathbb{N})$ and $c(\sqrt{\xi_k},l)=(-1)^k$ $(k\in\mathbb{N}_0)$.

Proof. The representations

$$s(\lambda,l)=\frac{\sin\lambda l}{\lambda}-Q\frac{\cos\lambda l}{2\lambda^2}+\int_0^l B_1(t)\frac{\cos\lambda t}{\lambda^2}\,dt, \tag{6.3.30}$$

$$c'(\lambda,l)=-\lambda\sin\lambda l+\frac{1}{2}Q\cos\lambda l+\int_0^l B_2(t)\cos\lambda t\,dt, \tag{6.3.31}$$

where $B_1,B_2\in L_2(0,l)$ are real-valued functions, see, e.g., [96, (12.2.22) and (12.2.25)], show that $\lambda\mapsto\lambda s(\sqrt{\lambda},l)$ and $\lambda\mapsto -c'(\sqrt{\lambda},l)$ have the same leading terms. Therefore the asymptotic representation (6.3.21) implies that the zeros of the latter function can be written as a sequence $(\xi_k)_{k=0}^\infty$ of complex numbers satifying the asymptotic representation (6.3.29). For each $k\in\mathbb{N}$, the function $\lambda\mapsto c^2(\sqrt{\lambda},l)-1$ has the value -1 at the endpoints of the interval $[\zeta_k,\zeta_{k+1}]$. Hence this function has an even number of zeros in (ζ_k,ζ_{k+1}), counted with multiplicity.

By (6.3.20), λ_k is the only zero of $\lambda \mapsto s(\sqrt{\lambda}, l)$ in (ζ_k, ζ_{k+1}). Hence it follows from (6.3.5) that $\lambda \mapsto c'(\sqrt{\lambda}, l)$ has an odd number of zeros in (ζ_k, ζ_{k+1}) $(k \in \mathbb{N})$. Also, the function $\lambda \mapsto c^2(\sqrt{\lambda}, l) - 1$ has an odd number of zeros, counted with multiplicity, in $(-\infty, \zeta_1)$ because of $c(\sqrt{\lambda}, l) \to \infty$ as $\lambda \to -\infty$, which has been shown in the proof of Proposition 6.3.6. Since $\lambda_1 > \zeta_1$ (see (6.3.20)) it is clear that $s(\sqrt{\lambda}, l) \neq 0$ for all $z \in (-\infty, \zeta_1)$. Again from (6.3.5) it follows that $\lambda \mapsto c'(\sqrt{\lambda}, l)$ has an odd number of zeros in the interval $(-\infty, \zeta_1)$. Hence $\lambda \mapsto c'(\sqrt{\lambda}, l)$ has at least one zero ξ_0 in $(-\infty, \zeta_1)$ and at least one zero ξ_k in (ζ_k, ζ_{k+1}) $(k \in \mathbb{N})$. In view of $\zeta_k = \zeta_{0,k}$ and the asymptotic representations (6.3.24) and (6.3.29) it follows that the zeros of $\lambda \mapsto c'(\sqrt{\lambda}, l)$ are real, simple and can be indexed such that (6.3.28) holds. Since $c(\sqrt{\lambda}, l) \to \infty$ as $\lambda \to -\infty$ and since the zeros of $\lambda \mapsto c(\sqrt{\lambda}, l)$ are simple, it is clear that $c(\sqrt{\lambda}, l) > 0$ when $\lambda \in (-\infty, \zeta_1)$ and that $(-1)^k c(\sqrt{\lambda}, l) > 0$ for $k \in \mathbb{N}$ and $\lambda \in (\zeta_k, \zeta_{k+1})$. In view of $c^2(\sqrt{\xi_k}, l) = 1$ this shows that $c(\sqrt{\xi_k}, l) = (-1)^k$ $(k \in \mathbb{N}_0)$. □

In view of Lemma C.5.5 the spectrum of the weighted adjacency matrix $\tilde{A}$ consists of p eigenvalues $-1 \le \alpha_1 \le \alpha_2 \le \cdots \le \alpha_{p-1} < \alpha_p = 1$, counted with multiplicity, where $\alpha_2 > -1$ and $\alpha_1 = -1$ if and only if G is bipartite.

We are now in a position to describe the asymptotic behaviour of the eigenvalues of $\mathcal{L}$. It is convenient to state two theorems, one for bipartite graphs and one for nonbipartite graphs. But both theorems will be proved together.

We will use the notation for the sequences considered in Propositions 6.3.5, 6.3.6 and 6.3.7 with $\lambda_k^{(j)} := \lambda_k$ $(k, j \in \mathbb{N})$.

Theorem 6.3.8. *Let G be a bipartite connected simple graph with at least two edges. Then the spectrum of the operator $\mathcal{L}$ on G given by the differential equation* (6.3.1), *the Neumann conditions* (6.2.1), (6.2.2) *and the conditions* (6.2.3), (6.2.4), *consists of g sequences of eigenvalues, counted with multiplicity, as follows:*

(i) *the sequence $(\xi_k)_{k=0}^\infty$ with the asymptotic representation* (6.3.29);

(ii) *if G is not a tree, the sequences $(\lambda_k^{(j)})_{k=1}^\infty$ $(j = 1, \dots, g - p + 1)$ with the asymptotic representation* (6.3.21);

(iii) *the sequences $(\zeta_{\alpha_j,k})_{k=1}^\infty$ $(j = 2, \dots, p-1)$ with the asymptotic representation* (6.3.24).

Theorem 6.3.9. *Let G be a nonbipartite connected simple graph with at least two edges. Then the spectrum of the operator $\mathcal{L}$ on G given by the differential equation* (6.3.1), *the Neumann conditions* (6.2.1), (6.2.2) *and the conditions* (6.2.3), (6.2.4), *consists of $g + 1$ sequences of eigenvalues, counted with multiplicity, as follows:*

(i) *the sequence $(\xi_{2k})_{k=0}^\infty$ with the asymptotic representation* (6.3.29) *and the sequence $(\lambda_{2k})_{k=1}^\infty$ with the asymptotic representation* (6.3.21);

(ii) *if $g > p$, the sequences $(\lambda_k^{(j)})_{k=1}^\infty$ $(j = 1, \dots, g - p)$ with the asymptotic representation* (6.3.21);

(iii) *the sequences* $(\zeta_{\alpha_j,k})_{k=1}^{\infty}$ $(j = 1,\ldots,p-1)$ *with the asymptotic representation* (6.3.24).

Proof. Theorems 6.3.2 and 6.3.3 describe all eigenvalues of $\mathcal{L}$. Taking Proposition 6.3.6 into account, it is clear that part (iii) is a representation of the eigenvalues λ for which $|c(\sqrt{\lambda}, l)| < 1$. All other eigenvalues λ satisfy $|c(\sqrt{\lambda}, l)| = 1$.

Therefore, let now λ be an eigenvalue of $\mathcal{L}$ with $c^2(\sqrt{\lambda}, l) = 1$.

First consider the case that G is bipartite. If $c'(\sqrt{\lambda}, l) = 0$ and $s(\sqrt{\lambda}, l) \neq 0$, then λ is a simple eigenvalue by Theorem 6.3.2, and if $c'(\sqrt{\lambda}, l)=0$ and $s(\sqrt{\lambda}, l)=0$, then λ is an eigenvalue by Theorem 6.3.3, part 1. This proves that the sequence in part (i) of Theorem 6.3.8 contributes to the spectrum. It is therefore clear from Theorem 6.3.3 parts 1 and 2 that all remaining eigenvalues are described in part (ii) of Theorem 6.3.8.

Finally, consider the case that G is not bipartite. If $c'(\sqrt{\lambda}, l) = 0$ and $s(\sqrt{\lambda}, l) \neq 0$, then λ is a simple eigenvalue by Theorem 6.3.2 when $c(\sqrt{\lambda}, l) = 1$. If $c'(\sqrt{\lambda}, l) = 0$ and $s(\sqrt{\lambda}, l) = 0$, then λ is an eigenvalue by Theorem 6.3.3, part 1, when $c(\sqrt{\lambda}, l) = 1$. Furthermore, apart from this sequence, also $(\lambda_{2k})_{k=1}^{\infty}$ contributes to the eigenvalues by Theorem 6.3.3, parts 1 and 2, and Proposition 6.3.7. This proves that the two sequences in part (i) of Theorem 6.3.9 contribute to the spectrum. It is therefore clear from Theorem 6.3.3 that all remaining eigenvalues are described in part (ii) of Theorem 6.3.9. □

For a connected simple graph G with adjacency matrix A and matrix of degree T let

$$P_G(z) := \det(zT - A) \quad (z \in \mathbb{C}). \tag{6.3.32}$$

The zeros of P_G are the eigenvalues of the weighted adjacency matrix $\widetilde{A}$, counted with multiplicity. It might appear to be more natural to consider $\det(\widetilde{A} - z) = \det(-T^{-1})\det(zT - A)$, but the function P_G has integer coefficients and a positive leading coefficients, which looks nicer when P_G is calculated explicitly. The characteristic function ϕ of the operator $\mathcal{L}$ is defined by

$$\phi_{\mathcal{L}}(z) := s^{g-p}(\sqrt{\lambda}, l) P_G(c(\sqrt{\lambda}, l)) \quad (z \in \mathbb{C}). \tag{6.3.33}$$

Corollary 6.3.10. *Let G be a simple connected graph. Then the eigenvalues of the operator $\mathcal{L}$ on G given by the differential equation* (6.3.1), *the Neumann conditions* (6.2.1), (6.2.2) *and the conditions* (6.2.3), (6.2.4), *are the zero of the entire function $\phi_{\mathcal{L}}$, counted with multiplicity.*

Proof. If G is bipartite, we use Theorem 6.3.8 for the proof. The eigenvalues in part (i) are the zeros of $\lambda \mapsto c'(\sqrt{\lambda}, l)$, the eigenvalues in part (ii) are the zeros of $\lambda \mapsto s^{g-p+1}(\sqrt{\lambda}, l)$, and the eigenvalues in part (iii) are the zeros of $\lambda \mapsto P_G(c(\sqrt{\lambda}, l))(c^2(\sqrt{\lambda}, l) - 1))^{-1}$ since -1 and 1 are simple eigenvalues of $\widetilde{A}$ by Lemma C.5.5. In view of (6.3.5),

$$c'(\sqrt{\lambda}, l) s^{g-p+1}(\sqrt{\lambda}, l) P_G(c(\sqrt{\lambda}, l))(c^2(\sqrt{\lambda}, l) - 1))^{-1} = s^{g-p}(\sqrt{\lambda}, l) P_G(c(\sqrt{\lambda}, l)),$$

which proves the statement for bipartite graphs.

If G is not bipartite, we use Theorem 6.3.9 for the proof. In view of (6.3.5) and Propositions 6.3.5 and 6.3.7, the eigenvalues in part (i) are the zeros of $\lambda \mapsto c^2(\sqrt{\lambda}, l) - 1$ with $c(\sqrt{\lambda}, l) = 1$, that is, the zeros of $\lambda \mapsto c(\sqrt{\lambda}, l) - 1$. The eigenvalues in part (ii) are the zeros of $\lambda \mapsto s^{g-p}(\sqrt{\lambda}, l)$, and the eigenvalues in part (iii) are the zeros of $\lambda \mapsto P_G(c(\sqrt{\lambda}, l))(c(\sqrt{\lambda}, l) - 1))^{-1}$ since -1 is not an eigenvalue and 1 a simple eigenvalue of $\widetilde{A}$ by Lemma C.5.5. Then the identity

$$(c(\sqrt{\lambda}, l) - 1)s^{g-p}(\sqrt{\lambda}, l)P_G(c(\sqrt{\lambda}, l))(c(\sqrt{\lambda}, l) - 1))^{-1} = s^{g-p}(\sqrt{\lambda}, l)P_G(c(\sqrt{\lambda}, l))$$

completes the proof. □

Remark 6.3.11. 1. It was convenient to consider g sequences in Theorem 6.3.8 and $g+1$ sequences in Theorem 6.3.9. However, if we combine the two sequences in Theorem 6.3.9, part (i), into one sequence, then one arrives in both cases at g sequences where the k-th term, say a_k, satisfies the asymptotic estimate

$$\left|\sqrt{a_k} - \frac{\pi k}{l}\right| \le \frac{\pi}{l} + o(1) \text{ as } k \to \infty.$$

Denoting the nondecreasing sequence of the eigenvalues of $\mathcal{L}$, counted with multiplicity, by $(\lambda_{\mathcal{L},k})_{k=1}^{\infty}$, Weyl's law

$$g = \frac{\pi}{l} \lim_{k\to\infty} \frac{k}{\sqrt{\lambda_{\mathcal{L},k}}}$$

follows (see [3], [100] or [10, Corollary 1]).

2. The smallest eigenvalue is ξ_0, which is the smallest zero of $\lambda \mapsto c(\sqrt{\lambda}, l)$, and this eigenvalue is simple.

Example 6.3.12. In this example we will find the characteristic functions as defined in (6.3.33) for the graphs associated with the platonic solids tetrahedron, cube and octahedron. All three graphs are planar graphs.

1. The tetrahedral graph is not a bipartite graph since it has cycles with three edges.

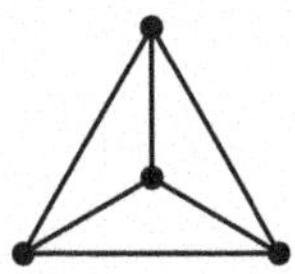

It has 4 vertices and 6 edges, that is, $g - p = 2$, and

$$zT - A = \begin{pmatrix} B_4(z) & -A_4 \\ -A_4 & B_4(z) \end{pmatrix},$$

where

$$B_4(z) = \begin{pmatrix} 3z & -1 \\ -1 & 3z \end{pmatrix}, \quad A_4 = \begin{pmatrix} 1 & 1 \\ 1 & 1 \end{pmatrix}.$$

Denoting the polynomial P_G in (6.3.32) by P_4, it follows for $\det B_4(z) \neq 0$ from the Schur factorization (B.4.2) that

$$P_4(z) = \det(zT - A) = \det B_4(z) \det(B_4(z) - A_4 B_4^{-1}(z) A_4).$$

A straightforward calculation shows that

$$A_4 B_4^{-1}(z) A_4 = \frac{6z+2}{9z^2-1} A_4,$$

and hence

$$\begin{aligned} P_4(z) &= (9z^2-1)\left(\left(3z - \frac{6z+2}{9z^2-1}\right)^2 - \left(1 + \frac{6z+2}{9z^2-1}\right)^2\right) \\ &= (9z^2-1)(3z+1)\left(3z - 1 - \frac{4(3z+1)}{9z^2-1}\right) \\ &= 3^4(z-1)\left(z + \frac{1}{3}\right)^3. \end{aligned}$$

This shows that the characteristic function as defined in (6.3.33) for the tetrahedral graph is

$$\phi_4(\lambda) = 3^4 s^2(\sqrt{\lambda}, l)(c(\sqrt{\lambda}, l) - 1)\left(c(\sqrt{\lambda}, l) + \frac{1}{3}\right)^3.$$

2. The octahedral graph is not a bipartite graph since it has cycles with three edges.

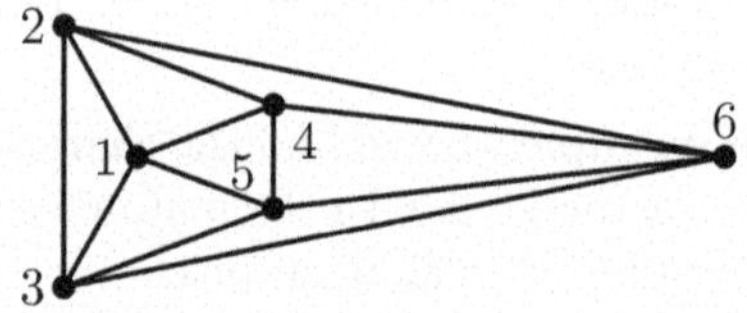

It has 6 vertices and 12 edges, that is, $g - p = 6$. With respect to the indexing of the vertices as indicated in the figure above we have

$$zT - A = \begin{pmatrix} B_6(z) & -A_6 \\ -A_6 & B_6(z) \end{pmatrix},$$

where

$$B_6(z) = \begin{pmatrix} 4z & -1 & -1 \\ -1 & 4z & -1 \\ -1 & -1 & 4z \end{pmatrix}, \quad A_6 = \begin{pmatrix} 1 & 1 & 0 \\ 1 & 0 & 1 \\ 0 & 1 & 1 \end{pmatrix}.$$

Denoting the polynomial P_G in (6.3.32) by P_6, it follows for $\det B_6(z) \neq 0$ from the Schur factorization (B.4.2) that

$$P_6(z) = \det(zT - A) = \det B_6(z) \det(B_6(z) - A_6 B_6^{-1}(z) A_6).$$

A straightforward calculation shows that $\det B_6(z) = 2(2z-1)(4z+1)^2$,

$$B_6^{-1}(z) = \frac{1}{2(4z+1)(2z-1)} \begin{pmatrix} 4z-1 & 1 & 1 \\ 1 & 4z-1 & 1 \\ 1 & 1 & 4z-1 \end{pmatrix}$$

and then

$$A_6 B_6^{-1}(z) A_6 = \frac{1}{(4z+1)(2z-1)} \begin{pmatrix} 4z & 2z+1 & 2z+1 \\ 2z+1 & 4z & 2z+1 \\ 2z+1 & 2z+1 & 4z \end{pmatrix}.$$

The diagonal elements of the matrix $(4z+1)(2z-1)(B_6(z) - A_6 B_6^{-1}(z) A_6)$ are $4z(4z+1)(2z-1) - 4z = 8z(4z^2 - z - 1)$, whereas its off-diagonal elements are $-(4z+1)(2z-1) - (2z+1) = -8z^2$. Then elementary row operations give

$$\begin{aligned} &\det(B_6(z) - A_6 B_6^{-1}(z) A_6) \\ &= \frac{2^9 z^3 (4z^2-1)}{(4z+1)^3(2z-1)^3} \det \begin{pmatrix} 4z^2 - z - 1 & -z & 0 \\ -2z & 4z^2 - 2z - 1 & 0 \\ -z & -z & 1 \end{pmatrix} \\ &= \frac{2^9 z^3 (4z^2-1)}{(4z+1)^3(2z-1)^3} \left[(4z^2 - z - 1)(4z^2 - 2z - 1) - 2z^2\right] \\ &= \frac{2^9 z^3 (2z+1)}{(4z+1)^3(2z-1)^2} (4z^2-1)(4z^2 - 3z - 1) \\ &= \frac{2^9 z^3 (2z+1)^2}{(4z+1)^2(2z-1)} (z-1), \end{aligned}$$

which shows that

$$P_6(z) = 4^6 (z-1) z^3 \left(z + \frac{1}{2}\right)^2.$$

Therefore the characteristic function as defined in (6.3.33) for the octahedral graph is

$$\phi_6(\lambda) = 4^6 s^6(\sqrt{\lambda}, l)(c(\sqrt{\lambda}, l) - 1) c^3(\sqrt{\lambda}, l) \left(c(\sqrt{\lambda}, l) + \frac{1}{2}\right)^2.$$

3. The cube graph is a bipartite graph since every edge has one black end and one white end.

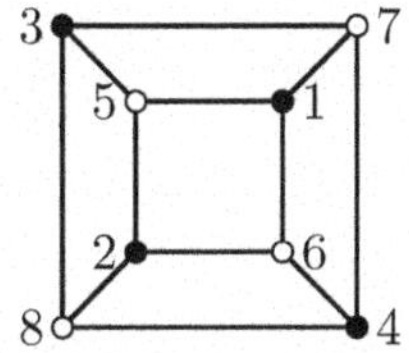

It has 8 vertices and 12 edges, that is, $g-p=4$. With respect to the indexing of the vertices as indicated in the figure above we have

$$zT - A = \begin{pmatrix} 3zI_4 & -A_8 \\ -A_8 & 3zI_4 \end{pmatrix},$$

where

$$A_8 = \begin{pmatrix} 1 & 1 & 1 & 0 \\ 1 & 1 & 0 & 1 \\ 1 & 0 & 1 & 1 \\ 0 & 1 & 1 & 1 \end{pmatrix}.$$

The Schur factorization (B.4.2) shows that

$$P_8(z) := \det(zT - A) = \det(9z^2 I_4 - A_8^2).$$

But

$$A_8^2 = \begin{pmatrix} 3 & 2 & 2 & 2 \\ 2 & 3 & 2 & 2 \\ 2 & 2 & 3 & 2 \\ 2 & 2 & 2 & 3 \end{pmatrix},$$

and letting $\zeta = \frac{1}{2}(3z^2 - 1)$ it follows that $9z^2 I_4 - A_8^2$ equals $2(\zeta T - A)$ with A and T from the tetrahedral graph. Therefore part 1 leads to

$$\begin{aligned} P_8(z) = 2^4 P_4(\zeta) &= 3^4(2\zeta - 2)\left(2\zeta + \frac{2}{3}\right)^3 \\ &= 3^8(z^2-1)\left(z^2 - \frac{1}{9}\right)^3, \end{aligned}$$

which gives the characteristic function as defined in (6.3.31) for the cube graph as

$$\phi_8(\lambda) = 3^8 s^4(\sqrt{\lambda}, l)(c^2(\sqrt{\lambda}, l) - 1)\left(c^2(\sqrt{\lambda}, l) - \frac{1}{9}\right)^3.$$

Example 6.3.13. For a star graph with $g \geq 2$ edges we have

$$zT - A = \begin{pmatrix} gz & \mathbb{1}_g^\top \\ \mathbb{1}_g & zI_g \end{pmatrix},$$

and the Schur factorization gives

$$\det(zT - A) = z^g \left(gz - z^{-1}\mathbb{1}_g^\top \mathbb{1}_g\right) = gz^{g-1}(z^2 - 1).$$

It follows in view of (6.3.5) that the characteristic function as defined in (6.3.31) is

$$\phi(\lambda) = gs^{-1}(\sqrt{\lambda}, l)(c^2(\sqrt{\lambda}, l) - 1)c^{g-1}(\sqrt{\lambda}, l) = gc'(\sqrt{\lambda}, l)c^{g-1}(\sqrt{\lambda}, l).$$

6.4 Connected graphs with Dirichlet conditions at pendant vertices

In this section it is assumed that Dirichlet conditions are imposed at r pendant vertices of the connected simple graph G with at least two edges. The graph $\widehat{G}$ is defined by removing the pendant vertices with Dirichlet conditions and their incident edges in G. For convenience, the vertices of $\widehat{G}$ will be denoted by $v_1, \dots, v_{p-r}$, whereas the remaining vertices of G will be denoted by $v_{p-r+1}, \dots, v_p$. Let $\widehat{A}$ be the adjacency matrix of $\widehat{G}$, let $\widehat{T} = \operatorname{diag}(\widehat{d}(v_1), \dots, \widehat{d}(v_{p-r}))$ and let $\widehat{T}_G = \operatorname{diag}(d(v_1), \dots, d(v_{p-r}))$, where $\widehat{d}(v_i)$ is the degree of the vertex v_i in $\widehat{G}$, whereas $d(v_i)$ is the degree of the vertex v_i in G.

Proposition 6.4.1. *Let $r > 0$. Then the zeros of the polynomial $P_{G,\widehat{G}}$ defined by*

$$P_{G,\widehat{G}}(z) := \det(z\widehat{T}_G - \widehat{A}) \quad (z \in \mathbb{C})$$

are real and lie in the interval $(-1, 1)$. If G is bipartite, then $P_{G,\widehat{G}}$ is an even or an odd polynomial.

Proof. It is clear that $\widehat{G}$ is a connected simple graph and that $\widehat{G}$ is bipartite if and only if G is bipartite. Furthermore, when $\widehat{G}$ is an isolated vertex v, then $P_{G,\widehat{G}}(z) = d(v)z$, which proves the statement in this case. From now on we consider the case that $\widehat{G}$ is not an isolated vertex. Since $\widehat{A}$ is a symmetric matrix, also $\widehat{T}_G^{-\frac{1}{2}} \widehat{A} \widehat{T}_G^{-\frac{1}{2}}$ is a symmetric matrix, and all its eigenvalues are therefore real. From Lemma C.5.4 applied to $\widehat{G}$ we conclude for $y \in \mathbb{C}^{p-r}$ that

$$-(\widehat{T}_G y, y) \le -(\widehat{T} y, y) \le (\widehat{A} y, y) \le (\widehat{T} y, y) \le (\widehat{T}_G y, y).$$

This shows that the zeros of $P_{G,\widehat{G}}$ lie in $[-1, 1]$. Furthermore, when $\widehat{G}$ is not bipartite and $y \neq 0$, then $-(\widehat{T} y, y) < (\widehat{A} y, y)$, which shows that -1 is not a zero. However, when $\widehat{G}$ is bipartite, then it follows as in the proof of Lemma C.5.4 that the zeros are symmetric, and it therefore remains to show that 1 is not a zero. Observe that $(\widehat{A} - \widehat{T})\mathbb{1}_{p-r} = 0$. Since 1 is a simple eigenvalue of the weighted adjacency matrix it follows that $(\widehat{A} y, y) < (\widehat{T} y, y)$ whenever y is not a multiple of $\mathbb{1}_{p-r}$. On the other hand, $\widehat{d}(v_i) \le d(v_i)$ for all $i = 1, \dots, p - r$ and $\widehat{d}(v_i) < d(v_i)$ for at least one i, and therefore it is clear that $(\widehat{T}\mathbb{1}_{p-r}, \mathbb{1}_{p-r}) < (\widehat{T}_G \mathbb{1}_{p-r}, \mathbb{1}_{p-r})$. Altogether, it follow that 1 is not a zero of $P_{G,\widehat{G}}$. □

Theorem 6.4.2. *Let G be a connected simple graph with at least two edges. Assume that all edges have the same length and the same symmetric potential. Also assume that $\mathcal{L}$ is the operator corresponding to Dirichlet conditions at r pendant vertices. Then the eigenvalues of $\mathcal{L}$, counted with multiplicity, are the zeros of the characteristic function $\phi_{\mathcal{L}}$ defined by*

$$\phi_{\mathcal{L}}(\lambda) = s^{g-p+r}(\sqrt{\lambda}, l) P_{G,\widehat{G}}(c(\sqrt{\lambda}, l)). \tag{6.4.1}$$

Proof. When $r = 0$, this theorem is exactly Corollary 6.3.10. Now let $r > 0$ and let $\lambda \in \mathbb{C}$ such that $s(\sqrt{\lambda}, l) \neq 0$. We are going to adapt the proof of Theorem 6.3.2 to the case $r > 0$. Then the Dirichlet conditions give $f_{p-r+1} = \cdots = f_p = 0$, whereas that remaining conditions from (6.2.1), (6.2.2) and (6.2.4) can be written as (6.3.8) and (6.3.9) restricted to $1 \leq i \leq p - r$ and

$$\sum_{j \in \widehat{I}^+(v_i)} f_{j,0} + \sum_{j \in \widehat{I}^-(v_i)} f_{j,l} - d(v)c(\sqrt{\lambda}, l) f_i = 0 \quad (v \in V^{\text{int}}),$$

where $\widehat{I}^+(v_i)$ and $\widehat{I}^-(v_i)$ are the indegrees and outdegrees with respect to the graph $\widehat{G}$. But these are exactly the $p - r$ components of the vector equation

$$\widehat{A}\widehat{F} - c(\sqrt{\lambda}, l)\widehat{T}_G \widehat{F} = 0,$$

where $\widehat{F} = (f_1, \ldots, f_{p-r})^\mathsf{T}$. Hence $\lambda \in \mathbb{C}$ such that $s(\sqrt{\lambda}, l) \neq 0$ is an eigenvalue of $\mathcal{L}$ if and only if λ is a zero of $\lambda \mapsto P_{G,\widehat{G}}(c(\sqrt{\lambda}, l))$, and the multiplicities are equal. Here we have to observe that $P_{G,\widehat{G}}(c(\sqrt{\lambda}, l)) = 0$ implies $c^2(\sqrt{\lambda}, l) < 1$ by Proposition 6.4.1 and therefore $s(\sqrt{\lambda}, l) \neq 0$ by (6.3.5) and that the zeros of $\lambda \mapsto c(\sqrt{\lambda}, l) - a$ ($|a| < 1$) are simple by Proposition 6.3.6.

Finally let $r > 0$ and let $\lambda \in \mathbb{C}$ such that $s(\sqrt{\lambda}, l) = 0$. Adapting the proof of Theorem 6.3.3 to this situation we have $f_{p-r+1} = \cdots = f_p = 0$ and in particular $F = 0$. The remaining conditions (6.2.1), (6.2.2) and (6.2.4) can then be written as

$$\widehat{U}_\varepsilon F' = 0,$$

where $\widehat{U}_\varepsilon$ consists of the first $q - r$ rows of U_ε. From the proof of Proposition C.5.1 it is clear that $\widehat{U}_\varepsilon \widehat{U}_\varepsilon^* = \widehat{T}_G + \varepsilon\widehat{A}$. By Proposition 6.4.1, the matrix $\widehat{T}_G + \varepsilon\widehat{A}$ is invertible both for $\varepsilon = 1$ and $\varepsilon = -1$, so that

$$\operatorname{rank} \widehat{U}_\varepsilon = \operatorname{rank} \widehat{U}_\varepsilon \widehat{U}_\varepsilon^* = \operatorname{rank}(\widehat{T}_G + \varepsilon\widehat{A}) = p - r.$$

Since the zeros of $\lambda \mapsto s(\sqrt{\lambda}, l)$ are simple (see Proposition 6.3.5), it follows that each $\lambda \in \mathbb{C}$ with $s(\sqrt{\lambda}, l) = 0$ is an eigenvalue of $\mathcal{L}$ of multiplicity $g - (p - r)$. □

6.5 Spectral problems on perfect binary trees

6.5.1 The spectral problem

In this section we consider equilateral perfect binary trees, which will be denoted by T_n ($n \in \mathbb{N}$). The trees T_n will be rooted and directed towards the root and can be constructed recursively as follows. The tree T_1 consists of two edges joined at the root. T_n will be obtained from T_{n-1} by attaching two edges to each of the pendant vertices of T_{n-1}. Thus, the degree of the root is 2 and the degree of each other interior vertex is 3. It is easy to see that the number of pendant vertices of T_n is $p_n^{\text{pen}} = 2^n$, that the number of all vertices of T_n is $p_n = 2^{n+1} - 1$, and that the number of edges of T_n is $g_n = 2^{n+1} - 2$. We enumerate the vertices by $v_0, \ldots, v_g$ in such a way that the distance to the root is a nondecreasing function of the edge index. The enumeration is arbitrary in other respects. In particular, v_0 is the root. Each vertex v_j with $j = 1, \ldots, g_n$ has a unique outgoing edge, which will be indexed as e_j.

In the following, n will be a generic natural number unless specified otherwise. Recall that the local coordinate on T_n identifies each directed edge e_j of T_n with the interval $[0, l]$ and that the coordinate increases in the direction of the edge.

On each edge of T_n consider the Sturm-Liouville equation

$$-y_j'' + qy_j = \lambda y_j, \quad (j = 1, \ldots, g_n), \tag{6.5.1}$$

where $q \in L_2(0, l)$ is a symmetric real-valued function. We impose the Dirichlet boundary conditions

$$y_j(0) = 0 \quad (j = g_{n-1} + 1, \ldots, g_n) \tag{6.5.2}$$

at each pendant vertex. At each interior vertex v_j which is not the root we impose the continuity conditions

$$y_{j_1}(l) = y_{j_2}(l) = y_j(0) \quad (j = 1, \ldots, g_{n-1}) \tag{6.5.3}$$

and Kirchhoff's condition

$$y_{j_1}'(l) + y_{j_2}'(l) - y_j'(0) = 0 \quad (j = 1, \ldots, g_{n-1}). \tag{6.5.4}$$

Here j is the index of the edge outgoing from the vertex v_j and j_1 and j_2 are the indices of the edges incoming into v_j. Note that j_1 and j_2 depend on j and that $j_1 > j$ and $j_2 > j$. For definiteness we may take $j_1 < j_2$. At the root, the continuity and Kirchhoff conditions are

$$y_1(l) = y_2(l), \tag{6.5.5}$$
$$y_1'(l) + y_2'(l) = 0. \tag{6.5.6}$$

With the notation from the first paragraph of Section 6.4 we have $\widehat{T}_n = T_{n-1}$ ($n \in \mathbb{N}$), where T_0 denotes the subgraph of T_1 consisting of the isolated vertex v_0.

6.5.2 Characteristic functions of perfect binary trees

For the tree T_n we denote by $\phi_N^{(n)}$ the corresponding Neumann characteristic function and by $\phi_D^{(n)}(\lambda)$ the corresponding Dirichlet characteristic function with respect to the root. This means, that $\phi_N^{(n)}$ is given by the problem (6.5.1)–(6.5.6), whereas $\phi_D^{(n)}$ is given by (6.5.1)–(6.5.4) and $y_1(l) = 0$, $y_2(l) = 0$.

Recall that the characteristic function $\phi_N^{(n)}$ has been defined in Theorem 6.4.2. The characteristic function $\phi_D^{(n)}$ is not directly covered by previous work. However, one can split the tree T_n at the root into two subtrees, say $T_{n,1}$ and $T_{n,2}$, where e_j becomes an edge of $T_{n,j}$ for $j = 1, 2$. Clearly, the problem with Dirichlet condition at the root splits into two separate problems for the trees $T_{n,1}$ and $T_{n,2}$. Note that the trees $T_{n,1}$ and $T_{n,2}$ are identical. Hence we can define

$$\phi_D^{(n)} := \phi_{D,n}^2, \tag{6.5.7}$$

where $\phi_{D,n}$ is the characteristic function for the problem on $T_{n,1}$ with Dirichlet conditions at its pendant vertices and continuity and Kirchhoff conditions at its interior vertices. For $n > 1$ let $T_{n,0}$ be the graph obtained by removing all pendant vertices and their adjacent edges from $T_{n,1}$. Clearly, $T_{n,0}$ is an isolated vertex when $n = 2$ and $T_{n,0}$ is a copy of T_{n-2} when $n > 2$. In the notation from the first paragraph of Section 6.4 we can also write $T_{n,0} = \widehat{T}_{n,1}$.

Proposition 6.5.1. *For each $n \in \mathbb{N}$,*

$$\phi_N^{(n)}(\lambda) = s^{2^n-1}(\sqrt{\lambda}, l) P_{T_n,T_{n-1}}(c(\sqrt{\lambda}, l)) \tag{6.5.8}$$

$$\phi_D^{(n)}(\lambda) = s^{2^n}(\sqrt{\lambda}, l) P_{T_{n,1},T_{n,0}}^2(c(\sqrt{\lambda}, l)), \tag{6.5.9}$$

where $P_{T_{1,1},T_{1,0}} := 1$.

Proof. The identities (6.5.8) and (6.5.9) for $n > 1$ follow immediately from the above discussion and Theorem 6.4.2. When $n = 1$, then $T_{n,1}$ is an edge with Dirichlet conditions at both ends, so that $\phi_{D,1}(\lambda) = s(\sqrt{\lambda}, l)$. □

For $k \in \mathbb{N}$ let A_k be the adjacency matrix of T_k and let $A_0 = 0$ and $p_0 = 1$. For $\alpha = 2, 3$ and $k \in \mathbb{N}$ define the $(1, p_{k-1}, p_{k-1})$ block matrix

$$\Phi_{k,\alpha}(z) := \begin{pmatrix} \alpha z & -\mathbf{e}_1^{\mathsf{T}} & -\mathbf{e}_1^{\mathsf{T}} \\ -\mathbf{e}_1 & 3zI - A_{k-1} & 0 \\ -\mathbf{e}_1 & 0 & 3zI - A_{k-1} \end{pmatrix} \quad (z \in \mathbb{C}), \tag{6.5.10}$$

where $\mathbf{e}_1$ is the first unit vector in $\mathbb{C}^{p_{k-1}}$. Rearranging the indexing of the vertices of T_n in such a way that v_0 and v_1 are still the first two vertices, but then followed by all vertices connected to v_1 by a path which does not pass through v_0, then the vertex v_2 and then all remaining vertices, it is clear that $\Phi_{k,\alpha}(0) = -A_k$ for $n \geq 2$ and that the first row and column in A_{k-1} in (6.5.10) is taken with respect to the root of T_{k-1}. Hence we have

$$P_{T_n,T_{n-1}} = \det \Phi_{n-1,2} \quad (n \geq 2), \tag{6.5.11}$$

$$P_{T_{n,1},T_{n,0}} = \det \Phi_{n-2,3} \quad (n \geq 3). \tag{6.5.12}$$

We note that P_{T_1,T_0} and $P_{T_{2,1},T_{2,0}}$ correspond to isolated vertices obtained by removing two or three pendant vertices from T_1 and $T_{2,1}$, respectively. Hence it follows that

$$P_{T_1,T_0}(z) = 2z, \quad P_{T_{2,1},T_{2,0}}(z) = 3z \quad (z \in \mathbb{C}). \tag{6.5.13}$$

Proposition 6.5.2. *The functions* $\det \Phi_{k,\alpha}$ *satisfy the following recurrence relations:*

$$\det \Phi_{1,\alpha}(z) = 9\alpha z^3 - 6z, \tag{6.5.14}$$

$$\det \Phi_{2,\alpha}(z) = (27z^3 - 6z)(\alpha z(27z^3 - 6z) - 18z^2), \tag{6.5.15}$$

$$\det \Phi_{k,\alpha}(z) = \det \Phi_{k-1,3}(z)[\alpha z \det \Phi_{k-1,3}(z) - 2(\det \Phi_{k-2,3}(z))^2] \ (k \geq 3). \tag{6.5.16}$$

Proof. For $k = 1$ we find

$$\det \Phi_{1,\alpha}(z) = \det \begin{pmatrix} \alpha z & -1 & -1 \\ -1 & 3z & 0 \\ -1 & 0 & 3z \end{pmatrix} = 9\alpha z^3 - 6z,$$

which proves (6.5.14).

Now let $k \geq 2$. From the Schur factorization (B.4.2) it follows for $z \in \mathbb{C}$ with $\det(3zI - A_{k-1}) \neq 0$ that

$$\begin{aligned} \det \Phi_{k,\alpha}(z) &= (\det(3zI - A_{k-1}))^2 \times \\ &\times \left[\alpha z - \begin{pmatrix} \mathbf{e}_1^\mathsf{T} & \mathbf{e}_1^\mathsf{T} \end{pmatrix} \begin{pmatrix} 3zI - A_{k-1} & 0 \\ 0 & 3zI - A_{k-1} \end{pmatrix}^{-1} \begin{pmatrix} \mathbf{e}_1 \\ \mathbf{e}_1 \end{pmatrix} \right] \\ &= (\det(3zI - A_{k-1}))^2 \left[\alpha z - 2\mathbf{e}_1^\mathsf{T}(3zI - A_{k-1})^{-1}\mathbf{e}_1 \right]. \end{aligned}$$

We observe that $3zI - A_{k-1} = \Phi_{k-1,3}(z)$ and that

$$\det(3zI - A_{k-1})\mathbf{e}_1^\mathsf{T}(3zI - A_{k-1})^{-1}\mathbf{e}_1$$

is the minor of the matrix $3zI - A_{k-1}$ with respect to its top left entry. We therefore have

$$\det(3z - A_{k-1})\mathbf{e}_1^\mathsf{T}(3zI - A_{k-1})^{-1}\mathbf{e}_1 = \begin{cases} 9z^2 & \text{when } k = 2, \\ (\det \Phi_{k-2,3}(z))^2 & \text{when } k > 2. \end{cases}$$

This proves (6.5.15) and (6.5.16). □

We introduce the numbers

$$b_k = \frac{1}{3}(2^k - (-1)^k), \quad \delta_k = \begin{cases} 1 & \text{if } k \text{ is odd}, \\ 0 & \text{if } k \text{ is even}, \end{cases} \quad (k \in \mathbb{N}_0) \tag{6.5.17}$$

and for $\alpha = 2, 3$ the polynomials

$$P_{-1,3}(z) = -1, \quad P_{0,\alpha}(z) = \alpha, \tag{6.5.18}$$

$$P_{k,\alpha}(z) = P_{k-1,3}(z)[\alpha z^{\delta_k} P_{k-1,3}(z) - 2P_{k-2,3}^2(z)] \quad (k \in \mathbb{N}). \tag{6.5.19}$$

Proposition 6.5.3. *For $k \in \mathbb{N}$ and $\alpha = 2, 3$, the functions* $\det \Phi_{k,\alpha}$ *are odd polynomials and have the representation*

$$\det \Phi_{k,\alpha}(z) = z^{b_{k+1}} P_{k,\alpha}(z^2). \tag{6.5.20}$$

The polynomial $P_{k,\alpha}$ has degree $b_{k+1} + \delta_k - 1$, $(-1)^k P_{k,\alpha}(0) > 0$, and the zeros of $P_{k,\alpha}$ are located in the interval $(0, 1)$.

Proof. We are going to prove the statement by induction on k. For $k = 1$ and $k = 2$, the result is obvious from (6.5.14) and (6.5.15).

Now let $k \geq 3$ and assume that the statement is true for indices less than k. Then (6.5.16) gives

$$\det \Phi_{k,\alpha}(z) = z^{b_k} P_{k-1,3}(z^2)[\alpha z^{b_k+1} P_{k-1,3}(z^2) - 2z^{2b_{k-1}} P_{k-2,3}^2(z^2)].$$

We calculate

$$\begin{aligned} b_k + 1 - 2b_{k-1} &= \frac{1}{3}(2^k - (-1)^k) + 1 - \frac{2}{3}(2^{k-1} - (-1)^{k-1}) = 1 - (-1)^k = 2\delta_k, \\ b_k + 2b_{k-1} &= \frac{1}{3}(2^k - (-1)^k) + \frac{2}{3}(2^{k-1} - (-1)^{k-1}) = b_{k+1}. \end{aligned}$$

Hence the representation (6.5.20) holds in view of (6.5.19) with an odd number b_{k+1}. Furthermore,

$$(-1)^k P_{k,\alpha}(0) = (-1)^{k-1} P_{k-1,3}(0)[\alpha(1 - \delta_k)(-1)^{k-1} P_{k-1,3}(0) + 2P_{k-2,3}^2(0)] > 0.$$

Since all zeros of $\det \Phi_{k,\alpha}$ lie in $(-1, 1)$, it follows that the zeros of $P_{k,\alpha}$ lie in $(0, 1)$. Finally, by definition of $\Phi_{k,a}$ in (6.5.10) it is clear that the degree of the polynomial $\det \Phi_{k,\alpha}$ is $2^{k+1} - 1$, and

$$b_{k+1} + 2(b_{k+1} + \delta_k - 1) = 2^{k+1} + (-1)^k + 1 - (-1)^k - 2 = 2^{k+1} - 1$$

shows that the degree of $P_{k,\alpha}$ is $b_{k+1} + \delta_k - 1$. □

Theorem 6.5.4. *With the notation from Proposition* 6.5.3, *the characteristic functions on T_n ($n \in \mathbb{N}$) have the representations*

$$\phi_N^{(n)}(\lambda) = s^{2^n-1}(\sqrt{\lambda}, l) c^{b_n}(\sqrt{\lambda}, l) P_{n-1,2}(c^2(\sqrt{\lambda}, l)), \tag{6.5.21}$$
$$\phi_D^{(n)}(\lambda) = s^{2^n}(\sqrt{\lambda}, l) c^{2b_{n-1}}(\sqrt{\lambda}, l) P_{n-2,3}^2(c^2(\sqrt{\lambda}, l)), \tag{6.5.22}$$

where the functions $P_{k,\alpha}$ are defined in (6.5.18) *and* (6.5.19). *The zeros of the functions $P_{k,\alpha}$ are located in the interval* $(0, 1)$.

Proof. For $n = 1$, (6.5.21) follows from (6.5.8), (6.5.13) and (6.5.18). For $n \geq 2$, (6.5.21) follows from (6.5.8), (6.5.11) and (6.5.20). For $n = 1$, (6.5.22) follows from (6.5.9) and (6.5.18). For $n = 2$, (6.5.22) follows from (6.5.9), (6.5.11) and (6.5.18). For $n \geq 3$, (6.5.22) follows from (6.5.9), (6.5.12) and (6.5.20). The location of the zeros of the polynomials $P_{k,\alpha}$ has already been stated in Proposition 6.5.3. □

6.5.3 The eigenvalues

The results of this chapter so far allow us to give a representation of the eigenvalues of the problem for perfect binary trees. We will make use of the notation (6.5.17)–(6.5.19).

Remark 6.5.5. We observe that for $n = 1$ we have $\phi_N^{(1)}(\lambda) = 4s(\sqrt{\lambda}, l)$, and therefore all eigenvalues of (6.5.1)–(6.5.6) are simple and are given by λ_k ($k \in \mathbb{N}$) from Proposition 6.3.5.

The next theorem is concerned with the general case $n \geq 2$.

Theorem 6.5.6. *Let $n \in \mathbb{N}$, $n \geq 2$. Then there is a natural number $p_0(n)$ such that the spectrum of problem (6.5.1)–(6.5.6) on T_n with a real-valued symmetric function $q \in L_2(0,l)$ can be represented as the set of the terms of increasing sequences $(\lambda_k^{(1)})_{k=1}^\infty$, $(\lambda_k^{(2)})_{k=1}^\infty$, $(\lambda_k^{(j,+)})_{k=1}^\infty$, $(\lambda_k^{(j,-)})_{k=1}^\infty$ ($j = 1, \dots, p_0(n)$) of pairwise distinct real numbers with the following properties.*

1. *The eigenvalues $\lambda_k^{(1)}$ ($k \in \mathbb{N}$) have multiplicity $2^n - 1$ and have the asymptotic representation*

$$\lambda_k^{(1)} = \frac{\pi^2 k^2}{l^2} + \frac{Q}{l} + \beta_k^{(1)} \tag{6.5.23}$$

with $Q = \frac{1}{2}\int_0^l q(x)dx$ and $(\beta_k^{(1)})_{k=1}^\infty \in l_2$.

2. *The eigenvalues $\lambda_k^{(2)}$ ($k \in \mathbb{N}$) have multiplicity b_n and have asymptotic representation*

$$\lambda_k^{(2)} = \left(\frac{\pi(k-1/2)}{l}\right)^2 + \frac{Q}{l} + \beta_k^{(2)}, \tag{6.5.24}$$

with $(\beta_k^{(2)})_{k=0}^\infty \in l_2$.

3. *The eigenvalues $\lambda_k^{(j,+)}$ and $\lambda_k^{(j,-)}$ ($k \in \mathbb{N},\ j = 1, \dots, p_0(n)$) have multiplicity $p_j(n) > 0$ with*

$$\sum_{j=1}^{p_0(n)} p_j(n) = b_n + \delta_{n-1} - 1$$

and have the asymptotic representation

$$\lambda_k^{(j,\varepsilon)} = \left(\frac{k\pi - \arccos((-1)^k \varepsilon \sqrt{z_j})}{l}\right)^2 + \frac{Q}{l} + \beta_k^{(j,\varepsilon)}, \tag{6.5.25}$$

where $\varepsilon \in \{+,-\}$, $z_1, \dots, z_{p_0(n)}$ are the distinct zeros of the polynomial $P_{n-1,2}$, and $(\beta_k^{(j,\varepsilon)})_{k=1}^\infty \in l_2$.

4. *For $j = 1, \dots, p_0(n)$ and $\varepsilon \in \{+, -\}$, the eigenvalues $\lambda_k^{(1)}$, $\lambda_k^{(2)}$ and $\lambda_k^{(j,\varepsilon)}$ $(k \in \mathbb{N})$ are interlaced as follows:*

$$\lambda_1^{(j,+)} < \lambda_1^{(2)} < \lambda_1^{(j,-)} < \lambda_1^{(1)} < \lambda_2^{(j,-)} < \lambda_2^{(2)} < \lambda_2^{(j,+)} < \lambda_2^{(1)} < \lambda_3^{(j,+)} < \dots. \tag{6.5.26}$$

Proof. The eigenvalues of problem (6.5.1)–(6.5.6), counted with multiplicity, are the zeros of the characteristic function $\phi_N^{(n)}$. By (6.5.21) the eigenvalues are as stated, with $(\lambda_k^{(1)})_{k=1}^\infty$ being the increasing sequence of zeros of $\lambda \mapsto s(\sqrt{\lambda}, l)$, $(\lambda_k^{(2)})_{k=1}^\infty$ being the increasing sequence of zeros of $\lambda \mapsto c(\sqrt{\lambda}, l)$, and $(\lambda_k^{(j,\varepsilon)})_{k=1}^\infty$ being the increasing sequence of zeros of $\lambda \mapsto c(\sqrt{\lambda}, l) - \varepsilon\sqrt{z_p}$, where $0 < z_1 < \dots < z_{p_0(n)} < 1$ are the distinct zeros of $P_{n-1,2}$. The asymptotics in part 3 and the interlacing properties have been shown in Propositions 6.3.5 and 6.3.6. To prove part 4 we recall from Proposition 6.3.7 that $c\left(\sqrt{\lambda_k^{(1)}}, l\right) = (-1)^k$, using the notation of this theorem. Also, for $a \in (-1, 1)$ the function $\lambda \mapsto c(\sqrt{\lambda}, l) - a$ has a unique zero in each of the intervals $(-\infty, \lambda_1)$ and $(\lambda_k, \lambda_{k+1})$ $(k \in \mathbb{N})$ by Proposition 6.3.6. The interlacing (6.5.26) immediately follows from the above observations. □

For solving the inverse problem, we observe that there is a huge degree of redundancy in the spectrum of problem (6.5.1)–(6.5.6). We will consider both cases, that only the spectrum is known and that the eigenvalues, with multiplicity, are known. We will start with the latter case.

Theorem 6.5.7. *Let $(\lambda_k)_{k=1}^\infty$ be the nondecreasing sequence of eigenvalues of problem (6.5.1)–(6.5.6) on T_n $(n \in \mathbb{N},\ n \geq 2)$. Then the largest segment of equal numbers in $(\lambda_k)_{k=1}^\infty$ is of length $2^n - 1$. The sequences $(\lambda_k^{(j)})_{k=1}^\infty$ $(j = 1, 2)$ in Theorem 6.5.6 are given by $\lambda_k^{(1)} = \lambda_{k(2^{n+1}-2)}$ and $\lambda_k^{(2)} = \lambda_{(k-1)(2^{n+1}-2)+2^{n-1}}$ $(k \in \mathbb{N})$. Furthermore,*

$$\psi_0(\lambda) := \lambda^m \prod_{k=1}^{\infty}{}' \left(1 - \frac{\lambda}{\lambda_k^{(2)}}\right) \qquad (\lambda \in \mathbb{C}),$$

defines an entire function, where $\prod'$ means that the term with k_0 is omitted in case $\lambda_{k_0}^{(2)} = 0$ for some k_0, and in this case $m = 1$, whereas $m = 0$ otherwise. With $C_r := \psi_0(2\pi r l^{-1})$ $(r \in \mathbb{N})$ the limit

$$C := \lim_{r\to\infty} C_r \neq 0$$

exists, and

$$\psi := C^{-1}\psi_0 \tag{6.5.27}$$

satisfies $\psi(\lambda_k^{(1)}) = (-1)^k$ $(k \in \mathbb{N})$. The sequence $(\lambda_k)_{k=1}^\infty$ consists of $2^n - 1$ copies of the sequence $(\lambda_k^{(1)})_{k=1}^\infty$ with $\lambda_{k(2^{n+1}-2)-2^n+2} = \lambda_k^{(1)}$ $(k \in \mathbb{N})$, b_n copies of the *sequence* $(\lambda_k^{(2)})_{k=1}^\infty$ *and all zeros of the function* $P_{n-1,2} \circ \psi^2$, *counted with multiplicity.*

Proof. By (6.5.26), a segment of equal numbers in the sequence of eigenvalues consists of copies of one of $\lambda_k^{(1)}$, $\lambda_k^{(2)}$, $\lambda_k^{(j,\varepsilon)}$ $(k \in \mathbb{N}, j \in \{1, \dots, p_0(n)\}, \varepsilon \in \{+,-\}$. Because of $2^n - 1 > b_n \geq b_n + \delta_{n-1} - 1$ the largest segment of equal numbers in $(\lambda_k)_{k=1}^\infty$ is of length $2^n - 1$, and exactly each segment consisting of all copies of $\lambda_k^{(1)}$ has this maximal length.

The eigenvalues are grouped into segments $\lambda_{(k-1)(2^{n+1}-2)+1}, \dots, \lambda_{k(2^{n+1}-2)}$ $(k \in \mathbb{N})$. By (6.5.26), the first $b_n + \delta_{n-1} + 1$ eigenvalues in such a segment are of the form $\lambda_k^{(j,\pm)}$, followed by b_n copies of $\lambda_k^{(2)}$, then again $b_n + \delta_{n-1} + 1$ eigenvalues of the form $\lambda_k^{(j,\pm)}$, followed by $2^n - 1$ copies of $\lambda_k^{(1)}$. In particular, the first and last term of the latter are $\lambda_{k(2^{n+1}-2)-2^n+2} = \lambda_{k(2^{n+1}-2)} = \lambda_k^{(1)}$. It is now clear that there are $2^{n+1} - 2 - (2^n - 1) = 2^n - 1$ consecutive terms starting at $\lambda_{(k-1)(2^{n+1}-2)+1}$ with $\lambda_k^{(2)}$ in the middle, which shows that $\lambda_k^{(2)} = \lambda_{(k-1)(2^{n+1}-2)+2^{n-1}}$.

The function $c(\cdot, l)$ is a sine type function, see [96, Corollary 12.2.11]. Then [96, Lemma 11.2.29] shows that the function $\lambda \mapsto \psi_0(\lambda^2)$ is a constant multiple of $c(\cdot, l)$. Hence (6.3.25) shows that $c(\lambda, l) = \psi(\lambda^2)$ $(\lambda \in \mathbb{C})$. This completes the proof in view of Theorem 6.5.4. □

6.5.4 The inverse problem

Here we will show that the properties of the eigenvalues of (6.5.1)–(6.5.6) obtained in Theorem 6.5.7, together with the asymptotic properties (6.3.21) and (6.3.24) are also sufficient to realize such a sequence of real numbers as the eigenvalues, counted with multiplicity, of a problem (6.5.1)–(6.5.6) on an equilateral perfect binary tree with symmetric potential. The size of the tree, the lengths of the edges and the potential on the edges are then uniquely determined by the eigenvalues.

Theorem 6.5.8. *Let* $(\lambda_k)_{k=1}^\infty$ *be a nondecreasing sequence of real numbers such that there is a number* $n \in \mathbb{N}$, $n \geq 2$, *with*

$$\lambda_{k(2^{n+1}-2)-2^n+1} < \lambda_{k(2^{n+1}-2)-2^n+2} = \lambda_{k(2^{n+1}-2)} < \lambda_{k(2^{n+1}-2)+1} \quad (k \in \mathbb{N}). \tag{6.5.28}$$

Define

$$\lambda_k^{(1)} = \lambda_{k(2^{n+1}-2)}, \ \lambda_k^{(2)} = \lambda_{(k-1)(2^{n+1}-2)+2^{n-1}} \quad (k \in \mathbb{N}). \tag{6.5.29}$$

Assume that

$$l := \lim_{k \to \infty} \frac{\pi k}{\sqrt{\lambda_k^{(1)}}} \tag{6.5.30}$$

exists and is positive. Furthermore, assume that there is a real number Q *such that* $(\lambda_k^{(1)})_{k=1}^\infty$ *satisfies the asymptotic representation* (6.3.21) *and that* $(\lambda_k^{(2)})_{k=1}^\infty$ *satisfies the asymptotic representation* (6.3.24). *Let the function* ψ *defined by* (6.5.27)

satisfy $\psi(\lambda_k^{(1)}) = (-1)^k$ $(k \in \mathbb{N})$. *Assume that the terms of the sequence* $(\lambda_k^{(2)})$ *occur* b_n *times in the sequence* $(\lambda_k)_{k=1}^\infty$ *and assume that the terms of the sequence* $(\lambda_k)_{k=1}^\infty$ *which are different from* $\lambda_s^{(j)}$ $(j = 1, 2,\ s \in \mathbb{N})$ *are exactly the zeros of the the function* $\lambda \mapsto P_{n-1,2}(\psi^2(\lambda))$, *counted with multiplicity.*
Then there is a real-valued symmetric $q \in L_2(0, l)$ *such that* $(\lambda_k)_{k=1}^\infty$ *is the sequence of the eigenvalues of* (6.5.1)–(6.5.6) *on the equilateral perfect binary tree* T_n *with edge length* l. *The numbers* n *and* l *and the function* q *are uniquely determined by* $(\lambda_k)_{k=1}^\infty$.

Proof. Formula (6.5.28) implies that each of the numbers $\lambda_k^{(1)}$ $(k \in \mathbb{N})$ occurs exactly $2^n - 1$ times in the sequence $(\lambda_k)_{k=1}^\infty$. The remaining eigenvalues are therefore located in nonadjacent segments of length $2^{n+1} - 2 - (2^n - 1) = 2^n - 1$. Each of these segments contains a value $\lambda_k^{(2)}$ of multiplicity $b_n < 2^n - 1$. Hence also the remaining terms have multiplicity less than $2^n - 1$.

By [96, Theorem 12.6.2] there is a unique real-valued function $q \in L_2(0, l)$ such that the sequences $(\lambda_k^{(1)})_{k=1}^\infty$ and $(\lambda_k^{(2)})_{k=1}^\infty$ are the zeros of the functions $\lambda \mapsto s(\sqrt{\lambda}, l)$ and $\lambda \mapsto c(\sqrt{\lambda}, l)$ associated with (6.3.1), respectively.

Next we are going to show that q is symmetric. In the proof of Theorem 6.5.7 we have shown that $c(\lambda, l) = \psi(\lambda^2)$ $(\lambda \in \mathbb{C})$ (the fact that q is symmetric in Theorem 6.5.7 is irrelevant for this identity). Using the Lagrange identity

$$c(\lambda, l)s'(\lambda, l) - c'(\lambda, l)s(\lambda, l) = 1$$

we arrive at $c\left(\sqrt{\lambda_k^{(1)}}, l\right) s'\left(\sqrt{\lambda_k^{(1)}}, l\right) = 1$. Then the assumption $c\left(\sqrt{\lambda_k^{(1)}}, l\right) = \psi(\lambda_k^{(1)}) = (-1)^k$ shows that $c\left(\sqrt{\lambda_k^{(1)}}, l\right) = s'\left(\sqrt{\lambda_k^{(1)}}, l\right)$. Hence all singularities of the function

$$h := \frac{c(\cdot, l) - s'(\cdot, l)}{s(\cdot, l)}$$

are removable singularities, which means that h can be identified with an entire function. By [96, Corollary 12.2.11], the functions $\lambda \mapsto \lambda s(\lambda, l)$, $c(\cdot, l)$ and $s'(\cdot, l)$ are sine type functions of type l, which implies by [96, Lemma 11.2.6, part 2] that there is a number $M > 0$ such that $|c(\lambda, l) - s'(\lambda, l)| < Me^{l|\operatorname{Im}\lambda|}$ $(\lambda \in \mathbb{C})$. Furthermore, by [96, Remark 11.2.21], for each $\delta > 0$ there is a constant $c_\delta > 0$ such that the estimate $|\lambda s(\lambda, l)| \geq c_\delta e^{l|\operatorname{Im}\lambda|}$ holds for all $\lambda \in \mathbb{C} \setminus \Lambda_\delta$, where

$$\Lambda_\delta = \{\lambda \in \mathbb{C} : \exists \zeta \in \mathbb{C},\ \zeta s(\zeta, l) = 0 \text{ and } |\lambda - \zeta| < \delta\}.$$

It follows that $|h(\lambda)| \leq Mc_\delta^{-1}|\lambda|$ for $\lambda \in \mathbb{C} \setminus \Lambda_\delta$. When $\delta > 0$ is sufficiently small, then all components of Λ_δ are bounded, and the maximum principle shows that there are positive numbers c and d such that $|h(\lambda)| \leq c|\lambda| + d$ for $\lambda \in \mathbb{C}$. Therefore the entire function h is a polynomial of degree at most 1.

Observing the representations (6.3.25) and

$$s'(\lambda, l) = \cos\lambda l + Q\frac{\sin\lambda l}{2\lambda} + \int_0^l B_3(t)\frac{\sin\lambda t}{\lambda}\, dt$$

with $B_3 \in L_2(0,l)$ (see [96, 12.2.23)], it follows that

$$\lambda\left(c(\cdot,l)-s'(\cdot,l)\right)=\int_0^l (B(t)-B_3(t))\sin\lambda t\,dt=o(1) \text{ for } \lambda\in\mathbb{R} \text{ with } \lambda\to\infty$$

by the Riemann-Lebesgue lemma. On the other hand, the representation (6.3.30) shows that

$$\left(2r+\frac{1}{2}\right)\pi l^{-1}s\left(\left(2r+\frac{1}{2}\right)\pi l^{-1}\right)=1+O(r^{-1}) \text{ for } r\in\mathbb{N} \text{ with } r\to\infty.$$

These last two estimates show that $h\left(\left(2r+\frac{1}{2}\right)\pi l^{-1}\right)=o(1)$. Since h is a polynomial, it follows that $h=0$ and therefore $c(\cdot,l)=s'(\cdot,l)$.

In particular, $c(\cdot,l)$ and $s'(\cdot,l)$ have the same (simple) zeros. This means that the spectrum of the eigenvalue problem for

$$-y''+qy=\lambda y \tag{6.5.31}$$

with Neumann-Dirichlet boundary conditions $y'(0)=y(l)=0$ coincides with the spectrum of the eigenvalue problem for (6.5.31) with Dirichlet-Neumann boundary conditions $y(0)=y'(l)=0$. Putting $\widehat{q}(x)=q(l-x)$, the latter problem can be identified with the spectral problem

$$-y''+\widehat{q}y=\lambda y,\ y'(0)=y(l)=0.$$

Clearly, the Dirichlet-Dirichlet problem $y(0)=y(l)=0$ is the same for $-y''+qy=\lambda y$ and for $-y''+\widehat{q}y=\lambda y$. We have already used at the beginning of this proof that the potential q is uniquely determined by the spectra of the Dirichlet-Dirichlet problem and the Dirichlet-Neumann problem. This shows that $q=\widehat{q}$, which means that the function q is symmetric.

The statement about the remaining terms of the sequence $(\lambda_k)_{k=1}^\infty$ follows from Theorem 6.5.7.

By Theorem 6.5.6, the length of the largest segments of equal numbers in $(\lambda_k)_{k=1}^\infty$ is 2^n-1 and therefore uniquely determines n. Hence also $\lambda_k^{(1)}$ and $\lambda_k^{(2)}$ are uniquely determined. This shows that l and q are unique as well. □

We do not need to know the multiplicity of the eigenvalues. All we need to know is the spectrum. For this it is advantageous to have a more explicit representation of the polynomials $P_{k,\alpha}$. We note that (6.5.19) is essentially a three term recurrence relation for the polynomials $P_{k,3}$; $P_{k,2}$ is then obtained from $P_{k-1,3}$ and $P_{k-2,3}$. We will now rewrite this recurrence relation as a formally more complex two term recurrence relation for the pairs of polynomials $(P_{k,2},P_{k-1,3})$.

Proposition 6.5.9. *For $k\in\mathbb{N}$, the recurrence relations*

$$P_{k,2}(z)=2\left[P_{k-1,2}(z)+z^{\delta_{k-1}}P_{k-2,3}^2(z)\right]\left[z^{\delta_k}P_{k-1,2}(z)+(z-1)P_{k-2,3}^2(z)\right], \tag{6.5.32}$$

$$P_{k-1,3}(z)=P_{k-1,2}(z)+z^{\delta_{k-1}}P_{k-2,3}^2(z) \tag{6.5.33}$$

hold, and the initial terms are $P_{0,2}(z)=2$ and $P_{-1,3}(z)=-1$.

Proof. The initial terms have been stated in (6.5.18), and then (6.5.33) follows for $k = 1$ because $\delta_0 = 0$. The recurrence relation (6.5.19) shows for $k \geq 2$ that

$$P_{k-1,3}(z) - P_{k-1,2}(z) = P_{k-2,3}(z) z^{\delta_{k-1}} P_{k-2,3}(z),$$

which proves (6.5.33) for $k \geq 2$. Similarly, (6.5.19) and (6.5.33) give for $k \in \mathbb{N}$ that

$$\begin{aligned} P_{k,2}(z) &= P_{k-1,3}(z)[2z^{\delta_k} P_{k-1,3}(z) - 2P_{k-2,3}^2(z)] \\ &= [P_{k-1,2}(z) + z^{\delta_{k-1}} P_{k-2,3}^2(z)][2z^{\delta_k} P_{k-1,2}(z) + 2zP_{k-2,3}^2(z) - 2P_{k-2,3}^2(z)] \\ &= 2[P_{k-1,2}(z) + z^{\delta_{k-1}} P_{k-2,3}^2(z)][z^{\delta_k} P_{k-1,2}(z) + (z-1)P_{k-2,3}^2(z)], \end{aligned}$$

which is (6.5.32). □

Theorem 6.5.10. *For $n \in \mathbb{N}_0$ the polynomials $P_{n,2}$ and $P_{n-1,3}$ can be written in the form*

$$P_{n,2} = 2A_n^2 B_n C_n, \tag{6.5.34}$$
$$P_{n-1,3} = A_n B_n, \tag{6.5.35}$$

where A_n, B_n, and C_n are polynomials which satisfy the recurrence relations

$$A_{n+1}(z) = A_n^2(z) B_n(z), \tag{6.5.36}$$
$$B_{n+1}(z) = 2C_n(z) + z^{\delta_n} B_n(z), \tag{6.5.37}$$
$$C_{n+1}(z) = 2z^{\delta_{n+1}} C_n(z) + (z-1)B_n(z), \tag{6.5.38}$$

with the initial terms being

$$A_0(z) = -1, \quad B_0(z) = 1, \quad C_0(z) = 1. \tag{6.5.39}$$

Proof. It is clear from Proposition 6.5.9 and (6.5.39) that (6.5.34) and (6.5.35) hold for $n = 0$. Now let $n \in \mathbb{N}_0$ and assume that (6.5.34) and (6.5.35) hold for this n. Then, in view of (6.5.33),

$$\begin{aligned} A_{n+1}(z)B_{n+1}(z) &= A_n^2(z)B_n(z)(2C_n(z) + z^{\delta_n} B_n(z)) \\ &= 2A_n^2(z)B_n(z)C_n(z) + z^{\delta_n} A_n^2(z)B_n^2(z) \\ &= P_{n,2}(z) + z^{\delta_n} P_{n-1,3}^2(z) \\ &= P_{n,3}(z), \end{aligned}$$

which proves (6.5.35) for $n + 1$, and together with

$$\begin{aligned} A_{n+1}(z)C_{n+1}(z) &= A_n^2(z)B_n(z)(2z^{\delta_{n+1}} C_n(z) + (z-1)B_n(z)) \\ &= (2z^{\delta_{n+1}} A_n^2(z)B_n^2(z)C_n(z) + (z-1)A_n^2(z)B_n^2(z)) \\ &= z^{\delta_{n+1}} P_{n,2}(z) + (z-1)P_{n-1,3}^2(z) \end{aligned}$$

it follows in view of (6.5.32) that

$$
\begin{aligned}
2A_{n+1}^2(z)B_{n+1}(z)C_{n+1}(z) \\
&= 2[P_{n,2}(z) + z^{\delta_n} P_{n-1,3}^2(z)][z^{\delta_{n+1}} P_{n,2}(z) + (z-1)P_{n-1,3}^2(z)] \\
&= P_{n+1,2}(z),
\end{aligned}
$$

which proves (6.5.34) for $n+1$. □

Corollary 6.5.11. *The explicit forms of the polynomials A_n, B_n and C_n for $n \in N_0$ are:*

$$B_{2n}(z) = F(z)(\alpha_2^n(z) - \alpha_1^n(z)) + \frac{1}{2}(\alpha_1^n(z) + \alpha_2^n(z)), \tag{6.5.40}$$

$$B_{2n+1}(z) = (F(z) + 2G(z))(\alpha_2^n(z) - \alpha_1^n(z)) + \frac{3}{2}(\alpha_1^n(z) + \alpha_2^n(z)), \tag{6.5.41}$$

$$C_{2n}(z) = G(z)(\alpha_2^n(z) - \alpha_1^n(z)) + \frac{1}{2}(\alpha_1^n(z) + \alpha_2^n(z)), \tag{6.5.42}$$

$$C_{2n+1}(z) = H(z)(\alpha_2^n(z) - \alpha_1^n(z)) + \frac{3z-1}{2}(\alpha_1^n(z) + \alpha_2^n(z)), \tag{6.5.43}$$

$$A_n(z) = \prod_{k=1}^{n-1} B_k^{2^{n-1-k}}(z) \quad (n > 0), \tag{6.5.44}$$

where

$$F(z) = \frac{9z}{2(\alpha_2(z) - \alpha_1(z))}, \quad G(z) = \frac{9z-6}{2(\alpha_2(z) - \alpha_1(z))}, H(z) = \frac{27z^2 - 21z}{2(\alpha_2(z) - \alpha_2(z))},$$

$$\alpha_1(z) = \frac{9z - 4 - 3\sqrt{9z^2 - 8z}}{2}, \quad \alpha_2(z) = \frac{9z - 4 + 3\sqrt{9z^2 - 8z}}{2}.$$

Proof. The recurrence relations (6.5.37), (6.5.38) can be written in matrix form

$$\begin{pmatrix} B_{n+1}(z) \\ C_{n+1}(z) \end{pmatrix} = \begin{pmatrix} z^{\delta_n} & 2 \\ z-1 & 2z^{\delta_{n+1}} \end{pmatrix} \begin{pmatrix} B_n(z) \\ C_n(z) \end{pmatrix}. \tag{6.5.45}$$

With

$$D(z) := \begin{pmatrix} z & 2 \\ z-1 & 2 \end{pmatrix} \begin{pmatrix} 1 & 2 \\ z-1 & 2z \end{pmatrix} = \begin{pmatrix} 3z-2 & 6z \\ 3(z-1) & 6z-2 \end{pmatrix}$$

it follows that

$$\begin{pmatrix} B_{2n+2}(z) \\ C_{2n+2}(z) \end{pmatrix} = D(z) \begin{pmatrix} B_{2n}(z) \\ C_{2n}(z) \end{pmatrix}$$

for $n \in \mathbb{N}_0$, and this together with (6.5.39) gives

$$\begin{pmatrix} B_{2n}(z) \\ C_{2n}(z) \end{pmatrix} = D^n(z) \begin{pmatrix} 1 \\ 1 \end{pmatrix}.$$

A straightforward calculation shows that the eigenvalues of $D(z)$ are $\alpha_1(z)$ and $\alpha_2(z)$. For the generic case when the eigenvalues are distinct, a diagonalization of $D(z)$ has the form

$$D(z) = E(z)\Delta(z)E^{-1}(z),$$

with

$$\Delta(z) = \operatorname{diag}(\alpha_1(z), \alpha_2(z))$$

and where a matrix of eigenvectors $E(z)$ and its inverse $E^{-1}(z)$ are given by

$$E(z) = \begin{pmatrix} 4z & 4z \\ z - \sqrt{9z^2 - 8z} & z + \sqrt{9z^2 - 8z} \end{pmatrix},$$

$$E^{-1}(z) = \frac{1}{8z\sqrt{9z^2 - 8z}} \begin{pmatrix} z + \sqrt{9z^2 - 8z} & -4z \\ -z + \sqrt{9z^2 - 8z} & 4z \end{pmatrix}.$$

Then

$$\begin{aligned} \begin{pmatrix} B_{2n}(z) \\ C_{2n}(z) \end{pmatrix} &= E(z)\Delta^n(z)E^{-1}(z)\begin{pmatrix} 1 \\ 1 \end{pmatrix} \\ &= \frac{1}{8z\sqrt{9z^2 - 8z}} E(z) \begin{pmatrix} \alpha_1^n(z) & 0 \\ 0 & \alpha_2^n(z) \end{pmatrix} \begin{pmatrix} -3z + \sqrt{9z^2 - 8z} \\ 3z + \sqrt{9z^2 - 8z} \end{pmatrix}, \end{aligned}$$

and a straightforward calculation gives (6.5.40) and (6.5.42). Then (6.5.41) and (6.5.43) easily follow from (6.5.45). Observing that $A_1(z) = A_0^2(z)B_0(z) = 1$, the identity (6.5.44) easily follows by induction. □

For a polynomial P with at least one real zero let w_P be its largest real zero.

Corollary 6.5.12. *The polynomials B_n and C_n defined in Theorem* 6.5.10 *for $n = 1$ and $n = 2$ are*

$$B_1(z) = 3, \; C_1(z) = 3z - 1, \; B_2(z) = 9z - 2, \; C_2(z) = 9z - 5.$$

For $n \geq 2$, the zeros of B_n and C_n are real, lie in the interval $(0, 1)$ and satisfy $w_{C_n} > w_{B_n}$ and $w_{C_n} > w_{C_{n-1}}$.

Proof. The representations for $n = 1$ and $n = 2$ immediately follow from the recurrence relations (6.5.37) and (6.5.38). From $w_{C_2} = \frac{5}{9} > \frac{2}{9} = w_{B_2}$ and $w_{C_2} = \frac{5}{9} > \frac{1}{3} = w_{C_1}$ we see that the inequalities are satisfied for $n = 2$. Since the zeros of B_n and C_n are also zeros of $P_{n,2}$ by (6.5.34), the statement on the location of the zeros of B_n and C_n follows from Proposition 6.5.3. We are going to show by induction that $w_{C_n} > w_{B_n}$ and that $w_{C_n} > w_{B_n}$ implies $w_{C_{n+1}} > w_{C_n}$ for $n \geq 2$. The recurrence relations (6.5.37) and (6.5.38) show that the leading coefficients of B_n and C_n are positive. Now let $n \geq 2$ and assume that $w_{C_n} > w_{B_n}$. Then B_{n+1} is positive on $[w_{C_n}, \infty)$ by (6.5.37), which means that $w_{B_{n+1}} < w_{C_n}$. On the other hand, $(z-1)B_n(z) < 0$ for $z \in [w_{C_n}, 1)$, and then (6.5.38) shows that $C_{n+1}(w_{C_n}) < 0$. Therefore C_{n+1} has a least one zero in (w_{C_n}, ∞), which proves $w_{C_{n+1}} > w_{C_n}$. In particular, $w_{C_{n+1}} > w_{C_n} > w_{B_{n+1}}$. □

Corollary 6.5.13. *For $n \in \mathbb{N}$, $w_{P_{n,2}} < w_{P_{n+1,2}}$.*

Proof. We know from (6.5.34), (6.5.44) and Corollary 6.5.12 that

$$w_{P_{n,2}} = \max\{w_{B_2}, \dots, w_{B_n}, w_{C_n}\} = w_{C_n},$$

and another application of Corollary 6.5.12 completes the proof. □

In the last theorem of this section we will show that the data of the string problem on a perfect tree are uniquely determined by its spectrum.

Theorem 6.5.14. *For $n \geq 2$, the numbers n and l and the symmetric function q are uniquely determined by the spectrum of problem* (6.5.1)–(6.5.6).

Proof. We enumerate the spectrum as an increasing sequence of real numbers $(\sigma_k)_{k=1}^{\infty}$. Since $p_0(n)$ is the number of distinct zeros of $P_{n-1,2}$, it is clear from (6.5.27) that $\lambda_k^{(1)} = \sigma_{2kN}$ and $\lambda_k^{(2)} = \sigma_{(2k-1)N}$ where $N = p_0(n) + 1$. Hence it follows as in Theorem 6.5.6 that l and q are uniquely determined by the sequence $(\sigma_k)_{k=1}^{\infty}$ and the number n.

We still have to show the uniqueness of n. Therefore let $n \in \mathbb{N}$, $n \geq 2$, be an integer for which $(\sigma_k)_{k=1}^{\infty}$ enumerates the spectrum on the perfect tree T_n. Let $(\lambda_k)_{k=1}^{\infty}$ be the nondecreasing sequence of the eigenvalues of the problem under consideration, counted with multiplicity. Then Theorem 6.5.6, parts 3 and 4, shows that

$$\sigma_{2kN-1} = \frac{k^2\pi^2}{l^2} - \frac{2k\pi\alpha_n}{l} + O(1) \quad \text{as } k \to \infty,$$

where $\alpha_n = \arccos\sqrt{\tilde{z}}$ and $\tilde{z}$ is the largest zero of $P_{n-1,2}$. If n_1 and n_2 are two numbers for which $(\sigma_k)_{k=1}^{\infty}$ enumerates the spectrum on T_{n_1} and T_{n_2}, and denoting the numbers N and l for n_1 and n_2 by l_1, l_2 and N_1, N_2, respectively, it follows that

$$\sigma_{2kN_1N_2-1} = \frac{(kN_2)^2\pi^2}{l_1^2} - \frac{2kN_2\pi\alpha_{n_1}}{l_1} + O(1) \quad \text{as } k \to \infty,$$

$$\sigma_{2kN_1N_2-1} = \frac{(kN_1)^2\pi^2}{l_2^2} - \frac{2kN_1\pi\alpha_{n_2}}{l_2} + O(1) \quad \text{as } k \to \infty.$$

This shows that $N_2 l_1^{-1} = N_1 l_2^{-1}$ and then $\alpha_{n_1} = \alpha_{n_2}$. But by Corollary 6.5.13, α_n is strictly decreasing when n increases, and therefore $n_1 = n_2$. □

Remark 6.5.15. A combination of Theorems 6.5.6, 6.5.7, 6.5.8 and 6.5.14 gives necessary and sufficient conditions for a set of real numbers to be the spectrum of problem (6.5.1)–(6.5.6) on a perfect equilateral tree with more than two edges and equal symmetric potentials.

6.6 Sturm-Liouville problems on star graphs

In this section we consider a plane star graph of g strings ($g \geq 2$), joined at the central vertex called the root and with all g pendant vertices fixed. We recall that spectral problems for a star graph of Stieltjes strings have been considered in Section 2.1.

In what follows we need a slightly modified notion of majorization.

Definition 6.6.1 (see [94, p. 12]). Let $x = (x_i)_{i=1}^s$ and $y = (y_i)_{i=1}^t$ be two vectors with nonnegative entries ordered nonincreasingly, $x_1 \geq x_2 \geq \cdots \geq x_s \geq 0$ and $y_1 \geq y_2 \geq \cdots \geq y_t \geq 0$. If $s = t$, then x is said to *weakly majorize* y if

$$x \succ_w y :\Leftrightarrow \sum_{i=1}^{\tau} x_i \geq \sum_{i=1}^{\tau} y_i \quad (\tau = 1, \dots, t). \tag{6.6.1}$$

If $s \neq t$, we fill up the shorter vector with zeros, i.e., we set $\widetilde{x} = (x_i)_{i=1}^{\max\{s,t\}}$ and $\widetilde{y} = (y_i)_{i=1}^{\max\{s,t\}}$ with $x_i = 0$ for $i = s+1, \dots, \max\{s,t\}$, $y_i = 0$ for $i = t+1, \dots, \max\{s,t\}$. Then x is said to *weakly majorize* y, $x \succ_w y$, if $\widetilde{x}$ majorizes $\widetilde{y}$, $\widetilde{x} \succ_w \widetilde{y}$.

We will otherwise use the same notation as in Subsection 2.1.3.

6.6.1 Multiplicities and location of eigenvalues below a fixed value

In this subsection we revisit the string problem on a star graph whose edges are labelled by $j = 1, \dots, g$ such that the j-th edge carries $n_j > 0$ beads with mass $m_k^{(j)}$ in its interior ($k = 1, \dots, n_j$) and $n_1 \geq n_2 \geq \cdots \geq n_g$. Recall that $(\mu_k)_{k=1}^n$ denotes the nondecreasing sequence of the characteristic values of the Neumann problem (N1) on the star graph.

The results of Subsection 2.1.3 can be interpreted in the following way.

We notice that by Theorem 2.1.5, μ_k is a characteristic value of problem (N1) of multiplicity $p_k \geq 2$ if and only if it is a characteristic value of the problem (D1) of multiplicity $p_k + 1$. This means that μ_k is a characteristic value of the Dirichlet-Dirichlet problem on exactly $p_k + 1$ amongst the g edges.

Let $(\nu_k^{(j)})_{k=1}^{n_j}$ be the increasing sequence of the (positive) characteristic values of the Dirichlet-Dirichlet problem on the edges ($j = 1, \dots, g$). For $E \in (0, \infty]$ let

$$n_j(E) := \#\{k \in \{1, \dots, n_j\} : \nu_k^{(j)} \leq E\}.$$

Clearly, $n_j(E) = 0$ if and only if $\nu_1^{(j)} > E$, otherwise, $n_j(E)$ is the largest integer $k \in \{1, \dots, n_j\}$ for which $\nu_k^{(j)} \leq E$, i.e., $n_j(E) = \max\{k \in \{1, \dots, n_j\} : \nu_k^{(j)} \leq E\}$

if $\nu_1^{(j)} \le E$, and $n_j(\infty) = n_j$. We define

$$n(E) := \sum_{j=1}^{g} n_j(E), \tag{6.6.2}$$

$$N_i(E) := \#\{j \in \{1, \dots, g\} : n_j(E) \ge i\}, \tag{6.6.3}$$

$$n_N(E) := \#\{k \in \{1, \dots, n\} : \mu_k \le E\}. \tag{6.6.4}$$

Then $N_1(E) \ge N_2(E) \ge \dots \ge N_{n_1}(E)$.

With the above notation, the following theorem immediately follows from Theorems 2.1.5 and 2.1.19 and the proof of Theorem 2.1.19.

Theorem 6.6.2. *Let $(\mu_k)_{k=1}^n$ be the nondecreasing sequence of the characteristic values of the Neumann problem* (N1)*, counted with multiplicity, let $E > 0$, let $r(E)$ be the number of distinct characteristic values of* (N1) *which are less or equal E, let $(p_i)_{i=1}^{r(E)}$ be the vector of their multiplicities, and let $p^{\downarrow}(E) = (p_i^{\downarrow})_{i=1}^{r(E)}$ be the corresponding vector of multiplicities in nonincreasing order. Then:*

1. $n_N(E) = \sum_{i=1}^{r(E)} p_i$; $n_N(E) = n(E)$ *or* $n_N(E) = n(E) + 1$;

2. $0 < \mu_1 < \mu_2 \le \mu_3 \le \dots \le \mu_{n_N(E)-1} \le \mu_{n_N(E)} \le E$;

3. *if* $p_i > 1$, *then* $p_{i-1} = p_{i+1} = 1$ $(i = 2, \dots, r(E) - 1)$, *and if* $p_{r(E)} > 1$, *then* $p_{r(E)-1} = 1$;

4. $(N_1(E) - 1, N_2(E) - 1, \dots, N_{n_1(E)}(E) - 1) \succ (p_1^{\downarrow}, p_2^{\downarrow}, \dots, p_{r(E)-n_1(E)}^{\downarrow})$ *in case* $n_N(E) = n(E)$;

4′. $(N_1(E) - 1, N_2(E) - 1, \dots, N_{n_1(E)}(E) - 1) \succ (p_1^{\downarrow}, p_2^{\downarrow}, \dots, p_{r(E)-n_1(E)-1}^{\downarrow})$ *in case* $n_N(E) = n(E) + 1$.

Although Theorem 6.6.2 is formulated for the characteristic values of a star graph of Stieltjes strings, in its essence it is purely algebraic, only depending on the properties of two sequences as stated in Theorem 2.1.5. Therefore being independent of the initial physical meaning, it can be extended to problems which have infinitely many characteristic values.

6.6.2 Sturm-Liouville problems on star graphs with finite edges

The following problem appears in quantum mechanics as well as in the theory of small transverse vibrations of a star graph of smooth strings with g edges of lengths $a_j > 0$ $(j = 1, \dots, g)$ (see, e.g., [96, p. 238]):

$$y_j'' + \lambda y_j - q_j(x)y_j = 0, \quad (j = 1, \dots, g, \; x \in [0, a_j]), \tag{6.6.5}$$

$$y_j(0) = 0 \quad (j = 1, \dots, g), \tag{6.6.6}$$

$$y_1(a_1) = \dots = y_g(a_g), \tag{6.6.7}$$

$$\sum_{j=1}^{g} y_j'(a_j) = 0. \tag{6.6.8}$$

We assume that $q_j \in L_2(0, a_j)$ $(j = 1, \dots, n)$ are real-valued functions and that $q_j \geq Q_j$ for some $Q_j \in \mathbb{R}$. The problem (6.6.5)–(6.6.8) is called the Neumann problem on the star graph.

We also consider the corresponding Dirichlet problem (6.6.5), (6.6.6),

$$y_1(a_1) = \dots = y_g(a_g) = 0. \tag{6.6.9}$$

With the notation from Section 6.3 and $s_j(\sqrt{\lambda}, \cdot)$ being the solution of

$$y_j'' + \lambda y_j - q_j(x)y_j = 0, \quad (x \in [0, a_j]),$$
$$y_j(0) = y_j(a_j) = 0,$$

the increasing sequence $(\nu_k^{(j)})_{k=1}^{\infty}$ of the eigenvalues of this problem is the increasing sequence of the zeros of $\lambda \mapsto s_j(\sqrt{\lambda}, a_j)$ $(j = 1, \dots, g)$. Also, let $\widehat{s}_j(\sqrt{\lambda}, \cdot)$ be the solution of this problem with q_j replaced by Q_j and let $(\widehat{\nu}_k^{(j)})_{k=1}^{\infty}$ be the increasing sequence of the corresponding eigenvalues. The well-know Sturm Comparison Theorem shows the inequalities $\widehat{\nu}_k^{(j)} \leq \nu_k^{(j)}$ $(j = 1, \dots, g, \; k \in \mathbb{N})$; these inequalities are, e.g., a special case of [145, Theorem 4.9.1(1)], or easily follow from [139, Theorem 13.1].

Let $E \in \mathbb{R}$ and define

$$n_j(E) := \#\{k \in \mathbb{N} : \nu_k^{(j)} \leq E\}, \quad \widehat{n}_j(E) := \#\{k \in \mathbb{N} : \widehat{\nu}_k^{(j)} \leq E\} \quad (j = 1, \dots, g).$$

It is clear that

$$n_j(E) \leq \widehat{n}_j(E) := \begin{cases} \lfloor \frac{a_j}{\pi} \sqrt{E - Q_j} \rfloor & \text{if } E > Q_j, \\ 0 & \text{if } E \leq Q_j, \end{cases} \tag{6.6.10}$$

where $\lfloor b \rfloor$ means the integer part of a nonnegative number b. The nondecreasing union of the sequences $(\nu_k^{(j)})_{j=1}^{\infty}$ $(j = 1, \dots, g)$ will be denoted by $(\nu_k)_{k=1}^{\infty}$.

In view of [96, (6.4.5)], the characteristic function of the Neumann problem (6.6.5)–(6.6.8) is

$$\phi_N(\lambda) = \sum_{j=1}^{g} s_j'(\sqrt{\lambda}, a_j) \prod_{\substack{r=1 \\ r \neq j}}^{g} s_r(\sqrt{\lambda}, a_r). \tag{6.6.11}$$

Proposition 6.6.3. *The eigenvalues of problem* (6.6.5)–(6.6.8) *are real and can be arranged in a nondecreasing sequence* $(\mu_k)_{k=1}^{\infty}$. *This sequence satisfies*

$$\mu_1 < \nu_1 \le \mu_2 \le \nu_2 \le \cdots \le \mu_{k-1} \le \nu_{k-1} \le \mu_k \le \cdots. \tag{6.6.12}$$

Proof. With

$$\psi(\lambda) := \frac{\phi_N(\lambda)}{\phi_D(\lambda)},\ \phi_D(\lambda) := \prod_{r=1}^{g} s_r(\sqrt{\lambda}, a_r),\ \psi_j(\lambda) := \frac{s_j'(\sqrt{\lambda}, a_j)}{s_j(\sqrt{\lambda}, a_j)}\ (j = 1, \dots, g),$$

it is clear that

$$\psi(\lambda) = \sum_{j=1}^{g} \psi_j(\lambda).$$

The zeros of $\lambda \mapsto s_j(\sqrt{\lambda}, a_j)$ are the eigenvalues of the Dirichlet-Dirichlet problem on the j-th string, whereas the zeros of $\lambda \mapsto s_j'(\sqrt{\lambda}, a_j)$ are the eigenvalues of the Dirichlet-Neumann problem on the j-th string. Reversing the direction of the strings, it follows from Proposition 6.3.5 that

$$\xi_1^{(j)} < \nu_1^{(j)} < \xi_2^{(j)} < \nu_2^{(j)} < \dots \quad (j = 1, \dots, g),$$

where $(\xi_k^{(j)})_{k=1}^{\infty}$ is the increasing sequence of the zeros of $\lambda \mapsto s_j'(\sqrt{\lambda}, a_j)$. By [96, Corollary 12.2.11], the functions $\lambda \mapsto \lambda s_j(\lambda, a_j)$ and $s_j(\cdot, a_j)$ are sine type functions. From the representation [96, Lemma 11.2.29], from [96, Remark 11.1.6] and from $s_j(\sqrt{\lambda}, a_j) \to \infty$ and $s_j'(\sqrt{\lambda}, a_j) \to \infty$ as $\lambda \to -\infty$ it is clear that $-\psi_j$ $(j = 1, \dots, g)$ are Nevanlinna functions since their smallest zero is smaller than their smallest pole. Hence also their sum $-\psi$ is a Nevanlinna function. Let $\nu_1^{(0)} := \min\{\nu_1^{(j)} : j = 1, \dots, g\}$ and $\xi_1^{(0)} := \min\{\xi_1^{(j)} : j = 1, \dots, g\}$. Then $-\psi < 0$ on $(-\infty, \xi_1^{(0)})$ and $\nu_1^{(0)}$ is the smallest pole of $-\psi$. Since Nevanlinna functions are locally increasing, $\xi_1^{(0)} \le \mu_1 < \nu_1^{(0)} = \nu_1$ holds. The remaining inequalities in (6.6.12) follow as the zeros and poles of the Nevanlinna function $-\psi$ interlace. □

For $E \in \mathbb{R}$ let $r(E)$ be the number of distinct eigenvalues of the Neumann problem which are less or equal E. Let $(p_1, \dots, p_{r(E)})$ be the vector and $(p_1^{\downarrow}, \dots, p_{r(E)}^{\downarrow})$ the ordered vector of multiplicities of the distinct eigenvalues of problem (6.6.7) which are less or equal E.

Theorem 6.6.4. *Let* $E \in \mathbb{R}$. *Then the eigenvalues of* (6.6.7) *have the following properties:*

1. $\mu_1 < \mu_2 \le \mu_3 \le \dots \le \mu_{n(E,N)-2} \le \mu_{n(E,N)-1} \le \mu_{n(E,N)} \le E < \mu_{n(E,N)+1}$, *where* $n(E, N) \le \widehat{n}(E) + 1 := \sum_{j=1}^{q} \widehat{n}_j(E) + 1$;
2. *if* $p_i > 1$, *then* $p_{i-1} = p_{i+1} = 1$ $(i = 2, \dots, r(E) - 1)$, *and if* $p_{r(E)} > 1$, *then* $p_{r(E)-1} = 1$;

3. $(N_1(E)-1, N_2(E)-1, \dots, N_{n_1(E)}(E)-1) \succ_w (p_1^\downarrow, p_2^\downarrow, \dots, p_{r(E)-n_1(E)-1}^\downarrow)$.

Proof. 1. Part 1 immediately follows from (6.6.10) and (6.6.12).

2. Let $k \in \mathbb{N}$ be such that μ_k is the i-th distinct eigenvalue of the Neumann problem and assume that $p_i > 1$. Then $\mu_k = \nu_{k-1}$ or $\mu_k = \nu_k$ so that μ_k is a pole of at least one ψ_j. As we have seen for ν_1 in the proof of Proposition 6.6.3, a pole of some ψ_j is also a pole of ψ, which means that $m(\phi_N, \mu_k) < m(\phi_D, \mu_k)$ and the interlacing in part 1 implies that $\nu_{k-1} = \mu_k = \nu_k$. It is clear that in view of the interlacing from part 1, no two adjacent distinct eigenvalue μ_k can have this property. This proves part 2.

3. If $\mu_k \le E < \nu_k$, then $\mu_{k-1} \le \nu_{k-1} < \mu_k$ by the proof of part 2. In the notation of Subsection 2.1.3 we identify n_j with $n_j(E)$. Then the results of Subsection 2.1.3 are applicable, and part 3 of Theorem 2.1.19 completes the proof. Here we note that the weak majorization $\succ_w$ is the majorization $\succ$ if $\mu_k \le E < \nu_k$ for some k. Otherwise, majorization is obtained by part 3 of Theorem 2.1.19 if $p_{r(E)-n_1(E)}^\downarrow$ is added as the last term in the vector on the right-hand side of the statement of part 3. □

6.6.3 Sturm-Liouville problems on star graphs with infinite edges

In this subsection we consider a star graph with g infinite edges, with Sturm-Liouville problems on the edges and with continuity and Kirchhoff conditions at the central vertex. Thus, we have

$$y_j'' + \lambda y_j - q_j(x) y_j = 0, \quad (j = 1, \dots, g, \ x \in [0, \infty]), \tag{6.6.13}$$

$$y_1(0) = \cdots = y_g(0), \tag{6.6.14}$$

$$\sum_{j=1}^{g} y_j'(0) = 0, \tag{6.6.15}$$

where q_j $(j = 1, \dots, g)$ are real valued measurable functions on $(0, \infty)$ satisfying $\int_0^\infty (1+x)|q_j(x)|\,dx < \infty$. Problem (6.6.13)–(6.6.15) is called Neumann problem on the star graph. Let $Q = \operatorname{diag}(q_1, \dots, q_g)$.

The operator $\mathcal{T}_N$ associated with the Neumann problem on the star graph is defined by

$$\mathcal{T}_N Y = -Y'' + QY,$$

where its domain $D(\mathcal{T}_N)$ is the set of all $Y = (y_1, \dots, y_g) \in (L_2(0,\infty))^g$ which satisfy $-Y'' + QY \in (L_2(0,\infty))^g$, (6.6.14) and (6.6.15).

We also consider the auxiliary Dirichlet problem (6.6.13),

$$y_1(0) = \cdots = y_g(0) = 0, \tag{6.6.16}$$

where the differential operator $\mathcal{T}_D$ associated with the Dirichlet problem on the star graph is defined by

$$\begin{aligned}
&\mathcal{T}_D Y = -Y'' + QY, \\
&D(\mathcal{T}_D) = \{Y \in (L_2(0,\infty))^g : -Y'' + QY \in (L_2(0,\infty))^g,\ Y(0) = 0)\},
\end{aligned}$$

The Dirichlet problem on the star graph is the direct product of the Dirichlet problems on the edges, where the differential operator $\mathcal{T}_{D,j}$ in $L_2(0,\infty)$ on the j-th edge is defined by

$$\begin{aligned}
&\mathcal{T}_{D,j} y = -y'' + q_j y, \\
&D(\mathcal{T}_{D,j}) = \{y \in L_2(0,\infty) : -y'' + q_j y \in L_2(0,\infty),\ y(0) = 0)\}.
\end{aligned}$$

Furthermore, the Neumann problem on the j-th edge is realized by the operator operator $\mathcal{T}_{N,j}$ in $L_2(0,\infty)$ defined by

$$\begin{aligned}
&\mathcal{T}_{N,j} y = -y'' + q_j y, \\
&D(\mathcal{T}_{N,j}) = \{y \in L_2(0,\infty) : -y'' + q_j y \in L_2(0,\infty),\ y'(0) = 0)\}.
\end{aligned}$$

Here we note that y as well as y' are continuous on $[0,\infty)$ for each $y \in L_2(0,\infty)$ which satisfies $-y'' + q_j y \in L_2(0,\infty)$.

Next we will state three propositions and one theorem, which will be proved together after the statement of the theorem.

Proposition 6.6.5. *For $j = 1,\dots,g$, the operator $\mathcal{T}_{D,j}$ is selfadjoint, its essential spectrum is $[0,\infty)$, and it has n_j eigenvalues, all simple and located in $(-\infty,0)$, where*

$$0 \le n_j \le \int_0^\infty x q_j^-(x)\,dx < \infty \tag{6.6.17}$$

and $q_j^-(x) := \max\{0, -q(x)\}$.

Proposition 6.6.6. *For $j = 1,\dots,g$, the operator $\mathcal{T}_{N,j}$ is selfadjoint, its essential spectrum is $[0,\infty)$, and it has at most $n_j + 1$ eigenvalues, all simple and located in $(-\infty,0)$.*

Proposition 6.6.7. *The operator $\mathcal{T}_D$ is selfadjoint, its essential spectrum is $[0,\infty)$, and it may have finitely many eigenvalues, all located in $(-\infty,0)$. The total multiplicity n of the eigenvalues satisfies*

$$0 \le n = \sum_{j=1}^g n_j.$$

Theorem 6.6.8. *The operator $\mathcal{T}_N$ is selfadjoint, its essential spectrum is $[0,\infty)$, and it may have finitely many eigenvalues, all located in $(-\infty,0)$.*

Proof of Propositions 6.6.5, 6.6.6, 6.6.7 *and Theorem* 6.6.8. All statements follow from the summary given in the introduction of [137], which also summarizes results from [135] and [1], taking [136] and [138] into account. In particular, [135, Theorem 6.3] states that the essential spectrum of any such selfadjoint operator is $[0, \infty)$, that the operator is bounded below and that its eigenvalues, if any, are located in $(-\infty, 0)$.

For Propositions 6.6.5 and 6.6.6 we may also use, e.g., [8, Corollaries 6.3.16 and 6.17], except for the finiteness of the eigenvalues and that 0 is not an eigenvalue.

Clearly, Proposition 6.6.7 follows immediately from Proposition 6.6.5 since $\mathcal{T}_D$ is the direct product of the operators $\mathcal{T}_{D,j}$ $(j = 1, \dots, g)$.

To complete the proof we still have to show that the operator $\mathcal{T}_N$ is selfadjoint, which will be accomplished by verifying the conditions (1.2)–(1.4) in [137]. In the notation [137, (1.2)], the boundary conditions (6.6.14) and (6.6.15) can be written as $-B^*Y(0) + A^*Y'(0) = 0$, where $A = (a_{j,k})_{j,k=1}^g$ and $B = (b_{j,k})_{j,k=1}^g$ are $g \times g$ matrices whose nonzero entries are given by $a_{j,j} = 1$ and $a_{j+1,j} = -1$ for $j = 1, \dots, g-1$ and $b_{j,g} = 1$ for $j = 1, \dots, g$. A straightforward calculation gives $B^*A = 0$ and therefore the condition $-B^*A + A^*B$ is satisfied, which is [137, (1.3)]. Furthermore, the matrix $A^*A + B^*B$ is positive semidefinite, and a straightforward calculation shows that it is a Jacobi matrix with diagonal elements $2, \dots, 2, g$ and the elements above and below the diagonal are $-1, \dots, -1, 0$. By Proposition B.3.7 it is now easy to see that its determinant is g^2. Therefore, $A^*A + B^*B > 0$, which is the condition [137, (1.4)]. □

In [137, Theorem 1.1] an upper bound for the total multiplicity of the eigenvalues of $\mathcal{T}_N$ in $(-\infty, 0)$ is given, but we will obtain a better estimate in our main theorem below.

For $\lambda \in \mathbb{C} \setminus \{0\}$ let $\sqrt{\lambda}$ be the unique square root with $0 \le \arg \sqrt{\lambda} < \pi$. For each $j = 1, \dots, g$ there is a unique solution of (6.6.13), called the Jost solution and denoted by $e_j(\sqrt{\lambda}, \cdot)$, which satisfies

$$e_j(\sqrt{\lambda}, x) = e^{i\sqrt{\lambda}x}(1 + o(1)) \text{ as } x \to \infty, \tag{6.6.18}$$

$$e_j'(\sqrt{\lambda}, x) = i\sqrt{\lambda}e^{i\sqrt{\lambda}x}(1 + o(1)) \text{ as } x \to \infty, \tag{6.6.19}$$

see, e.g., [8, Lemma 6.13.1]. For each $\lambda \in \mathbb{C} \setminus [0, \infty)$ the multiples of the Jost solution are those solutions of (6.6.13) which belong to $L_2(0, \infty)$.

The construction of e_j shows that the functions $\lambda \mapsto e_j(\sqrt{\lambda}, 0)$ and $\lambda \mapsto e_j'(\sqrt{\lambda}, 0)$ are analytic in $\mathbb{C} \setminus [0, \infty)$.

Proposition 6.6.9. *For $j = 1, \dots, g$, the eigenvalues of $T_{D,j}$ and $T_{N,j}$ in $(-\infty, 0)$ are the zeros of $\lambda \mapsto e_j(\sqrt{\lambda}, 0)$ and $\lambda \mapsto e_j'(\sqrt{\lambda}, 0)$ in $(-\infty, 0)$, respectively, these zeros are simple and interlace, and the functions ϕ_j defined by*

$$\phi_j(\lambda) := \frac{e_j'(\sqrt{\lambda}, 0)}{e_j(\sqrt{\lambda}, 0)}$$

are Nevanlinna functions. If ϕ_j has a pole in $(-\infty, 0)$, then ϕ_j has at least one zero there, and the smallest zero of ϕ_j is smaller than the smallest pole of ϕ_j.

Proof. Let $u_1(\lambda, \cdot)$ and $u_2(\lambda, \cdot)$ be the solutions of (6.6.13) satisfying the initial conditions $u_1(\lambda, 0) = 1$, $u_1'(\lambda, 0) = 0$, $u_2(\lambda, 0) = 0$, $u_2'(\lambda, 0) = 1$. By [8, (6.13.9)], the Weyl function M is the unique function such that $u_1(\lambda, \cdot) + M(\lambda)u_2(\lambda, \cdot) \in L_2(0, \infty)$ $(\lambda \in \mathbb{C} \setminus \mathbb{R})$. Hence there is a function α such that

$$u_1(\lambda, \cdot) + M(\lambda)u_2(\lambda, \cdot) = \alpha(\lambda)e_j(\sqrt{\lambda}, \cdot) \quad (\lambda \in \mathbb{C} \setminus \mathbb{R}). \tag{6.6.20}$$

Substituting the initial value 0 in (6.6.20) gives that $1 = \alpha(\lambda)e_j(\sqrt{\lambda}, 0)$. Differentiating (6.6.20) and then substituting the initial value 0 gives $M(\lambda) = \alpha(\lambda)e_j'(\sqrt{\lambda}, 0)$. This shows that $M = \phi_j$, and therefore ϕ_j is a Nevanlinna function by [8, Corollary 2.3.7].

It is clear that the eigenvalues of $\mathcal{T}_{D,j}$ are the zeros of $\lambda \mapsto e_j(\sqrt{\lambda}, 0)$ and that the eigenvalues of $\mathcal{T}_{N,j}$ are the zeros of $\lambda \mapsto e_j'(\sqrt{\lambda}, 0)$. Although [96, Lemma 11.1.3] is stated for meromorphic functions on $\mathbb{C}$, the proof shows that it is true for functions which are for example meromorphic on $\mathbb{C} \setminus [0, \infty)$. Hence the zeros and poles of ϕ_j in $(-\infty, 0)$ are simple and interlace. Finally, since solutions y of (6.6.13) are uniquely determined by $y(0)$ and $y'(0)$, it is clear that $\lambda \mapsto e_j(\sqrt{\lambda}, 0)$ and $\lambda \mapsto e_j'(\sqrt{\lambda}, 0)$ cannot have common zeros. The simplicity of the poles and zeros of ϕ_j therefore shows that the zeros of $\lambda \mapsto e_j(\sqrt{\lambda}, 0)$ and $\lambda \mapsto e_j'(\sqrt{\lambda}, 0)$ are simple and interlace.

By [8, (6.13.18)], the representation

$$\phi_j(\lambda) = \frac{i\sqrt{\lambda} - \int_0^\infty e^{i\sqrt{\lambda}t} q_j(t)u_1(\lambda, t)\, dt}{1 + \int_0^\infty e^{i\sqrt{\lambda}t} q_j(t)u_2(\lambda, t)\, dt}$$

holds, and by [8, (6.13.18)], the integral in the denominator tends to 0 as $\lambda \to -\infty$ and the integral in the numerator remains bounded when $\lambda \to -\infty$. Hence $\phi_j(\lambda) \to -\infty$ as $\lambda \to -\infty$, and therefore the smallest zero of ϕ_j, if any, is smaller than its smallest pole, if any, because Nevanlinna functions are increasing on each interval in their domain (see [96, Lemma 11.1.3]). □

Let

$$\phi_N(\lambda) := \sum_{j=1}^{g} e_j'(\sqrt{\lambda}, 0) \prod_{\substack{r=1 \\ r \neq j}}^{g} e_r(\sqrt{\lambda}, 0), \tag{6.6.21}$$

$$\phi_D(\lambda) := \prod_{r=1}^{g} e_r(\sqrt{\lambda}, 0). \tag{6.6.22}$$

For $\lambda \in (-\infty, 0)$ let $n_D(\lambda)$ and $n_N(\lambda)$ be the multiplicity of λ as an eigenvalue of $\mathcal{T}_D$ and $\mathcal{T}_N$, respectively.

Theorem 6.6.10. *The eigenvalues, counted with multiplicity, of $\mathcal{L}_D$ and $\mathcal{L}_N$ lie in $(-\infty,0)$ and are the zeros of ϕ_D and ϕ_N, counted with multiplicity. They can be written as finite (possibly void) nondecreasing sequences $(\zeta_k)_{k=1}^n$ for the Dirichlet problem and $(\lambda_k)_{k=1}^{n'}$ for the Neumann problem with $n := \sum_{j=1}^g n_j$ and $n' = n$ or $n' = n+1$ and satisfy*

$$\lambda_1 < \zeta_1 \le \lambda_2 \le \cdots \le \zeta_n < 0 \tag{6.6.23}$$

when $n' = n > 0$ and

$$\lambda_1 < \zeta_1 \le \lambda_2 \le \cdots \le \zeta_n < \lambda_{n+1} < 0 \tag{6.6.24}$$

when $n' = n+1$. The multiplicities satisfy $n_N(\zeta_j) = n_D(\zeta_j) - 1$.

Proof. The function

$$\phi := \frac{\phi_N}{\phi_D} = \sum_{j=1}^{n} \phi_j$$

is a sum of Nevanlinna functions and therefore a Nevanlinna function. A reasoning as in the proof of Proposition 6.6.3 shows that $\lambda_1 < \zeta_1$. Observing that we already know the zeros of $e_j(\cdot,0)$ and $e_j'(\cdot,0)$ are simple and distinct, all the statements follow from the proof of Theorem 2.1.4 with $R_{2n_j}^{(j)}(z)$ replaced with $e_j(\sqrt{\lambda},0)$ and $R_{2n_j-1}^{(j)}(z)$ replaced with $e_j'(\sqrt{\lambda},0)$. □

Let $(p_1,p_2,\dots,p_r)$ be the vector and $(p_1^\downarrow,p_2^\downarrow,\dots,p_r^\downarrow)$ the nonincreasingly ordered vector of distinct eigenvalues of problem (6.6.13)–(6.6.15).

Theorem 6.6.11. *Let*

$$N_i := \#\{j \in \{1,\dots,g\} : n_j(D) \ge i\} = \max\{j \in \{1,\dots,g\} : n_j(D) \ge i\}.$$

Then $(N_1-1, N_2-1,\dots,N_{n_1}-1) \succ_w (p_1^\downarrow,p_2^\downarrow,\dots,p_{r-n_1-1}^\downarrow)$. If $p_i > 1$, then $p_{i-1} = p_{i+1} = 1$ $(i = 2,\dots,r-1)$.

Proof. The last statement immediately follows from Theorem 6.6.10. The weak majorization can be proved as for part 2 of Theorem 6.6.4. □

Example 6.6.12. Let $q_j(x) = -\gamma_j e^{-\alpha_j x}$ with $\alpha_j > 0$, $\gamma_j > 0$ for $j = 1,\dots,g$. Then $\int_0^\infty x q_j^-(x)\,dx = \gamma_j\alpha_j^{-2}$ and therefore

$$n_j \le \lfloor \gamma_j\alpha_j^{-2} \rfloor. \tag{6.6.25}$$

It follows from Theorem 6.6.10 that the number of eigenvalues of $\mathcal{T}_N$, counted with multiplicity, is less or equal $1 + \sum\limits_{j=1}^{g} n_j \le 1 + \sum\limits_{j=1}^{g} \lfloor \gamma_j\alpha_j^{-2} \rfloor$.

6.7 Notes

Usually the term 'quantum graphs' means metric graphs considered as quasi-one-dimensional domains with differential equations defined on these domains (see, e.g., [67], [11], [82], [83], [5], [43]). In quantum mechanics the Sturm-Liouville and the Dirac equation and in vibration theory the string equation are considered on the edges of a graph subject to matching and boundary conditions at the vertices. These are Dirichlet or Neumann or Robin conditions at pendant vertices and continuity conditions together with Kirchhoff conditions at interior vertices. Such models are often used in problems of free-electron theory of conjugate molecules in chemistry and in the theory of quantum wires and thin wave-guides. The differential equations together with the matching and boundary conditions define an operator which is usually called continuous Laplacian (see, e.g., [33]).

There are different definitions of the so-called discrete (combinatorial) Laplacian (see [38], [12]); we have used the one in [38] (see (6.2.6)). The notion of discrete Laplacian is closely related to the notion of adjacency matrix of the classical spectral graph theory (see, e.g., [44]), namely, for a simple connected graph with no loops the discrete Laplacian is

$$L = I - T^{-1/2}AT^{-1/2}.$$

The question of connection between continuous and discrete Laplacian was pointed out in [2] and developed in a rigorous manner in [51], [30], [10], [57]. This connection exists in the case of the free continuous Laplacian or under restrictive conditions on the potentials of the Sturm-Liouville equations on the edges of the graph; the edges must be of the same length and the potential must be the same.

The correspondence between the spectra for the discrete (combinatorial) and continuous Laplacians for a large class of the equilateral graphs was given in [6], [33]. It was proved there that if the potentials on all edges are zero, then the spectrum of the continuous Laplacian on the graph consists of the Dirichlet spectrum $\frac{\pi^2 n^2}{l^2}$ ($n = 1, 2, \ldots$, and l is length of an edge) and those numbers z for which $\cos\sqrt{z}$ belongs to the spectrum of the corresponding discrete Laplacian.

Some of the results of Section 6.4 were obtained in [115].

Amongst the variety of graphs an important role is played by the perfect trees (see, e.g., [127]). The results of Section 6.5 are based on [113], but equations (6.5.37), (6.5.38) are given in [113] with errors.

The publication [102] considered the case of the same L_2 potential on each edge and δ-type boundary conditions at the vertices. The spectrum of the quantum graph is the union of the Dirichlet spectrum on an edge and of the set $\eta^{-1}(\sigma(2\Delta))$, where $\sigma(2\Delta)$ is the spectrum of 2Δ, Δ is the corresponding discrete Laplacian, and η is the discriminant (Lyapunov function) of the Hill equation on an edge.

The estimate (6.6.17) of the number of eigenvalues of the Sturm-Liouville problem on the semiaxis is called the Bargmann-Birman-Schwinger inequality. Therefore the estimate of the number of eigenvalues of the Sturm-Liouville problem on a star graph is called a Bargmann-Birman-Schwinger type inequality.

Appendix A

Rational Nevanlinna and Stieltjes Functions

In this appendix we will state and prove results for the class of rational Nevanlinna functions and its Stieltjes subclass. The results on the Stieltjes class are a slight generalization of the results in [59, Supplement II]. The interested reader can find corresponding results for analytic or meromorphic Nevanlinna functions for example in [96].

A.1 Rational functions

Definition A.1.1. A rational function f is the quotient of two polynomials q and p with $p \neq 0$:

$$f(z) = \frac{q(z)}{p(z)} \quad (z \in \mathbb{C},\ p(z) \neq 0).$$

Remark A.1.2. By the fundamental theorem of algebra, a nonconstant polynomial q assumes each value, i.e., for each $b \in \mathbb{C}$ there is at least one complex number $a \in \mathbb{C}$ such that $q(a) = b$. It follows that any polynomial q of degree $m \in \mathbb{N}$ has a representation

$$q(z) = C \prod_{j=1}^{m} (z - \beta_j) \quad (z \in \mathbb{C}) \tag{A.1.1}$$

with $\beta_j \in \mathbb{C}$ $(j = 1, \ldots, m)$ and $C \in \mathbb{C} \setminus \{0\}$. This representation extends to nonzero constant polynomials, which have degree 0, if one observes the convention that the product over an empty index set is defined to be 1.
A zero of q is a complex number z_0 such that $q(z_0) = 0$, i.e., $z_0 = \beta_{j_0}$ for some $j_0 \in \{1, \ldots, m\}$. If $z = \beta_{j_0}$ and $\beta_j \neq \beta_{j_0}$ for all $j \neq j_0$, then the zero z_0 of q is called a simple zero of q.

M. Möller, V. Pivovarchik, *Direct and Inverse Finite-Dimensional Spectral Problems on Graphs*, Operator Theory: Advances and Applications 283, https://doi.org/10.1007/978-3-030-60484-4

Remark A.1.3. Let f be a nonzero rational function with the representation $f = \frac{q}{p}$. If q and p would have common zeros, then we can cancel corresponding factors from q and p in the product representations (A.1.1) of q and p, so that we may assume that q and p do not have common zeros, that is, that there are no $z \in \mathbb{C}$ such that $q(z) = p(z) = 0$. Then $z_0 \in \mathbb{C}$ is called a (simple) zero of f if z_0 is a (simple) zero of q, whereas $z_0 \in \mathbb{C}$ is called a (simple) pole of f if z_0 is a (simple) zero of p. There are $m, n \in \mathbb{N} \cup \{0\}$, complex numbers $\alpha_1, \dots, \alpha_n$, $\beta_1, \dots, \beta_m$ with $\alpha_j \neq \beta_k$ for $j = 1, \dots, n$ and $\beta_k = 1, \dots, m$, and $C \neq 0$ such that

$$f(z) = C \prod_{j=1}^{m} (z - \beta_j) \prod_{j=1}^{n} (z - \alpha_j)^{-1} \quad (z \in \mathbb{C},\ p(z) \neq 0). \tag{A.1.2}$$

The polynomials q and p are unique up to a constant factor. When $z_0 \in \mathbb{C}$ is a zero of f, the number of indices j for which $\beta_j = z_0$ is called the multiplicity of the zero z_0 of f and denoted by $m(f, z_0)$. Similarly, when z_0 is a pole of f, the number of indices j for which $\alpha_j = z_0$ is called the pole order of the pole z_0 of f. In any case, for each $z_0 \in \mathbb{C}$ there is a unique integer k such that $f(z) = (z - z_0)^k h(z)$ where h is a rational function for which z_0 is neither a zero nor a pole. The complex number z_0 is a zero of f if and only if $k > 0$, and then k is the multiplicity of the zero z_0 of f. The complex number z_0 is a pole of f if and only if $k < 0$, and then $|k|$ is the order of the pole z_0 of f. Defining the function $\overline{f}$ by $\overline{f}(z) = \overline{f(\overline{z})}$ it follows from (A.1.2) that

$$\overline{f}(z) = \overline{C} \prod_{j=1}^{m} (z - \overline{\beta_j}) \prod_{j=1}^{n} (z - \overline{\alpha_j})^{-1}.$$

Hence also $\overline{f}$ is a rational function.

Definition A.1.4. A rational function f is called a real rational function if it maps real numbers to real numbers, i.e., if $f(z) \in \mathbb{R}$ for $z \in \mathbb{R}$ which are not poles of f.

Proposition A.1.5. *If f is a nonzero real rational function, then its zeros and poles are symmetric with respect to the real axis, $\overline{f} = f$, and f can be written as $f = \frac{q}{p}$, where q and p are real polynomials without common zeros.*

Proof. From the definition of $\overline{f}$ one concludes that $\overline{f}(z) = \overline{f(z)} = f(z)$ for real z which are not poles of $\overline{f}$ or f. Hence the rational function $\overline{f} - f$ has infinitely many (real) zeros, and $\overline{f} = f$ follows. In particular, if β_j is a nonreal zero of f, also $\overline{\beta_j}$ is a zero of f, and if α_j is a nonreal pole of f, also $\overline{\alpha_j}$ is a pole of f. Then the factors $(z - \beta_j)(z - \overline{\beta_j})$ and $(z - \alpha_j)(z - \overline{\alpha_j})$ are real for real z. If, however, β_j or α_j are real, then the factors $z - \beta_j$ and $z - \alpha_j$ are real for real z. When all these factors are exhausted, we see that also C is real. Defining

$$q(z) = C \prod_{j=1}^{m} (z - \beta_j), \quad p(z) = \prod_{j=1}^{n} (z - \alpha_j),$$

it follows that q and p are real polynomials such that $f = \frac{p}{q}$. □

It is sometimes convenient to consider a rational function f on $\mathbb{C}$ as a function from $\mathbb{C}$ into $\overline{\mathbb{C}} = \mathbb{C} \cup \{\infty\}$ by putting $f(z) = \infty$ for poles z of f.

Lemma A.1.6. *Let f be a nonconstant rational function. Then $\overline{\mathbb{C}} \setminus f(\mathbb{C})$ has at most one element.*

Proof. If $\infty \notin f(\mathbb{C})$, then f has no poles, so that f is a nonconstant polynomial, and $f(\mathbb{C}) = \mathbb{C}$ follows from Remark A.1.2. Now let $\infty \in f(\mathbb{C})$ but $b \notin f(\mathbb{C})$ for some $b \in \mathbb{C}$. Writing $f = \frac{q}{p}$ with polynomials q and p without common zeros it follows that $q(z) \neq bp(z)$ for all $z \in \mathbb{C}$. Hence the polynomial $q - bp$ has no zeros and must therefore be constant by the fundamental theorem of algebra, i.e., there is $c \in \mathbb{C} \setminus \{0\}$ such that $q(z) - bp(z) = c$ for all $z \in \mathbb{C}$. Then it follows for all $w \in \mathbb{C}$ with $w \neq b$ that

$$q(z) - wp(z) = (b - w)p(z) + c.$$

Since f has poles, p is nonconstant. Hence there is $z_0 \in \mathbb{C}$ such that $p(z_0) = \frac{c}{w-b}$ and thus $q(z_0) - wp(z_0) = 0$. Therefore $p(z_0) \neq 0$ and

$$f(z_0) = \frac{q(z_0)}{p(z_0)} = \frac{q(z_0) - wp(z_0)}{p(z_0)} + w = w$$

shows that all $w \in \mathbb{C} \setminus \{b\}$ belong to $f(\mathbb{C})$. □

Proposition A.1.7. *Let f be a nonconstant rational function, let $a \in \mathbb{C}$ be a zero of f or a pole of f, and let $\varphi_1, \varphi_2 \in \mathbb{R}$ with $\varphi_1 < \varphi_2$. For the representation $f(z) = (z - a)^j h(z)$ with $h(a) \neq 0$ as in Remark* A.1.3 *let $\psi_0 = \arg h(a)$. Let $\psi_1 \in \mathbb{R}$ be such that $j\psi_1 \in (\varphi_1 - \psi_0, \varphi_2 - \psi_0)$. Then there is $r_0 > 0$ such that for $r \in (0, r_0)$ the complex number $w = a + re^{i\psi_1}$ is neither a zero nor a pole of f and satisfies $f(w) = |f(w)|e^{i\varphi_0}$ with $\varphi_0 \in (\varphi_1, \varphi_2)$.*

Proof. Recall that $j \neq 0$ in the representation of f where $j > 0$ if a is a zero of f and $j < 0$ if a is a pole of f. Let $\eta_0 = \min\{j\psi_1 - \varphi_1 + \psi_0, \varphi_2 - \psi_0 - j\psi_1, \frac{\pi}{6}\}$ and put $\rho = |h(a)| \sin \eta_0$.

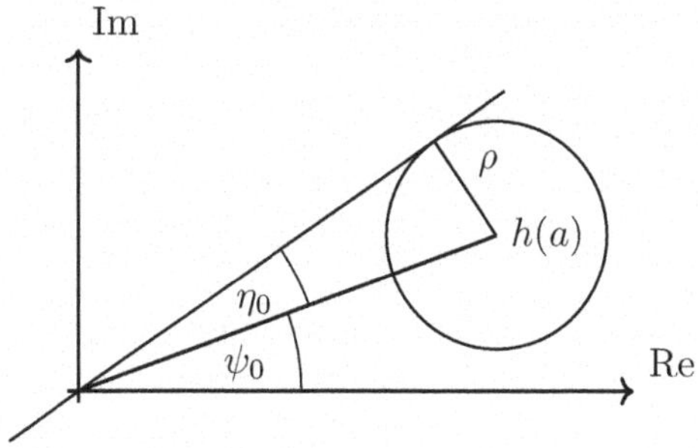

Due to the continuity of h at a there is $r_0 > 0$ such that $|h(z) - h(a)| < \rho$ and $h(z) = (|h(a)| + \delta(z))e^{i(\psi_0 + \eta(z))}$ whenever $0 < |z - a| < r_0$, where $\delta(z) \in (-\rho, \rho)$

and $\eta(z) \in (-\eta_0, \eta_0)$. For $r \in (0, r_0)$ and $w = a + re^{i\psi_1}$ one has

$$f(w) = r^j e^{ij\psi_1}(h(a) + \delta(w))e^{i(\psi_0+\eta(w))} = |f(w)|e^{i\varphi_0},$$

where

$$\varphi_0 = j\psi_1 + \psi_0 + \eta(w) \in (j\psi_1 + \psi_0 - \eta_0, j\psi_1 + \psi_0 + \eta_0) \subset (\varphi_1, \varphi_2). \qquad \square$$

A.2 Rational Nevanlinna functions

Definition A.2.1. A rational function f is called a *Nevanlinna function* if

(i) f has no poles in $\mathbb{C}^+ \cup \mathbb{C}^-$,

(ii) $f(\overline{z}) = \overline{f(z)}$ for $\operatorname{Im} z \neq 0$,

(iii) $\operatorname{Im} z \cdot \operatorname{Im} f(z) \geq 0$ for $\operatorname{Im} z \neq 0$.

Note that property (ii) is equivalent to f being a real rational function. In this monograph only rational functions are considered. We recall the general definition of a Nevanlinna function: f is called a Nevanlinna function if f is analytic on $\mathbb{C}^+$ and $\mathbb{C}^-$ and if f has properties (ii) and (iii) of definition A.2.1. Nevanlinna functions are called R-functions in [74, §1].

Lemma A.2.2. *Let p and q be real polynomials without common nonreal zeros. Consider the functions*

$$\omega = p + iq, \quad f = \frac{q}{p}, \quad g = \frac{\omega}{\overline{\omega}}.$$

1. *The following statements are equivalent:*

   (i) $|g(z)| < 1$ *for all* $z \in \mathbb{C}^+$ *such that* $\overline{\omega}(z) \neq 0$;

   (ii) ω *has no zeros in the open lower half-plane, and* $|g(z)| < 1$ *for all* $z \in \mathbb{C}^+$;

   (iii) $p \neq 0$ *and* $\operatorname{Im} f(z) > 0$ *for all* $z \in \mathbb{C}^+$ *with* $p(z) \neq 0$;

   (iv) *All zeros of p and q are real,* $f(\mathbb{C}^+) \subset \mathbb{C}^+$ *and* $f(\mathbb{C}^-) \subset \mathbb{C}^-$.

2. *Assume that $p \neq 0$. Then the following statements are equivalent:*

   (i) $|g(z)| \leq 1$ *for all* $z \in \mathbb{C}^+$ *such that* $\overline{\omega}(z) \neq 0$;

   (ii) ω *has no zeros in the open lower half-plane, and* $|g(z)| \leq 1$ *for all* $z \in \mathbb{C}^+$;

   (iii) $\operatorname{Im} f(z) \geq 0$ *for all* $z \in \mathbb{C}^+$ *with* $p(z) \neq 0$;

   (iv) *All zeros of p are real, $q = 0$ or all zeros of q are real,* $f(\mathbb{C}^+) \subset \overline{\mathbb{C}^+}$ *and* $f(\mathbb{C}^-) \subset \overline{\mathbb{C}^-}$.

Proof. 1. (i)⇒(ii): From

$$\begin{pmatrix}\omega\\ \overline{\omega}\end{pmatrix} = \begin{pmatrix}1 & i\\ 1 & -i\end{pmatrix}\begin{pmatrix}p\\ q\end{pmatrix}$$

and the assumption that p and q do not have common nonreal zeros it follows that also ω and $\overline{\omega}$ have no common nonreal zeros. In particular, $\omega \neq 0$. Since g is bounded in the open upper half-plane, g does not have a pole there, and hence $\overline{\omega}$ has no zeros in the open upper half-plane, which means that ω has no zeros in the open lower half-plane.

(ii)⇒(iii): Observing that p and q are real polynomials it follows for $z \in \mathbb{C}$ that

$$\omega(z) + \overline{\omega}(z) = p(z) + iq(z) + p(z) - iq(z) = 2p(z).$$

Since $|g(z)| < 1$ for $z \in \mathbb{C}^+$ gives $\omega(z) \neq -\overline{\omega}(z)$, it follows that $p(z) \neq 0$ for $z \in \mathbb{C}^+$. In particular, $p \neq 0$. Similarly, $\omega - \overline{\omega} = 2iq$ so that

$$f = \frac{q}{p} = i\frac{\overline{\omega} - \omega}{\overline{\omega} + \omega} = i\frac{1-g}{1+g}. \tag{A.2.1}$$

Since the right-hand side is a Möbius transformation in g which maps the interior of the unit disk onto the upper half-plane, it is clear that $|g(z)| < 1$ implies $\operatorname{Im} f(z) > 0$ for $z \in \mathbb{C}^+$ with $p(z) \neq 0$.

(iii)⇒(iv): Since $f(z) \neq 0$ for all $z \in \mathbb{C}^+$ with $p(z) \neq 0$, q has no zeros in the open upper half-plane. Assume that p has a zero a in $\mathbb{C}^+$. Then f has a pole (of order k) at a. By Proposition A.1.7 with $\varphi_1 = -\pi$, $\varphi_2 = 0$ and ψ_0 as given there we can find $0 < r < \operatorname{Im} a$ and $\psi_1 \in ((\psi_0 - \varphi_2)k^{-1}, (\psi_0 - \varphi_1)k^{-1})$ such that $f(w) = |f(w)|e^{i\varphi_0} \neq 0$ with $w = a + re^{i\psi_1}$ and $\varphi_0 \in (-\pi, 0)$. But then $w \in \mathbb{C}^+$ and $f(\omega) \in \mathbb{C}^-$, which contradicts the assumption that $\operatorname{Im} f(w) > 0$. Therefore p has no zeros in $\mathbb{C}^+$. By assumption (iii) it follows that $f(\mathbb{C}^+) \subset \mathbb{C}^+$. Since f is a real rational function we have $\overline{f} = f$, which gives $f(\mathbb{C}^-) \subset \mathbb{C}^-$. In particular, f has only real zeros or poles, if any, and therefore all zeros of p and q are real.

(iv)⇒(i) is an immediate consequence of (A.2.1), see the proof of (ii)⇒(iii).

2. The difference to part 1 is that $|g(z)| = 1$ and $\operatorname{Im} f(z) = 0$ is possible for some $z \in \mathbb{C}^+$. In this case, f and, equivalently, g are constant. Indeed, assume that there is $a \in \mathbb{C}^+$ such that $f(a) \in \mathbb{R}$ but that f is not constant. Then

$$\tilde{f}(z) = \frac{1}{f(a) - f(z)}$$

defines a rational function $\tilde{f}$ which has a pole at a and which satisfies $\operatorname{Im} \tilde{f}(z) \geq 0$ for all $z \in \mathbb{C}^+$ which are not poles of $\tilde{f}$. As in the proof of part 1, (iii)⇒(iv), this leads to a contradiction. But if f is a constant function, then also p and q are constant functions whose values are real. In this case, (i)–(iv) are satisfied. □

Remark A.2.3. Let f be a rational Nevanlinna function. Due to Definition A.2.1 (i), $\overline{f} = f$ on $\mathbb{C} \setminus \mathbb{R}$ and therefore everywhere. Hence every rational Nevanlinna function is a real rational function and is thus characterized by property (iv) in part 2 of Lemma A.2.2. Indeed, in the proof of Lemma A.2.2 we have seen that a rational function as in part 2 is as in part 1 if and only if it is not constant. The equivalence of (iii) and (iv) in part 1 shows that a nonconstant rational function is a rational Nevanlinna function if and only if it is a real rational function and maps $\mathbb{C}^+$ into itself. Clearly, a constant function is a rational Nevanlinna function if and only if its value is real.

Lemma A.2.4. *If $f \neq 0$ is a rational Nevanlinna function, then so are the functions $-\frac{1}{f}$ and $(\frac{1}{f}+c)^{-1}$ for each real constant c unless $f = -\frac{1}{c}$.*

Proof. For each $w \in \mathbb{C} \setminus \{0\}$ the identity $\frac{1}{w} = \frac{\overline{w}}{|w|^2}$ shows that

$$\operatorname{Im}\left(-\frac{1}{f(z)}\right) = \frac{\operatorname{Im} f(z)}{|f(z)|^2} \quad \text{and} \quad \operatorname{Im}\left(\frac{1}{f(z)}+c\right)^{-1} = \frac{\operatorname{Im}(-\frac{1}{f(z)})}{\left|\frac{1}{f(z)}+c\right|^2}.$$

Thus the statement immediately follows from Remark A.2.3. □

Lemma A.2.5. *Let f be a real rational function which satisfies $f(\mathbb{C}^+) \subset \mathbb{C}^+$. Then f has at least one zero or pole, $f(\mathbb{C}^+) = \mathbb{C}^+$, and all zeros and poles of f lie on the real axis, are simple and interlace, i.e., between two poles of f lies a zero of f and between two zeros of f lies a pole of f. For each $z \in \mathbb{R}$ which is not a pole of f, the inequality $f'(z) > 0$ holds.*

Proof. Since $f(\mathbb{C}^+) \subset \mathbb{C}^+$, f has neither poles nor zeros in the open upper half-plane, and hence also not in the open lower half-plane since f is real. Therefore all zeros and poles of f are real.

If f would have no zeros or poles, then f would be constant, see Remark A.1.3. Since f is real on the real axis, $f(z)$ would be real for all $z \in \mathbb{C}$, which contradicts $f(\mathbb{C}^+) \subset \mathbb{C}^+$.

Assume that f has a pole or zero a which is not simple. Then there are an integer j with $|j| > 1$ and a rational function h with $h(a) \neq 0$ such that $f(z) = (z-a)^j h(z)$. Since f is real on the real axis, also h is real on the real axis, so that $h(a) = |h(a)|e^{i\psi_0}$ with $\psi_0 = 0$ or $\psi_0 = \pi$. If $j > 1$, put $\psi_1 = \frac{1}{j}(\frac{3\pi}{2} - \psi_0)$. Then

$$0 < \frac{\pi}{2j} \leq \psi_1 \leq \frac{3}{2j}\pi \leq \frac{3}{4}\pi < \pi$$

and

$$j\psi_1 = \frac{3}{2}\pi - \psi_0 \in (\pi - \psi_0, 2\pi - \psi_0).$$

If $j < -1$, put $\psi_1 = \frac{1}{|j|}(\frac{\pi}{2} + \psi_0)$. Then

$$0 < \psi_1 \leq \frac{3}{2|j|}\pi \leq \frac{3}{4}\pi < \pi$$

and

$$j\psi_1 = -\frac{\pi}{2} - \psi_0 \in (-\pi - \psi_0, -\psi_0).$$

Applying Proposition A.1.7 we obtain in both cases that there is $r > 0$ such that $w = a + re^{i\psi_1}$ gives a representation $f(w) = |f(w)|e^{i\varphi_0}$ with $\varphi_0 \in (\pi, 2\pi)$ if $j > 0$ and $\varphi_0 \in (-\pi, 0)$ if $j < 0$. Thus we have arrived at the contradiction $w \in \mathbb{C}^+$ and $f(w) \in \mathbb{C}^-$. Therefore all zeros and poles of f must be simple.

Now assume that the zeros and poles of f do not interlace. If there are two adjacent poles $z_1 < z_2$ of f without a zero of f between them, then f is either positive or negative on the interval (z_1, z_2). Hence either $f(z)$ tends to ∞ as z tends to z_1 from the right and as z tends to z_2 from the left, or $f(z)$ tends to $-\infty$ as z tends to z_1 from the right and as z tends to z_2 from the left. Therefore f must have a local extremum at some $a \in \mathbb{R}$ between these two poles. If there are two adjacent zeros of f without a pole of f between them, then f is continuous on the interval between these two zeros. Therefore f must have a local extremum at some $a \in \mathbb{R}$ between these two zeros. In either case we have a point $a \in \mathbb{R}$ with $f'(a) = 0$. Since f is real on the real axis, $f(a) \in \mathbb{R}$, and therefore the function f_1 defined by $f_1(z) = f(z) - f(a)$ is a real rational function with $f_1(a) = f_1'(a) = 0$ and with $f_1(\mathbb{C}^+) \subset \mathbb{C}^+$. This would mean that the zero a of f_1 were not simple. But by what we have already shown in the previous paragraph, zeros of f_1 have to be simple. This contradiction shows that zeros and poles of f interlace.

Assume there is $w \in \mathbb{C}^+ \setminus f(\mathbb{C})$. From Lemma A.1.6 it would follow that $\{w\} = \overline{\mathbb{C}} \setminus f(\mathbb{C})$; in particular, $\overline{w} = f(z)$ for some $z \in \mathbb{C}$. Then $\overline{w} \in \mathbb{C}^-$, f real and $f(\mathbb{C}^+) \subset \mathbb{C}^+$ would imply that $z \in \mathbb{C}^-$. Therefore $\overline{z} \in \mathbb{C}^+$ and $w = \overline{f(z)} = f(\overline{z})$ would lead to the contradiction $w \in f(\mathbb{C}^+)$. Hence $f(\mathbb{C}^+) = \mathbb{C}^+$.

Finally, when $a \in \mathbb{R}$ is not a pole of f, then also $f - f(a)$ is a real rational function which maps $\mathbb{C}^+$ into itself. Hence, by what we have already proved, a is a simple zero of $f - f(a)$. Therefore

$$f(z) = f(a) + (z - a)h(z),$$

where h is a real rational function and $h(a) \neq 0$. Since

$$\operatorname{Im} f(a + i\varepsilon) > 0 \quad \text{for } \varepsilon > 0,$$

it follows that

$$\operatorname{Re} h(a + i\varepsilon) = \frac{1}{\varepsilon} \operatorname{Im} f(a + i\varepsilon) > 0.$$

By continuity,

$$f'(a) = h(a) = \lim_{\varepsilon \searrow 0} \operatorname{Re} h(a + i\varepsilon) \geq 0,$$

and $f'(a) > 0$ follows since $h(a) \neq 0$. □

Theorem A.2.6. *A real rational function f satisfies $f(\mathbb{C}^+) \subset \mathbb{C}^+$ if and only if f is represented in the form*

$$f(z) = C \prod_{k=1}^{m} (z - \beta_k) \prod_{k=1}^{n} \frac{1}{z - \alpha_k}, \tag{A.2.2}$$

where $m, n \in \mathbb{N}_0$, $m + n > 0$, the sequences $(\alpha_1, \ldots, \alpha_n)$ and $(\beta_1, \ldots, \beta_m)$ are interlacing sequences of real numbers, $C > 0$ if f has at least one zero and the largest zero of f is larger than any pole of f, and $C < 0$ if f has at least one pole and the largest pole of f is larger than any zero of f.

Proof. If $f(\mathbb{C}^+) \subset \mathbb{C}^+$, then the representation (A.2.2) with $C \in \mathbb{R} \setminus \{0\}$ and the interlacing properties of its zeros and poles follows from (A.1.2) and Lemma A.2.5. Let a be the largest of the numbers $\alpha_1, \ldots, \alpha_n, \beta_1, \ldots \beta_m$. By Lemma A.2.5 we know that $f'(x) > 0$ for $x \in (a, \infty)$, i.e., f is increasing on the interval (a, ∞). It follows that f is positive on (a, ∞) if a is a zero of f and that f is negative on (a, ∞) if a is a pole of f. Since all factors $z - \alpha_k$ and $z - \beta_k$ in the product (A.2.2) are positive there, C must be positive if a is a zero of f and C must be negative if a is a pole of f.

Assume that f is of the form (A.2.2) with $m+n > 0$ and interlacing sequences of zeros and poles. Clearly, f is a real rational function. Without loss of generality we may also assume that $\alpha_1 < \alpha_2 < \cdots < \alpha_n$ and $\beta_1 < \beta_2 < \cdots < \beta_m$.

We will first consider the case that the largest zero of f, if any, is larger than the largest pole of f, if any. In case the number of poles equals the number of zeros, i.e., $m = n$, the interlacing property gives

$$\alpha_1 < \beta_1 < \alpha_2 < \beta_2 < \cdots < \alpha_n < \beta_n.$$

Denoting $\alpha = \arg(z - \alpha_k)$ and $\beta = \arg(z - \beta_k)$ for generic k, the sketch

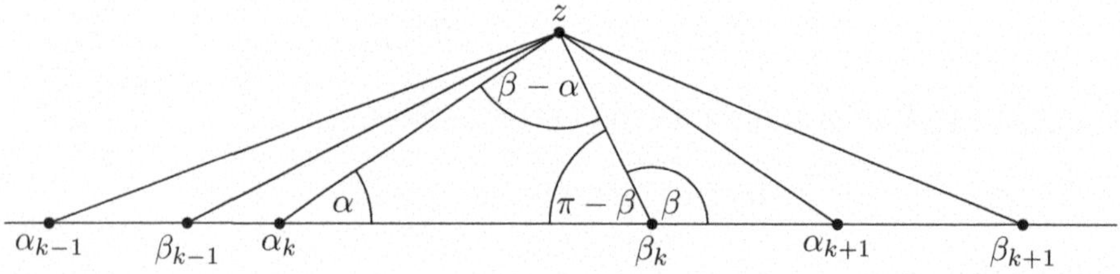

shows that for $\operatorname{Im} z > 0$ the difference $\beta - \alpha$ is the angle subtended by the segment $[\alpha_k, \beta_k]$ of the real axis from the point z. Since $C > 0$ in this case and since

$$\arg f(z) = \sum_{k=1}^{n} [\arg(z - \beta_k) - \arg(z - \alpha_k)],$$

it follows that $0 < \arg f(z) < \pi$ if $\operatorname{Im} z > 0$, i.e., $f(\mathbb{C}^+) \subset \mathbb{C}^+$.

If now the number of poles is different from the number of zeros, still in the case that β_m is larger than any pole of f, the interlacing property gives $m = n+1$. Observing that $0 < \arg(z - x_1) < \arg(z - x_2) < \pi$ for all $z \in \mathbb{C}^+$ and all real numbers x_1, x_2 with $x_1 < x_2$, it follows for $z \in \mathbb{C}^+$ that

$$\begin{aligned} 0 < \arg(z-\beta_1) &\le \arg(z-\beta_1) + \sum_{k=1}^{n}[\arg(z-\beta_{k+1}) - \arg(z-\alpha_k)] \\ &= \arg f(z) = \sum_{k=1}^{n}[\arg(z-\beta_k) - \arg(z-\alpha_k)] + \arg(z-\beta_m) \\ &\le \arg(z-\beta_m) < \pi. \end{aligned}$$

Therefore $f(\mathbb{C}^+) \subset \mathbb{C}^+$ also in this case.

It remains to consider the case that the largest pole of f is larger than the largest zero of f. Then $-\frac{1}{f}$ has zeros α_k and poles β_k interlacing as above, and the constant factor is $-\frac{1}{C} > 0$. Therefore $-\frac{1}{f}$ maps the upper half-plane into itself by what we have already shown. In view of Remark A.2.3 and Lemma A.2.4, also f maps the upper half-plane into itself. □

Lemma A.2.7. *Let $n \in \mathbb{N}$ and let f be a rational Nevanlinna function with n poles. Then there are real numbers $A \ge 0$, B, $\alpha_1 < \cdots < \alpha_n$, $C_j > 0$ $(j = 1, \ldots, n)$ such that*

$$f(z) = Az + B - \sum_{j=1}^{n} \frac{C_j}{z - \alpha_j}.$$

Proof. Referring to Remark A.2.3 and Lemma A.2.5 we know that f has $n-1$, n, or $n+1$ real zeros and n real simple poles. Hence the partial fraction decomposition is clear with $A, B \in \mathbb{R}$ and $C_j \in \mathbb{R} \setminus \{0\}$, $j = 1, \ldots, n$. Since $f(z)$ increases as $z \to \infty$ on the real axis, $A \ge 0$ follows, and similarly, since f increases on the real axis to the left and the right of α_j, it follows that $C_j > 0$ for $j = 1, \ldots, n$. □

A.3 Rational Stieltjes functions

In this section we are going to introduce two classes of Nevanlinna functions and we will show that these functions can be characterized by particular continued fraction expansions.

Definition A.3.1. A rational function f is called a rational Stieltjes function or *S-function* if it is a rational Nevanlinna function and satisfies in addition that

(i) f has no poles in $(-\infty, 0)$,

(ii) $f(z) > 0$ for $z \in (-\infty, 0)$.

A rational $\mathcal{S}$-function f is called a rational $\mathcal{S}_0$-*function* if

(iii) 0 is not a pole of f.

We also say that an $\mathcal{S}$-function ($\mathcal{S}_0$-function) f belongs to the class $\mathcal{S}$ ($\mathcal{S}_0$) and we write $f \in \mathcal{S}$ ($f \in \mathcal{S}_0$).

Proposition A.3.2. *A nonconstant rational Nevanlinna function is a rational $\mathcal{S}$-function if and only if it has at least one pole, and the smallest pole is nonnegative and smaller that all zeros, if any.*

Proof. Let f be a nonconstant rational $\mathcal{S}$-function. Then f has at least one zero or pole, and let a be the smallest of the zeros and poles. By assumption and Lemma A.2.5, $f(z) > 0$ for $z \in (-\infty, 0)$ and $f'(z) > 0$ for $z \in (-\infty, a)$, which shows that $a \geq 0$ and that a is a pole.

On the other hand, if f is a rational Nevanlinna function with at least one pole and such that the smallest pole is nonnegative and smaller than all zeros, then $f'(z) > 0$ and $f(z) \neq 0$ for all $z \in (-\infty, 0)$ implies $f(z) > 0$ for all $z \in (-\infty, 0)$. □

Corollary A.3.3. *A nonconstant real rational function f is a rational $\mathcal{S}$-function if and only if f is represented in the form*

$$f(z) = C_0 \prod_{k=1}^{m} (\beta_k - z) \prod_{k=1}^{n} \frac{1}{\alpha_k - z}, \tag{A.3.1}$$

where $n \in \mathbb{N}$, $m \in \mathbb{N}_0$, the increasing sequences $(\alpha_1, \dots, \alpha_n)$ and $(\beta_1, \dots, \beta_m)$ are interlacing sequences of real numbers, $\alpha_1 \geq 0$, $\alpha_1 < \beta_1$ if $m > 0$, and $C_0 > 0$.

Proof. This follows immediately from Proposition A.3.2 and Theorem A.2.6 with $C_0 = (-1)^{m+n}C$ if we observe that $m = n$ or $m = n - 1$ and that $m = n$ if and only if $C > 0$. □

Lemma A.3.4. *Let $r \in \mathbb{N}$ and let f be a rational $\mathcal{S}$-function with r poles. Then:*

1. *There are unique numbers a, b and a unique rational function g which is either identically zero or has at most $r - 1$ zeros such that*

$$f(z) = a + \frac{1}{-bz + g(z)}; \tag{A.3.2}$$

2. *$a = \lim_{z \to -\infty} f(z) \geq 0$, $b > 0$, and the number of zeros of f is r if $a > 0$ and $r - 1$ if $a = 0$.*

3. *The poles α_k and zeros β_k of f are all simple and interlace,*

$$0 \leq \alpha_1 < \beta_1 < \alpha_2 < \dots < \beta_{r-1} < \alpha_r < \beta_r \quad \text{if } a > 0, \tag{A.3.3}$$

$$0 \leq \alpha_1 < \beta_1 < \alpha_2 < \dots < \beta_{r-1} < \alpha_r \quad \text{if } a = 0. \tag{A.3.4}$$

4. *The rational function f is strictly increasing between its poles, i.e., in the intervals $(-\infty, \alpha_1)$, (α_k, α_{k+1}), $k = 1, \dots, r-1$, and (α_r, ∞).*
5. *If $r = 1$, then g is a nonnegative constant function, and $g = 0$ if and only if 0 is a pole of f.*
6. *If $r > 1$, then $g \neq 0$ and $h = \frac{1}{g}$ is a rational $\mathcal{S}$-function with $r-1$ poles and $r-1$ zeros $\beta_{h,k}$, $k = 1, \dots, r-1$ which are simple and interlace, 0 is a pole of h if and only if 0 is a pole of f, and*

$$\beta_k < \beta_{h,k} < \beta_{k+1} \quad \textit{for } k = 1, \dots, r-1 \textit{ if } a > 0,$$
$$\beta_k = \beta_{h,k} \quad \textit{for } k = 1, \dots, r-1 \textit{ if } a = 0.$$

Proof. By Lemma A.2.5 and Remark A.2.3 we know that the zeros and the poles of f are simple and interlace, and together with Proposition A.3.2 this shows that f has r or $r-1$ zeros. Hence f can be written as

$$f(z) = \frac{q(z)}{p(z)}$$

with

$$p(z) = z^r + \sum_{j=0}^{r-1} c_j z^j,$$
$$q(z) = \sum_{j=0}^{r} d_j z^j,$$

where p and q have no common zeros, their coefficients are real, and $d_r \neq 0$ or $d_{r-1} \neq 0$. Below we will show that $a = d_r$, which would complete the proof of part 3, again taking Proposition A.3.2 into account.

Since the function f is not constant, (A.3.2) is equivalent to

$$\begin{aligned} g(z) &= bz + \frac{1}{f(z) - a} \\ &= \frac{bz(q(z) - ap(z)) + p(z)}{q(z) - ap(z)} \\ &= \frac{p_g(z)}{q_g(z)} \end{aligned}$$

with

$$q_g(z) = q(z) - ap(z), \quad p_g(z) = bzq_g(z) + p(z).$$

Clearly, q_g is not identically zero since f is not constant. We observe that q_g and p_g do not have common zeros because $q_g(z) = p_g(z) = 0$ would lead to the contradiction $p(z) = 0$ and $q(z) = ap(z) = 0$.

In order to prove statement 1 it is therefore necessary and sufficient to find unique a and b such that the degree of p_g does not exceed $r-1$. This means that the coefficients of z^{r+1} and z^r of the Taylor expansion of p_g about 0 must be zero, that is, we must satisfy

$$b(d_r - a) = 0 \quad \text{and} \quad b(d_{r-1} - ac_{r-1}) + 1 = 0. \tag{A.3.5}$$

The second equation in (A.3.5) gives $b \neq 0$, and then the first equation has the unique solution $a = d_r$. It is clear that $a = d_r = \lim_{z\to-\infty} f(z) \geq 0$.

To solve for b in the second equation we first consider the case $d_r = 0$. Then $d_{r-1} \neq 0$ and $f(z) > 0$ for $z \in (-\infty, 0)$ gives $d_{r-1} < 0$. Hence we have a unique $b = -d_{r-1}^{-1}$, and $b > 0$ is satisfied.

Now let $d_r \neq 0$. Then

$$c_{r-1} = -\sum_{j=0}^{r} \alpha_j, \quad \frac{d_{r-1}}{d_r} = -\sum_{j=0}^{r} \beta_j,$$

and (A.3.3) gives

$$\frac{d_{r-1}}{d_r} < c_{r-1}.$$

Hence the second equation in (A.3.5) has a unique solution b, and $d_r = a > 0$ shows that

$$b = \frac{1}{a\left(c_{r-1} - \dfrac{d_{r-1}}{d_r}\right)} > 0.$$

This completes the proof of the statements 1 to 4.

Observe that the above proof shows that both q_g and p_g are polynomials of degree not exceeding $r-1$ and that q_g is not identically zero since f is not constant. Furthermore, $p_g(0) = 0$ if and only if $p(0) = 0$. If 0 is not a pole of f, then the fact that f increases on $(-\infty, 0]$ gives $f(0) > \lim_{x\searrow-\infty} f(x) = a$ and therefore

$$g(0) = \frac{1}{f(0) - a} > 0. \tag{A.3.6}$$

In case $r = 1$ both q_g and p_g are constant. Hence g is constant. Above we have seen that this constant is zero if 0 is a pole of f and positive otherwise, see (A.3.6). This proves statement 5.

Now we turn our attention to statement 6 and we henceforth assume that $r > 1$. From $q_g(\alpha_k) = q(\alpha_k)$ and the interlacing property of the zeros and poles of f it follows that q_g has opposite signs at consecutive α_k. Hence q_g has an odd number of zeros, counted with multiplicity, in each of the intervals (α_k, α_{k+1}), $k = 1, \ldots, r-1$. We already know that the degree of q_g does not exceed $r-1$, and therefore q_g has degree $r-1$ and all its zeros are real and simple, so that they can be indexed as $\beta_{h,k}$, $k = 1, \ldots, r-1$ and satisfy

$$\alpha_k < \beta_{h,k} < \alpha_{k+1}, \quad k = 1, \ldots, r-1. \tag{A.3.7}$$

If $a = 0$ then $q_g = q$, and hence $\beta_{h,k} = \beta_k$ for $j = 1, \dots, r-1$. If $a > 0$, then $q_g(\beta_k) = -ap(\beta_k)$, so that q_g has opposite signs at consecutive β_k, which means that q_g has an odd number of zeros, counted with multiplicity, in each of the intervals (β_k, β_{k+1}), $k = 1, \dots, r-1$. Hence it follows that $\beta_k < \beta_{h,k} < \beta_{k+1}$ for $k = 1, \dots, r-1$.

From $p_g(\alpha_k) = b\alpha_k q_g(\alpha_k)$ it follows that the sign of $p_g(\alpha_k)$ alternates with k (except for $\alpha_k = 0$). In any case, it follows that p_g has $r-1$ simple zeros. For $k = 1, \dots, r-1$, $p_g(\beta_{h,k}) = p(\beta_{h,k})$. Since the zeros of p are simple, p has opposite signs on adjacent intervals (α_{k-1}, α_k), (α_k, α_{k+1}). In view of (A.3.7), this means that $p_g(\beta_{h,k})$ has opposite signs for consecutive k. Hence between two zeros of q_g there is an odd number of zeros of p_g. Since q_g and p_g have the same number of zeros, counted with multiplicity, it follows that the zeros of q_g and p_g interlace.

If $\alpha_1 = 0$, then $p_g(0) = p(0) = 0$ and $0 = \alpha_1 < \beta_{h,1}$ shows that the smallest zero of p_g is smaller than the smallest zero of q_g. If $\alpha_1 > 0$, then $0 < \alpha_1 < \beta_{h,1} < \alpha_2$ shows that $p_g(0) = p(0)$ and $p_g(\beta_{h,1}) = p(\beta_{h,1})$ have opposite signs, that is, p_g has a positive zero which is smaller than the smallest zero $\beta_{h,1}$ of q_g. Hence it follows in any case that the smallest zero of p_g is nonnegative and is smaller than the smallest zero of q_g.

Altogether, we have shown so far that $h = \frac{1}{g}$ has a representation (A.3.1) with all properties as stated in Corollary A.3.3 satisfied except possibly $C_0 > 0$. That is, h or $-h$ is a rational $\mathcal{S}$-function. Therefore consider

$$g'(z) = b - \frac{f'(z)}{(f(z) - a)^2}.$$

The pole $\beta_{h,1}$ of g is a zero of $z \mapsto f(z) - a$, and therefore $\lim_{x \to \beta_{h,1}} (f(x) - a)^2 = 0$, whereas $f'(\beta_{h,1}) > 0$ in view of Lemma A.2.5. Hence $\lim_{x \to \beta_{h,1}} g'(x) = -\infty$, which gives that $h'(x)$ is positive for real x near $\beta_{h,1}$. In view of Lemma A.2.5, $-h$ cannot be a rational Nevanlinna function, and therefore h must be a rational $\mathcal{S}$-function. $\square$

Lemma A.3.5. *Let $n \in \mathbb{N}$ and let f be a rational S-function with n poles. Then:*

1. *f admits a unique continued fraction expansion*

$$f(z) = a_n + \cfrac{1}{-b_n z + \cfrac{1}{a_{n-1} + \cfrac{1}{-b_{n-1} z + \ddots + \cfrac{1}{a_1 + \cfrac{1}{-b_1 z + c_0}}}}} \tag{A.3.8}$$

   with $a_n = \lim_{z \to -\infty} f(z) \geq 0$, $a_k > 0$ for $k = 1, \dots, n-1$, $b_k > 0$ for $k = 1, \dots, n$, $c_0 \geq 0$ with $c_0 = 0$ if and only if 0 is a pole of f.

2. *The number of zeros of f is n if $a_n > 0$ and $n - 1$ if $a_n = 0$.*

3. *The poles* α_k *and zeros* β_k *of* f *are all simple and interlace,*

$$0 \le \alpha_1 < \beta_1 < \alpha_2 < \cdots < \beta_{n-1} < \alpha_n < \beta_n \quad \text{if } a_n > 0, \tag{A.3.9}$$
$$0 \le \alpha_1 < \beta_1 < \alpha_2 < \cdots < \beta_{n-1} < \alpha_n \quad \text{if } a_n = 0. \tag{A.3.10}$$

4. *The rational function* f *is strictly increasing between its poles, i.e., in the intervals* $(-\infty, \alpha_1)$, (α_k, α_{k+1}), $k = 1, \dots, n-1$, *and* (α_n, ∞).

5. *If* $f_j(z)$ *denotes the* $n-j$*-th tail of the continued fraction expansion* (A.3.8) *for* $j = 1, \dots, n$, *i.e.,*

$$f_j(z) = a_j + \cfrac{1}{-b_j z + \cfrac{1}{a_{j-1} + \cfrac{1}{-b_{j-1}z + \ddots + \cfrac{1}{a_1 + \cfrac{1}{-b_1 z + c_0}}}}} \tag{A.3.11}$$

then all zeros of f_j *are real and simple, written* $\beta_1^{(j)} < \cdots < \beta_{m_j}^{(j)}$ *with* $m_j = j$ *if* $j < n$ *or* $j = n$ *and* $a_n > 0$, *and* $m_n = n - 1$ *if* $a_n = 0$, *and the following holds:*

$$\beta_k^{(j)} < \beta_k^{(j-1)} < \beta_{k+1}^{(j)} \quad \text{for } j = 2, \dots, m_n,\ k = 1, \dots, j-1,$$
$$\beta_k^{(n)} = \beta_k^{(n-1)} \quad \text{for } k = 1, \dots, n-1 \text{ if } a_n = 0;$$

here we observe that $\beta_k = \beta_k^{(n)}$.

Proof. Parts 2, 3, and 4 follow from Lemma A.3.4 with $r = n$ and $a_n = a$.

We will prove parts 1 and 5 by induction on n. For $n = 1$, Lemma A.3.4 with $r = 1$, $a_1 = a$, $b_1 = b$ and $c_0 = g$ shows that indeed properties 1 and 5 are satisfied with $f_1 = f$.

Now let $n > 1$. From Lemma A.3.4 with $r = n$ we know that there are $a_n = a \ge 0$, $b_n = b > 0$ and a rational $\mathcal{S}$-function h with $n-1$ zeros and $n-1$ poles such that

$$f(z) = a_n + \cfrac{1}{-b_n z + \cfrac{1}{h(z)}}.$$

By induction hypothesis, h has the representation (A.3.8) with n replaced by $n-1$ and $a_{n-1} > 0$. Therefore $h = f_{n-1}$ as in (A.3.11). This proves part 1 for n. The remaining statements of part 5 follow by induction hypothesis for $j < n$ and from Lemma A.3.4 for $j = n$.

We still have to prove the uniqueness of the representation (A.3.8). To this end observe that for any such representation with (complex) numbers a_j, b_j and c_0 and corresponding representations (A.3.11) we have

$$f_j(z) = a_j + \cfrac{1}{-b_j z + \cfrac{1}{f_{j-1}(z)}} \tag{A.3.12}$$

for $j = 2, \dots, n$. We can write the rational functions f_j in the form $\frac{q_j}{p_j}$ with polynomials p_j and q_j without common zeros. Then, recursively, for $j = 2, \dots, n$,

$$f_j(z) = a_j + \cfrac{1}{-b_j z + \cfrac{p_{j-1}(z)}{q_{j-1}(z)}} = \frac{q_{j-1}(z) - a_j b_j z q_{j-1}(z) + a_j p_{j-1}(z)}{-b_j z q_{j-1}(z) + p_{j-1}(z)}.$$

Observing that f_1 has one pole and at most one zero, it follows by inductions that f_j has at most j poles and zeros. In particular, for $n > 1$, $f_n = f$ has n poles and f_{n-1} has at most $n-1$ poles. In view of Lemma A.3.4 the numbers a_n and b_n as well as the rational function f_{n-1} are unique, and f_{n-1} is a rational $\mathcal{S}$-function. By induction it follows that also the numbers $a_{n-1}, \dots, a_2, b_{n-1}, \dots, b_2$ as well as the rational $\mathcal{S}$-function f_1 are unique. Another application of Lemma A.3.4 finally shows that also a_1, b_1 and c_0 are unique. □

Lemma A.3.6. *Let $n \in \mathbb{N}$ and let f be given by (A.3.8) with $a_n \geq 0$, $a_k > 0$ for $k = 1, \dots, n-1$, $b_k > 0$ for $k = 1, \dots, n$ and $c_0 \geq 0$. Then f is a rational $\mathcal{S}$-function with n poles.*

Proof. For $j = 1, \dots, n$ define f_j by (A.3.11). We are going to show by induction that f_j is a rational $\mathcal{S}$-function. We have that

$$f_1(z) = a_1 + \frac{1}{-b_1 z + c_0}$$

has a pole at $z_0 = \frac{c_0}{b_1} \geq 0$. Then

$$f_1(z) = \begin{cases} \dfrac{1}{b_1} \dfrac{1}{z_0 - z} & \text{if } a_1 = 0, \\[2ex] a_1 \dfrac{z_0 + \frac{1}{a_1 b_1} - z}{z_0 - z} & \text{if } a_1 > 0, \end{cases}$$

shows by Corollary A.3.3 that f_1 is a rational $\mathcal{S}$-function.

Now let $n > 1$, $2 \leq j \leq n$ and assume that f_{j-1} is a rational $\mathcal{S}$-function. Then f_j has the representation (A.3.12). Lemma A.2.4 and the fact that a sum of Nevanlinna functions is a Nevanlinna function show that, in turn, with f_{j-1} also

$$-\frac{1}{f_{j-1}}, \quad z \mapsto b_j z - \frac{1}{f_{j-1}(z)}, \quad z \mapsto \cfrac{1}{-b_j z + \cfrac{1}{f_{j-1}(z)}}, \quad f_j$$

are rational Nevanlinna functions, and $f_{j-1}(z) > 0$ for $z < 0$ implies that also $f_j(z) > 0$ for $z < 0$. If k is the number of poles of f, then $n = k$ by Lemma A.3.5. □

Definition A.3.7. A rational function f is called a positive real function if

(i) $f(z)$ is real for real z,

(ii) $\operatorname{Re} f(z) > 0$ for $\operatorname{Re} z > 0$.

Lemma A.3.8. *If all zeros and poles of a nonconstant positive real function f lie on the imaginary axis, then the function h defined by $h(z) = if(-iz)$ is a rational Nevanlinna function and there is a rational $\mathcal{S}$-function g such that $f(z) = zg(-z^2)$. In this case, g belongs to $\mathcal{S}_0$ if and only if 0 is not a pole of f.*

Proof. By condition (i) of Definition A.3.7, f is a real rational function and hence $\overline{f} = f$ by Proposition A.1.5. In particular, zeros and poles of f are symmetric with respect to the real axis. Since zeros and poles of f lie on the imaginary axis, there are nonnegative integers m and n, an integer j, positive real numbers $\alpha_1, \dots, \alpha_n$ and $\beta_1, \dots, \beta_m$, and a nonzero complex number C such that

$$f(z) = Cz^j \prod_{k=1}^{m} \left(z^2 + \beta_k\right) \prod_{k=1}^{n} \left(z^2 + \alpha_k\right)^{-1}. \tag{A.3.13}$$

Without loss of generality we assume that $\alpha_k \le \alpha_{k+1}$ and $\beta_k \le \beta_{k+1}$. Observe that at least one of m, n, j is nonzero since f is nonconstant and that $C > 0$ since $f(z) > 0$ for $z > 0$.

Assume that $|j| > 1$. Then f has the representation $f(z) = z^j f_1(z)$ where f_1 is a rational function with $f_1(0) > 0$. Let $\psi_1 = \frac{3\pi}{4j} \in \left(-\frac{\pi}{2}, \frac{\pi}{2}\right)$ and observe that $j\psi_1 = \frac{3\pi}{4} \in \left(\frac{\pi}{2}, \pi\right)$. With $\psi_0 = 0$, $\varphi_1 = \frac{\pi}{2}$ and $\varphi_2 = \pi$ it follows from Proposition A.1.7 that there is $r > 0$ such that for $w = re^{i\psi_1}$ we have $f(w) = |f(w)|e^{i\varphi_0} \ne 0$ and $\varphi_0 \in \left(\frac{\pi}{2}, \pi\right)$. This would lead to the contradiction $\operatorname{Re} w > 0$ and $\operatorname{Re} f(w) < 0$.

If $j = 0$, then the representation (A.3.13) would give $f(z) = \overline{f(-\overline{z})}$ for all z which are not poles of f. When $\operatorname{Re} z < 0$ then $\operatorname{Re}(-\overline{z}) = -\operatorname{Re} \overline{z} = -\operatorname{Re} z > 0$, and hence condition (ii) of Definition A.3.7 would give $\operatorname{Re} f(z) = \operatorname{Re} f(-\overline{z}) > 0$. Hence $\operatorname{Re} f(z) \ge 0$ for all $z \in \mathbb{C}$ which are not poles of f. But this is impossible by Lemma A.1.6.

It follows from the previous two paragraphs that $j = 1$ or $j = -1$. In both cases, f maps the imaginary axis to itself, and h is therefore a rational function which maps the upper half-plane into itself and is real on the real axis. By Remark A.2.3 and Lemma A.2.5, h is a Nevanlinna function. In particular, all the numbers α_k, $k = 1, \dots, n$, and β_k, $k = 1, \dots, m$, are different and interlace. If $j = -1$, then 0 is a pole of f, and $n \le m$ as well as $\beta_1 < \alpha_1$ if $n > 0$ follows. If $j = 1$, then 0 is a zero of f, and $m \le n$ as well as $\alpha_1 < \beta_1$ if $m > 0$ follows.

Define

$$g(z) = C(-z)^{\frac{j-1}{2}} \prod_{k=1}^{m} (\beta_k - z) \prod_{k=1}^{n} \frac{1}{\alpha_k - z}.$$

Then poles and zeros of g interlace and the smallest pole is smaller than the smallest zero if g has any zeros. Hence g is a rational $\mathcal{S}$-function by Corollary A.3.3. The identity $f(z) = zg(-z^2)$ is clear by definition of g, and 0 is a pole of g if and only if $j = -1$. □

A.4 The classes $\mathcal{S}^r$

Definition A.4.1. Let $r \in \mathbb{N}$. A pair (q, p) of real polynomials is said to belong to $\mathcal{S}^r$ if $q(z) \neq 0$ and $p(z) \neq 0$ for all $z \in \mathbb{C} \setminus [0, \infty)$ and if there exist nonzero real polynomials $p_1, \ldots, p_r$ such that

$$\frac{q}{p_1} \in \mathcal{S}, \ \frac{p_1}{p_2} \in \mathcal{S}, \ \ldots, \ \frac{p_{r-1}}{p_r} \in \mathcal{S}, \ \frac{p_r}{p} \in \mathcal{S}.$$

A pair (q, p) of real polynomials is said to belong to $\mathcal{S}^0$ if $q(z) \neq 0$ and $p(z) \neq 0$ for all $z \in \mathbb{C} \setminus [0, \infty)$ and if $\frac{q}{p} \in \mathcal{S}$.

Remark A.4.2. 1. Let $r, s \in \mathbb{N}$ with $r < s$ and let $(q, p) \in \mathcal{S}^r$ with $p, q, q_1, \ldots, q_r$ as in Definition A.4.1. Then putting $p_s = p_{s-1} = \cdots = p_{r+1} = p_r$ and observing that the constant function 1 belongs to $\mathcal{S}$ it is clear that $(q, p) \in \mathcal{S}^s$. Thus we have seen that $\mathcal{S}^r \subset \mathcal{S}^s$ for $s > r$.

2. Starting with either p or q and observing that the zeros and poles of rational $\mathcal{S}$-functions are nonnegative real numbers, it follows by induction that the zeros of the polynomials $p_1, \ldots, p_r$ in Definition A.4.1 are also nonnegative real numbers.

Lemma A.4.3. *Let $X = (x_j)_{j=1}^n$ and $Y = (y_j)_{j=1}^n$ be two finite sequences of real numbers with $x_j \le x_{j+1}$ and $y_j \le y_{j+1}$ $(j = 1, \ldots, n-1)$. For $d \in \mathbb{R}$ let*

$$N(X, d) = \#\{j \in \{1, \ldots, n\} : x_j \le d\}.$$

Then $y_j \le x_j$ for $j = 1, \ldots, n$ if and only if $N(X, d) \le N(Y, d)$ for all $d \in \mathbb{R}$.

Proof. If $y_j \le x_j$ for $j = 1, \ldots, n$, and if $d \in \mathbb{R}$, then $x_j \le d$ implies $y_j \le d$, and therefore $N(X, d) \le N(Y, d)$. Conversely, if there is $j_0 \in \{1, \ldots, n\}$ such that $x_{j_0} < y_{j_0}$, then let $d = x_{j_0}$. In this case, $x_j \le x_{j_0} = d$ for $j = 1, \ldots, j_0$, so that $N(X, d) \ge j_0$. But $d = x_{j_0} < y_{j_0} \le y_j$ for $j = j_0, \ldots, n$ shows that $N(Y, d) < j_0$, and therefore $N(Y, d) < N(X, d)$. □

Lemma A.4.4. *Let $m, k \in \mathbb{N}$, let $X^{(0)} = (x_1^{(0)}, \ldots, x_m^{(0)})$ and $Y^{(0)} = (y_1^{(0)}, \ldots, y_m^{(0)})$ be sequences of real numbers such that*

$$y_1^{(0)} \le x_1^{(0)} \le y_2^{(0)} \le x_2^{(0)} \le \cdots \le y_m^{(0)} \le x_m^{(0)}.$$

Let $C = (c_1, \ldots, c_k)$ be a sequence of real numbers and let $X = (x_1, \ldots, x_{m+k})$ and $Y = (y_1, \ldots, y_{m+k})$ be the sequences consisting of the terms $x_1^{(0)}, \ldots, x_m^{(0)}, c_1, \ldots, c_k$

and $y_1^{(0)}, \dots, y_m^{(0)}, c_1, \dots, c_k$, *respectively, ordered in such a way that* $x_j \le x_{j+1}$ *and* $y_j \le y_{j+1}$ *for* $j = 1, \dots, k+m-1$, *respectively. Then*

$$y_1 \le x_1 \le y_2 \le x_2 \le \cdots \le y_{m+k} \le x_{m+k}.$$

Proof. Note that the statement of the lemma is equivalent to $y_j \le x_j$ for $j = 1, \dots, m+k$ and $x_j \le y_{j+1}$ for $j = 1, \dots, m+k-1$. Then $x_j \le x_{j+1}$ and $y_j \le y_{j+1}$ for $j = 1, \dots, k+m-1$ is satisfied as well so that these inequalities do not need to be shown separately. With the notation of Lemma A.4.3 we observe that $N(X,d) = N(X^{(0)},d) + N(C,d)$ and $N(Y,d) = N(Y^{(0)},d) + N(C,d)$ for all $d \in \mathbb{R}$. By Lemma A.4.3, $N(X^{(0)},d) \le N(Y^{(0)},d)$, which then implies $N(X,d) \le N(Y,d)$ for all $d \in \mathbb{R}$. Again by Lemma A.4.3, this shows $y_j \le x_j$ for $j = 1, \dots, m+k$. Choose numbers $x_0^{(0)} < \min\{y_1^{(0)}, c_1, \dots, c_k\}$ and $y_{m+1}^{(0)} > \max\{x_m^{(0)}, c_1, \dots, c_k\}$ and put $\overline{X}^{(0)} = (x_j^{(0)})_{j=0}^m$ and $\overline{Y}^{(0)} = (y_j^{(0)})_{j=1}^{m+1}$. Since $x_j^{(0)} \le y_{j+1}^{(0)}$ for $j = 0, \dots, m$, Lemma A.4.3 gives $N(\overline{Y}^{(0)},d) \le N(\overline{X}^{(0)},d)$ for all $d \in \mathbb{R}$, and $N(\overline{Y},d) \le N(\overline{X},d)$ follows. Then Lemma A.4.3 shows that $x_j \le y_{j+1}$ for $j = 1, \dots, k+m-1$. □

Theorem A.4.5. *Let* $r \in \mathbb{N}$ *and let* $(q,p) \in \mathcal{S}^r$ *where the polynomials* p *and* q *have the same degree* n. *Then the nondecreasing sequence of zeros* $(a_k)_{k=1}^n$ *of* q *and the nondecreasing sequence of zeros* $(b_k)_{k=1}^n$ *of* p *have the following properties:*

1. *For* $s = 1, \dots, n$ *the interval* $(-\infty, b_s)$ *contains at most* $s-1$ *terms of* $(a_k)_{k=1}^n$.
2. *For* $s = r+2, \dots, n$ *the interval* $(-\infty, b_s]$ *contains at least* $s-1-r$ *terms of* $(a_k)_{k=1}^n$.

Proof. Observe that it is allowed that two adjacent polynomials in the sequence $(q, p_1, \dots, p_r, p)$ have common zeros. Since quotients of adjacent polynomials are rational $\mathcal{S}$-functions, the degree of the polynomials in the above sequence is nondecreasing, and it follows that all polynomials have the same degree n. Hence the zeros of p_j $(j = 1, \dots, r)$ can be written as a nondecreasing sequence $(\alpha_k^{(j)})_{k=1}^n$. Now consider any consecutive pair of the sequences $(a_k)_{k=1}^n$, $(\alpha_k^{(1)})_{k=1}^n$, $\dots$, $(\alpha_k^{(r)})_{k=1}^n$, $(b_k)_{k=1}^n$. After removal of common terms, the remaining two sequences are sequences of the zeros and poles of a rational $\mathcal{S}$-function, which interlace like (A.3.3). Therefore, making use of Lemma A.4.4, statement 1 is obvious due to

$$b_s \le \alpha_s^{(r)} \le \cdots \le \alpha_s^{(1)} \le a_s.$$

Now let $s \ge r+2$ and $j \in \{1, \dots, s-1-r\}$. Observing $j+r+1 \le s$ it follows that

$$a_j \le \alpha_{j+1}^{(1)} \le \cdots \le \alpha_{j+r}^{(r)} \le b_{j+r+1} \le b_s,$$

which proves statement 2. □

Lemma A.4.6. *Let* $(q,p) \in \mathcal{S}^0$, *let* $l, m > 0$, *and define the polynomials* f *and* g *by*

$$\begin{pmatrix} f(z) \\ g(z) \end{pmatrix} = \begin{pmatrix} 1 & l \\ -mz & 1 - lmz \end{pmatrix} \begin{pmatrix} q(z) \\ p(z) \end{pmatrix}.$$

Then $(f,g) \in \mathcal{S}^0$ *and*

$$\frac{f(z)}{g(z)} = \cfrac{1}{-mz + \cfrac{1}{l + \cfrac{q(z)}{p(z)}}}. \tag{A.4.1}$$

If additionally $\frac{q}{p} \in \mathcal{S}_0$, *then also* $\frac{f}{g} \in \mathcal{S}_0$.

Proof. Clearly, f and g are real polynomials, and

$$\begin{aligned}\frac{f(z)}{g(z)} &= \frac{q(z) + lp(z)}{-mzq(z) + (1 - lmz)p(z)} \\ &= \cfrac{1}{-mz + \cfrac{1}{l + \cfrac{q(z)}{p(z)}}}.\end{aligned}$$

If $\frac{q}{p}$ is constant, then $\frac{f}{g} \in \mathcal{S}$ by Lemma A.3.6. Otherwise, if $\frac{q}{p}$ is not constant, then, by Lemma A.3.5, $\frac{q}{p}$ has a representation of the form (A.3.8), and it follows immediately that also $\frac{f}{g}$ has such a representation. Hence $\frac{f}{g} \in \mathcal{S}$ by Lemma A.3.6.

To complete the proof that $(f,g) \in \mathcal{S}^0$ let $z \in \mathbb{C} \setminus [0, \infty)$. We have to show that $f(z) \neq 0$ and $g(z) \neq 0$. This is obvious if $\frac{q}{p}$ is constant. Otherwise, if $\frac{q}{p}$ is not constant, assume that $f(z) = 0$. Then $\frac{q(z)}{p(z)} = -l$, which contradicts $\frac{q}{p} \in \mathcal{S}$. Hence $f(z) \neq 0$. Since $\frac{f}{g} \in \mathcal{S}$, this function has no pole at z, and hence also $g(z) \neq 0$.

If $\frac{q}{p} \in \mathcal{S}_0$, then we may assume that $p(0) \neq 0$, and hence $g(0) = p(0) \neq 0$ proves that $\frac{f}{g} \in \mathcal{S}_0$. □

A.5 Hermite-Biehler polynomials

Let ω be a polynomial. Defining $P = \frac{1}{2}(\omega + \overline{\omega})$ and $Q = \frac{1}{2i}(\omega - \overline{\omega})$ it is immediately clear that P and Q are real polynomials and that the identity $\omega = P + iQ$ holds. We call P the *real part of* ω and Q the *imaginary part* of ω.

Definition A.5.1. A nonconstant polynomial is said to be Hermite-Biehler (HB) if it has no zeros in the closed lower half-plane.

Theorem A.5.2. *Let* ω *be a polynomial with real and imaginary parts* P *and* Q, *i.e., real polynomials* P *and* Q *such that* $\omega = P + iQ$. *Then* ω *is a Hermite-Biehler polynomial if and only if the following two properties hold:*

(i) *the zeros of the polynomials* P *and* Q *are real, simple and interlace, i.e., between any two zeros of one of these polynomials there lies one zero of the other;*

(ii) *for some* $\lambda_0 \in \mathbb{R}$,

$$Q'(\lambda_0)P(\lambda_0) - Q(\lambda_0)P'(\lambda_0) > 0.$$

Proof. Let ω be a Hermite-Biehler polynomial. Let $n \in \mathbb{N}$ be its degree and let a_j $(j = 1, \ldots, n)$ be its zeros. Since ω is a Hermite-Biehler polynomial, $\operatorname{Im} a_j > 0$ $(j = 1, \ldots, n)$. There is a nonzero complex number a such that

$$\omega(\lambda) = a \prod_{j=1}^{n} (\lambda - a_j) \quad (\lambda \in \mathbb{C}).$$

For $\operatorname{Im} \lambda > 0$ we calculate

$$\begin{aligned} |\lambda - a_j|^2 &= (\operatorname{Re} \lambda - \operatorname{Re} a_j)^2 + (\operatorname{Im} \lambda - \operatorname{Im} a_j)^2 \\ &< (\operatorname{Re} \lambda - \operatorname{Re} a_j)^2 + (\operatorname{Im} \lambda + \operatorname{Im} a_j)^2 = |\lambda - \overline{a_j}|^2. \end{aligned}$$

We therefore conclude that

$$|\omega(\lambda)| = |a| \prod_{j=1}^{n} |\lambda - a_j| < |a| \prod_{j=1}^{n} |\lambda - \overline{a_j}| = |\overline{\omega}(\lambda)| \quad (\operatorname{Im} \lambda > 0).$$

Hence condition (ii) in part 1 of Lemma A.2.2 holds. Therefore, condition (iv) in part 1 of Lemma A.2.2 holds, which shows that $f = \frac{P}{Q}$ maps $\mathbb{C}^+$ into itself. Furthermore, since ω does not have real zeros, P and Q do not have common real zeros. Then the properties (i) and (ii) immediately follow from Lemma A.2.5.

Conversely, let properties (i) and (ii) be satisfied. By Theorem A.2.6, $f = \frac{P}{Q}$ or $-f$ maps $\mathbb{C}^+$ into itself. In view of (ii) and Lemma A.2.5, this is not true for $-f$. Therefore $f(\mathbb{C}^+) \subset \mathbb{C}^+$. Hence f satisfies property (iv) in part 1 of Lemma A.2.2, and property (ii) in part 1 of Lemma A.2.2 shows that ω has no zeros in the open lower half-plane. But since P and Q are real on the real axis, a real zero of ω would be a common real zero of P and Q, which is impossible since property (i) holds. □

Definition A.5.3. A polynomial ω is said to be symmetric if $\omega(-\overline{\lambda}) = \overline{\omega(\lambda)}$ for all $\lambda \in \mathbb{C}$. The polynomial ω is said to belong to the class SHB if it is symmetric and belongs to the Hermite-Biehler class.

For a symmetric polynomial ω, we have $\omega(-\lambda) = \overline{\omega(\overline{\lambda})} = \overline{\omega}(\lambda)$ $(\lambda \in \mathbb{C})$. Hence, with P and Q the real and imaginary parts of ω and $\lambda \in \mathbb{C}$ it follows that

$$P(-\lambda) = \frac{1}{2}\left(\omega(-\lambda) + \overline{\omega}(-\lambda)\right) = \frac{1}{2}\left(\overline{\omega}(\lambda) + \omega(\lambda)\right) = P(\lambda)$$

and

$$Q(-\lambda) = \frac{1}{2i}\left(\omega(-\lambda) - \overline{\omega}(-\lambda)\right) = \frac{1}{2i}\left(\overline{\omega}(\lambda) - \omega(\lambda)\right) = -Q(\lambda).$$

This means that P is an even real polynomial and Q is an odd real polynomial. Therefore there exist real polynomials $\widetilde{P}$ and $\widetilde{Q}$ such that $P(\lambda) = \widetilde{P}(\lambda^2)$ and $Q(\lambda) = \lambda\widetilde{Q}(\lambda^2)$ $(\lambda \in \mathbb{C})$. Hence we have

$$\omega(\lambda) = \widetilde{P}(\lambda^2) + i\lambda\widetilde{Q}(\lambda^2) \quad (\lambda \in \mathbb{C}). \tag{A.5.1}$$

Proposition A.5.4. *Let $n \in \mathbb{N}$ and let ω be a polynomial of degree n. Let $J = \{-k, \dots, k\}$ if $n = 2k+1$ and $J = \{-k, \dots, -1, 1, \dots, k\}$ if $k = 2n$. Then ω is of* SHB *class if and only if there are $C \in \mathbb{R} \setminus \{0\}$ and $\lambda_j \in \mathbb{C}$ $(j \in J)$ such that*

(i) $\operatorname{Im} \lambda_j > 0$ *for* $j \in J$;

(ii) $\lambda_{-j} = -\overline{\lambda_j}$ *for not pure imaginary* λ_{-j};

(iii) $$\omega(\lambda) = C \prod_{j \in J} \left(1 - \frac{\lambda}{\lambda_j}\right) \quad (\lambda \in \mathbb{C}).$$

If ω is of SHB *class, then, in the representation* (A.5.1), $\widetilde{P}(0) = C$ *and*

$$C^{-1}\widetilde{Q}(0) = \sum_{j \in J} \frac{i}{\lambda_j} > 0. \tag{A.5.2}$$

Proof. Under the conditions (i), (ii) and $C \in \mathbb{R}\setminus\{0\}$, the polynomial defined by (iii) belongs to the SHB class. Conversely, the zeros of a Hermite-Biehler polynomial ω satisfy (i) by definition, and the symmetry of ω implies that its zeros, counted with multiplicity, are symmetric with respect to the imaginary axis. Hence they can be indexed in such a way that (ii) holds. Finally, in the representation (iii) of ω we have $C = \omega(0) \in \mathbb{R} \setminus \{0\}$ due to (i) and the symmetry of ω.

Now let ω be of SHB class. The representations (A.5.1) and (iii) give

$$\widetilde{P}(0) = \omega(0) = C \quad \text{and} \quad \widetilde{Q}(0) = -i\omega'(0) = C \sum_{j \in J} \frac{i}{\lambda_j}. \tag{A.5.3}$$

Since $\widetilde{Q}$ is a real polynomial, it follows that $\widetilde{Q}(0)$ is real, and in view of (i),

$$\operatorname{Re} \frac{i}{\lambda_j} = \operatorname{Re} \frac{i\overline{\lambda_j}}{|\lambda_j|^2} = \frac{\operatorname{Im} \lambda_j}{|\lambda_j|^2} > 0, \quad (j \in J),$$

which gives $C^{-1}\widetilde{Q}(0) > 0$. □

Theorem A.5.5. *Let ω be a polynomial of* SHB *class with the representation* (A.5.1). *Then $\frac{\widetilde{Q}}{\widetilde{P}}$ is a rational $\mathcal{S}_0$-function.*

Proof. By Theorem A.5.2, the zeros of P and Q are real, simple, and interlace. From

$$\frac{Q(\lambda)}{P(\lambda)} = \frac{\lambda\widetilde{Q}(\lambda^2)}{\widetilde{P}(\lambda^2)} \tag{A.5.4}$$

we conclude that the zeros of $\widetilde{P}$ and $\widetilde{Q}$ are real, positive, simple, and interlace, and that the smallest zero of $\widetilde{P}$ is smaller than the smallest zero of $\widetilde{Q}$, if any. Since P and Q satisfy property (ii) of Theorem A.5.2, it follows that

$$\frac{\widetilde{Q}(0)}{\widetilde{P}(0)} = \left(\frac{Q}{P}\right)'(0) > 0.$$

By Theorem A.2.6, either $\frac{\widetilde{Q}}{\widetilde{P}}$ or $\frac{\widetilde{P}}{\widetilde{Q}}$ is a Nevanlinna function. But since Nevanlinna functions are locally increasing by Lemma A.2.5, this Nevanlinna function cannot have a zero between 0 and its smallest pole, and therefore indeed $\frac{\widetilde{Q}}{\widetilde{P}}$ is a Nevanlinna function, which is also positive on $(-\infty, 0)$ since it has no zeros there. Altogether, it follows that $\frac{\widetilde{Q}}{\widetilde{P}}$ is a rational $\mathcal{S}_0$ function. □

Proposition A.5.6. *Let ω be a monic symmetric polynomial of degree $2n$ $(n \in \mathbb{N})$ with $\omega(0) \neq 0$, where $\widetilde{Q}$ in the representation* (A.5.1) *of ω is a positive constant. If $\omega(0) > 0$, then ω has n zeros in the open upper half-plane and n zeros in the open lower half-plane, counted with multiplicity. If $\omega(0) < 0$, then ω has $n-1$ zeros in the open upper half-plane and $n+1$ zeros in the open lower half-plane, counted with multiplicity.*

Proof. Because ω is symmetric, it is evident that $\omega(0)$ is a nonzero real number. It is also clear from the assumptions and the representation (A.5.1) that $\omega(\lambda) \notin \mathbb{R}$ whenever $\lambda \in \mathbb{R} \setminus \{0\}$. For $\eta \in \mathbb{R}$ let

$$\omega_\eta(\lambda) = \lambda^{2n} + \omega(0) + i\eta\lambda.$$

We observe that also $\omega_\eta(\lambda) \notin \mathbb{R}$ whenever $\eta \neq 0$ and $\lambda \in \mathbb{R} \setminus \{0\}$.

First assume that $\omega(0) < 0$. Then ω_0 has two distinct simple real zeros. Let λ_0 be such a zero. Then $\omega(0) = -\lambda_0^{2n}$, and for sufficiently small $\eta > 0$, say $\eta \in (0, \eta_0]$, ω_η has a zero $\lambda_\eta \in \mathbb{C} \setminus \mathbb{R}$ which depends continuously on $\eta \in [0, \eta_0]$. We may write $\lambda_\eta = \rho(\eta)e^{i\psi(\eta)}$ where ρ and ψ are real functions depending continuously on η with $\rho(0) = \lambda_0$ and $\psi(0) = 0$. We may also assume that η_0 is chosen so small that $0 < |\psi(\eta)| < \frac{\pi}{2n}$ for $\eta \in (0, \eta_0)$. Then

$$0 = \omega_\eta(\lambda_\eta) = \lambda_\eta^{2n} - \lambda_0^{2n} + i\eta\lambda_\eta,$$

and taking imaginary parts gives

$$0 = \operatorname{Im} \lambda_\eta^{2n} + \eta \operatorname{Re} \lambda_\eta = \rho^{2n}(\eta) \sin(2n\psi(\eta)) + \eta\rho(\eta) \cos\psi(\eta).$$

Therefore

$$\begin{aligned}\operatorname{Im} \lambda_\eta &= \rho(\eta) \sin\psi(\eta) \\ &= -\rho^{2n}(\eta) \frac{\sin\psi(\eta) \sin(2n\psi(\eta))}{\eta \cos\psi(\eta)} \\ &< 0\end{aligned}$$

for $\eta \in (0, \eta_0)$. Of the $2n-2$ remaining zeros of ω_0, $n-1$ lie in the open upper half-plane and $n-1$ lie in the open lower half-plane. Altogether, the continuous dependence of the zeros of ω_η on η shows that $\omega_{\widetilde{Q}}$ has $n-1$ zeros in the open upper half-plane and $n+1$ zeros in the open lower half-plane when $\omega(0) < 0$.

When $\omega(0) > 0$, then ω_0 has n zeros in the open upper half-plane and n zeros in the open lower half-plane, and the same reasoning as above shows that $\omega_{\widetilde{Q}}$ has n zeros in the open upper half-plane and n zeros in the open lower half-plane when $\omega(0) > 0$.

Finally, observing that $\omega - \omega_{\widetilde{Q}}$ is real on the real axis and has degree at most $2n-2$, it follows that the zeros of $\omega_{\widetilde{Q}} + t(\omega - \omega_{\widetilde{Q}})$ depend continuously on $t \in [0,1]$ and do not cross the real axis. Therefore the number of zeros in the open upper (lower) half-plane of ω and $\omega_{\widetilde{Q}}$ are equal. □

Definition A.5.7. A nonconstant polynomial is said to belong to the generalized Hermite-Biehler class ($\overline{\mathrm{HB}}$) if it has no zeros in the open lower half-plane. The class of all symmetric generalized Hermite-Biehler polynomials is denoted by $\overline{\mathrm{SHB}}$.

Proposition A.5.8. *Let ω be be nonconstant symmetric polynomial. If the polynomials $\widetilde{P}$ and $\widetilde{Q}$ in the representation* (A.5.1) *have no common zeros in $\mathbb{C} \setminus [0, \infty)$ and if $\frac{\widetilde{Q}}{\widetilde{P}}$ is an $\mathcal{S}$-function, then ω is of $\overline{\mathrm{SHB}}$ class. If additionally $\widetilde{P}$ and $\widetilde{Q}$ have no common zeros and $\frac{\widetilde{Q}}{\widetilde{P}}$ is an $\mathcal{S}_0$-function, then ω is of* SHB *class.*

Proof. It is convenient to write the formula (A.3.1) for the $\mathcal{S}$-function $\frac{\widetilde{Q}}{\widetilde{P}}$ in the form

$$\frac{\widetilde{Q}(z)}{\widetilde{P}(z)} = (-1)^{m+n} C_0 \prod_{k=1}^{m} (z - b_k^2) \prod_{k=1}^{n} \frac{1}{z - a_k^2}$$

with $0 \le a_1 < b_1 < a_2 < \dots$ and $C_0 > 0$. Then the representation (A.5.4) gives

$$\frac{Q(\lambda)}{P(\lambda)} = (-1)^{m+n} C_0 \lambda \prod_{k=1}^{m} (\lambda - b_k)(\lambda + b_k) \prod_{k=1}^{n} \frac{1}{(\lambda - a_k)(\lambda + a_k)}.$$

Clearly, the zeros and poles of $\frac{Q}{P}$ are simple and interlace. If the largest pole is larger than the largest zero, then $m = n-1$, and in this case $(-1)^{m+n} C_0 < 0$. If the largest zero is larger than the largest pole, then $m = n$, and in this case $(-1)^{m+n} C_0 > 0$. Hence Remark A.2.3 and Theorem A.2.6 shows that $\frac{Q}{P}$ is a Nevanlinna function. The assumption that common zeros of $\widetilde{P}$ and $\widetilde{Q}$ can only be nonnegative real numbers implies that P and Q have no nonreal common zeros. Therefore, the equivalence of (ii) and (iii) in Lemma A.2.2, part 1 or part 2, shows that $\omega \in \overline{\mathrm{SHB}}$.

Under the additional assumption, the real functions P and Q have no common zeros. Hence $\omega \in \mathrm{SHB}$. □

A.6 Miscellaneous results

A straightforward calculation proves the following result.

Proposition A.6.1. *Let ψ_j, P_j, Q_j $(j = 1, 2)$ be polynomials such that $\psi_j(\lambda) = P_j(\lambda^2) + i\lambda Q_j(\lambda^2)$ $(\lambda \in \mathbb{C})$. Then*

$$\psi_2(\lambda)\psi_1(-\lambda) - \psi_2(-\lambda)\psi_1(\lambda) = 2i\lambda[P_1(\lambda^2)Q_2(\lambda^2) - P_2(\lambda^2)Q_1(\lambda^2)].$$

Proposition A.6.2. *Let $\eta \in \mathbb{C}$, let $k \in \mathbb{N}$ and let ω be a polynomial of degree $2k$ such that $\lambda \in \mathbb{C}$ is not a zero of ω whenever $-\lambda$ is a zero of ω. Then there is a unique polynomial ψ of degree not exceeding $2k$ such that $\omega(0)\psi(0) = 1$ and*

$$\omega(\lambda)\psi(-\lambda) - \omega(-\lambda)\psi(\lambda) = i\eta\lambda \quad (\lambda \in \mathbb{C}).$$

The coefficients of ψ depend linearly on η, and there is at most one value of $\eta \in \mathbb{C}$ for which the degree of ψ is less than $2k$.

Proof. Let ψ be any polynomial of degree $2k$. Decomposing each of ω and ψ into a sum of an even and odd polynomial, we find unique polynomials $\widetilde{P}$ of degree k, $\widetilde{Q}$ of degree at most $k-1$, $\widetilde{Y}$ of degree k and $\widetilde{Z}$ of degree at most $k-1$ such that

$$\omega(\lambda) = \widetilde{P}(\lambda^2) + i\lambda\widetilde{Q}(\lambda^2), \quad \psi(\lambda) = \widetilde{Y}(\lambda^2) + i\lambda\widetilde{Z}(\lambda^2) \quad (\lambda \in \mathbb{C}). \tag{A.6.1}$$

By the Euclidean algorithm, there are a unique constant $c_1 \in \mathbb{C}$, a unique polynomial $\widehat{Y}$ of degree at most $k-1$, a unique polynomial $\widehat{Z}_0$, and a unique polynomial $\widehat{Z}$ whose degree is less than the degree of $\widetilde{Q}$ such that

$$\widetilde{Y} = c_1\widetilde{P} + \widehat{Y}, \quad \widetilde{Z} = \widehat{Z}_0\widetilde{Q} + \widehat{Z}. \tag{A.6.2}$$

The polynomials $\widetilde{P}$ and $\widetilde{Q}$ do not have common zeros because otherwise there would be a complex number λ such that $\widetilde{P}(\lambda^2) = 0$ and $\widetilde{Q}(\lambda^2) = 0$, which would lead to the contradiction $\omega(\lambda) = 0 = \omega(-\lambda)$. Also, $\widetilde{P}(0) = \omega(0) \neq 0$. By Proposition A.6.1,

$$\omega(\lambda)\psi(-\lambda) - \omega(-\lambda)\psi(\lambda) = 2i\lambda[\widetilde{Q}(\lambda^2)\widetilde{Y}(\lambda^2) - \widetilde{P}(\lambda^2)\widetilde{Z}(\lambda^2)] \quad (\lambda \in \mathbb{C}).$$

Therefore we have to show that there are unique polynomials $\widetilde{Y}$ of degree at most k with $\widetilde{P}(0)\widetilde{Y}(0) = 1$ and $\widetilde{Z}$ of degree at most $k-1$ such that

$$\widetilde{Q}\widetilde{Y} - \widetilde{P}\widetilde{Z} = \frac{\eta}{2},$$

which is equivalent to

$$\frac{\widetilde{Y}}{\widetilde{P}} - \frac{\widetilde{Z}}{\widetilde{Q}} = \frac{\eta}{2\widetilde{P}\widetilde{Q}}. \tag{A.6.3}$$

The representation in (A.6.2) gives that (A.6.3) is equivalent to $c_1 = \widehat{Z}_0$ and

$$\frac{\widehat{Y}}{\widetilde{P}} - \frac{\widehat{Z}}{\widetilde{Q}} = \frac{\eta}{2\widetilde{P}\widetilde{Q}}. \tag{A.6.4}$$

In view of the uniqueness of the partial fraction decomposition and the fact that $\widetilde{P}$ and $\widetilde{Q}$ do not have common zeros it follows that (A.6.4) has a unique pair of solutions $\widehat{Y}$, $\widehat{Z}$. If Y and Z denote the solutions for $\eta = 1$, then clearly $\widehat{Y} = \eta Y$ and $\widehat{Z} = \eta Z$. Therefore

$$\omega(0)\psi(0) = \widetilde{P}(0)\widetilde{Y}(0) = c_1\widetilde{P}^2(0) + \eta\widetilde{P}(0)Y(0),$$

and the requirement $\omega(0)\psi(0) = 1$ gives a unique

$$c_1 = \frac{1 - \eta\widetilde{P}(0)Y(0)}{\widetilde{P}^2(0)}, \tag{A.6.5}$$

and hence also $\widehat{Z}_0 = c_1$ is unique. It is now obvious that all coefficients of $\widetilde{Y}$ and $\widetilde{Z}$ depend linearly on η. When $Y(0) = 0$, then $c_1 \neq 0$ for all $\eta \in \mathbb{C}$. However, when $Y(0) \neq 0$, then there is exactly one $\eta \in \mathbb{C}$ such that $c_1 = 0$. This completes the proof since the degree of ψ is $2k$ if and only if $c_1 \neq 0$. □

Corollary A.6.3. *In addition to the assumptions of Proposition* A.6.2 *assume that* $\eta \in \mathbb{R}$ *and* ω *is symmetric. Then also* ψ *is symmetric. For fixed* ω *let* J_ω *be the set of all* $\eta \in \mathbb{R}$ *for which the coefficient of* λ^{4k} *in* $\omega(\lambda)\psi(\lambda)$ *is positive. Then* J_ω *is an unbounded open interval with* $0 \in J_\omega$.

Proof. Since ω is symmetric, i.e., $\omega(-\lambda) = \overline{\omega}(\lambda)$, it follows that

$$\begin{aligned}
\omega(\lambda)\overline{\psi}(\lambda) - \omega(-\lambda)\overline{\psi}(-\lambda) &= -[\overline{\omega}(\lambda)\overline{\psi}(-\lambda) - \overline{\omega}(-\lambda)\overline{\psi}(\lambda)] \\
&= -\overline{[\omega(\overline{\lambda})\psi(-\overline{\lambda}) - \omega(-\overline{\lambda})\psi(\overline{\lambda})]} \\
&= -\overline{i\eta\overline{\lambda}} \\
&= i\eta\lambda.
\end{aligned}$$

The solution ψ of the functional equation in Proposition A.6.2 is unique, and therefore $\psi(-\lambda) = \overline{\psi}(\lambda)$, i.e., ψ is symmetric.

Since the polynomials $\widetilde{P}$ and $\widetilde{Q}$ are real when ω is symmetric (see p. 287), also the polynomial Y considered in the proof of Proposition A.6.2 is real. In particular, $\widetilde{P}(0)$ is a nonzero real number and $Y(0)$ is a real number. Then the statement on J_ω immediately follows from (A.6.5) since the coefficient of λ^{4k} in $\omega(\lambda)\psi(\lambda)$ equals the leading coefficient of λ^{2k} in $\widetilde{P}(\lambda)\widetilde{Y}(\lambda)$, which equals the leading coefficient of λ^{2k} in $c_1\widetilde{P}^2(\lambda)$ by (A.6.2). □

Definition A.6.4. Let U be a metric space, e.g., a subset of $\mathbb{C}^r$ ($r \in \mathbb{N}$). Then a rational function $f(\cdot, u)$ whose coefficients depend on $u \in U$ is said to be strongly continuous in $u \in U$ if it has a representation $f(\cdot, u) = \frac{g(\cdot,u)}{h(\cdot,u)}$, where $g(\cdot, u)$ and $h(\cdot, u)$ are polynomials with the following properties:

(i) the degrees of $g(\cdot, u)$ and $h(\cdot, u)$ are independent of $u \in U$;

(ii) the coefficients of the polynomials $g(\cdot, u)$ and $h(\cdot, u)$ depend continuously on $u \in U$;

(iii) for all $u \in U$, the polynomials $g(\cdot, u)$ and $h(\cdot, u)$ do not have common zeros.

Lemma A.6.5. *Let $f(\cdot, u)$ be a rational function which is strongly continuous in $u \in U$. Then the zeros and poles of $f(\cdot, u)$ depend continuously on $u \in U$.*

Proof. Due to property (iii) in Definition A.6.4 it suffices to consider g and h separately. Hence we only need to consider, say, the polynomial $g(\cdot, u)$. In this case, the statement of this lemma immediately follows from Rouché's theorem for simple zeros, and for multiple (branching) zeros if one clusters them as a group. More effort is needed to show that there are continuous branches, see, e.g., [96, Theorem 9.1.1]. □

Corollary A.6.6. *In addition to the assumptions of Proposition* A.6.2 *let U be a metric space and assume that the coefficients of ω depend continuously on $u \in U$. For $u \in U$ let $\eta_0(u) \in \overline{\mathbb{C}}$ be the exceptional value of η for which $c_1 = 0$ in* (A.6.5)*, with $\eta_0(u) := \infty$ if $c_1 \neq 0$ for all $\eta \in \mathbb{C}$. Then η_0 depends continuously on $u \in U$, and ψ is strongly continuous in $u \in U$ and $\eta \in \mathbb{C} \setminus \{\eta_0(u)\}$.*

Proof. It is well known that the coefficients in the partial fraction decomposition (A.6.4) can be obtained as solutions of a linear system of equations, which is obtained by comparing coefficients of powers of λ (after multiplying through by $\widetilde{P}\widetilde{Q}$). The given data of this linear system (for $\eta = 1$) are polynomials of the coefficients of $\widetilde{P}$ and $\widetilde{Q}$. Hence the solution of this linear system and therefore also Y and Z depend continuously on $u \in U$, and so does $\eta_0 \in \mathbb{C}$. In particular, the coefficients of the polynomial ψ satisfy property (ii) in Definition A.6.4 with respect to $u \in U$ and $\eta \in \mathbb{C}$, whereas property (i) is clear from the definition of $\eta_0(u)$. When $\eta_0(u_0) = \infty$ for some $u_0 \in U$, then $Y(0) = 0$ for this u_0, and the continuous dependence of $Y(0)$ and $\widetilde{P}(0)$ on $u \in U$ implies that $\eta_0(u_0) \to \infty$ as $u \to u_0$. □

Corollary A.6.7. *Let $f(\cdot, u)$ be a rational S-function which is strongly continuous in $u \in U$. Then the numbers $a_1, \dots, a_n$, $b_1, \dots, b_n$ and c_0 in the continued fraction expansion* (A.3.8) *depend continuously on $u \in U$.*

Proof. From the proof of Lemma A.3.4 we see that the numbers a and b there would depend continuously on $u \in U$ and that the function g there would be strongly continuous in $u \in U$. Mathematical induction completes this proof. □

Appendix B

Linear and quadratic matrix pencils

B.1 Matrix pencils

In this appendix we will state and prove results for quadratic matrix pencils of the form

$$T(\lambda) = \lambda^2 M - i\lambda K - A \quad (\lambda \in \mathbb{C}), \tag{B.1.1}$$

where M, K, A are $n \times n$ symmetric matrices with real entries ($n \in \mathbb{N}$), K being positive semidefinite and M being positive definite. We write $M > 0$ and $K \geq 0$. The trivial case $n = 1$ is included for convenience. We shall assume throughout this appendix that M is a diagonal matrix with positive diagonal entries since all pencils considered in this monograph are of this form.

A complex number λ is called an eigenvalue of T if the equation $T(\lambda)y = 0$ has a nontrivial solution $y \in \mathbb{C}^n \setminus \{0\}$, and each such y is called an eigenvector of T at λ. The set of all such solutions, including 0, is a linear subspace of $\mathbb{C}^n$ and its dimension is called the geometric multiplicity of the eigenvalue λ. It is well known that the equation $T(\lambda)y = 0$ has a nontrivial solution if and only if the determinant $\det T(\lambda) = 0$. The multiplicity of the zero λ of the function $\det T$ will be called the algebraic multiplicity of the eigenvalue λ and denoted by $m(\lambda, T)$. We may write $m(\lambda, T) = 0$ when λ is not an eigenvalue of T. Since M is invertible, $\det T$ is a polynomial of degree $2n$, and therefore the sum of the algebraic multiplicities of all eigenvalues of T equals $2n$.

When $K = 0$ one can set $\lambda^2 = z$, and (B.1.1) becomes the linear matrix pencil

$$L(z) = zM - A \quad (z \in \mathbb{C}). \tag{B.1.2}$$

The notion of eigenvalue and multiplicity as well as the notation $m(z, L)$ applies accordingly. The sum of the algebraic multiplicities of all eigenvalues of L equals n.

M. Möller, V. Pivovarchik, *Direct and Inverse Finite-Dimensional Spectral Problems on Graphs*, Operator Theory: Advances and Applications 283, https://doi.org/10.1007/978-3-030-60484-4

B.2 The linear pencil L

In this section we assume that $K = 0$, which leads to the pencil L defined in (B.1.2). Then $T(\lambda) = L(\lambda^2)$ $(\lambda \in \mathbb{C})$, and λ is an eigenvalue of T if and only if λ^2 is an eigenvalue of L; in this case also $-\lambda$ is an eigenvalue of T. Denoting the zeros of $\det L$, counted with multiplicity, with $z_1, \dots, z_n$, it follows that

$$\det L(z) = \det M \prod_{k=1}^{n} (z - z_k).$$

Clearly, $m(\lambda, T) = m(-\lambda, T) = m(\lambda^2, L)$ when $\lambda \neq 0$ and $m(0, T) = 2m(0, L)$. Hence we will only deal with the linear pencil L in this section.

Proposition B.2.1. *All eigenvalues of L are real and the geometric multiplicity of each eigenvalue equals its algebraic multiplicity. If A is positive semidefinite, then all eigenvalues of L are nonnegative, and if A is positive definite, then all eigenvalues of L are positive.*

Proof. Since M is a diagonal matrix with positive entries, there are positive numbers m_k $(k = 1, \dots, n)$ such that $M = \operatorname{diag}(m_1, \dots, m_n)$. With the matrix $M^{\frac{1}{2}} = \operatorname{diag}(m_1^{\frac{1}{2}}, \dots, m_n^{\frac{1}{2}})$ it is clear that

$$L(z) = zM - A = M^{\frac{1}{2}} \left(z - M^{-\frac{1}{2}} A M^{-\frac{1}{2}} \right) M^{\frac{1}{2}},$$

and therefore $L(z)y = 0$ for $y \in \mathbb{C}^n$ if and only $\left(M^{-\frac{1}{2}} A M^{-\frac{1}{2}} - z \right) M^{\frac{1}{2}} y = 0$. The matrix $M^{-\frac{1}{2}} A M^{-\frac{1}{2}}$ is symmetric, and it is positive (semi)definite if and only if A is positive (semi)definite. Hence the result follows from well-known properties of the eigenvalues of Hermitian matrices. □

B.3 The quadratic pencil T

For a matrix C let $\overline{C}$ be its conjugate complex and let C^* be its conjugate complex transpose, i.e., for $C = (c_{j,k})_{j=1,\dots,n;\, k=1,\dots,m}$ define $\overline{C} = (\overline{c_{j,k}})_{j=1,\dots,n;\, k=1,\dots,m}$ and $C^* = (\overline{c_{k,j}})_{j=1,\dots,m;\, k=1,\dots,n}$. Since the matrices M, K, A are real symmetric, $M^* = M$, $K^* = K$, $A^* = A$.

For $x, y \in \mathbb{C}^n$, $x = (x_j)_{j=1}^n$, $y = (y_j)_{j=1}^n$, the inner product is defined by

$$(x, y) = \sum_{j=1}^{n} x_j \overline{y_j}.$$

Observe that $(C^* x, y) = (x, Cy)$ for $n \times n$ matrices C.

Lemma B.3.1. *The geometric multiplicity of an eigenvalue of T does not exceed its algebraic multiplicity.*

Proof. Let λ_0 be an eigenvalue of T with geometric multiplicity k. Then the matrix $T(\lambda_0)$ has defect k and therefore there is an invertible $n \times n$ matrix C such that the first k columns of $CT(\lambda_0)$ are zero. Since the entries of CT are polynomials (of degree 2 or less), there are polynomial vector functions $a_1, \ldots, a_n$ in $\mathbb{C}^n$ such that

$$CT(\lambda) = ((\lambda - \lambda_0)a_1(\lambda), \ldots, (\lambda - \lambda_0)a_k(\lambda), a_{k+1}(\lambda), \ldots, a_n(\lambda)).$$

Hence

$$\det T(\lambda) = (\lambda - \lambda_0)^k (\det C)^{-1} \det(a_1(\lambda), \ldots, a_n(\lambda))$$

shows that $k \leq m(\lambda_0, T)$. □

An eigenvalue of T is called semisimple if its algebraic multiplicity equals its geometric multiplicity. An eigenvalue of T is called simple if its algebraic multiplicity is 1.

Lemma B.3.2. $T(\lambda)^* = \overline{T(\lambda)} = T(-\overline{\lambda})$ *for all* $\lambda \in \mathbb{C}$. *In particular, the eigenvalues of T are symmetric with respect to the imaginary axis, for each eigenvalue λ_0 of T, the geometric multiplicity of λ_0 equals the geometric multiplicity of the eigenvalue $-\overline{\lambda_0}$, and the algebraic multiplicity of λ_0 equals the algebraic multiplicity of $-\overline{\lambda_0}$.*

Proof. For all $\lambda \in \mathbb{C}$,

$$T(\lambda)^* = (\lambda^2 M - i\lambda K - A)^* = \overline{\lambda}^2 M + i\overline{\lambda} K - A = T(-\overline{\lambda}),$$

which proves $T(\lambda)^* = \overline{T(\lambda)} = T(-\overline{\lambda})$. It follows that $\overline{\det T(\lambda)} = \det T(-\overline{\lambda})$. Hence, if $\det T(\lambda) = (\lambda - \lambda_0)^k f(\lambda)$ with $k = m(\lambda_0, T)$ and some polynomial f, then

$$\det T(\lambda) = \overline{\det T(-\overline{\lambda})} = (-\lambda - \overline{\lambda_0})^k\, \overline{f(-\overline{\lambda})}.$$

Therefore $m(-\overline{\lambda_0}, T) \geq m(\lambda_0, T)$, and applying this inequality with λ_0 replaced by $-\overline{\lambda_0}$, it follows that $m(-\overline{\lambda_0}, T) = m(\lambda_0, T)$.

From $T(-\overline{\lambda_0})\overline{y} = \overline{T(\lambda_0)}\,\overline{y} = \overline{T(\lambda_0)y}$ for $y \in \mathbb{C}^n$ it is clear that $T(\lambda_0)y = 0$ if and only if $T(-\overline{\lambda_0})\,\overline{y} = 0$. Therefore the geometric multiplicity of an eigenvalue λ_0 equals the geometric multiplicity of the eigenvalue $-\overline{\lambda_0}$. □

Lemma B.3.3. *The eigenvalues of the pencil T are located in the closed upper half-plane and on the imaginary axis.*

Proof. Let y_0 be an eigenvector corresponding to an eigenvalue λ_0 of T. Then

$$(T(\lambda_0)y_0, y_0) = 0,$$

and consequently, taking real and imaginary parts,

$$((\operatorname{Re}\lambda_0)^2 - (\operatorname{Im}\lambda_0)^2)(My_0, y_0) + \operatorname{Im}\lambda_0(Ky_0, y_0) - (Ay_0, y_0) = 0 \qquad \text{(B.3.1)}$$

and

$$\operatorname{Re}\lambda_0[2\operatorname{Im}\lambda_0(My_0,y_0)-(Ky_0,y_0)]=0. \tag{B.3.2}$$

If $\operatorname{Re}\lambda_0\neq 0$, then (B.3.2) reduces to

$$2\operatorname{Im}\lambda_0(My_0,y_0)-(Ky_0,y_0)=0, \tag{B.3.3}$$

and $\operatorname{Im}\lambda_0\geq 0$ follows since $(My_0,y_0)>0$ and $(Ky_0,y_0)\geq 0$ in view of $M>0$ and $K\geq 0$. □

Lemma B.3.4. *If $K>0$, then the eigenvalues of the pencil T located in the closed lower half-plane lie on the imaginary axis.*

Proof. In view of Lemma B.3.3 we have to show that the pencil T has no nonzero real eigenvalues. Hence, assume that the pencil T has a nonzero real eigenvalue λ_0 with corresponding eigenvector y_0. Then (B.3.2) leads to $(Ky_0,y_0)=0$, which contradicts $K>0$. □

Lemma B.3.5.

1. *If $A\geq 0$, then the spectrum of the pencil T is located in the closed upper half-plane.*
2. *If $A>0$ and $K>0$, then the spectrum of the pencil T is located in the open upper half-plane.*
3. *If $A>0$ and $\lambda^2My-Ay\neq 0$ for all real λ and all nonzero $y\in N(K)$, then the spectrum of the pencil T is located in the open upper half-plane.*

Proof. 1. In view of Lemma B.3.3 we have to show that there are no eigenvalues on the negative imaginary semiaxis. Hence, let y_0 be an eigenvector corresponding to a pure imaginary eigenvalue λ_0. Then $\operatorname{Re}\lambda_0=0$, and in view of $M>0$, $K\geq 0$, $A\geq 0$, equation (B.3.1) would imply that $(My_0,y_0)=0$, $(Ky_0,y_0)=0$, and $(Ay_0,y_0)=0$ if $\operatorname{Im}\lambda_0<0$. But this contradicts $M>0$.

2. From $A>0$ it follows that 0 is not an eigenvalue of T, and hence the statement of part 2 follows in view of part 1 and Lemma B.3.4.

3. The same proof as in part 2 above shows that 0 is not an eigenvalue of the pencil T. If the pencil T would have a nonzero real eigenvalue λ_0 with corresponding eigenvector y_0, then (B.3.2) would imply $Ky_0=0$, and therefore

$$\lambda_0^2My_0-Ay_0=T(\lambda_0)y_0=0,$$

which contradicts the assumptions of this case. □

Proposition B.3.6. *Let $A\geq 0$. Then there are a unique integer $k\geq 0$, a unique monic real polynomial R with $R(0)\neq 0$ and a unique monic polynomial ω which is either 1 or a Hermite-Biehler polynomial such that*

$$\det T(\lambda)=(\det M)\lambda^kR(\lambda^2)\omega(\lambda). \tag{B.3.4}$$

If R is not constant, then all zeros of R are positive real numbers. If ω is not constant, then ω belongs to the SHB *class.*

Proof. Let $w_1, \ldots, w_m$ be the nonreal zeros of $\det T$, counted with multiplicity, and define

$$\omega(\lambda) = \prod_{j=1}^{m} (\lambda - w_j).$$

Let $z_1, \ldots, z_l$ be the positive real zeros of $\det T$, counted with multiplicity, and define

$$R(z) = \prod_{j=1}^{l} (z - z_j^2).$$

Putting $k = m(0, T)$ it is clear from the symmetry statement in Lemma B.3.2 and from $\det M$ being the coefficient of λ^{2n} in T that the representation (B.3.4) holds, that all zeros of R, if any, are positive, that $\omega = 1$ if $\det T$ has only real zeros, and that ω is an SHB polynomial on account of Lemma B.3.5, part 1, if $\det T$ has nonreal zeros.

For the uniqueness we first observe that $k = m(0, T)$ since $R(0) \neq 0$ and since a Hermite-Biehler polynomial does not have real zeros. Since $\det T$ has no zeros in the open lower half-plane by Lemma B.3.5, part 1, and since $\lambda \mapsto R(\lambda^2)$ is a real polynomial, all its zeros must be real because nonreal zeros would occur in conjugate complex pairs. On the other hand, all zeros of ω are nonreal, so that the monic polynomials P and ω are uniquely determined by the real and nonreal zeros of $\det T$, respectively. $\square$

Proposition B.3.7. *Let $(a_j)_{j=1}^{\infty}$ and $(b_j)_{j=1}^{\infty}$ be sequences of complex number. Then the determinants of the Jacobi matrices*

$$J_k = \begin{pmatrix} a_1 & b_1 & 0 & \cdots & \cdots & 0 & 0 \\ b_1 & a_2 & b_2 & 0 & \cdots & \cdots & 0 \\ 0 & b_2 & a_3 & b_3 & \ddots & & \vdots \\ \vdots & & \ddots & \ddots & \ddots & \ddots & \vdots \\ \vdots & & & \ddots & \ddots & \ddots & 0 \\ 0 & & & & b_{k-2} & a_{k-1} & b_{k-1} \\ 0 & 0 & \cdots & \cdots & 0 & b_{k-1} & a_k \end{pmatrix} \quad (k \in \mathbb{N})$$

satisfy the recursion relations

$$\det J_1 = a_1, \quad \det J_2 = a_1 a_2 - b_1^2, \quad \det J_k = a_k \det J_{k-1} - b_{k-1}^2 \det J_{k-2} \ (k \geq 3).$$

Proof. Clearly, $\det J_1 = a_1$ and $\det J_2 = a_1 a_2 - b_1^2$. For $k \geq 3$ we use the block decompositions

$$J_k = \left(\begin{array}{cccc|cc} & & & & 0 & 0 \\ & & J_{k-2} & & \vdots & \vdots \\ & & & & 0 & 0 \\ & & & & b_{k-2} & 0 \\ \hline 0 & \cdots & 0 & b_{k-2} & a_{k-1} & b_{k-1} \\ 0 & \cdots & 0 & 0 & b_{k-1} & a_k \end{array}\right)$$

and we define

$$\widehat{J}_{k-1} = \left(\begin{array}{cccc|c} & & & & 0 \\ & & J_{k-2} & & \vdots \\ & & & & 0 \\ \hline 0 & \cdots & 0 & b_{k-2} & b_{k-1} \end{array}\right).$$

Expanding the determinant of J_k with respect to the last row it follows that

$$\begin{aligned} \det J_k &= a_k \det J_{k-1} - b_{k-1} \det \widehat{J}_{k-1} \\ &= a_k \det J_{k-1} - b_{k-1}^2 \det J_{k-2}. \end{aligned}$$ □

B.4 Schur complements

Let A be a square matrix, say of size $n \times n$. For $0 < k < n$ we write A as an $((n-k)+k) \times ((n-k)+k)$ block matrix

$$A = \begin{pmatrix} A_{k,1} & A_{k,2} \\ A_{k,3} & A_{k,4} \end{pmatrix}.$$

If $A_{k,4}$ is invertible, then

$$S(A,k) := A_{k,1} - A_{k,2} A_{k,4}^{-1} A_{k,3} \tag{B.4.1}$$

is called the Schur complement of $A_{k,4}$ in A. In this case, the Schur factorization

$$A = \begin{pmatrix} I & A_{k,2}A_{k,4}^{-1} \\ 0 & I \end{pmatrix} \begin{pmatrix} S(A,k) & 0 \\ 0 & A_{k,4} \end{pmatrix} \begin{pmatrix} I & 0 \\ A_{k,4}^{-1} A_{k,3} & I \end{pmatrix} \tag{B.4.2}$$

holds. In particular,

$$\det A = \det S(A,k) \det A_{k,4}, \tag{B.4.3}$$

A is invertible when $S(A,k)$ is invertible, and then

$$A^{-1} = \begin{pmatrix} I & 0 \\ -A_{k,4}^{-1} A_{k,3} & I \end{pmatrix} \begin{pmatrix} S(A,k)^{-1} & 0 \\ 0 & A_{k,4}^{-1} \end{pmatrix} \begin{pmatrix} I & -A_{k,2}A_{k,4}^{-1} \\ 0 & I \end{pmatrix}. \tag{B.4.4}$$

The following lemma shows that the Schur complement can be iterated.

Lemma B.4.1. *Let A be an $n \times n$ matrix and let $0 < k, l$ such that $k + l < n$. If $A_{k,4}$ and $A_{k+l,4}$ are invertible, then $S(A, k + l) = S(S(A, k), l)$.*

Proof. Writing A as an $((n-k-l)+l+k) \times ((n-k-l)+l+k)$ block matrix

$$A = \begin{pmatrix} B_{11} & B_{12} & B_{13} \\ B_{21} & B_{22} & B_{23} \\ B_{31} & B_{32} & B_{33} \end{pmatrix}$$

it follows from the assumptions that $A_{k,4} = B_{33}$ and

$$A_{k+l,4} = \begin{pmatrix} B_{22} & B_{23} \\ B_{32} & B_{33} \end{pmatrix}$$

are invertible. By definition,

$$\begin{aligned} S(A,k) &= \begin{pmatrix} B_{11} & B_{12} \\ B_{21} & B_{22} \end{pmatrix} - \begin{pmatrix} B_{13} \\ B_{23} \end{pmatrix} B_{33}^{-1} \begin{pmatrix} B_{31} & B_{32} \end{pmatrix} \\ &= \begin{pmatrix} B_{11} - B_{13}B_{33}^{-1}B_{31} & B_{12} - B_{13}B_{33}^{-1}B_{32} \\ B_{21} - B_{23}B_{33}^{-1}B_{31} & B_{22} - B_{23}B_{33}^{-1}B_{32} \end{pmatrix} \end{aligned}$$

and

$$S(A, k+l) = B_{11} - \begin{pmatrix} B_{12} & B_{13} \end{pmatrix} A_{k+l,4}^{-1} \begin{pmatrix} B_{21} \\ B_{31} \end{pmatrix}.$$

Furthermore, the Schur factorization

$$A_{k+l,4} = \begin{pmatrix} I & B_{23}B_{33}^{-1} \\ 0 & I \end{pmatrix} \begin{pmatrix} B_{22} - B_{23}B_{33}^{-1}B_{32} & 0 \\ 0 & B_{33} \end{pmatrix} \begin{pmatrix} I & 0 \\ B_{33}^{-1}B_{32} & I \end{pmatrix}$$

and the invertibility of $A_{k+l,4}$ show that $B_{22} - B_{23}B_{33}^{-1}B_{32}$ is invertible and

$$A_{k+l,4}^{-1} = \begin{pmatrix} I & 0 \\ -B_{33}^{-1}B_{32} & I \end{pmatrix} \begin{pmatrix} (B_{22} - B_{23}B_{33}^{-1}B_{32})^{-1} & 0 \\ 0 & B_{33}^{-1} \end{pmatrix} \begin{pmatrix} I & -B_{23}B_{33}^{-1} \\ 0 & I \end{pmatrix}.$$

Therefore

$$\begin{aligned} S(A, k+l) &= B_{11} - B_{13}B_{33}^{-1}B_{31} \\ &\quad - (B_{12} - B_{13}B_{33}^{-1}B_{32})(B_{22} - B_{23}B_{33}^{-1}B_{32})^{-1}(B_{21} - B_{23}B_{33}^{-1}B_{31}) \\ &= S(S(A,k), l). \end{aligned}$$

□

B.5 Eigenvalues of real symmetric matrices

The necessity of (B.5.1) in the following well-known equivalence result is attributed to Cauchy [31], whereas its sufficiency was proved in [52]. A proof can be found for example in [124, Section 71]. However, since the interlacing (B.5.1) below is related to poles and zeros of rational Nevanlinna functions, which are extensively used in this monograph, we will give a short proof making use of properties of rational Nevanlinna functions.

In this section, let H be a real symmetric $(n+1)\times(n+1)$ matrix ($n \in \mathbb{N}$), and denote by $H_{1,1}$ the $n \times n$ first principal submatrix obtained from H by deleting its first row and first column.

Proposition B.5.1. 1. *Let $n \in \mathbb{N}$, let H be an $(n+1) \times (n+1)$ real symmetric matrix H such that the eigenvalues of H are $\mu_1 \le \mu_2 \le \dots \le \mu_{n+1}$ and the eigenvalues of the submatrix $H_{1,1}$ are $\nu_1 \le \nu_2 \le \dots \le \nu_n$. Then*

$$\mu_1 \le \nu_1 \le \mu_2 \le \nu_2 \le \dots \le \nu_n \le \mu_{n+1}. \tag{B.5.1}$$

2. *Let $(\mu_k)_{j=1}^{n+1}$ and $(\nu_k)_{k=1}^{n}$ be sequences of real numbers satisfying* (B.5.1). *Then there exists an $(n+1)\times(n+1)$ real symmetric matrix H such that the eigenvalues of H are $\mu_1, \dots, \mu_{n+1}$ and the eigenvalues of the submatrix $H_{1,1}$ are $\nu_1, \dots, \nu_n$.*

Proof. 1. Define the rational function χ_H by

$$\chi_H(\lambda) = -\frac{\det(H-\lambda)}{\det(H_{1,1}-\lambda)}.$$

Let U be an orthogonal matrix such that $U^\top H_{1,1} U = \operatorname{diag}\{\nu_1, \dots, \nu_n\}$. For the orthogonal block matrix

$$V = \begin{pmatrix} 1 & 0 \\ 0 & U \end{pmatrix}$$

it follows that $\widehat{H} := V^\top H V$ is a real symmetric matrix of the form

$$\widehat{H} = \begin{pmatrix} b_0 & b_1 & \cdots & b_n \\ b_1 & \nu_1 & 0 & 0 \\ \vdots & 0 & \ddots & \vdots \\ b_n & 0 & \cdots & \nu_n \end{pmatrix} \tag{B.5.2}$$

with real numbers $b_0, \dots, b_n$. Expanding the determinant of $(\widehat{H}-\lambda)$ with respect to the first row, we have

$$\det(H-\lambda) = \det(\widehat{H}-\lambda) = \prod_{k=1}^{n}(\nu_k-\lambda)\left[(b_0-\lambda) - \sum_{j=1}^{n}\frac{b_j^2}{\nu_j-\lambda}\right]$$

and then, in view of $\det(H_{1,1}-\lambda) = \det(\widehat{H}_{1,1}-\lambda)$,

$$\chi_H(\lambda) = \lambda - b_0 - \sum_{j=1}^{n}\frac{b_j^2}{\lambda-\nu_j}.$$

Because of Theorem A.2.6 and the fact that the sum of Nevanlinna functions is a Nevanlinna function it follows that χ_H is a Nevanlinna function. Denote the increasing sequence of the zeros of χ_H by $(\widehat{\mu}_k)_{k=1}^{n_0+1}$ and the increasing sequence of the poles of χ_H by $(\widehat{\nu}_k)_{k=1}^{n_0}$. Then there is a sequence $(\tau_k)_{k=1}^{n-n_0}$ such that $(\mu_k)_{k=1}^{n+1} = (\widehat{\mu}_k)_{k=1}^{n_0+1} \cup (\tau_k)_{k=1}^{n-n_0}$ and $(\nu_k)_{k=1}^{n} = (\widehat{\nu}_k)_{k=1}^{n_0} \cup (\tau_k)_{k=1}^{n-n_0}$. If $n_0 > 0$, then

$$\widehat{\mu}_1 < \widehat{\nu}_1 < \cdots < \widehat{\mu}_{n_0+1}$$

by Theorem A.2.6, and (B.5.1) follows in view of Lemma A.4.4.

2. With the notation from the proof of part 1 of this proof we define

$$\chi(\lambda) = \prod_{k=1}^{n+1}(\lambda - \mu_k)\prod_{k=1}^{n}(\lambda - \nu_k)^{-1} = \prod_{k=1}^{n_0+1}(\lambda - \widehat{\mu}_k)\prod_{k=1}^{n_0}(\lambda - \widehat{\nu}_k)^{-1}. \tag{B.5.3}$$

Since (B.5.1) holds, χ is a Nevanlinna function by Theorem A.2.6. In view of Lemma A.2.7 there are real numbers $b_0, \ldots, b_n$ such that

$$\chi(\lambda) = \lambda - b_0 - \sum_{j=1}^{n} \frac{b_j^2}{\lambda - \nu_j}, \tag{B.5.4}$$

where we put $b_j = 0$ when ν_j corresponds to some τ_k. Hence, defining H by the right-hand side of (B.5.2) it follows that $\chi_H = \chi$. Clearly

$$\det(H_{1,1} - \lambda) = \prod_{j=1}^{n}(\nu_k - \lambda),$$

and therefore

$$\det(H - \lambda) = -\chi(\lambda)\det(H_{1,1} - \lambda) = \prod_{j=1}^{n+1}(\mu_k - \lambda).$$

Hence the eigenvalues of H are $\mu_1, \ldots, \mu_{n+1}$ and the eigenvalues of $H_{1,1}$ are $\nu_1, \ldots, \nu_n$. □

Let $n \in \mathbb{N}$ and let σ be a permutation of $\{1, \ldots, n\}$, i.e., $\sigma : \{1, \ldots, n\} \to \{1, \ldots, n\}$ is a bijective map. The $n \times n$ matrix $P_\sigma := (\delta_{i,\sigma(j)})_{i,j=1}^{n}$ is called a permutation matrix. We will also write $P_\sigma =: (p_{\sigma,i,j})_{i,j=1}^{n}$.

Lemma B.5.2. *Let $n \in \mathbb{N}$ and let σ, τ be permutations. Then $P_\tau P_\sigma = P_{\tau\circ\sigma}$ and $P_\sigma^* = P_\sigma^\mathsf{T} = P_{\sigma^{-1}} = P_\sigma^{-1}$.*

Proof. For $i, j = 1, \ldots, n$ we calculate

$$\sum_{s=1}^{n} p_{\tau,i,s}p_{\sigma,s,j} = \sum_{s=1}^{n} \delta_{i,\tau(s)}\delta_{s,\sigma(j)} = \delta_{i,\tau(\sigma(j))} = p_{\tau\circ\sigma,i,j},$$

which proves $P_\tau P_\sigma = P_{\tau\circ\sigma}$.

Since the entries of P_σ are real, $P_\sigma^* = P_\sigma^\mathsf{T}$ is obvious. For $i, j = 1, \dots, n$ the identities $p_{\sigma,j,i} = \delta_{j,\sigma(i)} = \delta_{i,\sigma^{-1}(j)} = p_{\sigma^{-1},i,j}$ show that $P_\sigma^\mathsf{T} = P_{\sigma^{-1}}$. Finally, the first statement gives $P_{\sigma^{-1}} P_\sigma = I_n$, and $P_{\sigma^{-1}} = P_\sigma^{-1}$ follows. □

For an $n \times n$ matrix $A = (a_{i,j})_{i,j=1}^n$ and a permutation σ of $\{1, \dots, n\}$ let

$$P_\sigma^\mathsf{T} A P_\sigma =: A_\sigma =: (a_{\sigma,i,j})_{i,j=1}^n. \tag{B.5.5}$$

For $i, j = 1, \dots, n$, Lemma B.5.2 gives

$$\begin{aligned} a_{\sigma,i,j} &= \sum_{s=1}^n \sum_{t=1}^n \delta_{i,\sigma^{-1}(s)} a_{s,t} \delta_{t,\sigma(j)} = \sum_{s=1}^n \delta_{i,\sigma^{-1}(s)} a_{s,\sigma(j)} \\ &= a_{\sigma(i),\sigma(j)}. \end{aligned}$$

This shows that

$$A_\sigma = (a_{\sigma(i),\sigma(j)})_{i,j}^n. \tag{B.5.6}$$

Definition B.5.3. Let $m \in \mathbb{N}$, $m \geq 2$. An $m \times m$ matrix $A = (a_{i,j})_{i,j=1}^m$ is called reducible if there are two nonempty subsets Δ_1, Δ_2 of $\{1, \dots, m\}$ satisfying $\Delta_1 \dot\cup \Delta_2 = \{1, \dots, m\}$ such that $a_{i,j} = 0$ for $i \in \Delta_1$ and $j \in \Delta_2$. The matrix A is called irreducible if it is not reducible.

Proposition B.5.4. *Let $m \geq 2$. A Hermitian $m \times m$ matrix A is reducible if and only if there is a permutation σ of $\{1, \dots, m\}$ with $\sigma(1) = 1$ and a number $k \in \mathbb{N}$ with $1 \leq k \leq m-1$ such that A_σ has the representation*

$$A_\sigma = \begin{pmatrix} A_{\sigma,1} & 0 \\ 0 & A_{\sigma,2} \end{pmatrix} \tag{B.5.7}$$

with $A_{\sigma,1} \in M_k(\mathbb{C})$ and $A_{\sigma,2} \in M_{m-k}(\mathbb{C})$.

Proof. If (B.5.7) holds, put $\Delta_1 = \sigma(\{1, \dots, k\})$ and $\Delta_2 = \sigma(\{k+1, \dots, m\})$. For $i \in \Delta_1$ and $j \in \Delta_2$ we have $1 \leq \sigma^{-1}(i) \leq k$ and $k+1 \leq \sigma^{-1}(j) \leq m$, and then (B.5.6) gives $a_{i,j} = a_{\sigma^{-1}(i),\sigma^{-1}(j)} = 0$. Therefore A is reducible.

Conversely, let A be reducible. Since A is Hermitian, $a_{i,j} = 0$ for $i \in \Delta_1$ and $j \in \Delta_2$ implies $a_{j,i} = 0$ for $i \in \Delta_1$ and $j \in \Delta_2$, that is, one can interchange Δ_1 and Δ_2. Therefore we may assume that $1 \in \Delta_1$. Let $k = \#\Delta_1$. Then $1 \leq k \leq n-1$. Choose a permutation σ of $\{1, \dots, m\}$ such a that $\sigma(\{1, \dots, k\}) = \Delta_1$ and $\sigma(1) = 1$. Then $\sigma(\{k+1, \dots, m\}) = \Delta_2$. By (B.5.6) it follows that

$$a_{\sigma,i,j} = a_{\sigma(i),\sigma(j)} = 0 \quad (i = 1, \dots, k,\ j = k+1, \dots, m).$$

Since A_σ is Hermitian, (B.5.7) follows. □

Corollary B.5.5. *Under the assumptions of Proposition* B.5.1 *consider the special case that*

$$\mu_1 < \nu_1 < \mu_2 < \nu_2 < \dots < \nu_n < \mu_{n+1}. \tag{B.5.8}$$

Then the matrix H is irreducible.

Proof. Assume that H is reducible. By Proposition B.5.4 there are an $n \times n$ permutation matrix P, integers $1 \le m_1, m_2 \le n$ with $m_1 + m_2 = n + 1$, an $m_1 \times m_1$ matrix F and an $m_2 \times m_2$ matrix G such that

$$\begin{pmatrix} 1 & 0 \\ 0 & P^\top \end{pmatrix} H \begin{pmatrix} 1 & 0 \\ 0 & P \end{pmatrix} = \begin{pmatrix} F & 0 \\ 0 & G \end{pmatrix}.$$

Recalling $P^\top = P^{-1}$ from Lemma B.5.2 it follows that

$$\det(H - \lambda I_{n+1}) = \det(F - \lambda I_{m_1}) \det(G - \lambda I_{m_2}).$$

If $m_1 = 1$, then $H_{1,1} = PGP^\top$, and therefore

$$\det(H_{1,1} - \lambda I_n) = \det(G - \lambda I_n).$$

If $m_1 > 1$, then

$$H_{1,1} = P \begin{pmatrix} F_{1,1} & 0 \\ 0 & G \end{pmatrix} P^\top,$$

and

$$\det(H_{1,1} - \lambda I_n) = \det(F_{1,1} - \lambda I_{m_1 - 1}) \det(G - \lambda I_{m_2})$$

follows.

In any case, it would follow that the eigenvalues of G are eigenvalues of H as well as $H_{1,1}$, which contradicts (B.5.8). □

Appendix C

Graphs

C.1 Definitions

In this section we fix our notation and give some elementary definitions and facts about graphs, as far as they are needed for the main subject of this monograph. Nearly 40 years ago Tutte wrote in the first edition of his monograph [132, p. 1] “The terminology of graph theory is not yet standardized”, and the same is still true today and will most likely be true for a long time. Hence we have chosen to use notation which seems to be commonly used in the context of our subject. The notation is closely related to that of Tutte and Lovász [90], and in particular the first sentence of the next paragraph is almost verbally taken from [132, p. 1].

A *graph* G is defined by a set $V = V^G$ of elements called *vertices*, a set $E = E^G$ of elements called *edges*, and a relation of *incidence*, which associates with each edge either one or two vertices called its *ends*. A vertex v will be called *incident* with an edge e and the edge e will be called *incident* with the vertex v if v is any of the ends of e.

We will only consider finite graphs, that is, all graphs in this monograph will satisfy $\#V < \infty$ and $\#E < \infty$. Otherwise, we will allow arbitrary graphs and even the null graph, which has no vertices and no edges.

Definition C.1.1. A *walk* in a graph is a sequence

$$(v_0, e_1, v_1, e_2, v_2, \ldots, v_{k-1}, e_k, v_k)$$

of vertices v_j and edges e_j. Here v_{j-1} and v_j are the ends of e_j if e_j has two ends, whereas $v_{j-1} = v_j$ is the single end of e_j if e_j has only one end. A walk is called *open* when $v_0 \neq v_k$, and it is called *closed* when $v_0 = v_k$. A walk in which all the edges are distinct is called a *trail*. A trail in which all vertices $v_0, v_1, \ldots, v_k$ are distinct (except possibly for $v_k = v_0$) is called a *path*; an open path is called a *chain*; a closed path is called a *cycle*. Any edge joining a vertex v to itself, i.e., a cycle consisting of exactly one edge, is called a *loop*.

M. Möller, V. Pivovarchik, *Direct and Inverse Finite-Dimensional Spectral Problems on Graphs*, Operator Theory: Advances and Applications 283, https://doi.org/10.1007/978-3-030-60484-4

A graph is called *simple* if it has no loops and if for any two vertices there is is at most one edge which is incident with these two vertices.

We may briefly write $(v_0, \ldots, v_k)$ for a walk, trail, etc., if its edges are not essential in the corresponding context or uniquely determined by the vertices.

If there is a walk from a vertex v in G to a vertex $w \neq v$ in G which is not a path, then there are indices $j < k$ in the walk such that $v_j = v_k$. Removing $e_{j+1}, v_{j+1}, \ldots, e_k, v_k$ from the walk, one arrives at another walk starting at v and ending at w. Repeating this procedure, if necessary, finitely many times gives a chain from v to w.

A graph is called *connected* if for each pair of distinct vertices (v, w) there is a walk from v to w. A graph G' is called a *subgraph* of a graph G if $V^{G'} \subset V^G$, $E^{G'} \subset E^G$ and each edge of G' has the same ends in G' as in G. For a set V_1 of vertices of a graph G let $G|_{V_1}$ be the subgraph of G whose set of vertices is V_1 and whose edges are all edges in G whose ends are in V_1.

For vertices v, w in G define

$$v \sim w \Leftrightarrow (v = w \text{ or there is a walk from } v \text{ to } w).$$

It is clear that this is an equivalence relation and that each equivalence class of vertices together with the edges in G which connect these vertices defines a (maximal) connected subgraph, called a *component* of G. Each graph G is the disjoint union of its components, and G is connected if and only if it has exactly one component.

For each vertex v in G, the *degree* $d(v)$ is the number of edges incident with v. If v is the common endpoint of a loop, then this loop is counted twice in $d(v)$. A vertex v with $d(v) = 1$ is called a *pendant vertex* of G, and a vertex v with $d(v) \geq 2$ is called an *interior vertex* of G. For convenience we denote the set of pendant vertices of G by V^{pen} and the set of interior vertices of G by V^{int}. We let $\widehat{V} = V^{\text{pen}} \dot{\cup} V^{\text{int}}$ be the set of all vertices of G which are incident with at least one edge of G. We recall that we may allow *isolated vertices* v, i.e., $d(v) = 0$.

It is convenient to index the edges of G by e_j $(j = 1, \ldots, g)$, where g (or g^G) is the number of edges of G. Unless otherwise stated, the indexing of the edges may be chosen arbitrarily. A graph is called a *metric graph* if a length $l_j > 0$ is assigned to each of its edges e_j.

A graph with at least one edge is called a *digraph* if a direction is associated to each of its edges. If v and w are the vertices incident with an edge e and if e is directed from v to w, then v is called the *entrance* of e, denoted by $v_-(e)$, w is called the *exit* of e, denoted by $v_+(e)$, and e is called *outgoing from* v and *incoming into* w. For a vertex v of G let $I^+(v)$ be the set of indices $j \in \{1, \ldots, g\}$ such that e_j is incoming into v, and let $I^-(v)$ be the set of indices $j \in \{1, \ldots, g\}$ such that e_j is outgoing from v. Then $I(v) := I^-(v) \cup I^+(v)$ is the set of indices j for which e_j is an edge incident with v. It is clear that $j \in I^-(v) \cap I^+(v)$ if and only if e_j is a loop with end v. For each vertex v of G, the numbers $d^-(v) := \#I^-(v)$ and $d^+(v) := \#I^+(v)$ denote the number of outgoing and incoming edges which are incident with v, respectively. Hence $d(v) = d^-(v) + d^+(v)$ $(v \in V)$.

Since each edge in a digraph has exactly one vertex for which it is an incoming edge and exactly one vertex for which it is an outgoing edge, it is clear that

$$\dot{\bigcup}\{I^-(v) : v \in V\} = \dot{\bigcup}\{I^+(v) : v \in V\} = \{1, \ldots, g\},$$

i.e., $\{1, \ldots, g\}$ equals the disjoint union of the sets $I^-(v)$ taken over all vertices of G as well as the disjoint union of the sets $I^+(v)$ taken over all vertices of G.

Definition C.1.2. Two distinct vertices v and w of a graph G are said to be cyclically connected if there is a finite number of cycles $C_1, \ldots, C_k$ in G such that v is a vertex in C_1, w is a vertex in C_k, and each pair (C_j, C_{j+1}) $(j = 1, \ldots, k-1)$ possesses at least one common vertex.

Definition C.1.3. A graph with at least one edge is said to be cyclically connected if each pair of distinct vertices in it is cyclically connected.

It is easy to verify that a cyclically connected graph is connected. Also note that a graph with one vertex and at least one edge is cyclically connected and that all its edges are loops.

Definition C.1.4. A cyclically connected subgraph G' of a graph G is said to be maximal cyclically connected if it is not a nontrivial subgraph of another cyclically connected subgraph of G.

Definition C.1.5. A connected graph with at least one edge is said to be a tree if it has no cycles.

Let G be a graph and let $E' \subset E^G$. With $V' \subset V^G$ being the set of all vertices of G which are incident with at least one $e \in E$ one obtains a unique subgraph G' of G without isolated vertices such that $V^{G'} = V'$ and $E^{G'} = E'$. In particular, each subgraph G' of G without isolated vertices is uniquely determined by $E^{G'}$.

Definition C.1.6. Let G be a graph without isolated vertices and let G_1 be a subgraph of G without isolated vertices satisfying $\emptyset \neq E^{G_1} \neq E^G$. Then the unique subgraph G_2 of G without isolated vertices satisfying $E^{G_2} = E^G \setminus E^{G_1}$ is called the complement of G_1, and G_1 and G_2 are called complementary subgraphs of G.

From the above definition of complementary subgraphs, the following remark is obvious.

Remark C.1.7. Let G be a graph without isolated vertices and let G_1 and G_2 be subgraphs of G without isolated vertices. Then G_1 and G_2 are complementary subgraphs of G if and only if $E^{G_1} \neq \emptyset$, $E^{G_2} \neq \emptyset$ and $E^G = E^{G_1} \dot{\cup} E^{G_2}$.

Let G be a digraph without isolated vertices and let G_1 and G_2 be complementary subgraphs of G. Then a common vertex v of G_1 and G_2 is called an *entrance* of G_j $(j = 1, 2)$ if there is at least one edge in G_j for which v is an entrance. Similarly, v is called an *exit* of G_j if there is at least one edge in G_j for which v is an exit. Clearly, such a vertex could be both an entrance and an exit. If there is a loop with vertex v in G_j, then v is surely an entrance and an exit of G_j.

We say that the graph G is the disjoint union of the graphs G_1 and G_2, written $G = G_1 \dot{\cup} G_2$, if the edges and vertices of G are the disjoint unions of the edges and vertices of G_1 and G_2, respectively. Clearly, if $V^G = V_1 \dot{\cup} V_2$, then $G = G|_{V_1} \dot{\cup} G|_{V_2}$ if and only if there is no edge in G which has one incident vertex in V_1 and the other incident vertex in V_2.

We note that G is the disjoint union of two complementary subgraphs G_1 and G_2 if and only if the two complementary subgraphs have no common vertices.

Definition C.1.8. A connected graph with at least one edge is said to be a quasi-tree if it is neither a tree nor cyclically connected and if each of its cyclically connected subgraphs has at least two vertices in common with the complement of this subgraph.

Clearly, a quasi-tree does not have any loops.

To every cycle we ascribe any of its two possible orientations. In a digraph the direction of an edge in a cycle may be opposite to the orientation of that cycle.

Definition C.1.9. Let G be a graph with g edges and let s be the number of cycles of the graph G. A direction is assigned to each edge of the graph and an orientation is assigned to each cycle of the graph. Indexing the cycles from 1 to s in an arbitrary manner, the $s \times g$ matrix $M = (M_{k,j})_{\substack{k=1,\dots,s,\\ j=1,\dots,g}}$ called the matrix of cycles of G, is defined as follows.

1. For an edge e_j which does not belong to the k-th cycle, set $M_{k,j} = 0$.
2. For an edge e_j which belongs to the k-th cycle and whose direction coincides with the orientation of the cycle, set $M_{k,j} = 1$.
3. For an edge e_j which belongs to the k-th cycle and whose direction is opposite to the orientation of the cycle, set $M_{k,j} = -1$.

In the above definition it may not be completely clear what the number of cycles is. By Definition C.1.1 of a cycle as a path $(v_0, e_1, v_1, e_2, v_2, \dots, v_{k-1}, e_k, v_k)$ with $v_0 = v_k$ $(k \geq 2)$ we may ask if the path $(v_1, e_2, v_2, \dots, v_{k-1}, e_k, v_k, e_1, v_1)$ describes the same cycle or whether the cycle with opposite orientation is the same cycle. We take here the pragmatic view that it does not matter since any such 'modified' cycle would contribute the same row or the negative of it to the matrix M and therefore would not change the rank of M. Hence we will consider cycles as uniquely determined by their edges. Therefore no two rows of M are multiples of each other.

A square matrix is called a signed permutation matrix if it has exactly one entry in each row and each column whose value is 1 or -1 and if all its other entries have the value 0. If A is an $r \times s$ matrix and J an $r \times r$ permutation matrix, then JA is obtained from A by a signed permutation of the rows of A, that is, by a permutation of rows, and rows may be replaced by their negatives. Conversely, each matrix obtained from A by a signed permutation of the rows of A is of the form JA with a signed permutation matrix J. If A is an $r \times s$ matrix and J an $s \times s$ signed permutation matrix, then AJ is obtained from A by a signed permutation of columns of A. Conversely, each matrix obtained from A by a signed permutation of the columns of A is of the form AJ with a signed permutation matrix J. It is clear that a signed permutation of a matrix keeps its rank unchanged.

The matrix M introduced in Definition C.1.9 is unique up to signed permutations of rows and columns, and all properties of M we are going to use are invariant under these signed permutations. In particular, the rank of M is invariant under signed permutations. Therefore it is justified to call M 'the matrix of cycles'.

Definition C.1.10. If the graph G has at least one cycle, then the rank $\mu =: \mu^G$ of the matrix of cycles of a graph G is called the cyclomatic number of the graph G. A set of cycles in a graph G is said to be linearly independent if the corresponding set of rows in the matrix of cycles is linearly independent. Each set of μ linearly independent cycles is said to be fundamental. If the graph G has no cycles, then we set $\mu^G := 0$.

For a concrete representation of a matrix of cycles one needs a digraph and orientations of the cycles. However, each graph G can be written as a digraph with arbitrarily chosen directions of the edges, and as was seen above, the cyclomatic number is independent of the chosen directions and orientations. Hence the cyclomatic number is well-defined for any graph.

C.2 Basic properties of graphs

We generalize the construction of the rows of the cyclomatic matrix to walks in graphs with at least one edge. To this end let G be a graph with at least one edge, which is considered as a digraph by assigning an arbitrary but fixed direction to each edge. The edges will be indexed by $e_1, \dots, e_g$ in an arbitrary but fixed manner. Each walk $W = (v_0, \widetilde{e}_1, v_1, \dots, v_{k-1}, \widetilde{e}_k, v_k)$ has a natural orientation, with the edge $\widetilde{e}_j$ directed from the vertex v_{j-1} towards the vertex v_j if $v_{j-1} \neq v_j$. Only if $\widetilde{e}_j$ is a loop, the direction is not determined by the ends of this edge.

For an oriented walk W define the row vectors $m^\pm(W) = (m_j^\pm(W))_{j=1}^g$ as follows: $m_j^+(W)$ is the number of times the edge e_j is traversed in W in the same direction as in the digraph, and $m_j^-(W)$ is the number of times the edge e_j is traversed in W in the direction opposite to that in the digraph. Then we define $m(W) := m^+(W) - m^-(W)$ and $|m|(W) = m^+(W) + m^-(W)$. In particular, if C_k is the k-th cycle in G, then $m(C_k)$ is the k-th row of the matrix of cycles of G. For two walks W_1 and W_2 we write $|m|(W_1) \geq |m|(W_2)$ if this inequality holds for all components of the two row vectors, and $|m|(W_1) > |m|(W_2)$ means that additionally at least one component of $|m|(W_1)$ is larger than the corresponding component of $|m|(W_2)$.

Traversing the walk $W = (v_0, \widetilde{e}_1, v_1, \ldots, v_{k-1}, \widetilde{e}_k, v_k)$ with the opposite orientation gives the walk $W^- := (v_k, \widetilde{e}_k, v_{k-1} \ldots, v_1, \widetilde{e}_1, v_0)$. In particular, when e is a directed edge, then e^- will denote the edge with the opposite direction.

Let $W_1 = (v_{1,0}, e_{1,1}, \ldots, e_{1,j}, v_{1,j})$ and $W_2 = (v_{2,0}, e_{2,1}, \ldots, e_{2,k}, v_{2,k})$ be two walks in a graph G. If $v_{1,j} = v_{2,0}$ then

$$(W_1, W_2) := (v_{1,0}, e_{1,1}, \ldots, e_{1,j}, v_{1,j}, e_{2,1}, \ldots, e_{2,k}, v_{2,k})$$

is the walk obtained by first traversing the walk W_1 and then the walk W_2. Clearly, $m(W_1, W_2) = m(W_1) + m(W_2)$ and $|m|(W_1, W_2) = |m|(W_1) + |m|(W_2)$.

Clearly, a walk W is a trail if and only if each component of $|m|(W)$ has the value 0 or 1.

Lemma C.2.1. *Let W be a closed walk in a graph G with $m(W) \neq 0$ and assume that W is not a cycle. Then there are two closed walks W_1 and W_2 in G such that $W = (W_1, W_2)$, $m(W) = m(W_1) + m(W_2)$ and $|m|(W) = |m|(W_1) + |m|(W_2)$. If W is a trail, then also W_1 and W_2 are trails.*

Proof. There is a vertex v_0 of W which is traversed at least twice. Indeed, if W is a trail or contains a loop, then there is such a vertex because W is not a cycle. And if W is not a trail and does not contain loops, then there is at least one edge which is traversed twice. Since $m(W) \neq 0$ it follows that W has more than two edges and therefore at least one of the vertices of those edges which are traversed at least twice is also traversed at least twice.

Using this vertex as the starting point of the closed walk we have a representation $W = (v_0, \widetilde{e}_1, \ldots, \widetilde{e}_j, v_j, \widetilde{e}_{j+1}, \ldots, \widetilde{e}_k, v_k)$ with $1 \leq j < k$ and $v_0 = v_j = v_k$. Defining $W_1 = (v_0, \widetilde{e}_1, \ldots, \widetilde{e}_j, v_j)$ and $W_2 = (v_j, \widetilde{e}_{j+1}, \ldots, \widetilde{e}_k, v_k)$ it is clear that W_1 and W_2 are closed walks with $W = (W_1, W_2)$.

If W is a trail, then all components of $|m|(W)$ are 1 or 0, and hence also all components of $|m|(W_1)$ and $|m|(W_2)$ have this property. It follows that W_1 and W_2 are trails. □

Corollary C.2.2. *Let W be a closed walk in a graph G satisfying $m(W) \neq 0$. Then there are cycles $C_1, \ldots, C_k$ in G ($k \geq 1$) such that $m(W) = \sum_{j=1}^{k} m(C_j)$ and $|m|(W) \geq \sum_{j=1}^{k} |m|(C_j)$.*

Proof. We are going to prove the statement by induction on the number of edges in the walk W, that is, the sum of the components of the vector $|m|(W)$. If W has only one edge, then W is a loop and hence the statement follows with $k = 1$ and $C_1 = W$. Now let $n \in \mathbb{N}$ with $n > 1$ and assume that the statement holds for all closed walks with less than n edges. Let W be a closed walk in G with exactly n edges satisfying $m(W) \neq 0$. If W is a cycle, then the statement of this corollary follows with $k = 1$ and $C_1 = W$. If W is not a cycle, then by Lemma C.2.1 there are closed walks W_1 and W_2 in G with

$$m(W) = m(W_1) + m(W_2) \quad \text{and} \quad |m|(W) = |m|(W_1) + |m|(W_2). \tag{C.2.1}$$

In particular, n is the sum of the number of edges of W_1 and the number of edges of W_2, so that both the number of edges of W_1 and the number of edges of W_2 are less than n. Since $m(W) \neq 0$, it follows from (C.2.1) that at least one of $m(W_1)$ and $m(W_2)$ is different from the zero vector, say $m(W_1) \neq 0$. By induction hypothesis, there are cycles $C_1, \ldots, C_{k_1}$ such that

$$m(W_1) = \sum_{j=1}^{k_1} m(C_j) \quad \text{and} \quad |m|(W_1) \geq \sum_{j=1}^{k_1} |m|(C_j). \tag{C.2.2}$$

If $m(W_2) = 0$, then (C.2.1) implies $m(W) = m(W_1)$ and, together with (C.2.2),

$$|m|(W) = |m|(W_1) + |m|(W_2) > |m|(W_1) \geq \sum_{j=1}^{k_1} |m|(C_j),$$

which proves the statement for W in this case. If $m(W_2) \neq 0$, the induction hypothesis gives cycles $C_{k_1+1}, \ldots, C_k$ such that

$$m(W_2) = \sum_{j=k_1+1}^{k} m(C_j) \quad \text{and} \quad |m|(W_2) \geq \sum_{j=k_1+1}^{k} |m|(C_j). \tag{C.2.3}$$

Then the statement of this corollary for W in this case follows from (C.2.1), (C.2.2) and (C.2.3). □

Lemma C.2.3. *Let G be a graph, let p^G be the number of its vertices, and let κ^G be the number of its components. Then $g^G + \kappa^G = p^G + \mu^G$.*

Proof. We will prove the statement by induction on $g = g^G$. When $g = 0$, then G has no edges and therefore no cycles, which means that $g^G = 0 = \mu^G$. Also, every vertex is isolated, and therefore $\kappa^G = p^G$, which proves the formula for $g = 0$. Now let $g > 0$ and assume the formula holds for $g - 1$. We will consider three different cases, but in each case we will remove one edge e_0 from G to obtain a subgraph G' of G, which obviously satisfies $g^{G'} = g^G - 1$ and $p^{G'} = p^G$. Therefore the graph G' will satisfy the induction hypothesis.

Firstly assume that G contains a loop, and any such loop will be chosen as the edge e_0 to be removed. Clearly, $\kappa^{G'} = \kappa^G$. The loop cannot be part of any other cycle, and hence the loop corresponds to a row and a column in the matrix of cycles of G whose entries are 0 except for the entry where this row and this column meet. Removing this row and this column from the matrix of cycles of G, its rank decreases by 1 (with the obvious interpretation if the loop is the only cycle in G).

But this new matrix is the matrix of cycles of G', and therefore $\mu^{G'} = \mu^G - 1$. Altogether, it follows that $g^G + \kappa^G = g^{G'} + 1 + \kappa^{G'} = p^{G'} + 1 + \mu^{G'} = p^G + \mu^G$.

Next assume that G has no loops and choose any edge e_0 of G for removal. Let v_1 and v_2 be the ends of e_0. Obviously, each component of G which does not contain e_0 is also a component of G'. Let G_0 be the component of G which contains e_0 (and thus v_1 and v_2), and let G'_k be the component of G' containing v_k $(k = 1, 2)$. Let $v \neq v_2$ be a vertex in G_0 which does not belong to G'_1, if any. Since G_0 is connected, there is a chain from v_1 to v in G. But this chain must contain the edge e_0 because otherwise it would be a chain in G'. Therefore this chain must be of the form $(v_1, e_0, v_2, \dots, v)$, giving a chain $(v_2, \dots, v)$ in G'_2. This shows that $V^{G'_1} \cup V^{G'_2} = V^{G_0}$. We will consider two cases.

Here we assume that $G'_1 \neq G'_2$, in which case G'_1 and G'_2 are components of G'. Therefore $\kappa^{G'} = \kappa^G + 1$. Furthermore, e_0 is not an edge of any cycle in G. Because if there were such a cycle, say $(v_1, e_0, v_2, \dots, v_1)$, then $(v_2, \dots, v_1)$ would be a path connecting v_2 and v_1 in G'. Therefore $\mu^{G'} = \mu^G$, and hence $g^G + \kappa^G = g^{G'} + \kappa^{G'} = p^{G'} + \mu^{G'} = p^G + \mu^G$.

Finally, assume that $G'_1 = G'_2$. Then $\kappa^{G'} = \kappa^G$ and there is a path $(v_2, \dots, v_1)$ in G'. But then $(v_2, \dots, v_1, e_0, v_2)$ is a cycle in G. Assume that there are at least two cycles in G with edge e_0. With suitably chosen orientation two such cycles are of the form $C^{(1)} = (v_1, e_0, v_2, e_{1,1}, \dots, e_{1,j}, v_1)$ and $C^{(2)} = (v_1, e_{2,1}, \dots, e_{2,r}, v_2, e_0, v_1)$ and satisfy $m(C^{(1)}) + m(C^{(2)}) \neq 0$ if they are distinct. Removing the edge e_0 from each of the cycles the concatenation $W := (v_2, e_{1,1}, \dots, e_{1,j}, v_1, e_{2,1}, \dots, e_{2,r}, v_2)$ is a walk in G which satisfies $m(W) = m(C^{(1)}) + m(C^{(2)}) \neq 0$. By Corollary C.2.2 there are k cycles $C_1, \dots, C_k$ with $m(W) = \sum_{j=1}^{k} m(C_j)$ and $|m|(W) \geq \sum_{j=1}^{k} |m|(C_j)$. Since W does not traverse the edge e_0, the last inequality implies that none of the C_j traverses the edge e_0. Therefore a fundamental set of cycles of G can be chosen in such a way that exactly one of the cycles traverses the edge e_0. Removing this cycle then gives a fundamental set of cycles of G', that is, $\mu^{G'} = \mu^G - 1$, and therefore $g^G + \kappa^G = g^{G'} + 1 + \kappa^{G'} = p^{G'} + 1 + \mu^{G'} = p^G + \mu^G$. □

Corollary C.2.4. *A connected graph G has at most $g^G + 1$ vertices.*

Proposition C.2.5. *Let G be a graph with $g \geq 1$ edges and p vertices. Then the following properties are equivalent.*

(i) *G is a tree.*

(ii) *The graph does not have loops and for every two distinct vertices v and w of G there is a unique path from v to w.*

(iii) *G is connected and $g = p - 1$.*

(iv) *G is connected and $g < p$.*

Proof. Corollary C.2.4 states that $p - 1 \leq g$ for a connected graph, and therefore (iii) and (iv) are equivalent.

In the notation of Lemma C.2.3, (i) means $\kappa^G = 1$ and $\mu^G = 0$, whereas (iii) means $\kappa^G = 1$ and $g^G = p^G - 1$. In view of Lemma C.2.3 we conclude that (i) and (iii) are equivalent.

Now assume that (i) holds. Since trees do not have cycles, they do not have loops. Since a tree is connected, each two distinct vertices can be connected by a path. If there would be distinct vertices v and w in G which are connected by two different paths, then taking one path from v to w followed by the other path from w to v we would obtain a closed walk with $m(W) \neq 0$. But in view of Corollary C.2.2 this is impossible since G has no cycles. Hence (ii) has been shown.

Now assume that (ii) holds. Then G has no loops, and G is connected since each two distinct vertices can be connected by a path. If G would contain a cycle with at least two edges, then it has also at least two distinct vertices, and the cycle could be split to connect these vertices by two distinct paths, contradicting (ii). Therefore (i) is true. □

Lemma C.2.6. *Let T be a tree and let v_0 be an arbitrarily chosen fixed vertex of T, called the root of T. For each $v \in V^T \setminus \{v_0\}$ let P_v be the oriented path from v to v_0. Then the following statements hold.*

1. *Each edge of T has a unique direction such that the direction of each edge in P_v $(v \in V^T \setminus \{v_0\})$ coincides with its direction in T.*
2. *There is an injective function $w_T : V^T \to \mathbb{R}$ such that $w_T(v_-(e)) > w_T(v_+(e))$ for each edge e of T.*
3. *$I^-(v_0) = \emptyset$ and $\#I^-(v) = 1$ for each $v \in V^T \setminus \{v_0\}$.*

Proof. By Proposition C.2.5, the paths P_v are well-defined. For $v \in V^T \setminus \{v_0\}$, the number of edges in P_v will be denoted by $\delta(v)$, and for convenience we put $\delta(v_0) = 0$. Let $w_T : V^T \to [0, \infty)$ be an injective function such that $w_T(u) < w_T(v)$ whenever $u, v \in V^T$ satisfy $\delta(u) < \delta(v)$. For example, we may choose w_T such that $w_T(u) \in [\delta(u), \delta(u) + 1)$ with $w_T(u) \neq w_T(v)$ when $u \neq v$ and $\delta(u) = \delta(v)$. Let e be an edge of T and let u and v be its ends. Since $w_T(u) \neq w_T(v)$, we may label the ends in such a way that $w_T(u) < w_T(v)$. Directing e in such a way that $v_-(e) = v$ and $v_+(e) = u$ it follows that statement 2 holds.

If e is an edge which is incident with v_0, then $v_0 = v_+(e)$ by statement 2 since $w_T(v_0)$ is the smallest value of the function w_T in view of $\delta(v_0) = 0 < \delta(v)$ for all $v \in V^T \setminus \{v_0\}$. Hence each edge which is incident with the root is an incoming edge, which means that $I^-(v_0) = \emptyset$. Now let $v \in V^T \setminus \{v_0\}$. Then the first edge of P_v is outgoing from v, which shows that $\#I^-(v) \geq 1$. For any edge e which is outgoing from v consider the walk $W = ((v = v_-(e), e, v_+(e)), P_{v_+(e)})$. If W were not a path, then $v_-(e)$ would be a vertex of P_{v_+}, and therefore $P_{v_-(e)} \subset P_{v_+(e)}$. This would imply that $\delta(v_-(e)) < \delta(v_+(e))$ and therefore the contradiction $w_T(v_-(e)) < w_T(v_+(e))$ would follow. Hence W is a path and therefore $W = P_v$, which means that e is uniquely determined as the first edge of P_v. This proves statement 3.

To prove statement 1 let $v \in V^T \setminus \{v_0\}$ and let e be an edge of P_v. Then there are vertices $v_1, v_2 \in V^T$ such that $P_v = (v, \ldots, v_1, e, v_2, \ldots, v_0)$, where $v = v_1$ and $v_2 = v_0$ are allowed. The edge e is directed from v_1 to v_2 in P_v. On the other hand, it is clear from the representation of P_v that $\delta(v_1) = 1 + \delta(v_2)$, and therefore $w_T(v_1) > w_T(v_2)$, which means that e is also directed from v_1 to v_2 in T. Finally, if e is any edge of T, then it is the unique outgoing edge from $v_-(e)$, and in the previous paragraph we have shown that e is an edge of $P_{v_-(e)}$. Therefore the direction of e is uniquely determined by its direction in $P_{v_-(e)}$. □

C.3 Cyclically connected graphs and subgraphs

Lemma C.3.1. *Let C_1 and C_2 be two cycles in a graph G with $\#(V^{C_1} \cap V^{C_2}) \geq 2$ and assume there are $v \in V^{C_1} \setminus V^{C_2}$ and $w \in V^{C_2} \setminus V^{C_1}$. Then there is a cycle C in G with $E^C \subset E^{C_1} \cup E^{C_2}$ and $v, w \in V^C$.*

Proof. Choosing an orientation of C_1, we can write $C_1 = (v, \ldots, v_1, \ldots, v_2, \ldots, v)$, where v_1 is the first vertex of C_1 which is also a vertex of C_2 and v_2 is the last vertex of C_1 which is also a vertex of C_2, that is, there are vertices $v_1 \neq v_2$ in C_2 such that $C_1 = (P_{1,1}, P_{1,2}, P_{1,3})$, with chains $P_{1,1} = (v, \ldots, v_1)$, $P_{1,2} = (v_1, \ldots, v_2)$, $P_{1,3} = (v_2, \ldots, v)$ such that the only vertex of $P_{1,1}$ which is also a vertex of C_2 is v_1 and such that the only vertex of $P_{1,3}$ which is also a vertex of C_2 is v_2. With a suitable orientation, we can write $C_2 = (P_{2,1}, P_{2,2}, P_{2,3})$ with $P_{2,1} = (w, \ldots, v_2)$, $P_{2,2} = (v_2, \ldots, v_1)$, $P_{2,3} = (v_1, \ldots, w)$. Then $C := (P_{1,3}, P_{1,1}, P_{2,3}, P_{2,1})$ is a closed walk. By construction, $P_1 := (P_{1,3}, P_{1,1})$ and $P_2 := (P_{2,3}, P_{2,1})$ are chains and the only common vertices of P_1 and P_2 are v_1 and v_2 since all other vertices of P_1 are no vertices of C_2. Therefore each vertex of C is traversed exactly once, which means that C is a cycle. By construction, v and w are vertices of C. □

Lemma C.3.2. *Let G' be a cyclically connected subgraph of a graph G.*

1. *Let C be a cycle in G with $V^{G'} \cap V^C \neq \emptyset$. Then $G' \cup C$ is cyclically connected.*
2. *Let $P = (v, \ldots, w)$ be a chain in G such that $V^{G'} \cap V^P = \{v, w\}$. Then $G' \cup P$ is cyclically connected.*

Proof. 1. Let $v \in V^{G'} \cap V^C$. Then v is cyclically connected to each vertex in $G' \setminus \{v\}$ and to each vertex in $C \setminus \{v\}$. Therefore $G' \cup C$ is cyclically connected.

2. Since G' is connected, there is a chain $P' = (w, \ldots, v)$ in G'. By assumption, $V^{P'} \cap V^P \subset V^{G'} \cap V^P = \{v, w\}$. Therefore (P, P') is a cycle which satisfies $V^{G'} \cap V^{(P,P')} \neq \emptyset$. By statement 1, $G' \cup P = G' \cup (P, P')$ is cyclically connected. □

Definition C.3.3. Let v and w be two cyclically connected vertices of a graph G. A sequence $(C_1, \ldots, C_k)$ $(k \geq 1)$ of cycles in G is called a minimal sequence of cycles for (v, w) if

(i) $v \in C_1$, $w \in C_k$;

(ii) if $k \geq 2$, then $V^{C_j} \cap V^{C_{j+1}} \neq \emptyset$ $(j = 1, \ldots, k-1)$;

(iii) the number k is minimal amongst all sequences of cycles having the properties (i) and (ii).

By definition of cyclically connected, the conditions (i) and (ii) in Definition C.3.3 are satisfied for at least one sequence of cycles. Hence there is at least one minimal sequence of cycles for each cyclically connected pair of vertices.

Lemma C.3.4. *Let $(C_1, \ldots, C_k)$ $(k \geq 2)$ be a minimal sequence of cycles for a cyclically connected pair (v, w) of vertices in a graph G. Then:*

1. $v \notin C_2$ *and* $w \notin C_{k-1}$;
2. $V^{C_r} \cap V^{C_s} = \emptyset$ *for* $r, s = 1, \ldots, k$ *with* $r + 1 \leq s - 1$;
3. $\#(V^{C_j} \cap V^{C_{j+1}}) = 1$ $(j = 1, \ldots, k-1)$.

Proof. 1. Assume that $v \in C_2$. Then the sequence of cycles $(C_2, \ldots, C_k)$ would satisfy properties (i) and (ii) of Definition C.3.3, contradicting the minimality of k. Hence $v \notin C_2$ has been shown. Similarly one proves $w \notin C_{k-1}$.

Assume that statement 2 is false. Then there would be numbers $1 \leq r < s \leq k$ with $r + 1 \leq s - 1$ and $V^{C_r} \cap V^{C_s} \neq \emptyset$, and we could remove all cycles C_j with $r + 1 \leq j \leq s - 1$. Hence the sequence of cycles $(C_1, \ldots, C_r, C_s, \ldots, C_k)$ would satisfy properties (i) and (ii) of Definition C.3.3, contradicting the minimality of k. Therefore statement 2 is true.

Assume that statement 3 is false. Then there is some $j \in \{1, \ldots, k-1\}$ such that C_j and C_{j+1} have more that one common vertex. Put $v_0 = v$ and $v_k = w$. By property (ii) of Definition C.3.3 we can choose $v_r \in V^{C_r} \cap V^{C_{r+1}}$ for $r = 1, \ldots, k-1$. From statements 1 and 2 we know that $v_{j-1} \notin V^{C_{j+1}}$ and $v_{j+1} \notin V^{C_j}$. An application of Lemma C.3.1 would show that we can replace C_j and C_{j+1} with a single cycle still having vertices v_{j-1} and v_{j+1}. But this would contradict the minimality of k. □

Lemma C.3.5. *Let v and w be two cyclically connected vertices of a graph G. Then there is a closed trail W in G which has the following properties. There are a positive integer k, mutually distinct vertices $v_0 = v, v_1, \ldots, v_k = w$ and chains $P_{1,j} = (v_{j-1}, \ldots, v_j)$, $P_{2,j} = (v_j, \ldots, v_{j-1})$ $(j = 1, \ldots, k)$ such that $P_1 = (P_{1,1}, \ldots, P_{1,k})$ and $P_2 = (P_{2,k}, \ldots, P_{2,1})$ are chains with $V^{P_1} \cap V^{P_2} = \{v_0, \ldots, v_k\}$ and such that $W = (P_1, P_2)$.*

Proof. Let $(C_1, \ldots, C_k)$ $(k \geq 1)$ be a minimal sequence of cycles for (v, w). When $k = 1$, both v and w are vertices of the cycle C_1, which can be written in the form $C_1 = (P_1, P_2)$ with chains $P_1 = (v, \ldots, w)$ and $P_2 = (w, \ldots, v)$ where $V^{P_1} \cap V^{P_2} = \{v, w\}$. Hence $W = C_1$ is a trail of the requested form with $k = 1$.

Now let $k > 1$. We put $v_0 := v$, $v_k := w$ and let $v_j \in V^{C_j} \cap V^{C_{j+1}}$ for $j = 1, \ldots, k-1$. It is clear from Lemma C.3.4 that these $k + 1$ vertices are mutually distinct and uniquely determined. Due to the minimality of k, none of the cycles is a loop. Therefore each edge of the graph G occurs at most once in $C_1, \ldots, C_k$. Writing $C_j = (P_{1,j}, P_{2,j})$ $(j = 1, \ldots, k)$, where $P_{1,j}$ is a chain from v_{j-1} to v_j and $P_{2,j}$ is a chain from v_j to v_{j-1}, it is clear that no two distinct chains have a common edge. In view of Lemma C.3.4 the proof is complete. □

Proposition C.3.6. *Let G be a graph with at least two vertices and let v and w be two distinct vertices of G. Then v and w are cyclically connected if and only if there is a closed trail in G which traverses both v and w.*

Proof. If v and w are cyclically connected, then Lemma C.3.5 shows that there is a closed trail in G which traverses both v and w.

Conversely, let W be a closed trail in G such that v and w are vertices of W. We will prove that v and w are cyclically connected by induction on the number of edges in W. The minimal number of edges in a closed trail with at least two vertices is 2, and in this case the closed trail is a cycle. Now assume that W has more than 2 edges. If W is a cycle, nothing has to be shown. Therefore assume now that W is not a cycle. Then there is a vertex v_0 such that $W = (W_1, W_2)$ and W_1 and W_2 are closed trails starting and ending at v_0. If v and w are vertices of the same subtrail, say W_1, then v and w are cyclically connected by induction hypothesis since W_1 has fewer edges than W. If, on the other hand, v and w do not belong to the same subtrail, then the induction hypothesis shows that v and v_0 as well as v_0 and w are cyclically connected since both W_1 and W_2 have fewer edges that W. Therefore also v and w are cyclically connected. □

Lemma C.3.7. *Each cyclically connected graph with cyclomatic number* 1 *is a cycle.*

Proof. For a cyclically connected graph, its matrix of cycles M contains a nonzero entry in each column. But the rank of M is the cyclomatic number of the graph and therefore equal to 1. Hence M has exactly one row (see the discussion after Definition C.1.9), and all entries of this row are different from zero. But this means that the only cycle in the graph traverses all edges, that is, the graph is a cycle. □

C.4 Maximally connected subgraphs and quasi-trees

For each vertex v of G let $C(v)$ be the set of all $w \in V^G$ for which there is a closed trail W such that v and w are vertices of W. Obviously, if $C(v) \neq \emptyset$, then $v \in C(v)$. For any two distinct vertices v and w of G it is clear by Proposition C.3.6 that $C(v) = C(w)$ if v and w are cyclically connected and that $C(v) \cap C(w) = \emptyset$ if v and w are not cyclically connected. For each vertex v of G let G_v be the subgraph of G consisting of the vertices in $C(v)$ and those edges of G for which both ends are vertices in $C(v)$.

Lemma C.4.1. 1. *Let v be a vertex of the graph G for which G_v is not the null graph. Then G_v is a maximal cyclically connected subgraph of G.*

2. *Each cyclically connected subgraph of G is a subgraph of a maximal cyclically connected subgraph of G.*

Proof. 1. If $C(v) = \{v\}$, then all edges in G_v are loops, and therefore G_v is clearly cyclically connected. Now consider the case that $\#C(v) \geq 2$ and let $w \in C(v)$ with $w \neq v$. Let $(C_1, \ldots, C_k)$ be a minimal sequence of cycles for the pair (v, w). Then it is clear that each vertex in each of these cycles is cyclically connected to v in G. By definition of G_v it follows that with all vertices of these cycles being in G_v also all edges of these cycles are in G_v. Therefore v and w are cyclically connected in G_v. This proves that the graph G_v is cyclically connected.

Let G' be a cyclically connected subgraph of G with $G_v \subset G'$. Since no vertex in G_v is cyclically connected in G to a vertex in G which does not belong to $C(v)$, the graphs G_v and G' have the same vertices. But then edges in G' are edges in G with both ends being vertices in $C(v)$. It follows that edges in G' are also edges in G_v. Therefore $G' = G_v$, which shows that G_v is maximal.

2. Let G_1 be a cyclically connected subgraph of G and let v be a vertex in G_1. Then the vertices of G_1 belong to $C(v)$, and therefore also the edges of G_1 belong to G_v. This proves $G_1 \subset G_v$. □

Lemma C.4.2. *Let G be a quasi-tree and let G' be a maximal cyclically connected subgraph of G. Let $v_0, \ldots, v_n$ $(n \geq 1)$ be an arbitrary indexing of the common vertices of G' and its complement in G. The remaining vertices of G' (if any) are indexed as $v_{n+1}, \ldots, v_r$ in an arbitrary manner. Let G_0 be the null graph and put $w(v_0) = 0$. Then there is a sequence $(G_j)_{j=1}^{\infty}$ of cyclically connected subgraphs of G' and an injective function*

$$w : \left(\bigcup_{j=1}^{\infty} V^{G_j} \right) \setminus \{v_0\} \to (0, \infty)$$

with the following properties:

(i) *$G_{j-1} \subset G_j$ $(j \in \mathbb{N})$;*

(ii) *if $k \in \{1, \ldots, r\}$ and $j \in \mathbb{N}$ such that $v_k \in V^{G_j} \setminus V^{G_{j-1}}$, then there is a number $i \in \{0, \ldots, r\}$ such that $v_i \in V^{G_j}$, $w(v_k) > w(v_i)$ and such that there is an edge in G_j which is incident with v_k and v_i;*

(iii) *if $k \in \{n+1, \ldots, r\}$ and $j \in \mathbb{N}$ such that $v_k \in V^{G_j} \setminus V^{G_{j-1}}$, then there is $i \in \{0, \ldots, r\}$ such that $v_i \in V^{G_j}$, $w(v_k) < w(v_i)$ and such that there is an edge in G_j which is incident with v_k and v_i.*

Proof. For $j \in \mathbb{N}$, the subgraphs G_j and the function w on $G_j \setminus G_{j-1}$ satisfying (i), (ii) and (iii) will be constructed recursively. First let $j = 1$. In view of Lemma C.3.5 and Proposition C.3.6 there are two chains $P_1^{(1)} = (v_0 = v_{1,1}^{(1)}, \ldots, v_{1,r_1}^{(1)} = v_1)$ and $P_2^{(1)} = (v_0 = v_{2,1}^{(1)}, \ldots, v_{2,s_1}^{(1)} = v_1)$ and two increasing sequences of integers $t((1,\iota))_{\iota=0}^{k}$ and $t((2,\iota))_{\iota=0}^{k}$ such that $v_{1,t(1,\iota)}^{(1)} = v_{2,t(2,\iota)}^{(1)}$ $(\iota = 0, \ldots, k)$ as well as

$V^{P_1^{(1)}} \cap V^{P_2^{(1)}} = \{v_{1,t(1,\iota)}^{(1)} : \iota = 0, \dots, k\}$, and such that $(P_1^{(1)}, P_2^{(1)-})$ is a closed trail. Therefore also $G_1 := G|_{V^{P_1^{(1)}} \cup V^{P_2^{(1)}}}$ is cyclically connected. Positive weights will be attributed to the vertices $v_{1,\iota}^{(1)}$ $(\iota = 1, \dots, r_1)$ of $P_1^{(1)}$ such that

$$w(v_{1,1}^{(1)}) < w(v_{1,2}^{(1)}) < \cdots < w(v_{1,r_1}^{(1)}) = w(v_1).$$

For $i = 1, \dots, k$ with $t(2, i-1)+1 < t(2,i)$ and $\iota \in \mathbb{N}$ with $t(2,i-1) < \iota < t(2,i)$, weights for $v_{2,\iota}^{(1)}$ will be chosen such that that

$$w(v_{2,t(2,i-1)}^{(1)}) < w(v_{2,t(2,i-1)+1}^{(1)}) < \cdots < w(v_{2,t(2,i)}^{(1)})$$

and such that these weights are distinct from the already chosen weights. That is, the weights of the vertices of G_1 are chosen in such a way that $w(v_{1,\iota}^{(1)})$ and $w(v_{2,\iota}^{(1)})$ are increasing with ι and such that w is injective on G_1.

Clearly, (i) is satisfied for $j = 1$. For $k \in \{1, \dots, r\}$ with $v_k \in V^{G_1}$ there is $\iota \in \{1, \dots, r_1\}$ such that $v_k = v_{1,\iota}^{(1)}$ or there is $\iota \in \{1, \dots, s_1\}$ such that $v_k = v_{2,\iota}^{(1)}$. With $i = v_{1,\iota-1}^{(1)}$ or $i = v_{2,\iota-1}^{(1)}$, respectively, it is now clear that (ii) holds for $j = 1$. Similarly, for $k \in \{n+1, \dots, r\}$ with $v_k \in V^{G_1}$ there is $\iota \in \{1, \dots, r_1 - 1\}$ such that $v_k = v_{1,\iota}^{(1)}$ or there is $\iota \in \{1, \dots, s_1 - 1\}$ such that $v_k = v_{2,\iota}^{(1)}$. With $i = v_{1,\iota+1}^{(1)}$ or $i = v_{2,\iota+1}^{(1)}$, respectively, it is now clear that (iii) holds for $j = 1$.

Now assume that $G_1, \dots, G_{j-1}$ have been constructed for some $j > 1$. We consider three cases.

The first case is that there exists $l \in \{2, \dots, n\}$ such that v_l is not a vertex of G_{j-1}. As for $j = 1$ earlier in this proof we conclude from Lemma C.3.5 and Proposition C.3.6 that there are two chains $P_1^{(j)} = (v_0 = v_{1,1}^{(j)}, \dots, v_{1,r_j}^{(j)} = v_l)$ and $P_2^{(j)} = (v_0 = v_{2,1}^{(j)}, \dots, v_{2,s_j}^{(j)} = v_l)$ in G' without common edges and two increasing sequences of integers $t((1,\iota))_{\iota=0}^k$ and $t((2,\iota))_{\iota=0}^k$ such that $v_{1,t(1,\iota)}^{(j)} = v_{2,t(2,\iota)}^{(j)}$ $(\iota = 0, \dots, k)$ and $V^{P_1^{(j)}} \cap V^{P_2^{(j)}} = \{v_{1,t(1,\iota)}^{(j)} : \iota = 0, \dots, k\}$. Since v_0 is a vertex of G_{j-1} but v_l is not, the numbers $\widetilde{r}_j := \max\{\iota \in \{0, \dots, r_j\} : v_{1,\iota}^{(j)} \in V^{G_{j-1}}\}$ as well as $\widetilde{s}_j := \max\{\iota \in \{0, \dots, s_j\} : v_{2,\iota}^{(j)} \in V^{G_{j-1}}\}$ are well-defined and satisfy $\widetilde{r}_j < r_j$ and $\widetilde{s}_j < s_j$. Then $\widetilde{P}_1^{(j)} = (v_{1,\widetilde{r}_j}^{(j)}, \dots, v_{1,r_j}^{(j)})$ and $\widetilde{P}_2^{(j)} = (v_{2,\widetilde{s}_j}^{(j)}, \dots, v_{2,s_j}^{(j)})$ are subchains of $P_1^{(j)}$ and $P_2^{(j)}$, respectively. The first vertex of each of these two chains belongs to G_{j-1}, whereas all their other vertices do not belong to G_{j-1}. Define $G_j := G'|_{V^{G_{j-1}} \cup V^{\widetilde{P}_1^{(j)}} \cup V^{\widetilde{P}_2^{(j)}}}$. Let $\widetilde{k}$ be the smallest index $\iota \in \{1, \dots, k\}$ such that $t(1,\iota) > \widetilde{r}_j$ and $t(2,\iota) > \widetilde{s}_j$. For $v_{1,t(1,\iota)}^{(j)}$ with $\iota \in \{\widetilde{k}, \dots, k\}$ we assign weights

$$\max\{w(v_{1,\widetilde{r}_j}^{(j)}), w(v_{2,\widetilde{s}_j}^{(j)})\} < w(v_{1,t(1,\widetilde{k})}^{(j)}) < \cdots < w(v_{1,t(1,k)}^{(j)}).$$

With the already assigned weights we now define weights to all the remaining vertices in $\widetilde{P}_1^{(j)}$ and $\widetilde{P}_2^{(j)}$ such that the values $w(v_{1,\iota}^{(j)})$ $(\iota = \widetilde{r}_j, \dots, r_j)$ and

$w(v_{2,\iota}^{(j)})$ ($\iota = \widetilde{s}_j, \ldots, s_j$) are increasing in ι. Furthermore, the values of w on $G_j \setminus G_{j-1}$ can and will be chosen in such a way that w is injective on G_j. If $v_{1,\widetilde{r}_j}^{(j)} = v_{2,\widetilde{s}_j}^{(j)}$, then define $W^{(j)} := (\widetilde{P}_1^{(j)}, \widetilde{P}_2^{(j)-})$. If $v_{1,\widetilde{r}_j}^{(j)} \neq v_{2,\widetilde{s}_j}^{(j)}$, then, in view of the connectedness of G_{j-1}, there is a chain $\widetilde{P}_3^{(j)}$ from $v_{2,\widetilde{s}_j}^{(j)}$ to $v_{1,\widetilde{r}_j}^{(j)}$ and we define $W^{(j)} := (\widetilde{P}_1^{(j)}, \widetilde{P}_2^{(j)-}, \widetilde{P}_3^{(j)})$. In both cases, $W^{(j)}$ is a closed trail connecting each vertex in $G_j \setminus G_{j-1}$ to a vertex in G_{j-1}. Since G_{j-1} is cyclically connected, it follows that also G_j is cyclically connected. With the same reasoning as for G_1 one shows that the properties (ii) and (iii) holds.

The second case is that $\{v_0, \ldots, v_n\} \subset V^{G_{j-1}}$ and that there is a chain $P^{(j)} = (v_0^{(j)}, \ldots, v_k^{(j)})$ with $k \geq 2$ such that $V^{G_{j-1}} \cap V^{P^{(j)}} = \{v_0^{(j)}, v_k^{(j)}\}$. Since $v_0^{(j)} \neq v_k^{(j)}$, we may choose the indexing of the vertices of this chain in such a way that $w(v_0^{(j)}) < w(v_k^{(j)})$. Then we choose weights $w(v_\iota^{(j)})$ ($\iota = 1, \ldots, k-1$) such that $(w(v_\iota^{(j)}))_{\iota=0}^k$ is a increasing sequence and such that w is injective on $G_j := G'|_{V^{G_{j-1}} \cup V^{P^{(j)}}}$. It follows as above that G_j is cyclically connected and that properties (ii) and (iii) hold.

The final case is that $\{v_0, \ldots, v_n\} \subset V^{G_{j-1}}$ and that each chain in G' with both ends in G_{j-1} must have all vertices in G_{j-1}. In this case we set $G_j = G_{j-1}$. □

Lemma C.4.3. *Let G be a quasi-tree and let G' be a maximal cyclically connected subgraph of G. Let $v_0, \ldots, v_n$ ($n \geq 1$) be an arbitrary indexing of the common vertices of G' and its complement in G. The remaining vertices of G' (if any) are indexed as $v_{n+1}, \ldots, v_r$ in an arbitrary manner. Then the edges of G' can be directed and indexed such that G' becomes a digraph with the following properties for $k = 0, \ldots, r$:*

1. *$I^-(v_k) = \emptyset$ if and only if $k = 0$;*
2. *$I^+(v_k) \neq \emptyset$ for $k = n+1, \ldots, r$;*
3. *if $I^-(v_k) \neq \emptyset$, $I^+(v_k) \neq \emptyset$, $j \in I^-(v_k)$ and $i \in I^+(v_k)$, then $j < i$.*

Proof. Let a sequence $(G_j)_{j=1}^\infty$ of cyclically connected subgraphs of G' and an injective function w be chosen according to Lemma C.4.2. Since G' is a finite graph, there is an index j_0 such that $G'' := G_{j_0} = G_j$ for all $j > j_0$. Assume that $G'' \neq G'$. For $u \in V^{G'} \setminus V^{G''}$ and $v \in V^{G''}$ let $P(u)$ be a chain from v to u. Writing $P(u) = (v = \widetilde{v}_0, \ldots, \widetilde{v}_k = u)$, the number $\widetilde{k} := \max\{\iota \in \{0, \ldots, k\} : \widetilde{v}_\iota \in G''\}$ is well-defined and satisfies $\widetilde{k} < k$. Therefore $\widetilde{P}(u) = (\widetilde{v}_{\widetilde{k}}, \ldots, \widetilde{v}_k = u)$ is a well-defined chain whose first vertex $v(P(u)) := \widetilde{v}_{\widetilde{k}}$ belongs to $V^{G''}$, whereas all other vertices of $\widetilde{P}(u)$ are in $V^{G'} \setminus V^{G''}$. Let $P_1(u)$ and $P_2(u)$ be two such chains (with possibly distinct v) and assume that $v(P_1(u)) \neq v(P_2(u))$. Then $W = (\widetilde{P}_1(u), \widetilde{P}_2(u)^-)$ is a walk from $v(P_1(u))$ to $v(P_2(u))$ with its interior vertices belonging to $V^{G'} \setminus V^{G''}$. We know that there is a chain $P \subset W$ from $v(P_1(u))$ to $v(P_2(u))$. For $\iota = 1, 2$, the first edge $e_{\iota,1}$ of $\widetilde{P}_\iota(u)$ has one end $v(P_1(u))$ in $V^{G''}$ and the other end in $V^{G'} \setminus V^{G''}$, which implies that $e_{1,1} \neq e_{2,1}$. Hence $e_{1,1}$ and $e_{2,1}$ are edges of P, which therefore must have at least one interior vertex. But by construction of the G_j, such an interior vertex would belong to G_{j_0+1}, contradicting $G_{j_0} = G_{j_0+1}$. Therefore the vertex $v(P(u))$ is uniquely determined by u, and we write $v(u) := v(P(u))$.

Since G' is cyclically connected, Lemma C.3.5 gives a closed trail W with vertices u and $v(u)$ such that $W = (P_1(u), P_2(u)^-)$, where $P_1(u)$ and $P_2(u)$ are chains from $v(u)$ to u without common edges. Then $\widetilde{W} = (\widetilde{P}_1(u), \widetilde{P}_2(u)^-)$ is a closed trail whose only vertex in G'' is $v(u)$ and which passes through u.

Now choose any $u_0 \in V^{G'} \setminus V^{G''}$ and let $\widetilde{V}$ be the set of all $u \in V^{G'} \setminus V^{G''}$ with $v(u) = v(u_0)$. Then each $u \in \widetilde{V}$ is cyclically connected in $\widetilde{G} := G'|_{\widetilde{V} \cup \{v(u_0)\}}$ to $v(u_0)$, which means that $\widetilde{G}$ is cyclically connected. Since G is a quasi-tree, $\widetilde{G}$ must therefore have at least two vertices in common with its complement in G. Hence there is $u_1 \in \widetilde{V}$ such that u_1 is a vertex of the complement of $\widetilde{G}$. Then there is an edge e of G which is not an edge of $\widetilde{G}$ and which is incident with u_1. Let u_2 be the other end of e. Then $u_2 \notin \widetilde{V}$ by definition of $\widetilde{G}$. From $\widetilde{V} \subset V^{G'} \setminus V^{G''} \subset V^{G'} \setminus \{v_0, \ldots, v_n\}$ it follows that u_1 is not a vertex of the complement of G' in G, which implies that e is not an edge of this complement. Therefore e is an edge of G', which gives $u_2 \in V^{G'}$. Choose a chain $\widetilde{P}(u_1)$ from $v(u_1) = v(u_0)$ to u_1 whose vertices other than $v(u_1)$ belong to $V^{G'} \setminus V^{G''}$ and consider the chain $P = (\widetilde{P}(u_1), (u_1, e, u_2))$. Since $u_2 \notin \widetilde{V}$, it follows that $u_2 \in V^{G''}$ and P would be a chain from $v(u_0)$ to u_2 with u_1 and all other interior vertices (if any) belonging to $V^{G'} \setminus V^{G''}$. But since $G'' = G_{j_0} = G_{j_0+1}$, this contradicts the construction of G_{j_0+1} in the penultimate paragraph of the proof of Lemma C.4.2. This contradiction shows that $G'' = G'$.

We have therefore constructed an injective function $w : G' \to [0, \infty)$ with $0 = w(v_0) < w(v_i)$ for $i = 1, \ldots, r$. Recall that a quasi-tree does not have loops. Hence for each edge e of G' there are unique indices $\tau_{e,1}$ and $\tau_{e,2}$ in $\{0, \ldots, r\}$ such that the ends of e are $v_{\tau_{e,1}}$ and $v_{\tau_{e,2}}$ and such that $w(v_{\tau_{e,1}}) < w(v_{\tau_{e,2}})$. Then we choose an indexing of the edges with respect to the lexicographic order of the pairs $(w(v_{\tau_{e,1}}), w(v_{\tau_{e,2}}))$. For definiteness, letting $g' := \#E^{G'}$, we can (uniquely) index the edges of G' as e_j $(j = 1, \ldots, g')$ such that $w(v_{\tau_{e_{j+1},1}}) \geq w(v_{\tau_{e_j,1}})$ and $w(v_{\tau_{e_{j+1},2}}) > w(v_{\tau_{e_j,2}})$ if $w(v_{\tau_{e_{j+1},1}}) = w(v_{\tau_{e_j,1}})$ $(j = 1, \ldots, g' - 1)$. The edges e of G' will be directed from $v_{\tau_{e,2}}$ to $v_{\tau_{e,1}}$. This means that $v_-(e) = v_{\tau_{e,2}}$ and $v_+(e) = v_{\tau_{e,1}}$.

Finally, we are going to verify that properties 1, 2 and 3 are satisfied for this choice of the indexing of the edges of G'. Since $w(v_0) < w(v_i)$ for all $i \in \{1, \ldots, r\}$, it is clear that each edge which is incident with v_0 is an incoming edge, which shows that $I^-(v_0) = \emptyset$. On the other hand, let $k \in \{1, \ldots, r\}$. Since G_0 is the null graph and $G' = G_{j_0}$ for some $j_0 \in \mathbb{N}$, it follows that there is $j \in \mathbb{N}$ such that $v_k \in V^{G_j} \setminus V^{G_{j-1}}$. By property (ii) in Lemma C.4.2 there are $i \in \{0, \ldots, r\}$ and an edge e of G' such that v_k and v_i are ends of e and $w(v_k) > w(v_i)$. In particular, the edge e is outgoing from v_k, and therefore $I^-(v_k) \neq \emptyset$. Similarly, statement 2 follows from property (iii) in Lemma C.4.2. To prove statement 3, let $j \in I_k^-$ and $i \in I_k^+$. Then $\tau_{e_j,2} = k = \tau_{e_i,1}$. Therefore $w(v_{e_j,1}) < w(v_{e_j,2}) = w(v_k) = w(v_{e_i,1})$, which gives $j < i$. □

Let G be a graph and let

$$\mathcal{C}_{\mathcal{G}} := \{G_v : v \in V^G,\ C(v) \neq \emptyset\}$$

be the set of maximal cyclically connected subgraphs of G. If $\mathcal{C}_{\mathcal{G}} \neq \emptyset$, then the graph $G^{\mathcal{C}_0}$ is defined for any nonempty subset $\mathcal{C}_0$ of $\mathcal{C}_{\mathcal{G}}$ by contracting each maximal cyclically connected subgraph in $\mathcal{C}_0$ into a vertex, i.e., writing $\mathcal{C}_0 =: \{\Gamma_1, \ldots, \Gamma_r\}$, we identify each Γ_j as a new vertex $\widetilde{v}_j$ $(j = 1, \ldots, r)$ and put

$$V^{G^{\mathcal{C}_0}} = \left(V^G \setminus \bigcup_{j=1}^{r} V^{\Gamma_j}\right) \cup \{\widetilde{v}_1, \ldots, \widetilde{v}_r\}, \quad E^{G^{\mathcal{C}_0}} = \left(E^G \setminus \bigcup_{j=1}^{r} E^{\Gamma_j}\right). \tag{C.4.1}$$

We still have to define the incidence relation in $G^{\mathcal{C}_0}$. Let $e \in E^{G^{\mathcal{C}_0}}$ and let v be a vertex of G which is incident with e. If $v \in \left(V^G \setminus \bigcup_{j=1}^{r} V^{\Gamma_j}\right)$, then this end of e remains v in $G^{\mathcal{C}_0}$. If however $v \in V^{\Gamma_j}$ for some $j \in \{1, \ldots, r\}$, then the end v of e becomes $\widetilde{v}_j$ in $G^{\mathcal{C}_0}$. Since a vertex in G cannot belong to two different maximal cyclically connected subgraphs, the incidence relation is well-defined. Note that $\mathcal{C}_G = \{G\}$ and $G^{\mathcal{C}_G}$ is an isolated vertex when G is cyclically connected. The above construction can formally be extended to the case $\mathcal{C}_0 = \emptyset$, in which case $G^{\emptyset} = G$.

Proposition C.4.4. *Let G be a connected graph with at least one edge which is neither a tree nor cyclically connected. Let $\emptyset \neq \mathcal{C}_0 \subset \mathcal{C}_G$. Then $G^{\mathcal{C}_0}$ is a connected graph with at least one edge, $G^{\mathcal{C}_0}$ is not cyclically connected, and $\mathcal{C}_{\mathcal{G}} \setminus \mathcal{C}_0$ is the set of maximal cyclically connected subgraphs of $G^{\mathcal{C}_0}$.*

Proof. First consider the case $\mathcal{C}_0 = \{\Gamma_1\}$. Assume there is a cycle in $G^{\Gamma_1} := G^{\{\Gamma_1\}}$ which passes through $\widetilde{v}_1$. Such a cycle is represented as $(\widetilde{v}_1, \widetilde{e}_1, v_2, \ldots, v_3, \widetilde{e}_2, \widetilde{v}_1)$, where $(v_2, \ldots, v_3)$ is a chain in G with all vertices in $V^G \setminus V^{\Gamma_1}$. In G, the cycle with the same edges is a path $P_1 = (v_1, \widetilde{e}_1, v_2, \ldots, v_3, \widetilde{e}_2, v_4)$ with $v_1, v_4 \in V^{\Gamma_1}$. If $v_1 = v_4$, then P_1 is a cycle. If $v_1 \neq v_4$, then there is a path $P_2 = (v_4, \ldots, v_1)$ in Γ_1 since Γ_1 is connected. Hence (P_1, P_2) is a cycle in G. In either case, there would be a cycle in G with and edge $\widetilde{e}_1 \notin V^{\Gamma_1}$ but with one of its ends $v_1 \in V^{\Gamma_1}$. But this is impossible since Γ_1 is a maximal cyclically connected subgraph of G.

Therefore $\widetilde{v}_1$ is not cyclically connected to any other vertex in G^{Γ_1}, which implies that G^{Γ_1} is not cyclically connected. Furthermore, the cycles in G^{Γ_1} are exactly the cycles in G which are not cycles in Γ_1, and it follows that $\mathcal{C} \setminus \{\Gamma_1\}$ is the set of maximal cyclically connected subgraphs of G^{Γ_1}. In a similar way one can show that G^{Γ_1} is connected. Alternatively, $g^{G^{\Gamma_1}} = g^G - g^{\Gamma_1}$, $\mu^{G^{\Gamma_1}} = \mu^G - \mu^{\Gamma_1}$, $p^{G^{\Gamma_1}} = p^G - p^{\Gamma_1} + 1$, $\kappa^G = 1$, $\kappa^{\Gamma_1} = 1$ and Lemma C.2.3 give $\kappa^{G^{\Gamma_1}} = 1$.

Finally, since $\Gamma_1 \neq G$, there is $v_0 \in V^G \setminus V^{\Gamma_1}$, and since G^{Γ_1} is connected, there is a path from v_0 to $\widetilde{v}_1$ in G^{Γ_1}. This shows that G^{Γ_1} has at least one edge.

A recursive application of the case $r = 1$ gives the result for general r. □

Corollary C.4.5. *Let G be a connected graph which is not the null graph. Then $G^{\mathcal{C}_G}$ is not the null graph, is connected, and has no cycles.*

Proof. We have already noticed above that $G^{\mathcal{C}_G} = G$ when G has no cycles and that $G^{\mathcal{C}_G}$ is an isolated vertex if G is cyclically connected. In all other cases, the result follows from Proposition C.4.4 with $\mathcal{C}_0 = \mathcal{C}_G$. □

Let $\mathcal{C}_G = \{\Gamma_1, \ldots, \Gamma_r\}$ be the set of maximal cyclically connected subgraphs of a connected graph G. For $j = 1, \ldots, r$ let $\widehat{V}_j$ be the set of all vertices of Γ_j which are also vertices of the complements of Γ_j in G and let

$$\widehat{V}^G = \left(V^G \setminus \bigcup_{j=1}^{r} V^{\Gamma_j} \right) \cup \bigcup_{j=1}^{r} \widehat{V}_j. \tag{C.4.2}$$

If G is a neither the null graph nor cyclically connected, then $\widehat{V}^G \neq \emptyset$. Choosing and fixing any vertex $v \in \widehat{V}^G$, G will be called a rooted graph with root v.

Lemma C.4.6. *The set of all edges of any rooted quasi-tree G can be oriented and indexed in such a way that for each vertex v of G the following properties hold:*

1. $I^-(v) = \emptyset$ *if and only if v is the root;*
2. $I^+(v) = \emptyset$ *if and only if v is a pendant vertex but not the root;*
3. *if $j \in I^-(v)$ and $i \in I^+(v)$, then $j < i$.*

Proof. By Corollary C.4.5, $T := G^{\mathcal{C}_G}$ is a tree. Let $\mathcal{C}_G =: \{\Gamma_1, \ldots, \Gamma_r\}$. If the root of G is a vertex of some Γ_{j_0}, we set $v_0^* := \widetilde{v}_{j_0}$. Otherwise, let v_0^* be the root of G. Then T is a rooted tree with root v_0^*. The edges of T will be oriented according to Lemma C.2.6. Also let w_T be as in Lemma C.2.6. For each edge e in T let $(w_T(v_-(e)), 0)$ be the index of e. In view of statements 2 and 3 in Lemma C.2.6, indices of distinct edges are distinct.

For each $j = 1, \ldots, r$, the edges of Γ_j are indexed and directed according to Lemma C.4.3, and if the index of such an edge of Γ_j is i, then we give it the index $(w_T(\widetilde{v}_j), i)$ as an edge of G. In this way we have given each edge in G a unique index and orientation, and the edges in G will be ordered with respect to the lexicographic ordering of their indices. Of course, the indices can now be identified in an order preserving manner with the numbers $1, \ldots, g^G$, but for the purpose of this proof, we will use the above notation.

The vertices of Γ_j ($j = 1, \ldots, r$) will be indexed as $v_{j,k}$ ($k = 0, \ldots, r_j$), where the indices k are as in Lemma C.4.3. Here $v_{j,k}$ for $k = 0, \ldots, n_j$ are the common vertices of Γ_j and its complement. The vertices $v_{j,0}$ play a particular role, and we are now going to identify them. If the root of G belongs to Γ_j, we put $v_{j,0} := v_0$. If the root of G does not belong to Γ_j, then $\widetilde{v}_j$ is not the root of T, and by Lemma C.2.6 there is a unique outgoing edge from $\widetilde{v}_j$ in T, and its entrance in G is a vertex of Γ_j and will be the vertex $v_{j,0}$ of G.

We are now in a position to prove the statements of this lemma. To this end, we partition the set of vertices of G into five classes, and we will consider each of these classes in turn.

The first class is the set of all vertices $v_{j,k}$ with $j = 1, \ldots, r$, $k = n_j+1, \ldots, r_j$. In this case, all edges of G which are incident with such a vertex of G are also edges of Γ_j, and the statements follow from Lemma C.4.3 if we observe that a cyclically connected subgraph has no pendant vertices.

The second class only consists of the root. In case the root belongs to some Γ_j, it is $v_{j,0}$. But in any case, an edge incident with the root would be an incoming edge to v_0 by Lemma C.2.6 or Lemma C.4.3.

The third class consists of all vertices $v_{j,0}$ which are not the root. All edges which are incident with such $v_{j,0}$ and which are edges of Γ_j are incoming edges by statement 1 of Lemma C.4.3, their indices satisfy $(w_T(\widetilde{v}_j), \imath) > (w_T(\widetilde{v}_j), 0)$, and there is at least one such edge. Any remaining incoming edges would be edges of T and would therefore have indices $(w_T(v^*), 0) > (w_T(\widetilde{v}_j), 0)$. By choice of $v_{j,0}$, the only outgoing edge from it is the unique outgoing edge from $\widetilde{v}_j$ in T, which has index $(w_T(\widetilde{v}_j), 0)$. Hence the statements of the lemma also hold for this class of vertices.

The fourth class consists of all vertices $v_{j,k}$ with $j = 1, \ldots, r$ and $k = 1, \ldots, n_j$. By statement 1 of Lemma C.4.3 each such vertex has at least one outgoing edge in Γ_j. Furthermore, it follows from Lemma C.4.3 that there is a positive real number $i_{j,k}$ such that the index $(w_T(\widetilde{v}_j), \iota)$ of any edge of Γ_j which is incoming into $v_{j,k}$ is larger than $(w_T(\widetilde{v}_j), i_{j,k})$, whereas the index $(w_T(\widetilde{v}_j), \iota)$ of any edge of Γ_j which is outgoing from $v_{j,k}$ is smaller than $(w_T(\widetilde{v}_j), i_{j,k})$. As we have seen above when choosing $v_{j,0}$, there are no edges of T outgoing from $v_{j,k}$. But since $v_{j,k}$ has at least one incident edge in T, $v_{j,k}$ has at least one incoming edge, and each incoming edge in T has index $(w_T(v^*), 0)$ with $w_T(v^*) > w_T(\widetilde{v}_j)$, which implies that its index satisfies $(w_T(v^*), 0) > w_T(\widetilde{v}_j), i_{j,k})$. This shows that the statement of the lemma also holds for the fourth class.

The fifth class consists of all remaining vertices, that is, all vertices of G which are not vertices of any Γ_j $(j = 1, \ldots, r)$ and not the root. It is clear that all edges of G which are incident with such a vertex v are edges of T. Observe that this class includes all pendant vertices of G since a pendant vertex cannot be a vertex of any cycle. Each vertex v of T in this fifth class has exactly one outgoing vertex by Lemma C.2.6 and is therefore a pendant vertex if and only if $I^+(v) = \emptyset$. Finally, statement 3 for vertices in this fifth class which are not pendant vertices follows from statement 2 of Lemma C.2.6. □

Definition C.4.7. A connected graph with at least one edge is said to be a generalized quasi-tree if it is neither a tree nor cyclically connected and if each of its cyclically connected subgraphs with a vertex in $\widehat{V}^G$ has at least two vertices in common with the complement of this subgraph.

Lemma C.4.8. *The set of all edges of any rooted generalized quasi-tree G can be oriented and indexed in such a way that for each vertex v of G the following properties hold:*

1. $I^-(v) = \emptyset$ *if and only if v is the root;*
2. $I^+(v) = \emptyset$ *if and only if v is a pendant vertex but not the root;*
3. *if v is neither a pendant vertex nor the root of G, then* $\min I^-(v) < \min I^+(v)$ *and* $\max I^-(v) < \max I^+(v)$.

Proof. We are going to prove the statement by induction on the cyclomatic number μ. Since $\mu^G = 0$ implies that G is a tree, there are no generalized quasi-trees with $\mu^G = 0$, so that the statement is trivially true for $\mu = 0$. Now let $\mu > 0$ and assume that the statement is true for all generalized quasi-trees G' with $\mu^{G'} < \mu$. Let G be a generalized quasi-tree with $\mu^G = \mu$. If G is a quasi-tree, then the statements of this lemma are true by Lemma C.4.6. It remains to consider the case that G is not a quasi-tree, and in this case there is a cyclically connected subgraph $\widetilde{G}$ of G which has only one vertex w in common with its complement in G. Then $V^{\widetilde{G}} \cap \widehat{V}^G = \emptyset$. By Lemma C.4.1, $\widetilde{G}$ is a subgraph of a maximally connected subgraph Γ_1 of G. Observing $V^{\Gamma_1} \cap \widehat{V}^G \supset \widehat{V}_1 \neq \emptyset$ it follows that $V^{\widetilde{G}} \subsetneqq V^{\Gamma_1}$.

We will now cut the graph G at w into two vertices to arrive at a graph G' as follows. Each edge of G which is not incident with w will have the same ends in G'. Since $w \notin \widehat{V}^G$, it is clear that $w \notin \widehat{V}_1$ and therefore each edge of G incident with w must be an edge of Γ_1. Since $\widetilde{G}$ is cyclically connected, there is a closed trail $(w, \widetilde{e}_1, \ldots, \widetilde{e}_2, w)$ in $\widetilde{G}$, possibly $(w, \widetilde{e}_1, w)$, which will be identified with a trail $\widetilde{P} = (w_1, \widetilde{e}_1, \ldots, \widetilde{e}_2, w_2)$ in G', thus determining the ends of $\widetilde{e}_1$ and $\widetilde{e}_2$ in G'. If $\widetilde{G}$ has any other edges with end w, this end will become one of w_1 or w_2 in G' (this choice is arbitrary, for definiteness we may choose w_1). Now let $v_1 \in V^{\Gamma_1} \setminus V^{\widetilde{G}}$. Since Γ_1 is cyclically connected, there is a closed trail $P = (w, \widehat{e}_1, \ldots, v_1, \ldots \widehat{e}_2, w)$ in Γ_1. If this trail would traverse an edge of $\widetilde{G}$, then there would be a subtrail in $\widetilde{G}$ entering and exiting at the only common vertex w of $\widetilde{G}$ and $\widehat{G}$. After removing any such closed subtrail of P we may therefore assume that P is a trail in $\widehat{G}$. Since P is a closed trail with $v_1 \neq w$, it follows that $\widehat{e}_1$ and $\widehat{e}_2$ are edges of $\widehat{G}$ with $\widehat{e}_1 \neq \widehat{e}_2$. Then we identify P with the trail $\widehat{P} = (w_1, \widehat{e}_1, \ldots, v_1, \ldots, \widehat{e}_2, w_2)$ in G', thus determining the ends of $\widehat{e}_1$ and $\widehat{e}_2$ in G'. If $\widehat{G}$ has any other edges with end w, this end will become one of w_1 or w_2 in G'. This completes the construction of G'.

To show that G' is connected choose $v \in G' \setminus \{w_1, w_2\}$. Since G is connected, there is a path from v to w in G', which corresponds to a path from v to w_1 or from v to w_2 in G'. But we have already seen that there is a trail $\widetilde{P}$ from w_1 to w_2

in G'. Therefore G' is indeed connected. Since $g^{G'} = g^G$, $p^{G'} = p^G + 1$ and both G and G' are connected, it follows from Lemma C.2.3 that $\mu^{G'} = \mu^G - 1$.

Next we are going to show that G' is a generalized quasi-tree. To this end let Γ_1' be the subgraph of G' which is generated by all edges of Γ_1, that is, Γ_1' is the graph obtained from Γ_1 by cutting the vertex w in the manner described above. Let $\Gamma_2, \ldots, \Gamma_r$ be the other maximal cyclically connected subgraphs of G, if any. We are going to show that $\Gamma_1', \Gamma_2, \ldots, \Gamma_r$ are the maximal cyclically connected subgraphs of G'. Let P be a closed trail in G'. If P traverses w_1 or w_2, we replace all occurrences of w_1 and w_2 with w to arrive at a closed trail P_0 in G. But this trail must lie in one of $\Gamma_1, \ldots, \Gamma_r$. Since for $j = 2, \ldots r$ any closed trail in Γ_j is the same for G and G', it is clear that $\Gamma_2, \ldots, \Gamma_r$ are maximal cyclically connected in G'. Furthermore, any other maximal cyclically connected subgraph of G' must be a subgraph of Γ_1'. Finally, we are going to show that each vertex $v \in \Gamma_1' \setminus \{w_1\}$ is cyclically connected to w_1, which would show that Γ_1' is maximal cyclically connected. Since $\widetilde{P}$ and $\widehat{P}$ are trails from w_1 to w_2 in $\widetilde{G}$ and $\widehat{G}$, respectively, they have no common edges. Hence $(\widetilde{P}, \widehat{P}^-)$ is a closed trail, and therefore w_1 and w_2 are cyclically connected. It remains to show that each vertex v of Γ_1' which is different from w_1 and w_2 is cyclically connected to one of w_1 and w_2. If v is a vertex in $\widetilde{G}$, choose a closed trail through v and w. This becomes a trail P_1 from w_3 to w_4 in Γ_1' with $w_3, w_4 \in \{w_1, w_2\}$. If $w_3 = w_4$, then P_1 is a closed trail through v and w_1. If $w_3 \neq w_4$, we may assume that $w_3 = w_2$ and $w_4 = w_1$. In this case, the trail $(P_1, \widehat{P})$ is a closed trail through v, w_1 and w_2. The proof when v is not a vertex of $\widetilde{G}$ is similar. Since $w \notin \widehat{V}^G$ it is clear that $\widehat{V}^{G'} = \widehat{V}^G$.

By induction hypothesis, we may choose an indexing of the edges of the generalized quasi-tree G' with $\mu^{G'} = \mu - 1$ such that the statements of this lemma hold for G'. Hence the statements also hold for all vertices v of G which are different from w. From $\min I^-(w_j) < \min I^+(w_j)$ and $\max I^-(w_j) < \max I^+(w_j)$ $(j = 1, 2)$ it follows that

$$\min I^-(w) = \min(I^-(w_1) \cup I^-(w_2)) < \min(I^+(w_1) \cup I^+(w_2)) = \min I^+(w),$$
$$\max I^-(w) = \max(I^-(w_1) \cup I^-(w_2)) < \max(I^+(w_1) \cup I^+(w_2)) = \max I^+(w).$$

Therefore property 3 also holds for w. □

Let G be a digraph and denote by $V^\times$ the set of all vertices $v \in V^{\text{int}}$ such that $I^-(v) \neq \emptyset$ and $I^+(v) \neq \emptyset$ and denote by $V_0^\times$ the set of all $v \in V^\times$ satisfying $\min I^-(v) < \min I^+(v)$ and $\max I^-(v) < \max I^+(v)$.

Lemma C.4.9. *Let G be a cyclically connected graph and let w be a vertex of G. Then the edges of G can be oriented and indexed in such a way that $v \in V_0^\times$ for all $v \in V^G \setminus \{w\}$.*

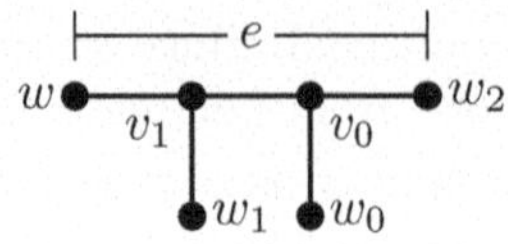

Figure C.1

Proof. Let e be any edge of G with is incident with w. We denote the other vertex of e by w_2, where $w_2 = w$ is allowed. We augment G to a graph $\widetilde{G}$ by adding two vertices v_0 and v_1 in the interior of the edge e and then two new vertices w_0 and w_1 and edges from v_0 to w_0 and v_1 to w_1, see Figure C.1. It is clear that G augmented by v_0 and v_1 is the only maximal cyclically connected subgraph of $\widetilde{G}$ and that $\widehat{V}^{\widetilde{G}} = \{v_0, v_1, w_0, w_1\}$. Clearly, if a cyclically connected subgraph of $\widetilde{G}$ traverses one of v_0 or v_1, then it traverses both. Hence $\widetilde{G}$ is a generalized quasi-tree with root at v_0. By Lemma C.4.8 the edges of $\widetilde{G}$ can be indexed and oriented in such a way that $v \in \widetilde{V}_0^{\times}$ for all $v \in V^{\widetilde{G}} \setminus \widehat{V}^{\widetilde{G}}$. In G, the edge e will be given the index and orientation of the edge from w_2 to v_0 (which is directed from w_2 to v_0 since v_0 has no outgoing edges by Lemma C.4.8, and therefore e is directed from w_2 to w.) All other edges of G will be given the indices and orientations they have in $\widetilde{G}$. For each $v \in V^G \setminus \{w\}$ it follows that $I^-(v) = \widetilde{I}^-(v)$ and $I^+(v) = \widetilde{I}^+(v)$ and therefore $v \in V_0^{\times}$. □

We have defined several special classes of connected graphs, namely trees, cyclically connected graphs, quasi-trees and generalized quasi-trees. Clearly, there are trees and cyclically connected graphs, and these two classes of graphs are disjoint. In the next proposition we will show that the classes of quasi-trees and generalized quasi-trees cannot be expressed in terms of the other classes. Recall that it is clear by definition that generalized quasi-trees are neither trees nor cyclically connected and that each quasi-tree is a generalized quasi-tree.

Proposition C.4.10. 1. *There are quasi-trees.*

2. *There are generalized quasi-trees which are not quasi-trees.*

3. *There are connected graphs which are neither trees nor cyclically connected nor generalized quasi-trees.*

Proof. We are going to present a simple example for each of the three statements. In each of these cases we will denote the graph by G. The edges of G will be denoted by numbers, and its vertices by bold numbers. For a subset E' of the set E of the edges of G, we let $G(E')$ denote the subgraph of G whose edges are the edges in E' and whose vertices are the ends of the edges in the set E'.

1. The graph G depicted below is a quasi-tree.

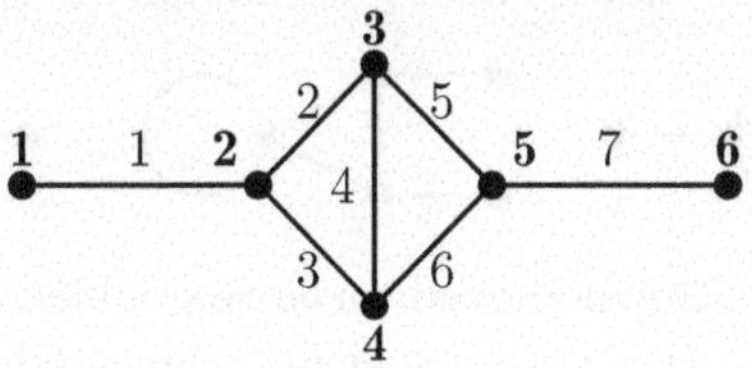

Indeed, the cyclically connected subgraphs if G are $G_j = G(E_j)$ $(j = 1, \ldots, 4)$ with $E_1 = \{2, 3, 4\}$, $E_2 = \{4, 5, 6\}$, $E_3 = \{2, 3, 5, 6\}$, $E_4 = \{2, 3, 4, 5, 6\}$. It is easy to see that the common vertices of G_j and its complement in G are $\{\mathbf{2}, \mathbf{3}, \mathbf{4}\}$ for $j = 1$, $\{\mathbf{3}, \mathbf{4}, \mathbf{5}\}$ for $j = 2$, $\{\mathbf{2}, \mathbf{3}, \mathbf{4}, \mathbf{5}\}$ for $j = 3$, and $\{\mathbf{2}, \mathbf{5}\}$ for $j = 4$. Since each of these sets contains at least two vertices, the graph G is a quasi-tree.

2. The graph G depicted below is a generalized quasi-tree but not a quasi-tree.

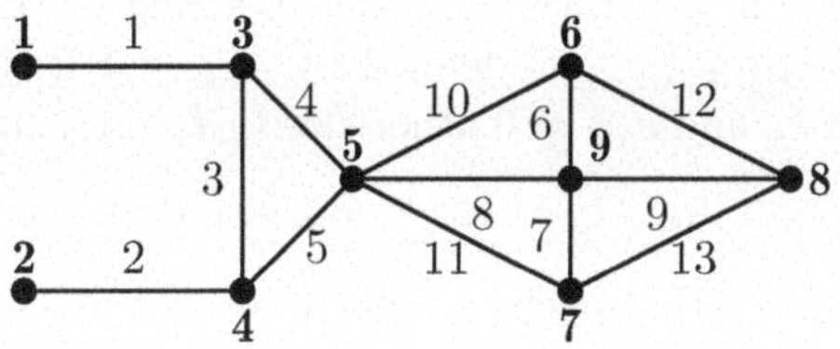

Indeed, the graph $G(E_1)$ with $E_1 = \{6, \ldots, 13\}$ is a cyclically connected graph, but **5** is the only common vertex of $G(E_1)$ and its complement in G. Hence G is not a quasi-tree. On the other hand, $G(E_2)$ with $E_2 = \{3, \ldots, 13\}$ is the only maximal cyclically connected subgraph of G, and **3** and **4** are the common vertices of $G(E_2)$ and its complement in G. Therefore, $\widehat{V} = \{\mathbf{1}, \mathbf{2}, \mathbf{3}, \mathbf{4}\}$. But any cyclically connected subgraph G' of G which has one of these vertices must contain the edge 3, and hence the vertices **3** and **4** are vertices of G'. This shows that G is a generalized quasi-tree.

3. The graph G depicted below is not a quasi-tree.

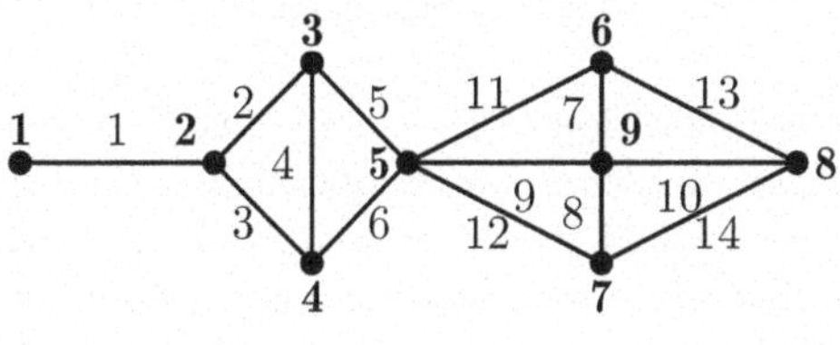

Indeed, the subgraph $G(E_1)$ of the graph G with $E_1 = \{2, \ldots, 14\}$ is the only maximal cyclically connected subgraph of G. Clearly, $\widehat{V} = \{\mathbf{1}, \mathbf{2}\}$ and **2** is the only vertex of $G(E_1)$ in $\widehat{V}$. □

In the above proposition we have chosen the examples in such a way that the graphs are simple plane graphs with interior vertices v satisfying $d(v) \geq 3$.

If these assumptions are relaxed, then even simpler graphs can be found for the three statements of Proposition C.4.10. For example, the graphs

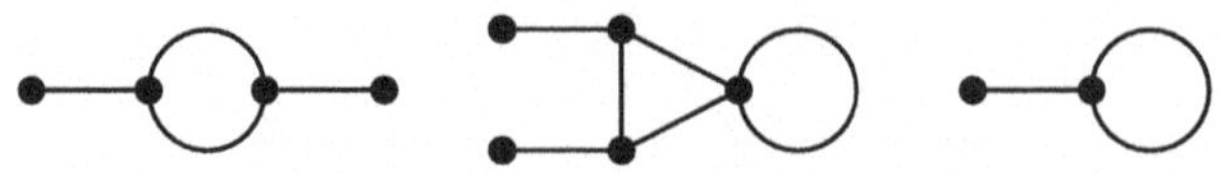

are, in turn, a quasi-tree, a generalized quasi-tree which is not a quasi-tree, and a graph which is neither a tree nor a cyclically connected graph nor a generalized quasi-tree.

C.5 Adjacency matrices and incidence matrices

In this section, G is a simple graph without isolated vertices. Therefore $p \geq 2$ and $g \geq 1$. Two vertices of G are called adjacent if they are the two ends of an edge of G. Let $v_1, \dots, v_p$ be an arbitrary but fixed indexing of the vertices of G. The *adjacency matrix* $A = (a_{j,k})_{j,k=1}^{p}$ is defined by $a_{j,k} = 1$ if there is an edge in G whose ends are v_j and v_k and $a_{j,k} = 0$ otherwise ($j, k = 1, \dots, p$). We observe that $d_j := d(v_j) = \sum_{k=1}^{p} a_{j,k} = \sum_{k=1}^{p} a_{k,j}$ ($j = 1, \dots, p$). Note that A is a symmetric matrix. Conversely, it is easy to see that each symmetric $p \times p$ matrix $A = (a_{j,k})_{j,k=1}^{p}$ with $a_{j,j} = 0$, $a_{j,k} \in \{0, 1\}$ and $\sum_{k=1}^{p} a_{j,k} > 0$ ($j, k = 1, \dots, p$) is the adjacency matrix of a simple graph G without isolated vertices. Let $T = \operatorname{diag}(d_1, \dots, d_p)$. Then $\widetilde{A} := T^{-\frac{1}{2}} A T^{-\frac{1}{2}}$ is called the weighted adjacency matrix.

Let $I_A := \{(j,k) : 1 \leq j < k \leq p,\ a_{k,j} = 1\}$ and let $\varepsilon \in \{-1, 1\}$ and $\varepsilon_{j,k}, \varepsilon_{k,j} \in \{-1, 1\}$ with $\varepsilon_{j,k}\varepsilon_{k,j} = \varepsilon$ for $(j,k) \in I_A$. Clearly, $\# I_A = g$. Define the $p \times g$ matrix U_ε by

$$u_{i,(j,k)} = \begin{cases} \varepsilon_{j,k} & \text{if } j = i,\ (j,k) \in I_A \\ \varepsilon_{k,j} & \text{if } k = i,\ (j,k) \in I_A \\ 0 & \text{otherwise.} \end{cases} \tag{C.5.1}$$

Up to the order of the enumeration of I_A, the matrix U_ε is well-defined. For $\varepsilon = 1$, the matrix U_1 is called the *incidence matrix* of G. For $\varepsilon = -1$, the matrix $U_\varepsilon = U_{-1}$ can be considered as the signed incidence matrix for a digraph structure on G, where $\varepsilon_{j,k} < 0$ means that the corresponding edge is directed from v_j to v_k.

A permutation of the indices of the vertices of G is equivalent to a permutation of the rows of U_ε and a simultaneous permutation of the rows and columns of A, T and $\widetilde{A}$. Since all results in this section are invariant under these permutations, we may assume particular enumerations of the vertices and edges of G if it is convenient to do so.

Proposition C.5.1. *For $\varepsilon \in \{-1, 1\}$, the identities $U_\varepsilon U_\varepsilon^* = T + \varepsilon A$ hold.*

Proof. Since the matrix $U_\varepsilon U_\varepsilon^*$ is symmetric, we only have to evaluate the entries

$$(U_\varepsilon U_\varepsilon^*)_{r,s} = \sum_{(j,k)\in I_A} u_{r,(j,k)} u_{s,(j,k)}$$

with $1 \le r \le s \le p$. For $r = s$ it is clear that

$$(U_\varepsilon U_\varepsilon^*)_{r,s} = \sum_{(j,k)\in I_A} (u_{r,(j,k)})^2 = \sum_{t=1}^{p} a_{r,t} = d_r.$$

For $r < s$ with $(r,s) \in I_A$, the sum contains exactly one nonzero element, which is $(U_\varepsilon U_\varepsilon^*)_{r,s} = \varepsilon_{r,s}\varepsilon_{s,r} = \varepsilon = \varepsilon a_{r,s}$. For $r < s$ with $(r,s) \notin I_A$, all terms in the sum are 0, which shows that $(U_\varepsilon U_\varepsilon^*)_{r,s} = 0 = a_{r,s}$ in this case. □

Definition C.5.2. A graph G is called bipartite if V^G is the disjoint union of two sets V_1 and V_2 such that each edge of G has one end belonging to V_1 and the other end belonging to V_2.

The length of a walk W in a graph G, denoted by $\operatorname{len}(W)$, is the number of edges traversed in that walk and therefore equals the sum of the components of $|m|(W)$.

Let G be a connected graph and choose any vertex v_0 of G as the root of G. For each $v \in V \setminus \{v_0\}$ let $\operatorname{Dist}_{v_0}(v)$ be the set of all numbers $\operatorname{len}(P)$, where P runs through all chains from v_0 to v. For convenience we put $\operatorname{Dist}_{v_0}(v_0) = \{0\}$. If G is a simple graph it is clear that $\#\operatorname{Dist}_{v_0}(v) = 1$ for all v if and only if G does not have cycles. The number $\operatorname{dist}_{v_0}(v) := \min \operatorname{Dist}_{v_0}(v)$ is called the *distance* of v from v_0.

Lemma C.5.3. *A graph is bipartite if and only if it has no cycle with an odd number of edges.*

Proof. Clearly, it suffices to consider the components of G separately. Therefore we may assume for this proof that G is a connected digraph which has a root v_0. First assume that G has no cycle with an odd number of edges. Let W be a closed walk in G. If $m(W) \neq 0$, then there are cycles $C_1, \dots, C_k$ such that $m(W) = \sum_{j=1}^{k} m(C_k)$, see Corollary C.2.2. Hence

$$|m|(W) = m(W) + 2m^-(W) = \sum_{j=1}^{k} |m|(C_k) - 2\sum_{j=1}^{k} m^-(C_k) + 2m^-(W),$$

and therefore $\operatorname{len}(W)$ is an even number, which is also true when $m(W) = 0$. Let $v \neq v_0$ be a vertex of G, let P_1 and P_2 be two chains from v_0 to v and put $W = (P_1, P_2^-)$. Then W is a closed walk with $\operatorname{len}(W) = \operatorname{len}(P_1) + \operatorname{len}(P_2)$ and therefore $\operatorname{len}(P_1) - \operatorname{len}(P_2) = \operatorname{len}(W) + 2\operatorname{len}(P_2)$ is even. Hence the set $\operatorname{Dist}_{v_0}(v)$ consists either of odd numbers or of even numbers. Define G_1 to be the set of all

$v \in G$ for which $\mathrm{Dist}_{v_0}(v)$ consists of odd numbers and G_2 to be the set of all $v \in G$ for which $\mathrm{Dist}_{v_0}(v)$ consists of even numbers. Let e be an edge of G with ends v and w. If one of the ends is the root, say $v = v_0$, then $v \in G_2$, whereas $\mathrm{len}(v_0, e, w) = 1$ shows that $w \in G_1$. Now let both v and w be distinct from v_0 and choose chains P_v and P_w from v_0 to v and w, respectively. Then $W = (P_v, (v, e, w), P_w^-)$ is a closed walk, and therefore $\mathrm{len}(P_v) + \mathrm{len}(P_w) = \mathrm{len}(W) - 1$ is an odd number, which implies that one end of e belongs to G_1, whereas the other end belongs to G_2. This shows that G is bipartite.

Now let G have a cycle with an odd number of edges and assume that G is bipartite. Traversing a cycle with an odd number of edges starting at a vertex, say, in G_1, the traversed vertices must be alternatively in G_1 and G_2, and the terminal vertex must therefore be in G_2. But since the terminal vertex is the starting vertex, this is impossible. □

Lemma C.5.4. *All eigenvalues of the matrix* $\widetilde{A}$ *lie in the interval* $[-1, 1]$ *and* 1 *is an eigenvalue of* $\widetilde{A}$. *If* G *is bipartite, then the eigenvalues, with multiplicities, are symmetric with respect to* 0. *If* $g < p$, *then* -1 *is an eigenvalue of* $\widetilde{A}$.

Proof. Since the matrix $\widetilde{A}$ is real and symmetric, its eigenvalues are real. The representations of $T - A$ and $T + A$ in Proposition C.5.1 show that $T - A \geq 0$ and $T + A \geq 0$, and therefore the eigenvalues of $I - \widetilde{A}$ and $I + \widetilde{A}$ are nonnegative. This means that all eigenvalues of the matrix $\widetilde{A}$ lie in the interval $[-1, 1]$. By definition of T it is clear that $(T - A)(1, \ldots, 1)^{\mathsf{T}} = 0$, which proves that 1 is an eigenvalue of $\widetilde{A}$.

If G is bipartite let $v_1, \ldots, v_{p_1}$ be the edges of G_1 and let $v_{p_1+1}, \ldots, v_p$ be the edges of G_2. Then there is a $p_1 \times (p - p_1)$ matrix A_1 such that

$$A = \begin{pmatrix} 0 & A_1 \\ A_1^* & 0 \end{pmatrix},$$

and therefore a $p_1 \times (p - p_1)$ matrix $\widetilde{A}_1$ such that

$$\widetilde{A} = \begin{pmatrix} 0 & \widetilde{A}_1 \\ \widetilde{A}_1^* & 0 \end{pmatrix}.$$

For $\lambda \neq 0$, the Schur factorization (B.4.2) gives

$$\widetilde{A} - \lambda I = \begin{pmatrix} I & -\lambda^{-1}\widetilde{A}_1 \\ 0 & I \end{pmatrix} \begin{pmatrix} \lambda^{-1}\widetilde{A}_1\widetilde{A}_1^* - \lambda I & 0 \\ 0 & -\lambda I \end{pmatrix} \begin{pmatrix} I & 0 \\ -\lambda^{-1}\widetilde{A}_1^* & I \end{pmatrix},$$

which shows that λ is an eigenvalue of $\widetilde{A}$ if and only if λ^2 is an eigenvalue of $\widetilde{A}_1\widetilde{A}_1^*$, with equal multiplicities. This proves that λ is an eigenvalue of $\widetilde{A}$ if and only if $-\lambda$ is an eigenvalue of $\widetilde{A}$, and their multiplicities are equal.

If $g < p$, then the rank of the $p \times g$ matrix U_1 is at most g. It follows that also the rank of $T + A = U_1U_1^*$ is at most g, which means that the $p \times p$ matrix $T + A$ is not invertible. Thus -1 is an eigenvalue of $\widetilde{A}$ in this case. □

Lemma C.5.5. *Let A be irreducible. Then 1 is a simple eigenvalue of $\widetilde{A}$. If G is bipartite, then -1 is a simple eigenvalue of $\widetilde{A}$, and if G is not bipartite, then -1 is not an eigenvalue of $\widetilde{A}$.*

Proof. It is well-known that the rank of $U_\varepsilon U_\varepsilon^* = T+\varepsilon A$ equals the rank of U_ε. Since A is assumed to be irreducible, the (di)graph which describes U_ε is connected. We are going to reduce the sizes of U_ε by deleting rows and columns until we arrive at a case where it is easy to prove the claim. Let us agree to still denote the reduced matrix U_ε by U_ε (it will correspond to subgraph of the (di)graph G) and to call the number $p - \operatorname{rank} U_\varepsilon$ the defect of U_ε.

First we consider the case that G has at least two edges and at least one pendant vertex. We may assume that v_1 is a pendant vertex and that the unique edge $\widetilde{e}$ which is incident with v_1 is represented by the first column in U_ε. Therefore the first row of U_ε has the value 1 or -1 in the first column and the value 0 everywhere else. Since each vertex $v \neq v_0$ is incident with at least one edge which is different from $\widetilde{e}$, all other rows have a nonzero entry outside the first column. It follows that the first row of U_ε is linearly independent from the other rows. Hence removing the vertex v_1 and its incident edge leads to a new connected graph G of reduced size and the same cyclomatic number, and the defect of the (signed) incidence matrix is unchanged.

If the starting graph G was a tree, then all subsequent graphs are trees, and after a finite number of steps we arrive at a graph with two vertices and an edge connecting them. But in this case the final U_ε, which we will denote by $\widetilde{U}_\varepsilon$, is $\widetilde{U}_\varepsilon = (\widetilde{\varepsilon}_1, \widetilde{\varepsilon}_2)^\mathsf{T}$ with $\widetilde{\varepsilon}_1, \widetilde{\varepsilon}_2 \in \{-1, 1\}$. Clearly, the defect of $\widetilde{U}_\varepsilon$ is 1, which shows in this case that 1 and -1 are simple eigenvalues of $\widetilde{A}$ in this case.

If the cyclomatic number μ is positive, then the above procedure stops at a graph with cyclomatic number μ and without pendant vertices. Choosing any cycle in G and removing one edge of the cycle, we arrive at a connected graph with cyclomatic number lowered by 1. Removal of that edge could only increase the defect of U_ε. If $\mu \geq 1$ and the new graph has pendant vertices, we repeat the first procedure and the second procedure and so on until we have arrived at a graph with $\mu = 0$. But then this graph is a tree, and we already know that the corresponding U_ε has defect 1. Hence the original U_{-1} has defect at most 1. By Lemma C.5.4, 1 is an eigenvalue of $\widetilde{A}$, and it follows that the eigenvalue 1 is simple. By Lemma C.5.4, -1 is a simple eigenvalue of $\widetilde{A}$ if the graph G is bipartite.

It remains to consider the case $\varepsilon = 1$ when G is not bipartite. By Lemma C.5.3, the graph G has a cycle C_1 with an odd number of edges. We follow the same procedure as above, except that we do not remove any edge of C_1 and that we stop once we have arrived at a graph $\widehat{G}$ with cyclomatic number $\widehat{\mu} = 1$ and with no pendant vertices. But then $\widehat{G} = C_1$. Indexing the vertices and edges in the order they are traversed, it follows that the corresponding incidence matrix is

$$\widehat{U}_1 = \begin{pmatrix} 1 & 0 & \cdots & \cdots & \cdots & 1 \\ 1 & 1 & 0 & \cdots & \cdots & 0 \\ 0 & 1 & 1 & 0 & \cdots & 0 \\ \vdots & & \ddots & \ddots & \ddots & \vdots \\ 0 & \cdots & \cdots & 1 & 1 & 0 \\ 0 & \cdots & \cdots & 0 & 1 & 1 \end{pmatrix},$$

which is a square matrix with an odd number of rows and columns. Expanding its determinant with respect to the first row shows that $\det \widehat{U}_1$ is the sum of the determinant of a lower triangular matrix and the determinant of an upper triangular matrix, both with all diagonal entries being 1. Thus $\det \widehat{U}_1 = 2$, which shows that $\widehat{U}_1$ and hence also U_1 has defect 0. This proves that -1 is not an eigenvalue of $\widetilde{A}$ when G is not bipartite. □

C.6 T-acyclic matrices

Definition C.6.1. [88] Let $n \in \mathbb{N}$, let T be a tree with $n+1$ vertices $v_1, \dots, v_{n+1}$ and let $A = (a_{j,k})_{j,k=1}^{n+1}$ be an $(n+1) \times (n+1)$ matrix (with entries $a_{j,k}$ from some ring). Then A is called *T-acyclic* if $a_{j,k} = a_{k,j} = 0$ whenever $j \neq k$ and the vertices v_j, v_k are not adjacent.

Let T be a tree with $n+1$ vertices, let A be T-acyclic matrix and let T' be a subgraph of T. Then $A_{T'}$ denotes the principal submatrix of A consisting of all rows and columns whose indices are indices of the vertices of T'. For $j, k \in \{1, \dots, n+1\}$, $j \neq k$, we denote by $T(j)$ the subgraph obtained from T by deleting the vertex v_j and all the edges incident with v_j, and by $T_k(j)$ the connected component of $T(j)$ that has v_k as a vertex. Put $N_j(T) := \{k \in \{1, \dots, n+1\} : v_k \text{ adjacent to } v_j \text{ in } T\}$. It follows for each $j \in \{1, \dots, n+1\}$ that the set of the vertices of T is the disjoint union of $\{v_j\}$ and the sets of vertices of $T_k(j)$ with $k \in N_j(T)$. Furthermore, each $T_k(j)$ is either a tree or an isolated vertex.

The proof of the following theorem is along the lines of the original proof. However, we are making use of properties of rational Nevanlinna functions rather than giving ad hoc proofs.

Theorem C.6.2. [88] *Let T be a tree with $n+1$ vertices, let $j \in \{1, \dots, n+1\}$ and write $N_j(T) = \{k_1, \dots, k_m\}$. Let $h_1, \dots, h_m$ be monic polynomials with real zeros and $\deg h_r$ equal to the number of vertices of $T_{k_r}(j)$ ($r \in \{1, \dots, m\}$). Denote the zeros of the product $h := h_1 \cdots h_m$, counted with multiplicity, by $\nu_1 \leq \cdots \leq \nu_n$ and let $\mu_1 \leq \mu_2 \leq \cdots \leq \mu_{n+1}$ be real numbers.*
Then there exists a real symmetric T-acyclic matrix A possessing the eigenvalues $\mu_1, \dots, \mu_{n+1}$ such that for each $k \in N_j(T)$ the eigenvalues of the submatrix $A_{T_k(j)}$ are the zeros of h_k if and only if

$$\mu_1 \leq \nu_1 \leq \mu_2 \leq \nu_2 \leq \cdots \leq \nu_n \leq \mu_{n+1} \tag{C.6.1}$$

holds. If all inequalities are strict, i.e., if

$$\mu_1 < \nu_1 < \mu_2 < \nu_2 < \cdots < \nu_n < \mu_{n+1}. \tag{C.6.2}$$

holds, then A is irreducible.

Proof. The statements of this theorem are independent of the indexing of the vertices. We may therefore choose the indices in such a way that the ordered set of vertices of the tree starts with v_j followed by the vertices of $T_{k_1}(j)$ and then up to those of $T_{k_m}(j)$. Furthermore, we assume that k_r becomes the smallest index amongst all indices of vertices in $T_{k_r}(j)$. In other words, we assume that $j = 1$, that the set of the vertices of $T_{k_r}(1)$ is $\{v_k : k_r \le k < k_{r+1}\}$ where we put $k_{m+1} := n+2$ for convenience.

To prove that (C.6.1) is necessary we observe that

$$A_{1,1} = \bigoplus_{r=1}^{m} A_{T_{k_r}(1)}\,. \tag{C.6.3}$$

By assumption, $\lambda_1, \dots, \lambda_{n+1}$ are the eigenvalues of A and $\mu_1, \dots, \mu_n$ are the eigenvalues of $A_{1,1}$. Hence Proposition B.5.1 shows that (C.6.1) holds. Furthermore, Corollary B.5.2 implies that A is irreducible if (C.6.2) is satisfied.

The sufficiency of (C.6.1) will be proved by induction on n. For $n = 1$, the tree T consists of two vertices and the edge connecting them, and trivially the two vertices of T are adjacent to each other. Any real symmetric matrix A chosen according to Proposition B.5.1 is therefore T-acyclic.

Now let $n \ge 2$ and assume that the sufficiency statement is true for trees with at most n edges. With the notation from the proof of Proposition B.5.1 we can write $h_r = \widehat{h}_r \widetilde{h}_r$ $(r = 1, \dots, m)$, where the $\widetilde{h}_r$ and $\widehat{h}_r$ are (possibly constant) monic polynomials such that the zeros of $\widehat{h} := \widehat{h}_1 \cdots \widehat{h}_m$ are $\widehat{\nu}_1, \dots, \widehat{\nu}_{n_0}$ and such that the zeros of $\widetilde{h} = \widetilde{h}_1 \cdots \widetilde{h}_m$ are $\tau_1, \dots, \tau_{n-n_0}$. Without loss of generality we may assume that the above reindexing is done in such a way that there is an integer m_0, $0 \le m_0 \le m$, such that $\widehat{h}_r$ is constant if and only if $r > m_0$. The function χ defined in (B.5.3) has the representation (B.5.4) and can be written in the form

$$\chi(\lambda) = \lambda - b_0 - \sum_{r=1}^{m_0} \gamma_r \frac{\widehat{f}_r(\lambda)}{\widehat{h}_r(\lambda)},$$

where the $\widehat{f}_r$ are (possibly constant) monic polynomials, the γ_r are nonzero real numbers, and the r-th summand is the sum of those summands in (B.5.4) whose poles are zeros of $\widehat{h}_r$. Since each summand in (B.5.4), including the minus sign, is a Nevanlinna function, also each summand in the above representation, including the minus sign, is a Nevanlinna function. Clearly, the degree of $\widehat{f}_r$ is smaller than the degree of $\widehat{h}_r$. In view of Theorem A.2.6 we conclude that the polynomial $\widehat{f}_r$ has $\deg \widehat{h}_r - 1$ zeros which interlace with the zeros of $\widehat{h}_r$, that the largest zero of

$\widehat{h}_r$ is larger than the largest zero of $\widehat{f}_r$, if any, and that $\gamma_r > 0$. Putting $f_r = \widehat{f}_r\widetilde{h}_r$ for $r = 1, \dots, m_0$, we consider two cases. If $\deg h_r \geq 2$, then $T_{k_r}(1)$ is a tree, and by induction hypothesis there is a $T_{k_r}(1)$-acyclic matrix A_r whose eigenvalues are the zeros of h_r such that the eigenvalues of $A_{r,1,1}$ are the zeros of f_r. Here we have to observe that we take a representation as in (C.6.3) for $T_{k_r}(1)$, which is split into the subgraphs $(T_{k_r}(1))_{\kappa_s}(k_r)$ for certain $\kappa_s \in \{k_r + 1, \dots, k_{r+1} - 1\}$. If $\deg h_r = 1$, then we define A_r to be the zero of h_r. For $r = m_0 + 1, \dots, m$ let A_r be a diagonal matrix whose diagonal entries are the zeros of h_r and let $\gamma_r = 0$. For $r = 1, \dots, m$, we let c_r be the vector $c_r = (\sqrt{\gamma_r}\delta_{1,l})_{l=1}^{\deg h_l}$. We now can define A as a block matrix,

$$A = \begin{pmatrix} b_0 & c_1^\mathsf{T} & \dots & c_m^\mathsf{T} \\ c_1 & A_1 & 0 & 0 \\ \vdots & 0 & \ddots & \vdots \\ c_m & 0 & \dots & A_m \end{pmatrix}.$$

It is clear by construction that A is T-acyclic and that the eigenvalues of $A_{T_{k_r}(1)} = A_r$ are the zeros of h_r for $r = 1, \dots, m$. Finally, using the multilinearity of the determinant, it follows that

$$\det(A - \lambda) = (b_0 - \lambda)\det(A_{1,1} - \lambda) + \sum_{r,s=1}^{m_0} \det B_{r,s}(\lambda),$$

where the matrices $B_{r,s}(\lambda)$ are obtained from $(A - \lambda)$ by replacing all entries in the first row and first column by 0, except for c_r and c_s^T. For $r = 1, \dots, m_0$ we find that

$$\det B_{r,r}(\lambda) = \det C_r(\lambda) \prod_{\substack{t=1 \\ t \neq r}}^{m} \det(A_t - \lambda)$$

with

$$C_r(\lambda) = \begin{pmatrix} 0 & c_r^\mathsf{T} \\ c_r & A_r - \lambda \end{pmatrix}.$$

Hence

$$\det C_r(\lambda) = -\gamma_r \det(A_{r,1,1} - \lambda).$$

For $r, s = 1, \dots, m_0$ with $r \neq s$ we have that

$$\det B_{r,s}(\lambda) = \pm \det C_{r,s}(\lambda) \prod_{\substack{t=1 \\ t \neq r,s}}^{m} \det(A_t - \lambda),$$

where

$$C_{r,s}(\lambda) = \begin{pmatrix} c_s^\mathsf{T} & 0 & 0 \\ A_s - \lambda & 0 & 0 \\ 0 & c_r & A_r - \lambda \end{pmatrix}.$$

Since $\begin{pmatrix} c_s^\mathsf{T} \\ A_s - \lambda \end{pmatrix}$ has more rows than columns, $\det C_{r,s}(\lambda) = 0$. Therefore

$$\begin{aligned}
\det(A-\lambda) &= (b_0-\lambda)\det(A_{1,1}-\lambda) - \sum_{r=1}^{m_0} \gamma_r \det(A_{r,1,1}-\lambda) \prod_{\substack{t=1\\ t\neq r}}^{m} \det(A_t-\lambda) \\
&= \det(A_{1,1}-\lambda)\left\{ b_0-\lambda-\sum_{r=1}^{m_0}\gamma_r \frac{\det(A_{r,1,1}-\lambda)}{\det(A_r-\lambda)} \right\} \\
&= (-1)^{n+1}h(\lambda)\left\{ \lambda-b_0-\sum_{r=1}^{m_0}\gamma_r\frac{f_r(\lambda)}{h_r(\lambda)} \right\} \\
&= (-1)^{n+1}h(\lambda)\chi(\lambda) \\
&= \prod_{r=1}^{n+1}(\mu_n-\lambda).
\end{aligned}$$

Hence the eigenvalues of A are $\mu_1, \dots, \mu_{n+1}$, and the proof is complete. □

In [101, Theorem 1.3] the above result with the stronger condition (C.6.2) was reproved by another method:

Corollary C.6.3. *Let the assumptions of Theorem* C.6.2 *be satisfied with strict inequalities* (C.6.2). *Then any matrix A satisfying the statement of Theorem* C.6.2 *has the property that $a_{j,k} = a_{k,j} = 0$ for $j, k = 1, \dots, n+1$ with $j \neq k$ if and only if v_j and v_k are adjacent.*

Proof. In view of Theorem C.6.2 we still have to prove that $a_{j,k} \neq 0$ whenever $j \neq k$ and v_j and v_k are adjacent. Hence let $j \neq k$ such that v_j and v_k are adjacent. Since $T_k(j)$ is a component of $T(j)$, it follows that $a_{r,s} = 0$ for all vertices v_r of $T_k(j)$ and all vertices v_s of T which do not belong to $T_k(j)$. But A is irreducible by Theorem C.6.2, and therefore $a_{j,k} = a_{k,j} \neq 0$. □

Bibliography

[1] T. Aktosun and R. Weder, *High-energy analysis and Levinson's theorem for the selfadjoint matrix Schrödinger operator on the half line*, J. Math. Phys. **54** (2013), no. 1, 012108, 27 pp.

[2] S. Alexander, *Superconductivity of networks. A percolation approach to the effects of disorder*, Phys. Rev. B (3) **27** (1983), no. 3, 1541–1557.

[3] F. Ali Mehmeti, *A characterization of a generalized C^∞-notion on nets*, Integral Equations Operator Theory **9** (1986), no. 6, 753–766.

[4] D. Z. Arov, *Realization of a canonical system with a dissipative boundary condition at one end of the segment in terms of the coefficient of dynamical compliance*, Sibirsk. Mat. Ž. **16** (1975), no. 3, 440–463, 643 (Russian); English transl., Siberian Math. J. **16** (1975), no. 3, 335–352.

[5] S. Avdonin and P. Kurasov, *Inverse problems for quantum trees*, Inverse Probl. Imaging **2** (2008), no. 1, 1–21.

[6] L. Barkwell and P. Lancaster, *Overdamped and gyroscopic vibrating systems*, Trans. ASME J. Appl. Mech. **59** (1992), no. 1, 176–181.

[7] L. Barkwell, P. Lancaster, and A. S. Markus, *Gyroscopically stabilized systems: a class of quadratic eigenvalue problems with real spectrum*, Canad. J. Math. **44** (1992), no. 1, 42–53.

[8] J. Behrndt, S. Hassi, and H. de Snoo, *Boundary value problems, Weyl functions, and differential operators*, Monographs in Mathematics, vol. 108, Birkhäuser/Springer, Cham, 2020.

[9] V. Belevitch, *Classical network theory*, Holden-Day, San Francisco, Calif.-Cambridge-Amsterdam, 1968.

[10] J. von Below, *Can one hear the shape of a network?*, Partial differential equations on multistructures (Luminy, 1999), Lecture Notes in Pure and Appl. Math., vol. 219, Dekker, New York, 2001, pp. 19–36.

[11] G. Berkolaiko and P. Kuchment, *Introduction to quantum graphs*, Mathematical Surveys and Monographs, vol. 186, American Mathematical Society, Providence, RI, 2013.

[12] N. Biggs, *Algebraic graph theory*, 2nd ed., Cambridge Mathematical Library, Cambridge University Press, Cambridge, 1993.

[13] D. Ī. Bodnar, *Vetvyashchiesya tsepnye drobi*, "Naukova Dumka", Kiev, 1986 (Russian).

[14] O. Boiko, O. Martynyuk, and V. Pivovarchik, *On maximal multiplicity of eigenvalues of finite-dimensional spectral problem on a graph*, Methods Funct. Anal. Topology **25** (2019), no. 2, 104–117.

[15] G. Borg, *Eine Umkehrung der Sturm-Liouvilleschen Eigenwertaufgabe. Bestimmung der Differentialgleichung durch die Eigenwerte*, Acta Math. **78** (1946), 1–96 (German).

M. Möller, V. Pivovarchik, *Direct and Inverse Finite-Dimensional Spectral Problems on Graphs*, Operator Theory: Advances and Applications 283, https://doi.org/10.1007/978-3-030-60484-4

[16] O. Boyko and V. Pivovarchik, *The inverse three-spectral problem for a Stieltjes string and the inverse problem with one-dimensional damping*, Inverse Problems **24** (2008), no. 1, 015019, 13 pp.

[17] O. Boyko and V. Pivovarchik, *Inverse problem for Stieltjes string damped at one end*, Methods Funct. Anal. Topology **14** (2008), no. 1, 10–19.

[18] O. Boyko and V. Pivovarchik, *Inverse spectral problem for a star graph of Stieltjes strings*, Methods Funct. Anal. Topology **14** (2008), no. 2, 159–167.

[19] L. de Branges, *Local operators on Fourier transforms*, Duke Math. J. **25** (1958), 143–153.

[20] ———, *Some mean squares of entire functions*, Proc. Amer. Math. Soc. **10** (1959), 833–839.

[21] ———, *Some Hilbert spaces of entire functions*, Proc. Amer. Math. Soc. **10** (1959), 840–846.

[22] ———, *Some Hilbert spaces of entire functions*, Trans. Amer. Math. Soc. **96** (1960), 259–295.

[23] ———, *Some Hilbert spaces of entire functions. II*, Trans. Amer. Math. Soc. **99** (1961), 118–152.

[24] ———, *Some Hilbert spaces of entire functions. III*, Trans. Amer. Math. Soc. **100** (1961), 73–115.

[25] ———, *Some Hilbert spaces of entire functions. IV*, Trans. Amer. Math. Soc. **105** (1962), 43–83.

[26] ———, *Hilbert spaces of entire functions*, Prentice-Hall Inc., Englewood Cliffs, N.J., 1968.

[27] B. M. Brown and R. Weikard, *A Borg-Levinson theorem for trees*, Proc. R. Soc. Lond. Ser. A Math. Phys. Eng. Sci. **461** (2005), no. 2062, 3231–3243.

[28] O. Brune, *Synthesis of a finite two-terminal network whose driving-point impedance is a prescribed function of frequency*, Doctoral thesis, MIT, 1931.

[29] R. Carlson, *Adjoint and self-adjoint differential operators on graphs*, Electron. J. Differential Equations (1998), No. 6, 10 pp. (electronic).

[30] C. Cattaneo, *The spectrum of the continuous Laplacian on a graph*, Monatsh. Math. **124** (1997), no. 3, 215–235.

[31] A. Cauchy, *Oeuvres complètes d'Augustin Cauchy publiées sous la direction Scientifique de l'Académie des sciences et sous les auspices de M. le ministre de l'instruction publique. (2) IX.*, Gauthier-Villars et Fils, Paris, 1891 (French).

[32] E. Cauer, W. Mathis, and R. Pauli, *Life and work of W. Cauer (1900–1945)*, Proc. MTNS 2000, Perpignan, France, June 19–23, 2000, 10 pp.

[33] W. Cauer, *Die Verwirklichung von Wechselstromwiderständen vorgeschriebener Frequenzabhängingkeit*, Arch. Elektrotech. **17** (1926), no. 4, 355–388 (German).

[34] ———, *Vierpole mit vorgeschriebenem Dämpfungsverhalten. I, II*, Telegraphen-Fernsprech-Funk-Fernseh-Techn. **29** (1940), 185-182, 228-235 (German).

[35] G. Chen and J. Zhou, *Vibration and damping in distributed systems. Vol. I*, Studies in Advanced Mathematics, CRC Press, Boca Raton, FL, 1993.

[36] Y. Chen, *Vibration of a string with attached concentrated masses*, J. Franklin Inst. **276** (1963), no. 3, 191–196.

[37] H. Cho and F. Udwadia, *Inverse Problem for Lagrangian Dynamics for Multi-Degree-of-Freedom Systems With Linear Damping*, Proceedings of the ASME 2012 International Mechanical Engineering Congress & Exposition IMECE2012 (2012).

[38] F. R. K. Chung, *Spectral graph theory*, CBMS Regional Conference Series in Mathematics, vol. 92, Published for the Conference Board of the Mathematical Sciences, Washington, DC; by the American Mathematical Society, Providence, RI, 1997.

[39] G. Cocci, *Rappresentazione di bipoli qualsiasi con quadripoli di pure reattanze chiusi su resistenze*, Alta Frequenza **9** (1940), 685–698 (Italian).

[40] R. Courant and D. Hilbert, *Methods of mathematical physics. Vol. I*, Interscience Publishers, Inc., New York, N.Y., 1953.

[41] S. J. Cox, M. Embree, and J. M. Hokanson, *One can hear the composition of a string: experiments with an inverse eigenvalue problem*, SIAM Rev. **54** (2012), no. 1, 157–178.

[42] S. Currie and B. A. Watson, *Eigenvalue asymptotics for differential operators on graphs*, J. Comput. Appl. Math. **182** (2005), no. 1, 13–31.

[43] ______, *The M-matrix inverse problem for the Sturm-Liouville equation on graphs*, Proc. Roy. Soc. Edinburgh Sect. A **139** (2009), no. 4, 775–796.

[44] D. M. Cvetković, M. Doob, I. Gutman, and A. Torgašev, *Recent results in the theory of graph spectra*, Annals of Discrete Mathematics, vol. 36, North-Holland Publishing Co., Amsterdam, 1988.

[45] S. Darlington, *Synthesis of reactance 4-poles which produce prescribed insertion loss characteristics including special applications to filter design*, J. Math. Phys. Mass. Inst. Tech. **18** (1939), 257–353.

[46] A. Dudko and V. Pivovarchik, *Three spectra problems for star graph of Stieltjes strings*, Methods Funct. Anal. Topology **25** (2019), no. 4, 311–323.

[47] R. J. Duffin, *A minimax theory for overdamped networks*, J. Rational Mech. Anal. **4** (1955), 221–233.

[48] H. Dym and H. P. McKean, *Gaussian processes, function theory, and the inverse spectral problem*, Academic Press, New York, 1976.

[49] J. Eckhardt, *An inverse spectral problem for a star graph of Krein strings*, J. Reine Angew. Math. **715** (2016), 189–206.

[50] P. Exner, *Weakly coupled states on branching graphs*, Lett. Math. Phys. **38** (1996), no. 3, 313–320.

[51] P. Exner, *A duality between Schrödinger operators on graphs and certain Jacobi matrices*, Ann. Inst. H. Poincaré Phys. Théor. **66** (1997), no. 4, 359–371 (English, with English and French summaries).

[52] K. Fan and G. Pall, *Imbedding conditions for Hermitian and normal matrices*, Canad. J. Math. **9** (1957), 298–304.

[53] W. E. Ferguson Jr., *The construction of Jacobi and periodic Jacobi matrices with prescribed spectra*, Math. Comp. **35** (1980), no. 152, 1203–1220.

[54] A. M. Filimonov, P. F. Kurchanov, and A. D. Myshkis, *Some unexpected results in the classical problem of vibrations of the string with n beads when n is large*, C. R. Acad. Sci. Paris Sér. I Math. **313** (1991), no. 13, 961–965 (English, with French summary).

[55] A. M. Filimonov and A. D. Myshkis, *On properties of large wave effect in classical problem of bead string vibration*, J. Difference Equ. Appl. **10** (2004), no. 13-15, 1171–1175.

[56] R. M. Foster, *A reactance theorem*, Bell System Techn. J. **III** (1924), 259–267.

[57] J. Friedman and J.-P. Tillich, *Wave equations for graphs and the edge-based Laplacian*, Pacific J. Math. **216** (2004), no. 2, 229–266.

[58] T. C. Fry, *The use of continued fractions in the design of electrical networks.*, Bull. Am. Math. Soc. **35** (1929), 153.

[59] F. P. Gantmacher and M. G. Krein, *Oscillation matrices and kernels and small vibrations of mechanical systems*, Revised edition, AMS Chelsea Publishing, Providence, RI, 2002.

[60] J. Genin and J. S. Maybee, *Mechanical vibration trees*, J. Math. Anal. Appl. **45** (1974), 746–763.

[61] F. Gesztesy and B. Simon, *On the determination of a potential from three spectra*, Differential operators and spectral theory, Amer. Math. Soc. Transl. Ser. 2, vol. 189, Amer. Math. Soc., Providence, RI, 1999, pp. 85–92.

[62] G. M. L. Gladwell, *Inverse problems in vibration*, 2nd ed., Solid Mechanics and its Applications, vol. 119, Kluwer Academic Publishers, Dordrecht, 2004.

[63] ———, *Matrix inverse eigenvalue problems*, Dynamical inverse problems: theory and application, CISM Courses and Lectures, vol. 529, SpringerWienNewYork, Vienna, 2011, pp. 1–28.

[64] B. J. Gómez, C. E. Repetto, C. R. Stia, and R. Welti, *Oscillations of a string with concentrated masses*, Eur. J. Phys. **28** (2007), no. 5, 961–975.

[65] G. M. Gubreev and V. Pivovarchik, *Spectral analysis of the Regge problem with parameters*, Funktsional. Anal. i Prilozhen. **31** (1997), no. 1, 70–74 (Russian); English transl., Funct. Anal. Appl. **31** (1997), no. 1, 54–57.

[66] E. A. Guillemin, *Synthesis of passive networks. Theory and methods appropriate to the realization and approximation problems*, John Wiley and Sons, Inc., New York, 1958.

[67] B. Gutkin and U. Smilansky, *Can one hear the shape of a graph?*, J. Phys. A **34** (2001), no. 31, 6061–6068.

[68] G. H. Hardy, J. E. Littlewood, and G. Pólya, *Inequalities*, Cambridge, at the University Press, 1952.

[69] H. Helmholtz, *Die Lehre von den Tonempfindungen als physiologische Grundlage für die Theorie der Musik*, Friedrich Vieweg und Sohn, Braunschweig, 1863 (German).

[70] ———, *On the sensations of Tone*, Dover Publications, New York, N. Y., 1954.

[71] H. Hochstadt and B. Lieberman, *An inverse Sturm-Liouville problem with mixed given data*, SIAM J. Appl. Math. **34** (1978), no. 4, 676–680.

[72] R. O. Hryniv and Ya. V. Mykytyuk, *Inverse spectral problems for Sturm-Liouville operators with singular potentials. III. Reconstruction by three spectra*, J. Math. Anal. Appl. **284** (2003), no. 2, 626–646.

[73] C. R. Johnson and A. Leal Duarte, *On the possible multiplicities of the eigenvalues of a Hermitian matrix whose graph is a tree*, Linear Algebra Appl. **348** (2002), 7–21.

[74] I. S. Kac and M. G. Kreĭn, *R-functions–analytic functions mapping the upper half-plane into itself.* Supplement I to the Russian edition of F. V. Atkinson: *Discrete and continuous boundary value problems*, Mir, Moscow, 1968 (Russian); English transl., Amer. Math. Transl. (2) **103** (1974), 1–18.

[75] ———, *On the spectral functions of the string.* Supplement II to the Russian edition of F. V. Atkinson: *Discrete and continuous boundary value problems*, Mir, Moscow, 1968 (Russian); English transl., Amer. Math. Transl. (2) **103** (1974), 19–102.

[76] I. Kac and V. Pivovarchik, *On multiplicity of a quantum graph spectrum*, J. Phys. A **44** (2011), no. 10, 105301, 14 pp.

[77] V. Kostrykin and R. Schrader, *Kirchhoff's rule for quantum wires*, J. Phys. A **32** (1999), no. 4, 595–630.

[78] M. G. Kreĭn, *On some new problems of the theory of oscillations of Sturmian systems*, Akad. Nauk SSSR. Prikl. Mat. Meh. **16** (1952), 555–568 (Russian).

[79] ———, *On a generalization of investigations of Stieltjes*, Doklady Akad. Nauk SSSR (N.S.) **87** (1952), 881–884 (Russian).

[80] M. G. Kreĭn and A. A. Nudel′man, *Direct and inverse problems for the frequencies of boundary dissipation of a nonuniform string*, Dokl. Akad. Nauk SSSR **247** (1979), no. 5, 1046–1049 (Russian).

[81] ———, *Some spectral properties of a nonhomogeneous string with a dissipative boundary condition*, J. Operator Theory **22** (1989), no. 2, 369–395 (Russian).

[82] P. Kuchment, *Quantum graphs. II. Some spectral properties of quantum and combinatorial graphs*, J. Phys. A **38** (2005), no. 22, 4887–4900.

[83] ———, *Quantum graphs: an introduction and a brief survey*, Analysis on graphs and its applications, Proc. Sympos. Pure Math., vol. 77, Amer. Math. Soc., Providence, RI, 2008, pp. 291–312.

[84] P. F. Kurchanov, A. D. Myshkis, and A. M. Filimonov, *Oscillations of a railroad train and Kronecker's theorem*, Prikl. Mat. Mekh. **55** (1991), no. 6, 989–995 (Russian); English transl., J. Appl. Math. Mech. **55** (1991), no. 6, 870–876 (1992).

[85] P. Lancaster and J. Maroulas, *Inverse eigenvalue problems for damped vibrating systems*, J. Math. Anal. Appl. **123** (1987), no. 1, 238–261.

[86] C. K. Law and V. Pivovarchik, *Characteristic functions of quantum graphs*, J. Phys. A **42** (2009), no. 3, 035302, 12 pp.

[87] C. K. Law, V. Pivovarchik, and W. C. Wang, *A polynomial identity and its application to inverse spectral problems in Stieltjes strings*, Oper. Matrices **7** (2013), no. 3, 603–616.

[88] A. Leal Duarte, *Construction of acyclic matrices from spectral data*, Linear Algebra Appl. **113** (1989), 173–182.

[89] B. M. Levitan and M. G. Gasymov, *Determination of a differential equation by two spectra*, Uspehi Mat. Nauk **19** (1964), no. 2 (116), 3–63 (Russian).

[90] L. Lovász, *Combinatorial problems and exercises*, 2nd ed., North-Holland Publishing Co., Amsterdam, 1993.

[91] J. A. Lubary, *Multiplicity of solutions of second-order linear differential equations on networks*, Linear Algebra Appl. **274** (1998), 301–315.

[92] V. A. Marchenko, *Sturm-Liouville operators and applications*, Revised edition, AMS Chelsea Publishing, Providence, RI, 2011.

[93] V. A. Marchenko, *Introduction to inverse problems of spectral analysis*, Akta, Kharkov, 2005 (Russian).

[94] A. W. Marshall, I. Olkin, and B. C. Arnold, *Inequalities: theory of majorization and its applications*, 2nd ed., Springer Series in Statistics, Springer, New York, 2011.

[95] J. Michor and G. Teschl, *Reconstructing Jacobi matrices from three spectra*, Spectral methods for operators of mathematical physics, Oper. Theory Adv. Appl., vol. 154, Birkhäuser, Basel, 2004, pp. 151–154.

[96] M. Möller and V. Pivovarchik, *Spectral theory of operator pencils, Hermite-Biehler functions, and their applications*, Operator Theory: Advances and Applications, vol. 246, Birkhäuser/Springer, Cham, 2015.

[97] M. Möller and V. Pivovarchik, *Damped star graphs of Stieltjes strings*, Proc. Amer. Math. Soc. **145** (2017), no. 4, 1717–1728.

[98] M. Möller and A. Zettl, *Symmetric differential operators and their Friedrichs extension*, J. Differential Equations **115** (1995), no. 1, 50–69.

[99] R. F. Muirhead, *Some methods applicable to identities and inequalities of symmetric algebraic functions of n letters*, Proc. Edinb. Math. Soc. **21** (1903), 143–157 (English).

[100] S. Nicaise, *Spectre des réseaux topologiques finis*, Bull. Sci. Math. (2) **111** (1987), no. 4, 401–413 (French, with English summary).

[101] P. Nylen and F. Uhlig, *Realizations of interlacing by tree-patterned matrices*, Linear and Multilinear Algebra **38** (1994), no. 1-2, 13–37.

[102] K. Pankrashkin, *Spectra of Schrödinger operators on equilateral quantum graphs*, Lett. Math. Phys. **77** (2006), no. 2, 139–154.

[103] S. Parmley, T. Zobrist, T. Clough, A. Perez-Miller, M. Makela, and R. Yu, *Vibrational properties of a loaded string*, Am. J. Phys. **63** (1995), 547–553.

[104] V. N. Pivovarchik, *Inverse problem for a smooth string with damping at one end*, J. Operator Theory **38** (1997), no. 2, 243–263.

[105] ______, *Direct and inverse problems for a damped string*, J. Operator Theory **42** (1999), no. 1, 189–220.

[106] ______, *An inverse Sturm-Liouville problem by three spectra*, Integral Equations Operator Theory **34** (1999), no. 2, 234–243.

[107] V. Pivovarchik, *Inverse problem for the Sturm-Liouville equation on a simple graph*, SIAM J. Math. Anal. **32** (2000), no. 4, 801–819 (electronic).

[108] ______, *Inverse problem for the Sturm-Liouville equation on a star-shaped graph*, Math. Nachr. **280** (2007), no. 13-14, 1595–1619.

[109] ______, *Existence of a tree of Stieltjes strings corresponding to two given spectra*, J. Phys. A **42** (2009), no. 37, 375213, 16 pp.

[110] ______, *Characteristic functions under series and parallel connection of quantum graphs*, J. Phys. A **48** (2015), no. 36, 365201, 13 pp.

[111] ______, *Corrigendum: Characteristic functions under series and parallel connection of quantum graphs (2015 J. Math. Phys. A: Math. Theor. 48 365201)*, J. Phys. A **49** (2016), no. 19, 199501, 1 pp.

[112] V. Pivovarchik and C. van der Mee, *The inverse generalized Regge problem*, Inverse Problems **17** (2001), no. 6, 1831–1845.

[113] V. Pivovarchik and N. Rozhenko, *Inverse Sturm-Liouville problem on equilateral regular tree*, Appl. Anal. **92** (2013), no. 4, 784–798.

[114] V. Pivovarchik, N. Rozhenko, and C. Tretter, *Dirichlet-Neumann inverse spectral problem for a star graph of Stieltjes strings*, Linear Algebra Appl. **439** (2013), no. 8, 2263–2292.

[115] V. Pivovarchik and O. Taystruk, *On characteristic functions of operators on equilateral graphs*, Methods Funct. Anal. Topology **18** (2012), no. 2, 189–197.

[116] V. Pivovarchik and C. Tretter, *Location and multiplicities of eigenvalues for a star graph of Stieltjes strings*, J. Difference Equ. Appl. **21** (2015), no. 5, 383–402.

[117] V. Pivovarchik and H. Woracek, *Sums of Nevanlinna functions and differential equations on star-shaped graphs*, Oper. Matrices **3** (2009), no. 4, 451–501.

[118] Yu. V. Pokornyĭ, O. M. Penkin, V. L. Pryadiev, A. V. Borovskikh, K. P. Lazarev, and S. A. Shabrov, *Differentsialnye uravneniya na geometricheskikh grafakh*, Fiziko-Matematicheskaya Literatura, Moscow, 2005 (Russian, with Russian summary).

[119] Yu. V. Pokornyi and V. L. Pryadiev, *The qualitative Sturm-Liouville theory on spatial networks*, J. Math. Sci. (N. Y.) **119** (2004), no. 6, 788–835.

[120] W. Pong, *Principal frequencies of a double spring-mass system*, Am. J. Phys. **21** (1953), 546–548.

[121] J. W. S. Rayleigh, *The Theory of Sound, Volume I*, Mcmillan and Co., London, 1877.

[122] ______, *The Theory of Sound*, Dover Publications, New York, N. Y., 1945.

[123] F. C. Santos, Y. A. Coutinho, L. Ribeiro-Pinto, and A. C. Tort, *The motion of two masses coupled to a finite mass spring*, Eur. J. Phys. **27** (2006), 1037–1051.

[124] G. E. Shilov, *An introduction to the theory of linear spaces*, Translated from the Russian by Richard A. Silverman, Prentice-Hall, Inc., Englewood Cliffs, N.J., 1961.

[125] M. A. Shubov and M. Rojas-Arenaza, *Four-branch vibrational spectrum of double-walled carbon nanotube model*, Proc. R. Soc. Lond. Ser. A Math. Phys. Eng. Sci. **467** (2011), no. 2125, 99–126.

[126] V. Ya. Skorobogat′ko, *Branched continued fractions and convergence acceleration problems*, Rational approximation and applications in mathematics and physics (Łańcut, 1985), Lecture Notes in Math., vol. 1237, Springer, Berlin, 1987, pp. 46–50.

[127] M. Solomyak, *On the spectrum of the Laplacian on regular metric trees*, Waves Random Media **14** (2004), no. 1, S155–S171.

[128] T.-J. Stieltjes, *Recherches sur les fractions continues*, Toulouse Ann. **8** (1894), j1–j122 (French).

[129] C. Texier, *On the spectrum of the Laplace operator of metric graphs attached at a vertex—spectral determinant approach*, J. Phys. A **41** (2008), no. 8, 085207, 9 pp.

[130] F. Tisseur and K. Meerbergen, *The quadratic eigenvalue problem*, SIAM Rev. **43** (2001), no. 2, 235–286.

[131] C.-W. Tsao and C.-K. Law, *The Stieltjes string and its associated nodal points*, Oper. Matrices **13** (2019), no. 2, 363–373.

[132] W. T. Tutte, *Graph theory*, Encyclopedia of Mathematics and its Applications, vol. 21, Cambridge University Press, Cambridge, 2001.

[133] K. Veselić, *On linear vibrational systems with one-dimensional damping*, Appl. Anal. **29** (1988), no. 1-2, 1–18.

[134] ______, *On linear vibrational systems with one-dimensional damping. II*, Integral Equations Operator Theory **13** (1990), no. 6, 883–897.

[135] R. Weder, *Scattering theory for the matrix Schrödinger operator on the half line with general boundary conditions*, J. Math. Phys. **56** (2015), no. 9, 092103, 24 pp.

[136] ______, *Erratum: Scattering theory for the matrix Schrödinger operator on the half line with general boundary conditions*, J. Math. Phys. **60** (2019), no. 1, 019901, 1 pp.

[137] ______, *The number of eigenvalues of the matrix Schrödinger operator on the half line with general boundary conditions*, J. Math. Phys. **58** (2017), no. 10, 102107, 11 pp.

[138] ______, *Erratum: The number of eigenvalues of the matrix Schrödinger operator on the half line with general boundary conditions*, J. Math. Phys. **60** (2019), no. 1, 019902, 1 pp.

[139] J. Weidmann, *Spectral theory of ordinary differential operators*, Lecture Notes in Mathematics, vol. 1258, Springer-Verlag, Berlin, 1987.

[140] M. R. Wohlers, *Lumped and distributed passive networks*, Academic Press, New York, 1969.

[141] L. Yang, Y. Guo, and G. Wei, *Inverse eigenvalue problems for a damped Stieltjes string with mixed data*, Linear Algebra Appl. **601** (2020), 55–71.

[142] ______, *Inverse problem for a Stieltjes string damped at an interior point*, Integral Equations Operator Theory **92** (2020), no. 4, 30.

[143] V. Yurko, *Inverse spectral problems for Sturm-Liouville operators on graphs*, Inverse Problems **21** (2005), no. 3, 1075–1086.

[144] E. E. Zajac, *The Kelvin-Tait-Chetaev Theorem and Extensions*, J.Astronaut. Sci. **11** (1964), 46–49.

[145] A. Zettl, *Sturm-Liouville theory*, Mathematical Surveys and Monographs, vol. 121, American Mathematical Society, Providence, RI, 2005.

[146] G. V. Zeveke, G. V. Ionkin, A. V. Netushil, and S. V. Strakhov, *Analysis of Electric Circuits*, Mir, Moscow, 1972.

Index

M. Möller, V. Pivovarchik, *Direct and Inverse Finite-Dimensional Spectral Problems on Graphs*, Operator Theory: Advances and Applications 283, https://doi.org/10.1007/978-3-030-60484-4

Index of Notation

$\mathbb{C}$	the set of complex numbers
$\overline{\mathbb{C}}$	$\mathbb{C} \cup \{\infty\}$
$\mathbb{C}^+$	the open upper half-plane
$\overline{\mathbb{C}^+}$	the closed upper half-plane
$\mathbb{C}^-$	the open lower half-plane
$\overline{\mathbb{C}^-}$	the closed lower half-plane
$\overline{C}$	the conjugate complex of a matrix C
C^*	the conjugate complex transpose of a matrix C
$\det M$	the determinant of a square matrix M
e_j	the j-th unit vector in $\mathbb{C}^n$
I	the identity operator or the identity matrix
I_r	the $r \times r$ identity matrix or a particular index set, depending on the context
$(\, , \,)$	the inner product
$L_p(a, b)$	the Banach space of all equivalence classes of measurable functions f on (a, b) such that $x \mapsto \|f(x)\|^p$ is Lebesgue integrable on (a, b)
$\#S$	the magnitude of a set S
$\mathbb{N}$	the set of positive integers
$\mathbb{N}_0$	$\mathbb{N} \cup \{0\}$, the set of nonnegative integers
$N(A)$	the nullspace (kernel) of an operator, or a matrix, A
$\mathbb{R}$	the set of real numbers
$\mathbb{Z}$	the set of integers
$\mathbb{Z}a$	the set of integer multiplies of the number a

Classes of functions

Special Symbols

M. Möller, V. Pivovarchik, *Direct and Inverse Finite-Dimensional Spectral Problems on Graphs*, Operator Theory: Advances and Applications 283, https://doi.org/10.1007/978-3-030-60484-4

The manufacturer's authorised representative in the EU is Springer Nature Customer Service Centre GmbH, Europaplatz 3, 69115 Heidelberg, Germany. If you have any concerns regarding our products, please contact ProductSafety@springernature.com

Printed and bound by CPI Group (UK) Ltd, Croydon, CR0 4YY
24/07/2026
02175922-0006